THE LIBRARY
ST. MARY'S COLLEGE OF MARYLAND
ST. MARY'S CITY, MARYLAND 20686

SMALL-SCALE FRESHWATER TOXICITY INVESTIGATIONS

Small-scale Freshwater Toxicity Investigations

Volume 1 – Toxicity Test Methods

Edited by

Christian Blaise
St. Lawrence Centre, Environment Canada,
Montreal, QC, Canada

and

Jean-François Férard
Paul Verlaine University, Metz,
Laboratoire Ecotoxicité et Santé Environnementale,
Metz, France

A C.I.P. Catalogue record for this book is available from the Library of Congress.

ISBN-10 1-4020-3119-X (HB)
ISBN-13 978-1-4020-3119-9 (HB)
ISBN-10 1-4020-3120-3 (e-book)
ISBN-13 978-1-4020-3120-5 (e-book)

Published by Springer,
P.O. Box 17, 3300 AA Dordrecht, The Netherlands.

www.springeronline.com

Cover design: Created by Patrick Bermingham (Montreal, Canada)

Printed on acid-free paper

All Rights Reserved
© 2005 Springer
No part of this work may be reproduced, stored in a retrieval system, or transmitted
in any form or by any means, electronic, mechanical, photocopying, microfilming, recording
or otherwise, without written permission from the Publisher, with the exception
of any material supplied specifically for the purpose of being entered
and executed on a computer system, for exclusive use by the purchaser of the work.

Printed in the Netherlands.

About the editors

Christian Blaise, D.Sc., is a senior research scientist at the Saint-Lawrence Centre, Environment Canada, Québec Region, where he heads the Aquatic Toxicology Unit (ATU), River Ecosystems Research Section. He also holds an adjunct professor status at UQAR (Université du Québec à Rimouski) where he contributes to teaching and (co)directs graduate students in the field of ecotoxicology. ATU strives to develop, validate, standardize, modernize (and promote the commercialization of) bioanalytical and biomarker techniques, making use of new instrumental technologies whenever possible, in order to determine the potential (geno)toxicity of chemicals and various types of environmental matrices (*e.g.*, effluents, sediments, pore/surface waters). ATU research output provides practical tools and approaches which facilitate decision-making for environmental management of aquatic ecosystems such as the Saint-Lawrence River. ATU also provides (inter)national technology transfer to interested professionals and agencies and promotes graduate student training by co-directing applied research projects with university collaborators.

Dr. Blaise obtained university diplomas from the U. of Montréal (B.A., 1967: biology and chemistry), U. of Ottawa (B.Sc., 1970: cell biology; M.Sc., 1973: environmental microbiology) and U. of Metz (D.Sc., 1984: ecotoxicology). He is a member of the editorial board for two scientific journals (*Environmental Toxicology*; *Ecotoxicology and Environmental Safety*) and holds membership in both the biologists' (Association des Biologistes du Québec) and microbiologists' (Association des Microbiologistes du Québec) associations of the province of Québec. He regularly attends and makes presentations during major venues held in the field of ecotoxicology (SETAC: Society of Environmental Toxicology and Chemistry; SECOTOX: Society of Ecotoxicology and Environmental Safety; ATW-Canada: Aquatic Toxicity Workshop-Canada; ISTA: International Symposium on Toxicity Assessment). Dr. Blaise has (co)authored over 100 scientific articles in internationally refereed journals, as well as having written several book chapters, reviews, and various government technical reports.

He recently co-edited, with Canadian colleagues, a book dedicated to small-scale toxicity testing (Wells, P., K. Lee and C. Blaise (eds.), 1998. *Microscale testing in Aquatic Toxicology Advances, Techniques and Practice*. CRC Lewis Publishers, Boca Raton, Florida, 679 pages). He was scientific organizer of the 10th International Symposium on Toxicity Assessment (ISTA 10), hosted by the Saint-Lawrence Centre where he works, and held in Quebec City, August 26-31, 2001. He further co-edited with another Canadian colleague, a special edition of *Environ. Toxicol.* (Volume 17 [3]: 2002, special issue) highlighting selected papers presented at the ISTA 10 venue.

Jean-François Férard, D.Sc., is a professor at the University of Metz (Lorraine province of France), where he heads a research team (RT) which is part of a C.N.R.S. (Centre National de la Recherche Scientifique) research unit for Ecotoxicity and Environmental Health (E.S.E). He also manages an undergraduate school program dedicated to Environmental Engineering. His teaching duties involve fundamental and applied Ecotoxicology, Physiology and Physiotoxicology, Cell Biology and other related disciplines.

In the field of ecotoxicology, his RT was involved in the behavior of metals, PAHs and complex mixtures in air, water and soil compartments and their effects on different organisms (bacteria, algae, crustaceans, plants, arthropods, etc.). His actual research endeavors are more specifically focused on the development of metal-resistance (*e.g.* phytochelatin) and genotoxic (*e.g.* comet assay) biomarkers. He also promotes knowledge and use of toxicity tests by organizing an annual course entitled "Ecotoxicity and carcinogenicity of chemicals" which provides a theoretical and practical view of numerous toxicity tests to decision-makers, industrialists and consultants. Since 1974, he has markedly contributed to numerous research programs that have successfully lead to i) development and validation of different ecotoxicity tools (*e.g.* new toxicity test methods, trophic chain models, biomarkers), ii) hazard/risk assessment schemes and iii) links between field and laboratory studies. These undertakings were financially supported by the European Economic

Community, the French ministry of the Environment, and agencies such as the French Water Agency of the Rhin-Meuse Watershed, the French Agency for Environment and Energy Ressources.

Professor Férard obtained university diplomas from the U. of Strasbourg (B.A., 1970: biology and chemistry; B.Sc., 1973: biochemistry) and U. of Metz (M.Sc., 1974: chemistry and environmental toxicology; D.Sc., 1978: environmental toxicology; State doctorate, 1986: environmental toxicology). He was European editor for *Environmental Toxicology and Water Quality* from 1992-1996 and holds membership in SETAC (Society of Environmental Toxicology and Chemistry). He regularly makes presentations during major symposia held in the field of ecotoxicology (*e.g.* SETAC meetings, Secotox conferences, International Symposia on Toxicity Assessment, Annual Aquatic Toxicity Workshops in Canada). Professor Férard has (co)authored over 50 scientific articles in (inter)nationally refereed journals, as well as having written several book chapters, reviews, and research reports. He also participates in several OECD (Organization for Economic Cooperation and Development) and AFNOR (*Association française de normalisation* - French standards association) initiatives to standardize and promote the use of biological tests.

Contributors

Merrin Adams Centre for Environmental Contaminants Research, CSIRO Energy Technology, Private Mail Bag 7, Bangor, NSW 2234, Australia Merrin.adams@csiro.au	Christian Blaise Centre Saint-Laurent, Environment Canada 105 McGill street, Montréal, Québec Canada, H2Y 2E7 christian.blaise@ec.gc.ca
Niels C. Bols Department of Biology University of Waterloo Waterloo, ON, Canada, N2L 3G1 ncbols@sciborg.uwaterloo.ca	Uwe Borgmann National Water Research Institute Environment Canada 867 Lakeshore Road, P.O. Box 5050, Burlington, Ontario Canada, L7R 4A6 uwe.borgmann@ec.gc.ca
Vivian R. Dayeh Department of Biology University of Waterloo Waterloo, Ontario, Canada, N2L 3G1 vrdayeh@sciborg.uwaterloo.ca	Ken Doe Environment Canada Environmental Science Centre P.O. Box 23005, Moncton, NB Canada, E1A 6S8 ken.doe@ec.gc.ca
Natasha Franklin Department of Biology McMaster University, 1280 Main Street West, Hamilton Canada, L8S 4K1 nfrank@univmail.cis.mcmaster.ca	François Gagné St. Lawrence Centre, Environment Canada 105 McGill Street, Montreal, Québec Canada, H2Y 2E7 francois.gagne@ec.gc.ca
Jeanne Garric Laboratoire d'écotoxicologie CEMAGREF, 3bis quai Chauveau CP 220, 69336 Lyon, France jeanne.garric@cemagref.fr	Guy Gilron Golder Associates Ltd. 2390 Argentia Road, Mississauga Ontario L5N 5Z7, Canada ggilron@golder.com
Douglas A. Holdway Faculty of Science, University of Ontario Institute of Technology 2000 Simcoe Street North Oshawa, Ontario L1H 7K4 Canada Douglas.Holdway@uoit.ca	Paula Jackman Environment Canada Environmental Science Centre P.O. Box 23005, Moncton, New Brunswick Canada, E1A 6S8 Paula.jackman@ec.gc.ca

Contributors

Emilia Jonczyk Stantec Consulting Ltd. R.R.2, Nicholas Beaver Road Guelph, Ontario Canada, N1H 6H9 ejonczyk@stantec.com	B. Thomas Johnson Environmental Microbiology, Columbia Environmental Research Center U. S. Geological Survey 4200 New Haven Road, Columbia Missouri 65201, U.S.A. b_thomas_johnson@usgs.gov
Lucy E.J. Lee Department of Biology Wilfrid Laurier University Waterloo, Ontario Canada, N2L 3C5 llee@wlu.ca	Don McLeay McLeay Environmental Ltd. 2999 Spring Bay Road Victoria, British Columbia Canada, V8N 5S4 mcleayenvir@islandnet.com
Jennifer A. Miller Miller Environmental Sciences Inc. Innisfil, Ontario L9S 3E9, Canada miller.smith@sympatico.ca	Raphaël Mons Laboratoire d'écotoxicologie CEMAGREF, 3bis quai Chauveau, CP 220, 69336 Lyon, France mons@lyon.cemagref.fr
Mary J. Moody Environment and Minerals Division Saskatchewan Research Council Saskatoon, Saskatchewan Canada, S7N 2X8 moody@src.sk.ca	Grzegorz Nałęcz-Jawecki Department of Environmental Health Sciences Medical University of Warsaw Banacha 1 str. 02-097 Warsaw, Poland grzes@farm.amwaw.edu.pl
Warren P. Norwood National Water Research Institute Environment Canada 867 Lakeshore Road, P.O. Box 5050, Burlington, Ontario Canada, L7R 4A6 warren.norwood@ec.gc.ca	Monica Nowierski Department of Biology University of Waterloo Waterloo, Ontario Canada, N2L 3G1, monica.nowierski@ene.gov.on.ca
Niels Nyholm Environment and Resources Technical University of Denmark, Bld 113 DK-2800 Lyngby, Denmark nin@er.dtu.dk	Alexandre R.R. Péry Laboratoire d'écotoxicologie CEMAGREF, 3bis quai Chauveau, CP 220 69336 Lyon, France alexandre.pery@cemagref.fr

Contributors

Hans G. Peterson WateResearch Corp. 11 Innovation Boulevard Saskatoon, Saskatchewan Canada, S7N 3H5 hanspeterson@sasktel.net	Norma Ruecker WateResearch Corp. 11 Innovation Boulevard Saskatoon, Saskatchewan Canada, S7N 3H5 N.Ruecker@provlab.ab.ca
Kristin Schirmer Department of Cell Toxicology UFZ-Centre for Environmental Research Permoserstrasse, 15 04318, Leipzig, Germany kristin.schirmer@ufz.de	Jerry C. Smrchek U.S. Environmental Protection Agency Office of Prevention, Pesticides and Toxic Substances 1200 Pennsylvania Ave., N.W. Washington, D.C., 20460, USA, smrchek.jerry@epamail.epa.gov
Terry W. Snell School of Biology Georgia Institute of Technology Atlanta, GA 30332-0230, U.S.A. terry.snell@biology.gatech.edu	Jennifer Stauber Centre for Environmental Contaminants Research CSIRO Energy Technology, Private Mail Bag 7 Bangor, NSW 2234, Australia Jenny.Stauber@csiro.au
Jane P. Staveley ARCADIS G&M, Inc. 4915 Prospectus Drive, Suite F Durham, North Carolina, 27713, USA jstaveley@arcadis-us.com	Paule Vasseur U. de Metz, EBSE Campus Bridoux, rue du Général Delestraint 57070 METZ, France vasseur@sciences.univ-metz.fr
Gary Wohlgeschaffen Fisheries and Oceans Canada Bedford Institute of Oceanography P.O. Box 1006, Dartmouth, NS Canada, B2Y 4A2 wohlgeschaffeng@mar.dfo-mpo.gc.ca	

Reviewers

Sharon G. Berk Water Resources Center P.O. Box 5033 Tennessee Technological University Cookeville, Tennessee 38505 U.S.A.	Denny R. Buckler Columbia Environmental Research Center U.S. Geological Survey 4200 New Haven Road Columbia, Missouri 65201 U.S.A.
G. Allen Burton Institute for Environmental Quality Wright State University Dayton, Ohio 45435 U.S.A.	Marc Crane Crane Consultants Chancel Cottage 23 London Street Farington Oxfordshire SN7 7AG, UK
Kenneth G. Doe Toxicology Laboratory Environment Canada, ECB Environmental Science Centre P.O. Box 23005 Moncton, New Brunswick Canada, E1A 6S8	James F. Fairchild U.S. Geological Survey 4200 New Haven Rd. Columbia, Missouri 65201 U.S.A.
Przemyslaw Fochtman Institute of Organic Industry, Dept. of Ecotoxicology Doswiadczalna 27 43-200 Pszczyna, Poland	Manon Harwood Centre Saint-Laurent, Env. Canada 105 McGill, Montreal, Quebec Canada, H2Y 2E7
Paula Jackman Environment Canada Environmental Science Centre P.O. Box 23005, Moncton NB, E1A 6S8, Canada	Anne Kahru National Institute of Chemical Physics and Biophysics Laboratory of Molecular Genetics Akadeemia tee 23 Tallinn 12618, Estonia
George Khachakourians College of Agriculture University of Saskatchewan 51 Campus Drive Saskatoon, Saskatchewan Canada, S7N 5A8	Henry O. Krueger Aquatic Toxicology/Terrestrial Plants and Insects Wildife International Ltd. 8598 Commerce Drive Easton, Maryland 21601 U.S.A.

Reviewers

Jussi V.K. Kukkonen University of Joensuu Dept. of Biology P.O. Box 111 FIN-80101 Joensuu Finland	Michael Lewis U.S. Environmental Protection Agency 1 Sabine Island Drive Gulf Breeze, Florida 32561 U.S.A.
Malcolm J. McCormick RMIT University GPO Box 2476V Melbourne 3001 Australia	Carmel Mothersill Medical Physics and Applied Radiation Sciences Unit Nuclear Research Building, room 228 McMaster University 1280, Main Street West Hamilton, Ontario Canada, L8S 4K1
Guido Persoone Microbiotests, Inc. Industriezone "De Prijkels" Venecoweg 19 9810 Nazareth, Belgium	Jim Petty USGS 373 McReynolds Hall Columbia, Missouri 65211 U.S.A.
David Poirier Laboratory Services Branch Ontario Ministry of the Environment 125 Resources Rd., Etobicoke, Ontario Canada, M9P 3V6	Brian Quinn Centre Saint-Laurent, Env. Canada 105 McGill, Montreal, Quebec Canada, H2Y 2E7
Pascal Radix Bayer CropScience AG Ecotoxicology Department Alfred Nobel Strasse 50 40789 Monheim am Rhein Germany	Hans Toni Ratte Aachen University Department of Biology V Ecology, Ecotoxicology, Ecochemistry Worringerweg 1 D-52056 Aachen, Germany
Roberto Rico-Martinez Universidad Autonoma de Aguascalientes Departmento de Quimica Avenida de Universidad 940 C.P. 20100 Aguascalientes, Ags. Mexico	Alicia Ronco CIMA, Facultad de Ciencias Exactas Universidad Nacional de la Plata La Plata, Argentina. 47 y 115, (1900) La Plata, Argentina

Reviewers

Kirstin Ross Department of Environmental Health Room 4E431.1 Flinders Medical Centre Flinders University GPO Box 2100 Adelaide SA 5001 Australia	Philippe Ross Environmental Science and Engineering Division Colorado School of Mines Golden, CO 80401 U.S.A.
Robert L. Roy Fisheries and Oceans Canada Institut Maurice-Lamontagne 850, route de la Mer, C.P. 1000 Mont-Joli, Quebec Canada, G5H 3Z4	Julie Schroeder Aquatic Toxicology Unit Laboratory Services Branch Ontario Ministry of the Environment 125 Resources Road Etobicoke, Ontario Canada, M9P 3V6
James Sherry Ecosystem Health Assessment Aquatic Ecosystem Protection Research Branch National Water Research Institute Burlington, Ontario Canada, L7R 4A6	Paul Sibley Department of Environmental Biology University of Guelph Guelph, Ontario Canada, N1G 2W1
Jennifer Stauber Centre for Environmental Contaminants Research CSIRO Energy Technology New Illawarra Rd Lucas Heights PMB 7, Menai NSW 2234 Australia	Jane P. Staveley ARCADIS G&M, Inc. 4915 Prospectus Drive, Suite F, Durham, North Carolina, 27713 U.S.A.
K. L. Tay Waste Management and Remediation Section Environment Canada 45 Alderney Drive Dartmouth, Nova Scotia Canada, B2Y 2N6	Sylvain Trottier Centre Saint-Laurent, Env. Canada 105 McGill, Montreal, Quebec Canada, H2Y 2E7
Don Versteeg The Proctor & Gamble Company Environmental Science Department Miami Valley Laboratory P.O. Box 538707, Rm. 1A04S BTF Cincinnati, Ohio 253-8707, U.S.A.	

Preface

Developed, developing and emerging economies worldwide are collectively contributing multiple stresses on aquatic ecosystems by the release of numerous contaminants. This in turn demands that basic toxicological information on their potential to harm living species be available. Hence, environmental protection programs aimed at preserving water quality must have access to comprehensive toxicity screening tools and strategies that can be applied reliably and universally.

While a good number of toxicity testing procedures and hazard assessment approaches have been published in the scientific literature over the past decades, many are wanting in that insufficient detail is available for users to be able to fully understand the test method or scheme and to be able to reproduce it successfully. Even standardized techniques published in recognized international standard organization documents are often lacking in thoroughness and *minutiae*. Paucity of information relating to biological test methods may be consequent and trigger several phenomena including generation of invalid data and resulting toxicity measurements, erroneous interpretation and decision-taking with regards to a particular chemical or environmental issue, or simply abandonment of testing procedures. Clearly, improperly documented toxicity testing methods can be detrimental to their promotion and use, as they open the doorway to unnecessary debate and criticism as to their *raison d'être*. Furthermore, this situation can indirectly contribute to delaying, minimizing or eliminating their application, thereby curtailing the important role toxicity testing plays in the overall protection and conservation of aquatic ecosystems.

The "cry for help" that we have often heard from people having encountered difficulties in properly conducting biological tests was the primary trigger that set off our desire to edit a book on freshwater toxicity testing procedures in the detailed manner described herein. We feel this book is rather unique in that it includes 1) a broad review on toxicity testing applications, 2) comprehensive small-scale toxicity test methods (Volume 1) and hazard assessment schemes (Volume 2) presented in a designated template that was followed by all contributors, and 3) a complete glossary of scientific/technical terms employed by editors/contributors in their respective chapters.

Indeed, the book provides information on the purposes of applying toxicity tests and regroups 15 validated toxicity test methods (Volume 1) and 11 hazard assessment schemes (Volume 2) for the benefit and use of the scientific community at large. Academia (students, professors), government (environmental managers, scientists, regulators) and consulting professionals (biologists, chemists, engineers) should find it of interest, because it encompasses, into a single document, comprehensive information on biological testing which is normally scattered and difficult to find. It should be, for example, very useful for (under)graduate courses in aquatic toxicology involving practical laboratory training. In this respect, it can be attractive, owing to some of its contents, as a laboratory manual for learning purposes or for undertaking applied research to assess chemical hazards. As a further example, it can also prove useful for environmentalists who wish to select the most appropriate test(s) or scheme(s) for future decision-taking with regards to protection of aquatic ecosystems. In short, all groups directly or indirectly involved with the protection and conservation of freshwater environments will find this book appealing, as will those who simply wish to become familiar with the field of toxicity testing.

We are grateful for the financial support given to us in the production of this book by Environment Canada (Centre Saint-Laurent, Québec region, Environmental Conservation), the University of Metz (Metz, France) and IDRC (International Development Research Centre, Ottawa, Ontario, Canada). For their assistance in many dedicated ways which facilitated our tasks and ensured the timely completion of our book, we extend our thanks to the following persons: Mr. Andrés Sanchez and Dr. Jean Lebel (IDRC); Ms. Jacinthe Leclerc, Dr. Alex Vincent and Dr. André Talbot (Centre Saint-Laurent); Ms. Sylvie Bibeau and Dr. Laura Pirastru (University of Québec in Montréal). We are also very appreciative of the dedicated professional help provided us by Anna Besse and Judith Terpos of **Springer Publishers** in guiding us through the editorial process.

Again, how could we not extend our appreciation to all of our devoted colleagues who accepted our invitation to contribute a chapter to this book? They number 54 in total and represent 11 countries including Argentina, Australia, Canada, Chile, Columbia, Denmark, France, Germany, Poland, Switzerland and the U.S.A. Needless to say that it is owing to their outstanding career experience and interest to promote their know-how that *Small-scale Freshwater Toxicity Investigations (Volume 1 and Volume 2)* has now become a reality. Last but not least, the ultimate acknowledgment must go to our other estimated colleagues who acted as peer-reviewers for all manuscript contributions and who significantly contributed to their final quality.

We are convinced that this book fills an important scientific gap that will stimulate international use and application of small-scale toxicity tests, whether for research, monitoring, or educational purposes. May the "blue planet" and its aquatic species ultimately profit from such endeavours!

Christian Blaise and Jean-François Férard

January, 2005

Foreword

Much has been said and done since the International Decade for Water and Sanitation of the 1980s to improve access to sufficient and safe drinking water in developing countries. Although we are nowhere near achieving universal access to this basic human need, progress has been accomplished. Technology has played an important role, but another critical legacy of the Decade has been a much better recognition and understanding of the social factors linked to sustainable access to safe drinking water for communities in developing countries.

One of the empowering factors has been the development of simple and affordable technologies for monitoring microbial water quality. Because they are inexpensive and are not dependent of sophisticated laboratories, such technologies have made their way into areas where electrical power has yet to reach and have allowed communities to perform their own water quality monitoring. The identification of specific micro-organisms are less important to rural inhabitants than an alarm system which they can depend on to consistently alert them to fecal contamination of their water supply. With water-borne diarrhea still causing the second highest mortality and morbidity toll in Third World countries (mainly infants and young children) the precautionary principle remains the only responsible strategy for poor communities.

Although fecal contamination of drinking water is still a serious problem in developing countries, it is not the only risk that need concern their populations and ecosystems. Both natural and anthropogenic processes are known to cause another kind, but no less dangerous contamination: recent surveys have shown for example that upwards of 36 million people in the Indian sub-continent are drinking water contaminated by arsenic; such contamination is also known to occur in the Southern Cone of Latin America and in areas of China. In Bangladesh, sadly, this problem has been compounded by altruistic efforts of AID agencies, digging wells to offer an alternative to fecally contaminated surface waters. Alas, the geologic makeup of the region has caused underground water to be heavily laced with Arsenic. Serious pathological manifestations have now been reported in affected areas. Some areas of India have also reported high fluoride concentration in well water leading to severe fluorosis in children and adults alike, with severe skeletal malformations and attendant physiological problems.

Human activity has also exacerbated this problem: Mercury contamination related to gold mining in frontier areas of South America; contamination of both surface and ground water by agricultural inputs such as pesticides and fertilizers; increased chemical pollution by recently implanted industries; global pollution by persistent chemicals used in industrialized countries such as PCBs and bromine-containing fire retardants. Unquestionably, the past and continuing release of toxicants of this nature to receiving waters, one of earth's crucial compartments, by way of numerous (non) point sources of pollution, have equally impaired the health of aquatic biota and even adversely affected the biodiversity of some of its communities (*e.g.*, invertebrates and fish). Indeed, while microbiological pollution poses predominantly a risk to human health, chemical contamination represents a much more global threat to all components of the ecosystem, with a potential for more profound and enduring consequences.

In most cases, laboratory analytical methods exist to detect such chemicals and to quantify them. However, they can be time consuming and very expensive. No one could even propose that screening programs could be set up for routine water testing which would be both timely and affordable. In fact, this would not be feasible for industrialized countries either. How is one to test water for safety from chemicals, then? One approach is to perform routine analysis for specific chemicals in a given area where they are presumed to exist. Therein lies a cautionary tale: in the early nineties, the British Geological Survey (BGS) carried out a survey of well waters in Bangladesh (in relation to the well digging program discussed earlier), seeking data on iron and phosphorus which were presumed to contaminate the water. No attempts were made to measure other toxic compounds such as arsenic, which we now know constituted a major contaminant. Following the appearance of severe arsenic poisoning in the affected area, Bangladesh sued the agency for failing to warn users that the toxic metal was present in well water. The BGS was cleared by a British court of any wrong doing, since the former had performed the assays for which their services had been retained – and which did not include assays for other contaminants. Could this situation have not been avoided if a test had been applied to evaluate the overall toxicity of water, irrespective of the contaminant present? What about waters which exhibit contamination by multiple chemicals: individual measurements may not give an assessment of the true toxicity if these chemicals act in synergy rather than in an additive fashion.

Thus, some environmental scientists suggest that tests be used that measure "toxicity" rather than individual contaminants. Toxic samples could then be further assayed for specific contaminants if necessary to identify point sources and/or water treatment procedures. Relatively rapid, affordable and dependable assays would be a boon for developing country communities, in the same way as earlier rapid tests were for fecal contamination. The latter have proven to be usable in a sustainable manner in developing country communities, empowering them to monitor water safety and to act appropriately when necessary.

Bioassays appeared to fit the bill to perform this service to monitor chemical contamination. They have been around for a while. Until relatively recently, however, they remained in the realm of the laboratory. Only over the last two decades have they found a niche in testing for toxic chemicals in water and sediment, but not yet specifically as a tool for routine water quality monitoring. As ***Small-scale Freshwater Toxicity Investigations (Volume 1 and Volume 2)*** amply demonstrates, the science has now come of age. Assays based on bacteria, microscopic or multicellular algae, protozoa, invertebrates and vertebrates (freshwater fish cell cultures) are discussed in Volume 1 of this book. Of equal importance to my mind, Volume 2 of the book describes hazard assessment schemes that are based on combinations of the various bioassays, the so-called "battery" of tests. Indeed, all organisms are not similarly sensitive to given toxics. For instance, algae are likely to be very sensitive to herbicides albeit at levels which are unlikely to represent a danger to humans, while vertebrate cells may be less so. Thus, testing the sample on a series of organisms is more likely to reflect an overall toxicity. Whether one is to assess the risk to aquatic organisms or human beings, it is important to monitor the toxicity of samples on more than one trophic level.

Another significant advance is the development of a number of schemes to combine the results of toxicity testing on multiple trophic levels into indices which could be used to standardize results from one sample to another, from one area to another. ***Small-scale toxicity testing for freshwater environments*** presents a number of such schemes, and for this the editors should be congratulated. Only through such approaches can we begin to promote the use of these techniques more generally, especially if we are to encourage their use by field workers who have at best a limited experience of analytic laboratory techniques. Along with the other excellent chapters on hazard assessment schemes described in this book, the paper by Ronco, Castillo and Diaz-Baez *et al.* is significant to my mind because these authors have been working with municipal governments of Latin America (Argentina, Chile and Mexico) to promote WaterTox$^©$. This is a battery of tests which they developed with colleagues elsewhere in Latin America, Canada, India and the Ukraine, with support from the International Development Research Centre (IDRC), the National Water Research Institute (Burlington, Ontario, Environment Canada) and the Saint-Lawrence Centre (Montreal, Quebec, Environment Canada). Results produced by this network of superb scientists have been extremely well received and, in some countries, governments are already incorporating batteries of bioassays in the national water quality testing programs (notably the Ukraine, Mexico and Chile).

All of this bodes very well for the future of bioassays, and for their transfer to poorer communities of the Third World where perhaps they are most needed.

Gilles Forget
Regional Director
In Central and West Africa
International Development Research Centre

Contents

PART 1. Introduction

Overview of contemporary toxicity testing..1
C. Blaise and J.-F. Férard, editors

PART 2. Toxicity test methods

Bacteria

Chapter 1. Microtox acute toxicity test 69
B. T. Johnson

Chapter 2. Solid-phase test for sediment toxicity using the luminescent
bacterium, *Vibrio fischeri*..107
K. Doe, P. Jackman, R. Scroggins, D. McLeay and G. Wohlgeschaffen

Micro-algae and aquatic macrophytes

Chapter 3. Algal microplate toxicity test..137
C. Blaise and P. Vasseur

Chapter 4. Algal toxicity test..181
J. P. Staveley and J. C. Smrchek

Chapter 5. Microalgal toxicity tests using flow cytometry........................203
J. Stauber, N. Franklin and M. Adams

Chapter 6. Algal microplate toxicity test suitable for heavy metals..........243
H. G. Peterson, N. Nyholm and N. Ruecker

Chapter 7. *Lemma minor* growth inhibition test......................................271
M. Moody and J. Miller

Protozoans

Chapter 8. Spirotox test – *Spirostomum ambiguum* acute toxicity test 299
G. Natęcz-Jawecki

Micro-invertebrates

Chapter 9. Rotifer ingestion test for rapid assessment of toxicity 323
T. W. Snell

Chapter 10. Acute and chronic toxicity testing with *Daphnia* sp. 337
E. Jonczyk and G. Gilron

Chapter 11. *Hydra* population reproduction toxicity test method 395
D. A. Holdway

Chapter 12. Amphipod (*Hyalella azteca*) solid-phase toxicity test using
High Water-Sediment ratios 413
U. Borgmann, W.P. Norwood and M. Nowierski

Chapter 13. *Chironomus riparius* solid-phase assay 437
A.R.R. Péry, R. Mons and J. Garric

Fish cells

Chapter 14. Acute toxicity assessment of liquid samples with primary
cultures of rainbow trout hepatocytes 453
F. Gagné

Chapter 15. Rainbow trout gill cell line microplate cytotoxicity test 473
V. R. Dayeh, K. Schirmer, L. E. Lee and N. C. Bols

Glossary...505

Index...539

OVERVIEW OF CONTEMPORARY TOXICITY TESTING

CHRISTIAN BLAISE
St. Lawrence Centre, Environment Canada
105 McGill Street, Montreal
Quebec H2Y 2E7, Canada
christian.blaise@ec.gc.ca

JEAN-FRANÇOIS FÉRARD
Université Paul Verlaine
Laboratoire Ecotoxicité et Santé Environnementale
CNRS FRE 2635, Campus Bridoux,
rue du Général Delestraint
57070 METZ, France
ferard@sciences.univ-metz.fr

Preamble

In co-editing this book on ***Small-scale Freshwater Toxicity Investigations (Volume 1 and Volume 2)*** we felt it would be of value to bring to light the numerous types of publications which have resulted from the development and use of laboratory bioassays over the past decades. Knowing why toxicity testing has been conducted is obviously crucial knowledge to grasp the importance and breadth of this field.

Our tracking of publications involving toxicity testing was carried out with several databases (Poltox, Current Contents, Medline, Biosis and CISTI: Canada Institute for Scientific and Technical Information) and key words tailored to our objectives. In undertaking our search of the literature, we exclusively circumscribed it to articles or reports dealing with toxicity testing performed in the context of freshwater environments – obviously the focus of this book. Excluded from this review are publications describing sub-cellular bioassays (*e.g.*, assays conducted with sub-mitochondrial particles or where specific enzymes are directly exposed to contaminants) and those carried out with recombinant DNA (micro)organisms (*e.g.*, promoter/reporter bacterial constructs) and biosensors. These essentially newer techniques are unquestionably of interest and will be called upon to play increasingly useful roles in the area of small-scale environmental toxicology in the future, but they are clearly beyond the primary aims of this book.

While this review cannot be judged exhaustive, it is nevertheless representative of toxicity tests developed and applied at different levels of biological organization to comprehend toxic effects associated with the discharge of xenobiotics to aquatic environments. In reading this chapter, it is our hope that readers will get a broad sense of the versatile ways in which bioassays have been used by the scientific community at large and of the genuine role they play - along with other tools and approaches in ecotoxicology - in ensuring the protection and conservation of the freshwater aquatic environment.

Introduction

Laboratory toxicity tests have been developed and conducted over the past decades to demonstrate adverse effects that chemicals can have on biological systems. Along with other complementary tools of ecotoxicology available to measure (potential or real) effects on aquatic biota (*e.g.*, microcosm, mesocosm and field study approaches with assessment of a variety of structural and/or functional parameters), they have been, and continue to be, useful to indicate exposure-effect relationships of toxicants under defined, controlled and reproducible conditions (Adams, 2003).

Among their multiple uses, acute and chronic bioassays have served, for example, to rank and screen chemicals in terms of their hazardous potential, to undertake biomonitoring studies, to derive water quality criteria for safe release of single chemicals into aquatic bodies and to assess industrial effluent quality in support of compliance and regulatory statutes.

Because of the pressing contemporary need to assess an ever-growing number of chemicals and complex environmental samples, the development and use of small-scale toxicity tests (also called "micro-scale toxicity tests" or "microbiotests") have increased because of their attractive features. Simply defined as "a test involving the exposure of a unicellular or small multicellular organism to a liquid or solid sample in order to measure a specific effect", small-scale tests are generally simple to execute and characterized by traits which can include small sample volume requirements, rapid turnaround time to results, enhanced sample throughput and hence cost-effectiveness (Blaise et al., 1998a).

Small-scale toxicity tests are numerous and their relative merits (and limitations) for undertaking environmental assessment have been amply documented (Wells et al., 1998; Persoone et al., 2000). The small-scale toxicity tests methods described in this book and the hazard assessment schemes into which they can be incorporated are certainly representative of the field of small-scale aquatic toxicology and of tests and approaches being applied actively in today's world.

Our scrutiny of publications identified in the literature search has enabled us to uncover the various ways in which laboratory toxicity tests have been applied, many of which are small-scale in nature. We have assembled papers based on their application affinities and classified them into specific sections, as shown in Figure 1. This classification scheme essentially comprises the structure of this chapter and each section is subsequently commented hereafter.

Main categories of aquatic bioassay applications based on representative publications involving toxicity testing

1. Liquid media toxicity assessment

- 1.1 Environmental samples
- 1.2 Chemical contaminants
- 1.3 Biological contaminants

2. Sediment toxicity assessment

- 2.1 Assessment of areas of concern
- 2.2 Critical body residues and links to (sub)lethal toxicity responses

3. Miscellaneous studies/initiatives linked to aquatic toxicity testing applications (liquid media and sediments)

- 3.1 Endeavors promoting development, validation and refinement of toxicity testing procedures
  - 3.1.1 Test method development
  - 3.1.2 Inter-calibration exercises
  - 3.1.3 Comparative studies
  - 3.1.4 Factors capable of affecting bioassay responses
- 3.2 Initiatives promoting the use of toxicity testing procedures
  - 3.2.1 Review articles, biomonitoring and HAS articles
  - 3.2.2 Standardized test methods and guidance documents

Figure 1. Presentation pathway for the overview on toxicity testing exposed in this chapter.

In discussing the developments and applications of bioassays to liquid media and to sediments, we have placed some emphasis on the types of chemicals and environmental samples that have been appraised, on the types and frequency of biotic level(s) employed, as well as on the relative use of single species tests as opposed to test battery approaches.

1. Liquid media toxicity assessment

1.1 ENVIRONMENTAL SAMPLES

Articles related to toxicity testing of waters, wastewaters and other complex media are separated into three groups: studies involving toxicity testing of wastewaters and solid waste leachates (Tab. 1); studies involving toxicity testing of specific receiving media and sometimes including wastewaters (Tab. 2); studies combining toxicity/chemical testing and sometimes integrating other disciplines to assess waters, wastewaters and solid waste leachates (Tab. 3). While some investigations have strictly sought to measure bioassay responses after exposure to (waste)waters (Tables 2 and 3), an equally important number have combined toxicity and chemical testing in an attempt to establish a link between observed effects and putative chemical stressors present in appraised samples (Tab. 3). In both cases, a wide variety of point source effluent wastewaters of diverse industrial and municipal origins, as well as solid matrix leachates and various receiving media have been assessed. On the industrial scene, pulp and paper wastewaters appear to have received more overall attention than other industrial sectors, very likely owing to the fact that the forestry industry is a major enterprise internationally. Historically, also, pulp and paper mills were notorious for their hazardous discharges to aquatic environments (Ali and Sreekrishnan, 2001), although secondary treatment application has greatly reduced their toxicity (Scroggins et al., 2002b).

Table 1. Studies involving toxicity testing of wastewaters and solid waste leachates.

Assessment category	Type of bioanalytical application[a]	Biotic levels employed[b,c] (and reference)
Industrial effluents		
Dyeing factory	TT	B (Chan et al., 2003)
Electrical utilities	TBA	B,F,I (Rodgers et al., 1996)
Metal plating	TT	P (Roberts and Berk, 1993)
	TBA	B,F,I (Choi and Meier, 2001)
Mining	TT	B,B,B (Gray and O'Neill, 1997); F (Gale et al., 2003)
	TBA	B,B,F,I,I,I,I (CANMET, 1996); A,A,B,F,F,I,L (CANMET, 1997b); I,F (CANMET, 1998); Bi,F,I,I (Milam and Farris, 1998); A,F,I,L (Scroggins et al., 2002a);
Oil refinery	TT	B (Riisberg et al., 1996)
	TBA	A,A,F (Roseth et al., 1996); A,B,F,F,I,I,I,L,S (Sherry et al., 1997)
Pulp and paper	TT	F (Gagné and Blaise, 1993); B (Oanh, 1996); F (Bennett and Farrell, 1998); F (Parrott et al., 2003); F (Sepúlveda et al., 2003); F (van den Heuvel and Ellis, 2002)
	TBA	A,B,F (Blaise et al., 1987); B,B,B,I (Rao et al., 1994); A,B,L (Oanh and Bengtsson, 1995); A,B,B,F,I (Ahtiainen et al., 1996); A,B,F,F (Priha, 1996); B,F,F,I,I,I,I (Côté et al., 1999); A,F,F,I (Scroggins et al., 2002b); B,I (Pintar et al., 2004)
Tannery	TT	B,B (Diaz-Baez and Roldan, 1996)
	TBA	A,B,I,I,I,I,I,I (Isidori, 2000)
Textile	TT	I (Villegas-Navarro et al., 1999)

Table 1 (continued). Studies involving toxicity testing of wastewaters and solid waste leachates.

Assessment category	Type of bioanalytical application[a]	Biotic levels employed[b,c] (and reference)
Industrial effluents		
Various effluents	TT	F (Blaise and Costan, 1987); B (Tarkpea and Hansson, 1989); B (Svenson et al., 1992); I (Seco et al., 2003)
	TBA	B,F,F,F,F,F,I (Williams et al., 1993); B,F,I (Gagné and Blaise, 1997); B,I,I (Jung and Bitton, 1997); B,I (Liu et al., 2002)
Wood industry	TT	F (Rissanen et al., 2003)
Municipal effluents	TT	B,B,B,B,B (Codina et al., 1994); I (Monda et al., 1995); Fc (Gagné and Blaise, 1998a); Fc (Gagné and Blaise, 1999); B (Sánchez-Mata et al., 2001)
	TBA	B,B,I (Arbuckle and Alleman, 1992); A,B,F,P (George et al., 1995); B,B,F,Fc (Dizer et al., 2002); F,I (Gerhardt et al., 2002a)
Municipal and industrial effluents	TT	B (Asami et al., 1996); Fc (Gagné and Blaise, 1998b); Fc,Fc,F (Gagné and Blaise, 1998c)
	TBA	F,F,I,I,I (Fisher et al., 1989); F,F,I,I,I (Fisher et al., 1998); B,I (Doherty et al., 1999); B,F,I,I,S (Castillo et al., 2000); A,A,B,I,I,P (Manusadžianas et al., 2003)
WWTP (waste water treatment plants)	TT	B (Hoffmann and Christofi, 2001); B (Paixão and Anselmo, 2002)
	TBA	B,F,I (Sweet et al., 1997)
Solid waste leachates	TT	A (McKnight et al., 1981); B (Bastian and Alleman, 1998); B (Coz et al., 2004)
	TBA	B,B,B,F,F,I,I (Day et al., 1993); A,B,I,I,I,I,L,P (Clément et al., 1996); A,B,I,I,Pl,Pl,Pl (Ferrari et al., 1999); A,I,I,P (Törökné et al., 2000); A,A,B,B,I,I,P,S (Sekkat et al., 2001)

a) <u>TT (toxicity testing)</u>: a study undertaken with test(s) at only one biotic level. <u>TBA (test battery approach)</u>: a study involving tests representing two or more biotic levels.
b) Levels of biological organization used in conducting (or describing) TT: A (algae), B (bacteria), Bi (bivalve), F (fish), Fc (fish cells), I (invertebrates), L (*Lemnaceae*, duckweed: small vascular aquatic floating plant), P (protozoans), Pl (plant), and S (seed germination test with various types of seeds, *e.g.*, *Lactuca sativa*).
c) A study reporting the use of more than one toxicity test at the same biotic level is indicated by additional lettering (*e.g.*, use of three different bacterial tests is coded as "B, B, B").

Table 2. Studies involving toxicity testing of specific receiving media and sometimes including wastewaters.

Assessment category	Type of bioanalytical application[a]	Biotic levels employed[b,c] (and reference)
Groundwater	TBA	A,B,B,I (Dewhurst et al., 2001)
Lake	TT	I (Kungolos et al., 1998)
	TBA	A,B,B,I,S (Okamura et al., 1996); A,I (Angelaki et al., 2000)
River/Stream	TT	I (Viganò et al., 1996); Bi,I (Stuijfzand et al., 1998); I (Jooste and Thirion, 1999); I (Lopes et al. 1999); I,I (Pereira et al., 1999); I (Sakai, 2001); I (Schulz et al., 2001); A (Okamura et al., 2002); I (Sakai, 2002a); I (Williams et al., 2003)
	TBA	A,B,F,I (Wilkes and Beatty-Spence, 1995); B,B,B,I,I (Dutka et al., 1996); A,F,F,I,L (CANMET, 1997c); A,I (Baun et al., 1998); B,B,I (Sabaliunas et al., 2000); A,B,I,I,I (Van der Wielen and Halleux, 2000)
Wetland	TT	B (Dieter et al., 1994)
Specific types of environmental samples		
Packaged water	TT	P (Sauvant et al., 1994)
Pond	TT	I,I,I (Lahr, 1998)
Rainwater	TT	I (Sakai, 2002b)
Rice field	TBA	A,I (Cerejeira et al., 1998)
Runoff water	TT	A (Wong et al., 2001); I (Boulanger and Nikolaidis, 2003)
	TBA	B,B,I (Marsalek et al., 1999); A,B (Heijerick et al., 2002)
Diverse types of environmental samples [d]	TT	B (Coleman and Qureshi, 1985); I (Samaras et al., 1998); I (Lechelt, 2000); A (Graff et al., 2003); Fc (Schweigert et al., 2002)

Table 2 (continued). Studies involving toxicity testing of specific receiving media and sometimes including wastewaters.

Assessment category	Type of bioanalytical application[a]	Biotic levels employed[b,c] (and reference)
Diverse types of environmental samples [d]	TBA	B,B,I (Cortes et al., 1996); B,I (Pardos et al., 1999a); A,I,I,L,P (Blinova, 2000); A,I,I,P (Czerniawska-Kusza and Ebis, 2000); A,I,I,P (Dmitruk and Dojlido, 2000); A,I,I,I (Isidori et al., 2000); B,I,I,P (Stepanova et al., 2000) A,I,I,S,S (Arkhipchuk and Malinovskaya,2002); A,I,I,S (Diaz-Baez et al., 2002); A,I,I (Mandal et al., 2002); A,I,I,S (Ronco et al., 2002)

a) TT (toxicity testing): a study undertaken with test(s) at only one biotic level. TBA test battery approach): a study involving tests representing two or more biotic levels.
b) Levels of biological organization used in conducting (or describing) TT: A (algae), B (bacteria), Bi (bivalve), F (fish), Fc (fish cells), I (invertebrates), L (*Lemnaceae*, duckweed: small vascular aquatic floating plant), P (protozoans), and S (seed germination test with various types of seeds, *e.g.*, *Lactuca sativa*).
c) A study reporting the use of more than one toxicity test at the same biotic level is indicated by additional lettering (*e.g.*, use of three different bacterial tests is coded as "B, B, B".
d) Includes samples such as potable/surface waters, as well as industrial effluents, soil/sediment/sludge extracts, landfill leachates and snow, where individual studies report testing one or more sample type(s).

Table 3. Studies combining toxicity/chemical testing and sometimes integrating other disciplines to assess waters, wastewaters and solid waste leachates.

Assessment category	Type of bioanalytical application[a]	Biotic levels employed[b,c] (and reference)
Industrial effluents		
Chemical plant	TT	B (Chen et al., 1997)
	TBA	B,I,I,I (Guerra, 2001)
Coal industry	TBA	A,I,I,I (Dauble et al., 1982); F,I,I (Becker et al., 1983)
Coke	TBA	A,B (Peter et al., 1995)
Complex munitions	TBA	A,A,A,A,F,F,F,F,I,I,I,I (Liu et al., 1983)
Mining	TT	I,I (Fialkowski et al., 2003)
	TBA	F,I (Erten-Unal et al., 1998); A,B (LeBlond and Duffy, 2001)
Pharmaceutical	TBA	A,B,B,B,F,I (Brorson et al., 1994); B,I (Tišler and Zagorc-Koncan, 1999)

Table 3 (continued). Studies combining toxicity/chemical testing and sometimes integrating other disciplines to assess waters, wastewaters and solid waste leachates.

Assessment category	Type of bioanalytical application[a]	Biotic levels employed[b,c] (and reference)
Industrial effluents		
Pulp and paper	TBA	B,I,F (Dombroski et al., 1993); B,F,I (Leal et al., 1997); B,F,I (Middaugh et al., 1997); A,B,B,F,I (Ahtiainen et al., 2000); B,I,I,P,P (Michniewicz et al., 2000)
Resin production	TBA	A,B,F,I (Tišler and Zagorc-Koncan, 1997)
Tannery	TT	I,I (Cooman et al., 2003)
	TBA	B,I (Fernández-Sempere et al., 1997); B,I (Font et al., 1998)
Tobacco plant	TBA	A,B,B,B,B,P,P (Sponza, 2001)
Water based drilling muds	TBA	A,I (Terzaghi et al., 1998)
Oily waste		
Olive oil	TBA	B,I,I (Paixão et al., 1999)
Oil refinery	TT	B (Aruldoss and Viraraghavan, 1998)
	TBA	A,B,B,F,F,I,I,I,L,S (Sherry et al., 1994); B,F,I (Bleckmann et al., 1995)
Oil-shale	TT	B,B,B (Kahru et al., 1996)
	TBA	B,B,I,I,I,P (Kahru et al., 1999); A,B,B,B,I,I,I,I,P (Kahru et al., 2000)
Composting oily waste	TBA	B,B,B,B,B,I,I,I,L,S (Juvonen et al., 2000)
Municipal effluents	TT	B (Pérez et al., 2001)
	TBA	B,B,Pl,Pl,S (Monarca et al., 2000)
WWTP (waste water treatment plant)	TT	B (Chen et al., 1999); I (Kosmala et al., 1999); B,B,B (Gilli and Meineri, 2000); B (Svenson et al., 2000); B (Wang et al., 2003)
	TBA	F,I (Fu et al., 1994); A,Fc,I (Pablos et al., 1996); B,B,B,B,P (Ren and Frymier, 2003)
Leachates		
From agricultural production solid waste	TT	B (Redondo et al., 1996)
From industrial solid waste	TT	L (Jenner and Janssen-Mommen, 1989); B (Coya et al., 1996); I,I (Rippon and Riley, 1996); I,I,I,I,I,I (Canivet and Gibert, 2002)

Table 3 (continued). *Studies combining toxicity/chemical testing and sometimes integrating other disciplines to assess waters, wastewaters and solid waste leachates.*

Assessment category	Type of bioanalytical application[a]	Biotic levels employed[b,c] (and reference)
Leachates		
From industrial solid waste	TBA	A,B,I (Lambolez et al., 1994); B,B,B,B,L,S,S,S (Joutti et al., 2000); A,B,I (Malá et al., 2000); A,B,B,I (Vaajasaari et al., 2000)
From municipal solid waste	TBA	A,A,B,I,I,S (Latif and Zach, 2000); A,B,B,F,I,I (Rutherford et al., 2000); A,B,I (Ward et al., 2002a)
Miscellaneous types of environmental samples [d]	TT	I (Gasith et al., 1988); I (Doi and Grothe, 1989) B (Bitton et al., 1992); I (Jop et al., 1992); A (Wong et al., 1995); B (Hao et al., 1996); I (Blaise and Kusui, 1997); B,B (Hauser et al., 1997); I (Eleftheriadis et al., 2000); F (Liao et al., 2003); I (Kszos et al., 2004); A,I,I,P,S (Latif and Licek, 2004)
	TBA	F,I,I (Tietge et al., 1997); A,B,I,I,I (Kusui and Blaise, 1999); A,A,I,I,P (Manusadžianas et al., 2000)
Natural waters		
Floodplain	TBA	B,I,I,I,I (de Jonge et al., 1999)
Groundwater	TBA	A,B,I,P,P,P (Helma et al., 1998); B,F,I (Gustavson et al., 2000)
Rivers and streams	TT	A (Guzzella and Mingazzini, 1994); Bi,I,I (Crane et al., 1995); I (Bervoets et al., 1996); A,A (O'Farrell et al., 2002)
Wetland	TT	B (Boluda et al., 2002)

a) TT (toxicity testing): a study undertaken with test(s) at only one biotic level. TBA (test battery approach): a study involving tests representing two or more biotic levels.
b) Levels of biological organization used in conducting (or describing) TT: A (algae), B (bacteria), Bi (bivalve), F (fish), Fc (fish cells), I (invertebrates), L (*Lemnaceae*, duckweed: small vascular aquatic floating plant), P (protozoans), Pl (plant), and S (seed germination test with various types of seeds, *e.g.*, *Lactuca sativa*).
c) A study reporting the use of more than one toxicity test at the same biotic level is indicated by additional lettering (*e.g.*, use of three different bacterial tests is coded as "B, B, B".
d) Includes samples such as storm waters, river waters, as well as industrial/municipal effluents, sludge extracts, where individual studies report testing one or more sample type(s).

While it is beyond our intent to discuss the main purpose(s) that prompted research groups to conduct individual investigations with particular toxicity tests, readers can access this information by consulting references of interest. Others are mentioned hereafter, however, to indicate bioanalytical endeavors that have taken place in past years. For example, Bitton et al. (1992), after developing a metal-specific bacterial toxicity assay, demonstrated its capacity to correctly pinpoint heavy-metal containing industrial wastewaters. In another venture, Roberts and Berk (1993) were motivated to undertake toxicity testing of a metal plating effluent and of a series of (in)organic chemicals in order to further validate a newly-developed protozoan chemo-attraction assay. Again, a test battery approach with chemical support to assess a coke plant effluent identified treatment methods that were superior for decontaminating the wastewater (Peter et al., 1995). In toxicity testing of tannery industry effluent samples, bacterial tests were shown to be sufficiently sensitive to act as screening tools for such wastewaters (Diaz-Baez and Roldan, 1996). In a study conducted on industrial, municipal and sewage treatment plants, toxicity testing identified chlorination as the most important contributor of toxic loading to the receiving environment (Asami et al., 1996). After a comprehensive assessment of pulp and paper mills, toxicity testing proved useful to ameliorate mill process control (Oanh, 1996). Another study conducted with three bacterial toxicity tests showed that oil-shale liquid wastes could be bio-degraded when activated sludge was pre-acclimated to phenolic wastewaters (Kahru et al., 1996). Petrochemical plant assessment using toxicity testing, chemical analysis and a TIE/TRE strategy combined to identify aldehydes as the main agent of effluent toxicity (Chen et al., 1997). Test battery assessment of a mine water discharge, which involved both toxicity testing and in-stream exposure of bivalves, helped to set a no-effect level criterion for a bioavailable form of iron (Milam and Farris, 1998). A comparison of laboratory toxicity testing and *in situ* testing of river sites downstream from an acid mine drainage demonstrated good agreement between the two approaches for the most contaminated stations (Pereira et al., 1999). A similar strategy to assess gold and zinc mining effluents confirmed the reliability of some chronic assays for routine toxicity monitoring (LeBlond and Duffy, 2001). Clearly, there are numerous reasons for conducting toxicity testing and/or chemical analysis of (waste)waters to derive relevant information that have eventually triggered enlightened decisions contributing to their improvement.

Of the 188 studies reported in Tables 1, 2 and 3, more than half (n = 101) were conducted with two or more tests representing at least two biotic levels (*i.e.*, test battery approach or TBA), as opposed to those performed with a single biotic level (n = 87). While test and biotic level selection may be based on a variety of reasons and study objectives (*e.g.*, practicality, cost, personnel availability), preference for TBAs can also be influenced by the need to assess hazard at different levels so as not to underestimate toxicity. Indeed, contaminants can demonstrate "trophic-level specificity" (*e.g.*, phytototoxic effects of herbicides) or they can exert adverse effects at multiple levels (*e.g.*, particular sensitivity of cladocerans toward heavy metals in contrast to bacteria). When TBAs are used, they are mostly conducted with two, three or four trophic levels (Tab. 4).

Whether TT (toxicity testing with single species tests at the same biotic level) or TBAs are performed, some test organisms have been more frequently used than others (Tab. 5). Invertebrates have been the most commonly employed, as had been pointed out in an earlier literature survey conducted between 1979 and 1987 (Maltby and Calow, 1989). Bacteria as well as fish and algal assays come next in frequency of use. Early standardization of invertebrate (*e.g.*, *Daphnia magna*) and bacterial test (*e.g.*, *Vibrio fischeri* luminescence assay) procedures, as well as increased miniaturization and cost-effectiveness, are likely factors explaining their popularity over the past decades. While some groups of small-scale toxicity tests (*i.e.*, fish cell, duckweed and protozoan tests) have thus far received less attention to appraise various environmental samples, recent efforts in test procedure validation and standardisation should effectively promote their use in the future (see Volume 1, Chapters 7, 8, 14 and 15).

Table 4. *Frequency of the number of biotic levels employed in test battery approaches (TBAs) for complex liquid media assessment based on the 101 TBA papers classified in Tables 1-3.*

TBA studies undertaken with:	Number and frequency (%)
Two biotic levels	39/101 (38.6)
Three biotic levels	38/101 (37.6)
Four biotic levels	19/101 (18.8)
Five biotic levels	3/101 (3)
Six biotic levels	2/101 (2)

Table 5. *Frequency of use of specific biotic levels employed in toxicity testing (TT) and test battery approaches (TBA) for complex liquid media assessment based on the 188 papers classified in Tables 1-3.*

TT and TBA studies undertaken with:	Number and frequency (%)
Algae	70/553* (12.7)
Bacteria	152/553 (27.5)
Bivalves	3/553 (< 1)
Fish	68/553 (12.3)
Fish cells	8/553 (1.5)
Invertebrates	199/553 (36.0)
Lemnaceae (duckweed)	10/553 (1.8)
Plants	3/553 (< 1)
Protozoans	23/553 (4.2)
Seeds	15/553 (2.7)

*Total number of single species tests reported in the 188 papers classified in Tables 1-3 (= sum of number of A, B, Bi, F, Fc, I, L, P, Pl, S tests indicated in the "Biotic levels employed" column).

1.2 CHEMICAL CONTAMINANTS

It has been estimated that as many as 250,000 man-made chemicals could possibly enter different compartments of the biosphere and cause adverse effects on ecosystem and human health (OSPAR, 2000). Out of concern for ensuring the protection of aquatic biota, a large number of scientists internationally have turned to bioassays as primary means of assessing the hazard (and risk) posed by these substances. Indeed, the scientific literature abounds with hundreds of publications dealing with toxicity testing of various classes of (in)organic chemicals. While it is beyond the intentions of this chapter to discuss all of these, papers have been selected that reflect the types of chemicals having undergone toxicity assessment. In general, published articles show that test organisms and biotic levels described are the same as those employed for assessing environmental samples.

Representative investigations involving toxicity assessment of metals, ions and oxidizing agents are highlighted in Table 6. Varied toxicological objectives have been pursued to evaluate metals singly or in groups of two or more with one toxicity test or with a test battery. The benefits of these initiatives to enhance our knowledge of undesirable effects that can be directed toward specific biotic levels (*e.g.*, Holdway et al., 2001), to identify useful sentinel species (*e.g.*, Madoni, 2000), or to promote useful (Couture et al., 1989) or potentially safer clean-up technologies (Leynen et al., 1998) should be fairly obvious.

Table 6. Studies involving toxicity assessment of metals, ions and oxidizing agents.

Assessment category	*Type of bioanalytical application*[a]	*Biotic levels employed*[b,c] *(and reference)*
One metal:		
Aluminium	TT: four species of invertebrates are exposed to Al over a pH range of 3.5 to 6.5.	I,I,I,I (Havas and Likens, 1985)
Cadmium	TT: a simple microcosm experiment associating two biotic levels conducted in a Petri dish allows measurement of reproduction effects on daphnids following Cd contamination of either their food source (algae) or of their water medium.	I (Janati-Idrissi et al., 2001)
Chromium (Cr^{+6})	TT: luminescent bacteria are exposed to assess the influence of pH speciation of chromium on toxicity response.	B (Villaescusa et al., 1997)

Table 6 (continued). Studies involving toxicity assessment of metals, ions and oxidizing agents.

Assessment category	Type of bioanalytical application[a]	Biotic levels employed[b,c] (and reference)
One metal:		
Copper	TT: comparison of effects occurring at molecular (DNA profiling) and population (ecological fitness parameters including acute and chronic toxicity) levels for *Daphnia magna*.	I (Atienzar et al., 2001)
Gallium	TT: assessment of inter-metallic elements used in making-high speed semiconductors such as gallium arsenic with *Cyprinus carpio*.	F (Yang and Chen, 2003)
Lead	TBA: assessment of toxicity, uptake and depuration of lead in fish and invertebrate species.	F,F,I,I (Oladimeji and Offem, 1989)
Manganese	TT: assessment at three levels of water hardness with *Ceriodaphnia dubia* and *Hyalella azteca*.	I,I (Lasier et al., 2000)
Mercury	TT: assessment of 10 mercury compounds to determine their relative toxicities to luminescent bacteria.	B (Ribo et al., 1989)
Nickel	TT: assessment with 12 species of freshwater ciliates to determine which could become, based on observed sensitivity, a good bio-indicator of waters polluted by heavy metals.	P (Madoni, 2000)
Selenium	TT: assessment of selenium compounds and relationships with uptake in an invertebrate species.	I (Maier and Knight, 1993)
Silver	TBA: assessment of toxicity to fish and invertebrates under a variety of water quality conditions.	F,I (La Point et al., 1996)
Uranium	TT: assessment of depleted uranium on the health and survival of *C. dubia* and *H. azteca*.	I,I (Kuhne et al., 2002)
Zinc	TT: assessment the influence of various ions and pH on phytotoxicity response.	A (Heijerick et al., 2002)

Table 6 (continued). Studies involving toxicity assessment of metals, ions and oxidizing agents.

Assessment category	Type of bioanalytical application[a]	Biotic levels employed[b,c] (and reference)
One metal:		
Zirconium	TBA: assessment of zirconium (ZrCl$_4$), considered of use as a P-precipitating agent to reduce the eutrophication potential of pig manure wastes to receiving environments.	A,B,F (Couture et al., 1989)
Two metals:		
Cadmium, Zinc	TT: assessment of their acute and chronic toxicity to two *Hydra* species.	I,I (Holdway et al., 2001)
Three metals:		
Arsenic, Cobalt, Copper	TT: assessment of relationships between acute toxicity and various experimental variables (*e.g.*, metal concentration in water, time of exposure, bioconcentration factor) with two fish species.	F,F (Liao and Lin, 2001)
Four metals or more:	TT: assessment of the adequacy of cultured fish cells (Bluegill BF-2) for toxicity testing of aquatic pollutants.	Fc (Babich and Borenfreund, 1987)
Ions:	TT: assessment of the phytotoxicity of high density brines (calcium chloride and calcium bromide) to *L. minor*.	L (Vujevic et al., 2000)
Rare earth elements:	TT: assessment of the aquatic toxicity of rare earth elements (La, Sm, Y, Gd) to a protozoan species.	P (Wang et al., 2000)
Oxidizing agents:	TBA: assessment of the acute toxicity of ozone, an alternative to chlorination to control biofouling in cooling water systems of power plants, to fish larvae of three species and to *D. magna*.	F,F,F,I (Leynen et al., 1998)

a) TT (toxicity testing): a study undertaken with test(s) at only one biotic level. TBA (test battery approach): a study involving tests representing two or more biotic levels.
b) Levels of biological organization used in conducting (or describing) TT: A (algae), B (bacteria), F (fish), Fc (fish cells), I (invertebrates), L (*Lemnaceae*, duckweed: small vascular aquatic floating plant), and P (protozoans).
c) A study reporting the use of more than one toxicity test at the same biotic level is indicated by additional lettering (*e.g.*, use of three different bacterial tests is coded as "B, B, B".

The toxicological properties of chemicals representing various classes and structures of organic substances have also been assessed by a series of bioassays at different levels of biological organization (Tab. 7). Featured in this table is but the tip of the iceberg in terms of the types of studies that have been conducted to further our knowledge about the hazards of anthropogenic molecules. While industrial progress has markedly enhanced the quality of life on this planet through production of countless xenobiotics synthesized for multiple human uses (*e.g.*, diverse household products and pharmaceuticals), it has also increased the risk linked to their discharge and fate in aquatic systems. Understanding their potential for adverse effects through the conduct of bioassays is clearly a first step in the right direction.

Table 7. Examples of studies involving toxicity assessment of organic substances.

Assessment category (and product tested)	Type of bioanalytical application^a, biotic levels employed^{b,c} (and reference)
Acaricide (Tetradifon)	TT: I (Villarroel et al., 1999)
Adjuvants (several used as surfactants for aquatic herbicide applications)	TT: F (Haller and Stocker, 2003)
Anti-fouling paint (TBT)	TBA: A,I (Miana et al., 1993)
Aromatic hydrocarbon (*para*-methylstyrene)	TBA: A,F,I (Baer et al., 2002)
Cationic fabric softener (DTDMAC)	TBA: A,B,B,I,I,I (Roghair et al., 1992)
Chelator ([S,S]-EDDS)	TBA: A,A,F,I (Jaworska et al., 1999)
Detergents and softeners (26 detergents and 5 softeners)	TT: I (Pettersson et al., 2000)
De-icing / anti-icing fluids	TT: B (Cancilla et al., 1997)
Disinfectant (Mono-chloramine)	TBA: F,I (Farrell et al., 2001)
Dyes (Fluorescein sodium salt, Phloxine B)	TT: I (Walthall and Stark, 1999)
Fatty acids (C_{14} to C_{18})	TT: A (Kamaya et al., 2003)
Fire control substances (Fire-Trol GTS-R and LCG-R, Phos-Chek D75-F and WD-881, Silv-Ex)	TBA: A,I (McDonald et al., 1996)
Flame retardant (Brominated diphenyl ether-99)	TBA: A,I (Evandri et al., 2003)
Fungicide (Ridomil plus 72)	TBA: F,I (Monkiédjé et al., 2000)
Herbicide (Atrazine)	TT: I (Dodson et al.,1999)
Household products (Abrasives, additives, disinfectants)	TBA: A,B,B,F,F,I (Bermingham et al., 1996)
Insecticide (Glyphosate)	TT: L (Lockhart et al., 1989)

Table 7 (continued). Examples of studies involving toxicity assessment of organic substances.

Assessment category (and product tested)	Type of bioanalytical application[a], biotic levels employed[b,c] (and reference)
Lubricant additives (Ashless dispersant A and B, Zinc dialkyldithiophosphate)	TT: A (Ward et al., 2002b)
(Tri *n*-butyl phosphate)	TBA: A,B (Michel et al., 2004)
Nitromusks (Ambrette, Setone, Moskene,Tibetene, Xylene)	TBA: A,B,I (Schramm et al., 1996)
Narcotics (*n*-alkanols)	TT: B (Gustavson et al., 1998)
Organochlorides (PCBs)	TT: B (Chu et al., 1997)
Organosulfur compounds (several benzothiophenes)	TBA: B,I (Seymour et al., 1997)
Pesticide (Cyromazine)	TT: I,I (Robinson and Scott, 1995)
Pharmaceutical compound (β-Blockers)	TBA: F,I,I,I (Huggett et al., 2002)
Phenolic compounds (Pentachlorophenol)	TBA: A,B,I,S (Repetto et al., 2001)
Phtalate esters (several)	TT: I,I,I (Call et al., 2001)
Solvents (Mono-, Di- and Tri PGEs)	TBA: A,B,F,F,F,I,I,L (Staples and Davis, 2002)
Surfactant (Genapol OX-80)	TT: A (Anastácio et al., 2000)
Volatilecompounds (*N*-nitrosodiethylamine, *N*-nitrosodimethylamine)	TBA: A,A,F,I,I (Draper III and Brewer, 1979)
Wood preservative (Bardac 2280)	TBA: F,F,F,F,I,I,I,I (Farrell et al., 1998)

a) <u>TT (toxicity testing)</u>: a study undertaken with test(s) at only one biotic level. <u>TBA (test battery approach)</u>: a study involving tests representing two or more biotic levels.
b) Levels of biological organization used in conducting (or describing) TT: A (algae), B (bacteria), F (fish), I (invertebrates), L (Lemnaceae, duckweed: small vascular aquatic floating plant) and S (seed germination test with various types of seeds, *e.g.*, Lactuca sativa).
c) A study reporting the use of more than one toxicity test at the same biotic level is indicated by additional lettering (*e.g.*, use of three different bacterial tests is coded as "B, B, B".

Several papers have also reported toxicity data for a variety of metals and organic substances simultaneously. Reasons for conducting such investigations include 1) establishing the concentrations at which chemicals exert their adverse effects (*e.g.*, at the ng/L, µg/L or mg/L levels), 2) estimating environmental risk based on measured toxicity endpoints and predicted environmental concentrations for specific chemicals and 3) defining toxicant concentrations harmful for specific biotic levels and/or assemblages of species within each level.

Studies have assessed the toxicological properties of one or more heavy metal(s) with one or more organic substance(s). Examples include copper and diazinon (van der Geest et al., 2000), cadmium and pentachlorophenol (McDaniel and Snell, 1999), several heavy metals (Cd, Cu, Ni, Pb, Zn) and organic (Chlorpyrifos, DDT, DDD, DDE, Dieldrin) toxicants (Phipps et al., 1995), and two metals (Cu, Zn) and eight surfactants (Dias and Lima, 2002). Again, test organisms employed for toxicity assessment are similar to those discussed previously and investigators make use of one or more biotic levels to undertake their evaluations.

Chemical toxicity assessment should also take into consideration the combined effects that groups of chemicals can have on living organisms. Indeed, contaminants are not discharged singly in aquatic systems but are joined by many others whose composition will depend on the origin of (non)point sources of pollution affecting particular reaches of receiving waters (*e.g.*, industrial, municipal and agricultural sources). The recognition that groups of chemicals can interact together to produce a resulting effect that can reduce (antagonistic effect) or exacerbate (synergistic effect) that of substances tested singularly has prompted scientists to appraise the toxicity characteristics of mixtures.

Published articles indicate that work has focussed on (binary, ternary, etc.) mixtures including metals, organics as well as metal/organic cocktails. For metals, examples include toxicity testing of various mixtures with algae (Chen et al., 1997), bacteria (Mowat and Bundy, 2002) and micro-invertebrates (Burba, 1999). For organics, mixtures have been assessed belonging to groups such as antifouling agents (Fernandés-Alba et al., 2002), herbicides (Hartgers et al., 1998), pesticides (Pape-Lindstrom and Lydy, 1997), and manufactured munitions (Hankenson and Schaeffer, 1991). For (in)organic mixtures, metal/pesticide (Stratton, 1987), metal/composted manure (Ghosal and Kaviraj, 2002), as well as metal/miscellaneous organic (Parrott and Sprague, 1993) combinations offer additional examples of interaction assessments. Because appraising mixtures of compounds (singularly and in binary, ternary or other combinations) is more laborious in time and effort than for single compounds, toxicity testing has, in most cases, been conducted with a single test organism, as opposed to the use of a test battery. Algal, bacterial and micro-invertebrate tests have thus far been favoured in this respect.

Another active field of research intended to estimate the toxic properties of organic compounds lies in the determination of their quantitative structure-activity relationships (QSAR). The rationale for this work is based on the fact that molecules will enter living organisms to exert adverse effects depending on their elemental composition and structure. In brief, QSARs are regression equations relating toxicological endpoints (*e.g.*, LC50s, EC50s, IC50s, NOECs) to physicochemical properties within a class of compounds. A good number of QSARs, for example, are determined with the octanol-water coefficient (K_{ow}), a well-known predictor of the tendency of a compound to be bio-accumulated. QSARs have several potential uses, some of which include 1) predicting the effects of newly-synthesized chemicals, 2) priority ranking of chemicals destined for more elaborate toxicity testing, 3) assistance in deriving water quality guidelines and 4) rapidly estimating toxicity for specific compounds when toxicity test data are unavailable (Environment Canada, 1999).

A quantitative structure-activity relationship (QSAR), for example, has been shown for aliphatic alcohols, where 96h-LC50s for fathead minnows are related to their K_{ow} status (Veith et al., 1983). Other QSARs based on K_{ow} have been reported for several classes of organics with test species including algae, invertebrates and fish (Suter, 1993). Hydrophobicity-based QSARs were also generated for fish and invertebrates with a set of 11 polar narcotics (Ramos et al., 1998) and for bacteria, fish and protozoan test organisms with a large set of (non)polar narcotic classes of chemicals (Schultz et al., 1998). QSARs were also employed to predict the biodegradation, bioconcentration and toxicity potential of more than 5000 xenobiotics (industrial chemicals, pesticides, food additives and pharmaceuticals) having a potential for release into the Great lakes basin (Walker et al., 2004). This study, in particular, illustrates the usefulness of QSARs as a cost-effective pre-screening adjunct to (significantly more expensive) monitoring studies that can then be prioritized towards those chemicals having the potential to persist and bio-accumulate in aquatic species. In these and other recent QSAR-based investigations of chemicals (Junghans et al., 2003; Choi et al., 2004; Schultz et al., 2004), it is noteworthy to mention that small-scale toxicity tests conducted with algae, bacteria, invertebrates and protozoans are used frequently.

1.3 BIOLOGICAL CONTAMINANTS

Besides the many hazards looming on aquatic life owing to the uncontrolled discharge of a myriad of chemicals, exposure to plants or microbes may also place it at risk. Indeed, toxicity tests conducted within the last decade on plant substances/extracts, and on microbes or their products (*e.g.*, metabolites), to investigate their biopesticide or toxicity potential, have indicated that species of different levels of biological organization can be adversely affected by such biological contaminants (Tab. 8). Since undesirable ecological effects to aquatic communities could result from exposure to naturally-produced chemicals or micro-organisms, documenting their toxicity potential via bioassays is fully justified.

As future applications with natural and/or genetically-modified plants and micro-organisms are expected to increase in the future (*e.g.*, for bioremediation treatments of contaminated soils, wastewaters, sediments), so will toxicity assessment programs to insure the protection of aquatic biota. In Canada, for example, information is now required to appraise new microbes (and their products) in terms of their toxicity potential toward aquatic organisms, and standardized toxicity test methods are being developed and recommended for this purpose (Environment Canada, 2004a). Risk assessment of biological contaminants is clearly an area that will receive sustained attention in the coming years.

Table 8. Examples of studies involving toxicity assessment of biological contaminants.

Assessment category and product tested	Type of bioanalytical application[a], biotic levels employed[b,c] (and reference)
Biopesticides	
Aquatic plant: essential oils from *Callicarpa americana*	TBA: A,A,A,B,B,B,B,B,S,S (Tellez et al., 2000)
Aquatic plant: phenanthrenoids from *Juncus acutus*	TT: A (DellaGreca et al., 2002)
Aquatic plant: essential oils from *Lepidium meyenii*	TBA: A,A,I,S,S (Tellez et al., 2002)
Aquatic plant: antialgal furano-diterpenes from *Potamogetonaceae*	TT: A (DellaGreca et al., 2001)
Aquatic plant: ent-labdane diterpenes from *Potamogetonaceae*	TBA: A,I,I,I,I (Cangiano et al., 2002)
Bacterium: *Bacillus thuringiensis*	TT: I (Manasherob et al., 1994); TT: I (Kondo et al., 1995)
Fungus: *Metarhizium anisopliae*	TT: B (Milner et al., 2002)
Biotoxins	
Cyanobacteria	
Microcystis aeruginosa	TBA: B,I (Campbell et al., 1994)
Anabaena sp., *M. aeruginosa*, *Microcystis* sp., *P. aghardii*, *P. rubenscens*	TT: I (Törökné, 2000; Törökné et al., 2000)
M. aeruginosa, M. wesenbergii	TBA: B,B,B,I,I,I,I,P (Maršálek and Bláha, 2000)
Cyanobacterial blooms	TBA: I,I,P,P (Tarczynska et al., 2000)
Pathogenic bacteria: *Aeromonas hydrophila*, *Flavobacter* spp., *Flexibacter columnaris*	TT: F (Geis et al., 2003)
Odor and taste compounds of microbial origin	
Geosmin, 2-methyliso-borneol	TT: Fc (Gagné et al. 1999)

a) TT (toxicity testing): a study undertaken with test(s) at only one biotic level. TBA (test battery approach): a study involving tests representing two or more biotic levels.
b) Levels of biological organization used in conducting (or describing) TT: A (algae), B (bacteria), F (fish), Fc (fish cells), I (invertebrates), P (protozoans), and S (seed germination test with various types of seeds, *e.g.*, *Lactuca sativa*).
c) A study reporting the use of more than one toxicity test at the same biotic level is indicated by additional lettering (*e.g.*, use of three different bacterial tests is coded as "B, B, B".

2. Sediment toxicity assessment

2.1 ASSESSMENT OF AREAS OF CONCERN

In today's world, sediment contamination continues to be a growing environmental issue. Indeed, the deposition of numerous (in)organic chemicals in aquatic systems stemming from various types of anthropogenic activities (urban, industrial, agricultural) has the potential to adversely affect aquatic biota. Once deposited, resuspension of contaminated sediment *via* both natural (*e.g.*, flood scouring) and man-made (*e.g.*, dredging, navigation, open water deposition) activities can further harm living organisms by increasing their contact with (and uptake of) deleterious chemicals. Integrated strategies to assess the toxic potential of contaminated sediments, such as the sediment quality triad approach (see Volume 2, Chapter 10) continue to favour the presence of a strong bioanalytical component within investigation schemes.

Our literature review has shown that sediment toxicity assessment has received marked attention over the past decades and that bioassays have been largely used for this purpose. Contaminated environments, for instance, have triggered many studies conducted to detect and quantify sediment toxicity, to determine the extent of its impact, and to enhance understanding of its short and long-term effects on aquatic communities.

To give readers a first insight into the ways in which toxicity tests have been applied for sediment assessment, we have regrouped publications dealing with sediments collected from areas of concern (Tab. 9) and those collected from other lotic and lentic environments, also impacted by pollutant discharges, where combined chemical-biological analyses were performed (Tab. 10). Sediments were collected from lakes and rivers to undertake initial assessment of sites, to study effects of diverse (in)organic contamination, as well as to investigate various toxicity aspects linked to oil spills and flooding events (Tab. 9). A number of studies also explored relationships between specific contaminants and observed toxicity effects (Tab. 10).

Table 9. Studies with field-collected sediments: assessment of areas of concern.

Assessment objective, type of bioanalytical application[a] and tested sediment phase(s)		Biotic levels employed[b,c] (and reference)
Areas impacted by wastewaters: with sediments potentially contaminated by (in)organic pollution		
Ammonia effects	TT: overlying water, pore water	I (Bartsch et al., 2003)
Initial/preliminary assessment of sites	TT: whole sediment	B (Onorati et al., 1998)
	TT: overlying water	I,I (Rediske et al., 2002)
	TT: whole sediment	I (Bettinetti et al., 2003)
	TT: whole sediment	I,I (Collier and Cieniawski, 2003)
	TBA: elutriate	A,B,I,I,I (Sloterdijk et al., 1989)
	TBA: pore water, whole sediment	B,I (Munawar et al., 2000)
Metal contamination	TT: overlying water	I,I,I (West et al., 1993)
	TT: spiked sediment, whole sediment	I,I (Dave and Dennegard, 1994)
	TT: pore water	I (Besser et al., 1995)
	TT: pore water	I (Deniseger and Kwong, 1996)
	TT: pore water	I (Call et al., 1999)
	TT: pore water	I (Hill and Jooste, 1999)
	TT: overlying water, pore water	I (Bervoets et al., 2004)
	TBA: pore water, whole sediment	B,F,F,I,I,I,I (Kemble et al., 1994)
	TBA: overlying water, pore water, whole sediment	B,I,I,I,I,S (Burton et al., 2001)
Metal and organic contamination	TT: whole sediment	I,I (Nebeker et al., 1988)
	TT: elutriate	A (Lacaze et al., 1989)
	TT: whole sediment	B,B (Kwan and Dutka, 1992)
	TT: whole sediment	I,I (Jackson et al., 1995)
	TT: elutriate	I (Bridges et al., 1996)
	TT: elutriate, pore water, whole sediment	I,I (Ristola et al., 1996)
	TT: whole sediment	B (Svenson et al., 1996)
	TT: pore, elutriate, whole sediment	I,I,I,I,I (Sibley et al., 1997b)
	TT: whole sediment	A (Blaise and Ménard, 1998)
	TT: OE[d], whole sediment	B (Salizzato et al., 1998)

Table 9 (continued). Studies with field-collected sediments: assessment of areas of concern.

Assessment objective, type of bioanalytical application[a] and tested sediment phase(s)		Biotic levels employed[b,c] (and reference)
Areas impacted by wastewaters: with sediments potentially contaminated by (in)organic pollution		
Metal and organic contamination	TT: overlying water	I (Call et al., 1999)
	TT: overlying water	I (Martinez-Madrid, 1999)
	TT: overlying water, whole sediment	I,I,I (Munawar et al., 1999)
	TT: overlying water	I,I,I (Cheam et al., 2000)
	TT: pore water	I (Kemble et al., 2002)
	TBA: pore water	B,I,I (Giesy et al., 1988)
	TBA: overlying water, whole sediment	A,B,B,B,B,I (Dutka et al., 1989)
	TBA: elutriate, whole sediment	A,I (Gregor and Munawar, 1989)
	TBA: pore water, whole sediment	B,I,I,I (Giesy et al., 1990)
	TBA: elutriate, pore water, whole sediment	A,B,B,F,I(8x) L, Pl (Ross et al., 1992)
	TBA: pore water, whole sediment	B,I,I,I (Hoke et al., 1993)
	TBA: elutriate, OE[d]	B,I,S (Lauten, 1993)
	TBA: elutriate, whole sediment	B,I,I (Moran and Chiles, 1993)
	TBA: elutriate, whole sediment	A,A,B,F,I,I (Naudin et al., 1995)
	TBA: pore water	B,B,I,I (Heida and van der Oost, 1996)
	TBA: overlying water, pore water	F,I,I (Watzin et al., 1997)
	TBA: pore water, whole sediment	A,B,I,I (Carter et al., 1998)
	TBA: pore water, whole sediment	A,B,B,B,I,I,I (Côté et al., 1998a)
	TBA: overlying water, whole sediment	B,I,I,I,S,S,S (Rossi and Beltrami, 1998)
	TBA: elutriate, OE[d]	B,I (Hong et al., 2000)
	TBA: pore water	A,B,I,I,I,I,P (Persoone and Vangheluwe, 2000)
	TBA: elutriate, OE[d]	A,B,B,I (Ziehl and Schmitt, 2000)

Table 9 (continued). Studies with field-collected sediments: assessment of areas of concern.

Assessment objective, type of bioanalytical application[a] and tested sediment phase(s)		Biotic levels employed[b,c] (and reference)
Areas impacted by wastewaters: with sediments potentially contaminated by (in)organic pollution		
Metal and organic contamination	TBA: whole sediment	B,I,I (Ingersoll et al., 2002)
	TBA: pore water	B,I,I,I,I (Lahr et al., 2003)
	TBA: pore water, whole sediment	B,I,I (Munawar et al., 2003)
Organic contamination	TBA: OE[d]	A,B,I (Santiago et al, 1993)
	TBA: pore water	B,I (Pastorok et al., 1994)
	TBA: elutriate, pore water	B,I (Hyötyläinen and Oikari, 1999)
Areas impacted by oil spill events		
Diesel fuel spill	TT: whole sediment	I,I (Keller et al., 1998)
Oil sands	TT: overlying water	F (Tetreault et al., 2003)
Oil pollution	TT: seepage water, whole sediment	I,I (Wernersson, 2004)
Simulated oil spill experiment	TT: whole sediment	B (Ramirez et al., 1996)
	TT: OE[d]	B (Johnson et al., 2004)
	TBA: whole sediment	B,B,B,I (Mueller et al., 2003)
	TBA: whole sediment	A,B,B,I,I (Blaise et al., 2004)
Areas impacted by flooding events		
Metal and organic contamination	TT: whole sediment	I (Kemble et al., 1998)
	TBA: overlying water, whole sediment	F,I,I (Hatch and Burton, 1999)

a) <u>TT (toxicity testing)</u>: a study undertaken with test(s) at only one biotic level. <u>TBA (test battery approach)</u>: a study involving tests representing two or more biotic levels.
b) Levels of biological organization used in conducting (or describing) TT: A (algae), B (bacteria), F (fish), I (invertebrates), L (*Lemnaceae*, duckweed: small vascular aquatic floating plant), P (protozoans), Pl (plant), and S (seed germination test with various types of seeds, *e.g.*, *Lactuca sativa*).
c) A study reporting the use of more than one toxicity test at the same biotic level is indicated by additional lettering (e.g., use of three different bacterial tests is coded as "B, B, B".
d) Organic (solvent) extract.

Table 10. Studies with field-collected sediments: assessment of areas of concern where combined toxicity and contaminant analysis studies were undertaken.

Assessment objective, type of bioanalytical application[a], tested sediment phase(s) and type of chemical analysis		Biotic levels employed[b,c] (and reference)
Lake sediments	TT: pore water Organic analysis	B (Guzzella et al., 1996)
	TT: elutriate, OE[d] Organic analysis	Fc (Gagné et al., 1999b)
	TT: whole sediment Organic analysis	I,I (Marvin et al., 2002)
River sediments	TT: whole sediment Heavy metal and organic analysis	I, Bc (Canfield et al., 1998)
	TT: overlying water, whole sediment Heavy metal and organic analysis	I,I,I,I (Bonnet, 2000)
	TT: pore water Heavy metal and organic analysis	I (Cataldo et al., 2001)
	TT: overlying water Heavy metal analysis	F (Bervoets and Blust, 2003)
	TT: whole sediment Organic analysis	I,I (Cieniawski and Collier, 2003)
	TBA: elutriate Organic analysis	A,B,F,I (Bradfield et al., 1993)
	TBA: elutriate Organic analysis	B,I (McCarthy et al., 1997)
	TBA: OE[d], pore water, whole sediment NH_3, heavy metal and organic analysis	A,B,B,B,B,Fc,I,I,I,I,I,I (Côté et al., 1998a,b)
	TBA: whole sediment Heavy metals	B,I,I,I (Richardson et al., 1998)

a) TT (toxicity testing): a study undertaken with test(s) at only one biotic level. TBA (test battery approach): a study involving tests representing two or more biotic levels.
b) Levels of biological organization used in conducting (or describing) TT: A (algae), B (bacteria), Bc (various benthic communities), F (fish), Fc (fish cells), and I (invertebrates).
c) A study reporting the use of more than one toxicity test at the same biotic level is indicated by additional lettering (*e.g.*, use of three different bacterial tests is coded as "B, B, B".
d) Organic (solvent) extract.

Of the 75 studies reported in Tables 9 and 10, less than half (n = 34) were conducted with two or more tests representing at least two biotic levels (*i.e.*, test battery approach or TBA), as opposed to those performed with a single biotic level (n = 41). This contrasts somewhat with bioassay applications for liquid media assessment, where TBAs comprised nearly 54% (101/188) of reported studies (Tables 1-3). Again, test and biotic level selection may be based on a variety of reasons and study objectives (*e.g.*, practicality, cost, personnel availability) and have influenced a preference for conducting TT assessments. Another factor may lie in that there were (and still are) less toxicity tests whose use is validated for undertaking sediment appraisals. With the exception of those conducted with several benthic invertebrates, most other tests conducted with other groups (*e.g.*, algae, bacteria, fish) were first developed and intended for liquid media assessment (*e.g.*, chemicals and polluted waters). Unlike invertebrate tests, their use to evaluate different liquid compartments associated with whole sediment (*i.e.*, interstitial waters, elutriates, organic extracts of whole sediment) was generally less frequent until the early 1990's when more small-scale assays were developed and validated for sediment toxicity assessment (Wells et al., 1998). Yet another factor is linked to the fact that sediments, unlike liquid samples, comprise several phases that can be assayed (pore waters, elutriates, whole sediment and organic extracts thereof). Ideally, all of these phases should be assessed with a relevant battery of tests for a comprehensive understanding of the sediment's full toxicity potential. In reality, however, scientists will make choices based on laboratory capability for testing and study objectives. When TBAs are used, they are mostly conducted with two or three trophic levels (Tab. 11), similarly to those TBAs performed to study liquid media (Tab. 4).

Table 11. Frequency of the number of biotic levels employed in test battery approaches (TBA) for sediment assessment based on the 34 TBA papers classified in Tables 9 and 10.

TBA studies undertaken with:	Number and frequency (%)
Two biotic levels	18/34 (52.9)
Three biotic levels	11/34 (32.4)
Four biotic levels	4/34 (11.8)
Five biotic levels	0/34 (0)
Six biotic levels	1/34 (2.9)

Whether TT (toxicity testing with single species tests at the same biotic level) or TBAs are performed, some test organisms have been more frequently used than others for sediment assessment (Tab. 12). With an overwhelming majority, invertebrates have unquestionably been the most commonly employed, even more so than for liquid media assessment (Tab. 5). The conduct of solid phase tests on whole sediment with invertebrate species explains their preferential selection as test organisms. Bacterial tests rank second in utilization, likely owing to the frequent use of sediment direct contact bioluminescence inhibition assays whose development began in the early 1990s (Brouwer et al., 1990). Algae and fish have also been used by some workers, in part to study the potential impact of contaminants on water column organisms owing to sediment resuspension.

Several phases associated with sediments are evaluated for their toxic potential as Tables 10 and 11 indicate. Whole sediment and pore water stand out as phases that are most frequently investigated (Tab. 13). Because sediments act as contaminant sinks where both readily-soluble and adsorbed toxicants can be present, it is not surprising that whole sediments should be the compartment to receive marked attention, as the (endo)benthic community lives in intimate contact with this matrix and therefore vulnerable to adverse effects. Man-made activities that cause sediments to move (*e.g.*, dredging) can spread contaminants back into the water column and pose a threat to pelagic organisms. Hence, testing sediment phases including elutriates, interstitial waters and overlying waters are fully justified and these have been amply tested as well. Organic extracts of whole sediment, purported by some to lack environmental relevance because they can extract persistent (lipophilic) compounds that would normally stay sequestered *ad infinitum* in sediments, can nevertheless indicate possible long-term effects for benthic organisms.

Table 12. Frequency of use of specific biotic levels employed in toxicity testing (TT) and test battery approaches (TBA) for sediment assessment based on the 75 papers classified in Tables 9 and 10.

TT and TBA studies undertaken with:	**Number and frequency (%)**
Algae	16/222* (7.2)
Bacteria	53/222 (23.9)
Fish	9/222 (4.1)
Invertebrates	136/222 (61.3)
Lemnaceae (duckweed)	1/222 (< 1)
Plant (*H. verticulata*)	1/222 (< 1)
Protozoans	1/222 (< 1)
Seeds	5/222 (2.3)

*Total number of single species tests reported in the 75 papers classified in Tables 9 and 10 (= sum of number of A,B,F,I,L,P,Pl,S tests indicated in the "Biotic levels employed" column).

Table 13. Testing frequency of specific sediment phases for sediment toxicity assessment based on the 75 papers classified in Tables 9 and 10.

Sediment phase	**Number and frequency (%)**
Elutriate	16/109* (14.7)
Overlying water/seepage water	17/109 (15.6)
Pore water	28/109 (25.7)
Organic extract	7/109 (6.4)
Whole sediment	41/109 (37.6)

*Total number of times different sediment phases have been assayed in the 75 papers classified in Tables 9 and 10 (= sum of number of sediment phases indicated in the "Assessment objective…" column).

2.2 CRITICAL BODY RESIDUE STUDIES AND LINKS TO (SUB)LETHAL TOXICITY RESPONSES

During exposure to contaminated sediments, test organisms can concentrate chemicals in their tissue and exhibit measurable (sub)lethal effects linked to accumulated substances. In the field of sediment toxicity assessment, it is noteworthy to mention that some studies have been conducted to characterize both exposure and biological effects in parallel. Exposure to contaminants can be gauged by measuring their concentrations in water/sediment and tissue, and effects can be estimated with endpoints such as survival and growth. These studies are important, for example, to detect threshold concentrations at which chemicals begin to exert adverse effects. As such, they can be useful to recommend effective chemical quality standards that will be protective of aquatic life.

CBR (critical body residue) studies include research on metals, organics and contaminants in mixtures. For instance, cadmium toxicity was appraised with the midge, *Chironomus tentans*, exposed to spiked-sediments that were stored for different periods of time (Sae-ma et al., 1998). Decreases in toxicity effects (lethality) and Cd accumulation in midge tissue with storage time suggested that decreased bioavailability of this metal had occurred. This work clearly illustrated the influence of sediment storage time on organism toxicity response and the impact it could have on test results. Effects of fluoranthene, a PAH (polycyclic aromatic hydrocarbon) congener, were appraised in benthic copepods exposed to dosed sediments for ten days (Lotufo, 1998). Relationships were found between organism health (survival, reproductive and grazing capacity) and fluoranthene concentration in both sediment and tissue. This study was therefore able to more closely pinpoint the NOEL (no observed effect level) concentration of this chemical for this group of biota. Another initiative in CBR studies sought to find out whether the AVS (acid-volatile sulphide) content of sediments collected in areas impacted by mining activities might influence the bioaccumulation of metals (Zn, Cu) and toxicity to the midge *C. tentans* (Besser et al., 1996). Results indicated differences in metal uptake in organisms based on AVS content and showed that growth inhibition was more markedly linked to Zn than Cu. Recommendations called for considering AVS concentrations in metal-contaminated sediments, because of the importance it can have on uptake by biota and subsequent toxicity responses. These investigations indeed confirm the usefulness of CBR-like approaches for evaluating hazard and risk to sediment-dwelling organisms from metals and organic pollutants.

3. Miscellaneous studies/initiatives linked to aquatic toxicity testing applications (liquid media and sediments)

3.1 ENDEAVOURS PROMOTING THE DEVELOPMENT, VALIDATION AND REFINEMENT OF TOXICITY TESTING PROCEDURES

There are literally hundreds of publications that, directly or indirectly, have contributed to the development, validation and refinement of bioassay techniques both for liquid and solid media assessment. These papers incorporate initiatives that have dealt with 1) test method development, 2) inter-calibration exercises, 3) comparative studies and 4) factors capable of affecting bioassay responses. Anyone familiar with the world of toxicity testing would likely not disagree with the statement that "the perfect bioassay is not of this world" and that developers of these instruments of ecotoxicology simply do their utmost to make each test "as least imperfect as possible". To reach this latter stage, assurance of reproducibility, demonstration of scope of use and understanding confounding factors capable of influencing toxicity responses are some of the issues that must be addressed. Hereunder, examples of such studies are given to reveal some of the ways in which they have contributed to the science of small-scale toxicity testing by enhancing its diagnostic tools.

3.1.1 Test method development
To guarantee that reliable procedures are consistently employed to generate toxicity data, it is first essential that sufficient effort be directed toward the development of reproducible toxicity test methods whose results will remain unchallenged. Those that are featured in this book are representative of dependable micro-assays presently in use internationally. Many other small-scale toxicity test methods have been developed at various levels of biological organization. These include bioassays conducted with **algae** (Daniels et al, 1989*; Radetski et al., 1995; St-Laurent and Blaise, 1995; Chen et al., 1997; Blaise and Ménard, 1998*; Persoone, 1998; Tessier et al., 1999; Geis et al., 2000), **bacteria** (Bitton et al., 1994; Blaise et al., 1994; Bulich and Bailey, 1995; Kwan, 1995*; Bulich et al., 1996; Botsford, 1998; Lappalainen et al., 1999*; Ulitzur et al., 2002; Gabrielson et al., 2003), **fish cells** (Ahne, 1985; Pesonen and Andersson, 1997; Sandbacka et al., 1999), **invertebrates** (Snell and Persoone, 1989; Oris et al., 1991; Kubitz et al., 1996*; Benoit et al., 1997*; Johnson and Delaney, 1998; Chial and Persoone, 2002*; Gerhardt et al., 2002b*; Tran et al., 2003), **Lemnaceae** (Bengtsson et al., 1999; Cleuvers and Ratte, 2002a), **protozoans** (Dive et al., 1991; Larsen et al., 1997; Berk and Roberts, 1998; Twagilimana et al., 1998; Gilron et al., 1999) and **yeast** (Ribeiro et al., 2000).

*(tests applying to sediment toxicity testing)

For freshwater solid media investigations, efforts have also been directed towards the development of formulated sediments (also called "artificial" or "synthetic" sediments) to assess their adequacy for conducting contaminant-spiked sediment toxicity studies (Suedel and Rodgers, 1994; Kemble et al., 1999). Among other uses, formulated sediments can be useful to recommend realistic sediment quality criteria for (in)organic substances. Different types of formulated sediments have been employed to evaluate both metal- spiked (Gonzalez, 1996; Harrahy and Clements, 1997; Chapman et al., 1999; Péry et al., 2003) and organic-spiked (Fleming et al., 1998; Besser et al., 2003; Lamy-Enrici et al., 2003) contaminants.

3.1.2 Inter-calibration exercises
Beyond test development and validation, inter-calibration exercises (also known as "round robin" or "inter-laboratory exercises") are mandatory steps that must be undertaken if a toxicity test method is intended for standardization. These exercises

further contribute to test validation by insuring reproducibility of results among different laboratories. In most cases, they also contribute to test method improvement and refinement (*e.g.*, Thellen et al., 1989; Dive et al., 1991; Persoone et al., 1993).

For example, inter-calibration exercises have been undertaken with **algae** (Thellen et al., 1989), **bacteria** (Ribo, 1997; Ross et al., 1999*), **fish cells** (Gagné et al., 1999a), **invertebrates** (Cowgill, 1986; Persoone et al., 1993; Burton et al., 1996*; Hayes et al., 1996), **protozoans** (Dive et al., 1990), and **test organisms of several biotic levels** (Rue et al., 1988; Ronco et al., 2002).

*(tests applying to sediment toxicity testing)

If toxicity tests fulfill the scientific criteria set out by inter-calibration exercises, they can then be considered for the standardization process. If this process is followed, an official toxicity test method document is eventually produced that ensures proper conduct of biological tests (see Section 3.2.1).

3.1.3 Comparative studies
Comparative studies involving toxicity tests abound in the scientific literature. There are many reasons compelling ecotoxicologists to conduct work of this nature, some of which are directed 1) to assess the performance, sensitivity and relevance of individual bioassays undertaken on various chemicals and (liquid and solid) media to specify their scope of use, 2) to optimize the diagnostic potential of bioassay batteries to broaden hazard detection (insure that tests in a battery are complementary and not redundant) and 3) to promote the application of novel assays capable of high throughput for cost-effective screening of (complex) environmental samples.

As an overview, **studies carried out with liquid media** have been launched to compare **bioassay responses** (Finger et al., 1985; Blaise et al., 1987; Kaiser and McKinnon, 1993; Ross, 1993; Isomaa et al., 1995; Dodard et al., 1999; Lucivjanskà et al., 2000; Brix et al., 2001a; Nalecz-Jawecki and Sawicki, 2002; Mummert et al., 2003; Sherrard et al., 2003; Tsui and Chu, 2003), **different endpoints** (Dunbar et al., 1983; Fernández-Casalderrey et al., 1993; Pauli and Berger, 1997; Froehner et al., 2000; Snell, 2000; Weyers and Vollmer, 2000; Jos et al., 2003), **responses of laboratory test organism species and endemic species and/or laboratory bioassay responses and field results** (Koivisto and Ketola, 1995; Traunspurger et al., 1996; van Wijngaarden et al., 1996; Jak et al., 1998; Crane et al., 1999; Tchounwou and Reed, 1999; Dyatlov, 2000; Milam et al., 2000; Pascoe et al., 2000; Bérard et al., 2003), **and bioassay and biomarker endpoints** (Gagné and Blaise, 1993; Nyström and Blanck, 1998; Connon et al., 2000; Perkins and Schlenk, 2000; De Coen and Janssen, 1997; Bierkens et al., 1998; Sturm and Hansen, 1999; den Besten and Tuk, 2000; Guilhermino et al., 2000; Maycock et al., 2003; Taylor et al., 2003).

In **studies conducted with sediments**, comparisons have been reported for **artificial (formulated) and natural sediments** (Barrett, 1995; Fleming et al., 1998), **bioassay and biomarker endpoints** (Gillis et al., 2002), **bioassay responses** (Ahlf et al., 1989; Becker et al., 1995; Day et al., 1995a; Kwan and Dutka, 1995; Suedel et al., 1996; Barber et al., 1997; Day et al., 1998; Fuchsman et al., 1998; Guzzella,

1998; Huuskonen et al., 1998; Côté et al., 1998a,b; Vanderbroele et al., 2000; Watts and Pascoe, 2000; Chial et al., 2003; Milani et al., 2003; Mueller et al., 2003; Petänen et al., 2003), **different endpoints** (Suedel et al., 1996; Watts and Pascoe, 1996; Sibley et al., 1997a; Pasteris et al., 2003; Landrum et al., 2004; Vecchi et al., 1999), **different sediment phases** (Harkey et al., 1994), **responses of laboratory test organism species and endemic species** (Conrad et al., 1999) **and/or laboratory bioassay responses and field results** (Reinhold-Dudok et al., 1999; Bombardier and Blaise, 2000; Peeters et al., 2001; den Besten et al., 2003) and **sediment collection techniques** (West et al., 1994).

3.1.4 Factors capable of affecting bioassay responses
Toxicity testing developers and users have also devoted significant energy to the understanding of specific factors capable of confounding (micro-) organism responses and/or interfering with data interpretation (*e.g.*, pH, temperature, light, growth medium, natural contaminants such as NH_3, H_2S, or grain size in case of solid phase tests).

In fact, any aspect of testing likely to impact toxicity results (*e.g.*, stimulatory effects in the case of algal toxicity assays, or sample colour interferences in the case of a toxicity endpoint measured by photometry) have been a focus of concern, as have been ways of minimizing, eliminating or circumventing particular problems or limitations that may be test-specific. In brief, seeking thorough understanding of a test's capabilities and limitations has been considered paramount for proper toxicity assessment (and final data interpretation) and marked efforts have been directed toward this goal.

With this purpose in mind, investigations have explored the influence of such factors as **acid volatile sulfides** (Sibley et al., 1996*; Long et al., 1998*), **alkalinity** (Lasier et al., 1997*), **ammonia** (Besser et al., 1998*; Newton et al., 2003*), **colored samples** (Cleuvers and Weyers, 2003), **equilibration time** (Lee et al., 2004*), **experimental design** (Naylor and Howcroft, 1997*; Bartlett et al., 2004*), **fluid dynamics** (Preston et al., 2001), **food** (Sarma et al., 2001; Gorbi et al., 2002; de Haas et al., 2002*; Antunes et al., 2004; de Haas et al., 2004*); **grain size** (Guerrero et al., 2003*), **genetic variability** (Baird et al., 1991; Barber et al., 1990; Barata et al., 1998), **gut contents** (Sibley et al., 1997c*), **heavy metal speciation** (Gunn et al., 1989*; Ankley et al., 1996*), **humic/fulvic acids** (Ortego and Benson, 1992; Alberts et al., 2001; Guéguen et al., 2003; Koukal et al., 2003; Ma et al., 2003), **intermittent or short exposures to contaminants** (Hickey et al., 1991; Brent and Herricks, 1998; Naddy and Klaine, 2001, Broomhall, 2002), **life-cycle stage/age** (Williams et al., 1986; Stephenson et al., 1991; Watts and Pascoe, 1998*; Hamm et al., 2001), **light regime** (Cleuvers and Ratte, 2002b), **organic matter content** (Ankley et al., 1994*; Lacey et al., 1999*; Besser et al., 2003*; Guerrero et al., 2003*; Lamy-Enrici et al., 2003*; Mäenpää et al., 2003*; VanGenderen et al., 2003), **pH** (Fisher and Wadleigh, 1986; Fu et al., 1991; Svenson and Zhang, 1995; Rousch et al., 1997; Franklin et al., 2000; Peck et al., 2002*; Long et al., 2004), **phosphorus** (Van Donk et al., 1992; Mkandawire et al., 2004), **potassium** (Bervoets et al., 2003*), **pre-exposure to contaminants** (Bearden et al., 1997; Muyssen and Janssen, 2001, 2002; Ristola et al., 2001*; Vidal and Horne, 2003*), **sand** (Thomulka et al., 1997), **sediment**

indigenous animals (Reynoldson et al., 1994*), **sediment processing** (Day et al., 1995b*), **sex** (Sildanchandra and Crane, 2000), **solvents** (Calleja and Persoone, 1993; Fliedner, 1997), **choice of statistical tests** (Isnard et al., 2001), **sulfates** (Brix et al., 2001c), **sulfur** (Jacobs et al., 1992*; Pardos et al., 1999b*), **suspended solids** (Herbrandson et al., 2003a,b), **temperature** (Fisher, 1986; Broomhall, 2002; Buchwalter et al., 2003; Heugens et al., 2003), **test exposure time** (Suedel et al., 1997; Naimo et al., 2000*; Froehner et al., 2002; Feng et al., 2003), **test medium** (Vasseur and Pandard, 1988; Guilhermino et al., 1997; Samel et al., 1999), **test organism inoculum density** (Moreno-Garrido et al., 2000; Franklin et al., 2002), **UV irradiation** (Bonnemoy et al., 2004), **water chemistry/quality** (Persoone et al., 1989; Jop et al., 1991; van Dam et al., 1998; Karen et al., 1999; Clément, 2000; Bury et al., 2002; Graff et al., 2003), **water hardness** (Fu et al., 1991; Baer et al., 1999; Verge et al., 2001; Charles et al., 2002; Gensemer et al., 2002; Naddy et al., 2003; Long et al., 2004), **water-sediment partitioning** (Stewart and Thompson, 1995*).

*(tests applying to sediment toxicity testing)

3.2 INITIATIVES PROMOTING THE USE OF TOXICITY TESTING PROCEDURES

For over three decades, the use of bioassays for toxicity testing has steadily increased and become an indispensable component of aquatic environmental assessment. In this section, specific types of publications are presented as important contributions that have 1) promoted the use of ecotoxicology testing in the biomonitoring, regulatory and compliance arena, 2) disseminated information and understanding relating to toxicity testing issues, 3) favoured technology transfer of test methods internationally and 4) provided overall sound scientific support to facilitate decision-making aimed at environmental protection and conservation.

3.2.1 Review, bio-monitoring and HAS articles

Review articles are particularly useful to synthesize research work that has been undertaken in different spheres relating to toxicity testing. By exposing the state of the art for a selective field, these articles will often circumscribe the limitations, advantages and scope of use of bioassays which then leads to their proper and effective application. Some examples of review articles include papers on **concept/management/policy** (MacGregor and Wells, 1984; U.S. EPA and Environment Canada, 1984; Sergy, 1987; Cairns and Pratt, 1989; Maltby and Callow, 1989; Blaise, 2003), as well as several others on specific trophic groups including **algae** (Blaise, 1993; Lewis, 1995; Sosak-Swiderska and Tyrawska, 1996; Blaise et al., 1998b; Blaise, 2002), **bacteria** (Bennett and Cubbage, 1992b*; Bitton and Koopman, 1992; Kross and Cherryholmes, 1993; Painter, 1993; Bitton and Morel, 1998; Ross, 1998; Doherty, 2001*), **fish cells** (Babich and Borefreund, 1991;Fentem and Balls, 1993; Denizeau, 1998; Fent, 2001; Castaño et al., 2003), **invertebrates** (Burton et al., 1992; Ingersoll et al., 1995*; Snell and Janssen, 1995, 1998; Chapman, 1998*; CANMET, 1999) and **protozoa** (Gilron and Lynn, 1998; Sauvant et al., 1999; Nicolau et al., 2001; Nalecz-Jawecki, 2004).

Other reviews have also encompassed **different levels of toxicity tests** (Giesy and Hoke, 1989*; Bennett and Cubbage, 1992a; CANMET, 1997a; Blaise et al., 1998a; de Vlaming et al., 1999; Blaise et al., 2000; Girling et al., 2000; Janssen et al., 2000; Repetto et al., 2000).

*applying to sediment toxicity assessment

Various papers expounding the value of **biomonitoring, routine and/or regulatory testing** have also advanced the practice of bioassays. Some of these include articles on **drinking water assessment** (Forget et al., 2000), **single chemical or mixture assessment** (Altenburger et al., 1996; Aoyama et al., 2000), **surface water assessment** (Canna-Michaelidou et al., 2000; Marsalek and Rojickova-Padrtova, 2000; Ruck et al., 2000), **wastewater assessment** (OECD, 1987; Blaise et al., 1988; Mackay et al., 1989; Hansen, 1993; Johnson et al., 1993; Stulhfauth, 1995; Kovacs et al., 2002), **sewage treatment plant performance assessment** (Fearnside and Hiley, 1993), and **sediment quality assessment** (Nipper, 1998).

Articles proposing new **hazard assessment schemes** (HAS) for liquid or sediment assessment have equally paved the way for the employment of test batteries in ecotoxicity appraisals. Some describe systems for evaluating **water/wastewater** (Blaise et al., 1985; Heinis et al., 2000; Ronco et al., 2000 ; Persoone et al., 2003), **chemicals** (Fochtman et al., 2000; Garay et al., 2000; Girling et al., 2000; Pica-Granados et al., 2000; Brix et al., 2001a,b,c) and **sediments** (Ingersoll et al., 1997; Côté et al., 1998b). These effects-based indices, varied in their concepts and objectives, demonstrate novel ways of utilizing groups of bioassays to deal with "real-life" environmental situations. As such, they highlight schemes that are complementary to the robust and validated HAS approaches described in Volume 2 of this book.

3.2.2 Standardized test methods and guidance documents
Finally, marked efforts have been undertaken nationally and internationally to publish **standardized toxicity test methods** and several standards organizations (*e.g.*, ASTM, ISO, OECD) have been very active in the production of documents too numerous to reproduce in this chapter. Publishing official test methods is not a simple task and can require a substantial amount of time and energy from dedicated scientists. Again, standardized toxicological method documents are crucial to environmental assessment as they ensure proper use of testing, (inter)national consistency and acceptance, as well as reliability of test results owing to the quality control and assurance components that are integrated in such protocols.

Test method standardization (TMS) calls for several actions that involve 1) preparation of a formal draft test method document for each bioassay intended for standardization, 2) a critical review by an expert subcommittee, 3) the preparation of a final draft test method, 4) an international peer review of each test method, 5) an inter-calibration exercise of the final draft test method, 6) finalization of each test method and 7) the formal publication of the toxicity test method document. Environment Canada (EC) has been particularly active in biological test method standardization and has thus far contributed 18 standardized aquatic and sediment toxicity methods, eight and three of which apply to acute/chronic freshwater liquid (tests with algae, bacteria, fish, invertebrates, and *Lemnaceae*) and solid (tests with bacteria and invertebrates) media assessment, respectively (IGETG, 2004). As a complement to TMS, EC has also produced several **guidance documents** that provide assistance on matters related to choice of reference toxicants (Environment Canada, 1990), sampling and spiking techniques for sediments (Environment Canada, 1994, 1995), interpretation of results (Environment Canada, 1999) and statistical considerations for toxicity tests (Environment Canada, 2004b).

Other **standardized/validated test methods** reported in the literature include acute/chronic tests performed with **algae** (*e.g*, OECD, 2002a; ISO, 2003), **fish cells** (Gagné and Blaise, 2001), **invertebrates** (Borgmann and Munawar, 1989*; Trottier et al., 1997; Pereira et al., 2000*; OECD, 2001*a,b), *Lemnaceae* (OECD, 2002b), and with **toxicity tests conducted at different trophic levels** (Nebeker et al., 1984*; U.S. EPA, 2002a,b).

*applying to sediment toxicity assessment

Additionally, **miscellaneous guidance/technical documents** have reported on various aspects linked to ecotoxicity that give advice on:
- choice of bioassays for general contaminant assessment (Calow, 1989);
- criteria to select tests for effluent testing (Grothe et al., 1996; Johnson, 2000);
- choice of species and endpoints for appraising pharmaceuticals (Länge and Deitrich, 2002);
- proper application of algal, bacterial and invertebrate tests (Santiago et al., 2002);
- approaches, design and interpretation of sediment tests (Ross and Leitman, 1995; Ingersoll et al., 2000; Wenning and Ingersoll, 2002; MacDonald and Ingersoll, 2002a,b).

4. Conclusion(s)

Small-scale freshwater toxicity testing is but a modest fraction of a diverse array of scientific activities connected to the field of ecotoxicology. Yet, within this still emerging discipline, few will argue the fact that tools and approaches developed to measure the undesirable effects that countless chemicals (alone or in mixtures) and complex (liquid and solid) media can exert on biota have markedly contributed to aquatic ecosystem preservation. Indeed, the breadth and scope of application of bioassays thus far directed toward obtaining relevant information aimed at problem-solving and prevention of contaminant-based issues has progressed well.

While many developed countries have been effective over past decades in eliminating acute toxicity from point source discharges owing to technological improvement of industrial processes and legislation, chronic effects on aquatic biota are still very much an issue. Furthermore, as the 21rst century unfolds, many emerging and developing countries active in joining the world economy are presently creating new contaminant burdens on aquatic systems that will contribute additional acute and chronic toxicity pressures until, once again, technology and legislation repress pollution. Hence, the techniques and hazard assessment schemes featured in this book can prove to be very relevant for use in all parts of the world. As editors of this book, it is our hope that readers will grasp that an effects-based approach is primordial to deal with hazard and risk assessment of pollutants and that use of toxicity tests is an essential cog in this respect. It is also our hope that many, directly or indirectly involved in ensuring the well-being of aquatic systems, will actually use (or suggest the use of) some of the toxicity testing methods and hazard assessment schemes described in subsequent sections.

Lastly, while acute and chronic (sub)lethal toxicity effects are basic concerns that must be first dealt with and eradicated, new demands will be made on ecotoxicology to address emerging issues. Indeed, several more subtle (and potentially deleterious) effects owing to long-term exposures to low concentrations of contaminants will merit investigation (Eggen et al., 2004). Genotoxicity, teratogenicity, immunotoxicity and endocrine disruption are some of the undesirable consequences of classical (*e.g.*, metals, pesticides, organochlorides) and more recent (*e.g.*, household products and pharmaceuticals) chemical discharges into receiving waters that require urgent comprehensive assessment. Here as well, reliable and relevant standardized tools and approaches will have to be developed and applied.

References

Adams, S.M. (2003) Establishing causality between environmental stressors and effects on aquatic ecosystems, *Human and Ecological Risk Assessment* **19**, 17-35.

Ahlf, W., Calmano, W., Erhard, J. and Förstner, U. (1989) Comparison of rive bioassay techniques for assessing sediment-bound contaminants, in M. Munawar, G. Dixon, C.I. Mayfield, T. Reynoldson and M.H. Sadar (eds.), *Environmental Bioassay Techniques and their Application: Proceedings of the 1st International Conference held in Lancaster, England, 11-14 July 1988*, Kluwer Academic Publishers, Dordrecht, Netherlands, pp. 285-289.

Ahne, W. (1985) Untersuchungen über die Verwendung von Fischzellkulturen fur Toxizitätsbestimmungen zur Einschränkung and Ersatz des Fishtests, *Zentralblatt Fur Bakteriologie, Mikrobiologie Und Hygiene. 1. Abt. Originale B, Hygiene* **180**, 480-504.

Ahtiainen, J., Nakari, T. and Silvonen, J. (1996) Toxicity of TCF and ECF pulp bleaching effluents assessed by biological toxicity tests, in M.R. Servos, K.R. Munkittrick, J.H. Carey and G.J. Van Der Kraak (eds.), *Environmental Fate and Effects of Pulp and Paper Mill Effluents,* St-Lucie Press, FL, pp. 33-40.

Ahtiainen, J., Nakari, T., Ruoppa, M., Verta, M. and Talka, E. (2000) Toxicity screening of novel pulp mill wastewaters in Finnish pulp mills, in G. Persoone, C. Janssen and W.M. De Coen (eds.), *New Microbiotests for Routine Toxicity Screening and Biomonitoring*, Kluwer Academic/Plenum Publishers, New York, pp. 307-317.

Alberts, J.J., Takács, M. and Pattanayek, M. (2001) Influence of IHSS standard and reference materials on copper and mercury toxicity to *Vibrio fischeri*, *Acta Hydrochimica et Hydrobiologica* **28**, 428-435.

Ali, M. and Sreekrishnan, T.R. (2001) Aquatic toxicity from pulp and paper mill effluents: a review, *Advances in Environmental Research* **5**, 175-196.

Altenburger, R., Boedeker, W., Faust, M. and Grimme, L.H. (1996) Regulations for combined effects of pollutants: consequences from risk assessment in aquatic toxicology, *Food and Chemical Toxicology* **34**, 1155-1157.

Anastácio, P.M., Lützhøft, H.C., Halling-Sørensen, B. and Marques, J.C. (2000) Surfactant (Genapol OX-80) toxicity to *Selenastrum capricornutum*, *Chemosphere* **40** (8), 835-838.

Angelaki, A., Sakellariou, M., Pateras, D. and Kungolos, A. (2000) Assessing the quality of natural waters in Magnesia prefecture in Greece using Toxkits, in G. Persoone, C. Janssen and W.M. De Coen (eds.), *New Microbiotests for Routine Toxicity Screening and Biomonitoring*, Kluwer Academic/Plenum Publishers, New York, pp. 281-288.

Ankley, G. T., Benoit, D. A., Balogh, J. C., Reynoldson, T. B., Day, K. E. and Hoke, R. A. (1994) Evaluation of potential confounding factors in sediment toxicity tests with three freshwater benthic invertebrates, *Environmental Toxicology and Chemistry* **13** (4), 627-635.

Ankley, G.T., Liber, K., Call, D.J., Markee, T.P., Canfield, T.J. and Ingersoll, C.G. (1996) A field investigation of the relationship between zinc and acid volatile sulfide concentrations in freshwater sediments, *Journal of Aquatic Ecosystem Health* **5** (4), 255-264.

Antunes, S.C., Castro, B.B. and Gonçalves, F. (2004) Effect of food level on the acute and chronic responses of daphnids to lindane, *Environmental Pollution* **127** (3), 367-375.

Aoyama, I., Okamura, H. and Rong, L. (2000) Toxicity testing in Japan and the use of Toxkit microbiotests, in G. Persoone, C. Janssen and W.M. De Coen (eds.), *New Microbiotests for Routine Toxicity Screening and Biomonitoring*, Kluwer Academic/Plenum Publishers, New York, pp. 123-133.

Arbuckle, W.B. and Alleman, J.E. (1992) Effluent toxicity testing using nitrifiers and Microtox™, *Water Environment Research* **64**, 263-267.

Arkhipchuk, V.V. and Malinovskaya, M.V. (2002) Quality of water types in Ukraine evaluated by WaterTox bioassays, *Environmental Toxicology* **17** (3), 250-257.

Aruldoss, J.A. and Viraraghavan, T. (1998) Toxicity testing of refinery wastewater using Microtox, *Bulletin of Environmental Contamination and Toxicology* **60** (3), 456-463.

Asami, M., Suzuki, N. and Nakanishi, J. (1996) Aquatic toxicity emission from Tokyo: wastewater measured using marine luminescent bacterium, *Photobacterium phosphoreum*, *Water Science and Technology* **33** (6), 121-128.

Atienzar, F.A., Cheung, V.V., Jha, A.N. and Depledge, M.H. (2001) Fitness parameters and DNA effects are sensitive indicators of copper-induced toxicity in *Daphnia magna*, *Toxicological Sciences* **59** (2), 241-250.

Babich, H. and Borenfreund, E. (1987) Cultured fish cells for the ecotoxicity testing of aquatic pollutants, *Toxicity Assessment* **2**, 119-133.

Babich, H. and Borenfreund, E. (1991) Cytotoxicity and genotoxicity assays with cultured fish cells: a review, *Toxicology in Vitro* **5**, 91-100.

Baer, K.N., Ziegenfuss, M.C., Banks, S.D. and Ling, Z. (1999) Suitability of high-hardness COMBO medium for ecotoxicity testing using algae, daphnids, and fish, *Bulletin of Environmental Contamination and Toxicology* **63** (3), 289-296.

Baer, K.N., Boeri, R.L., Ward, T.J. and Dixon, D.W. (2002) Aquatic toxicity evaluation of para-methylstyrene, *Ecotoxicology and Environmental Safety* **53** (3), 432-438.

Baird, D.J., Barber, I., Bradley, M., Soares, A.M.V.M. and Calow, P. (1991) A comparative study of genotype sensitivity to acute toxic stress using clones of *Daphnia magna* Straus, *Ecotoxicology and Environmental Safety* **21**, 257-265.

Barata, C., Baird, D.J. and Markich, S.J. (1998) Influence of genetic and environmental factors on the tolerance of *Daphnia magna* Straus to essential and non-essential metals, *Aquatic Toxicology* **42**(2), 115-137.

Barber, I., Baird, D.J. and Calow, P. (1990) Clonal variation in general responses of *Daphnia magna* Straus to toxic stress. II. Physiological effects, *Functional Ecology* **4**, 409-414.

Barber, T.R., Fuchsman, P.C., Chappie, D.J., Sferra, J.C., Newton, F.C. and Sheehan, P.J. (1997) Toxicity of hexachlorobenzene to *Hyallela azteca* and *Chironomus tentans* in spiked sediment bioassays, *Environmental Toxicology and Chemistry* **16** (8), 1716-1720.

Barrett, K.L. (1995) A comparison of the fate and effects of prochloraz in artificial and natural sediments, *Journal of Aquatic Ecosystem Health* **4** (4), 239-248.

Bartlett, A.J., Borgmann, U., Dixon, D.G., Batchelor, S.P. and Maguire, R.J. (2004) Tributyltin uptake and depuration in *Hyalella azteca*: implications for experimental design, *Environmental Toxicology and Chemistry* **23** (2), 426-434.

Bartsch, M.R., Newton, T.J., Allran, J.W., O'Donnell, J.A. and Richardson, W.B. (2003) Effects of pore-water ammonia on *in situ* survival and growth of juvenile mussels (*Lampsilis cardium*) in the St. Croix Riverway, Wisconsin, USA, *Environmental Toxicology and Chemistry* **22** (11), 2561-2568.

Bastian, K.C. and Alleman, J.E. (1998) Microtox characterization of foundry sand residuals, *Waste Management* **18** (4), 227-234.

Baun, A., Bussarawit, N. and Nyholm, N. (1998) Screening of pesticide toxicity in surface water from an agricultural area at Phuket Island (Thailand), *Environmental Pollution* **102** (2-3), 185-190.

Bearden, A.P., Gregory, B.W. and Schultz, T.W. (1997) Growth kinetics of preexposed and naive populations of *Tetrahymena pyriformis* to 2-decanone and acetone, *Ecotoxicology and Environmental Safety* **37** (3), 245-250.

Becker, C.D., Fallon, W.E., Crass, D.W. and Scott, A.J. (1983) Acute toxicity of water soluble fractions derived from a coal liquid (SRC-II) to three aquatic organisms, *Water, Air, and Soil Pollution* **19**, 171-184.

Becker, D.S., Rose, C.D. and Bigham, G.N. (1995) Comparison of the 10-day freshwater sediment toxicity tests using *Hyallela azteca* and *Chironomus tentans*, *Environmental Toxicology and Chemistry* **14** (12), 2089-2094.

Bengtsson, B.-E., Bongo, J.P. and Eklund, B. (1999) Assessment of duckweed *Lemna aequinoctialis* as a toxicological bioassay for tropical environments in developing countries, *Ambio* **28** (2), 152-155.

Bennett, J. and Cubbage, J. (1992a) *Evaluation of bioassay organisms for freshwater sediment toxicity testing*, Environmental Investigations and Laboratory Services, Washington State Department of Ecology, Washington, DC (December 19, 2003); http://www.nic.edu/library/superfund/refdocs%5Ccda0159.pdf.

Bennett, J. and Cubbage, J. (1992b) *Review and evaluation of Microtox test for freshwater sediments*, Washington State Department of Ecology, Washington, 28 pp.

Bennett, W.R. and Farrell, A.P. (1998) Acute toxicity testing with juvenile white sturgeon (*Acipenser transmontanus*), *Water Quality Research Journal of Canada* **33** (1), 95-110.

Benoit, D.A., Sibley, P.K., Juenemann, J.L. and Ankley, G.T. (1997) *Chironomus tentans* life-cycle test: design and evaluation for use in assessing toxicity of contaminated sediments, *Environmental Toxicology and Chemistry* **16** (6), 1165-1176.

Bérard, A., Dorigo, U., Mercier, I., Becker-van Slooten, K., Grandjean, D. and Leboulanger, C. (2003) Comparison of the ecotoxicological impact of the triazines Irgarol 1051 and atrazine on microalgal cultures and natural microalgal communities in Lake Geneva, *Chemosphere* **53** (8), 935-944.

Berk, S.G. and Roberts, R.O. (1998) Development of a protozoan chemoattraction inhibition assay for evaluating toxicity of aquatic pollutants, in P.G. Wells, K. Lee and C. Blaise (eds.) *Microscale Testing in Aquatic Toxicology: Advances, Techniques, and Practice*, CRC Press, Boca Raton, FL, pp. 337-348.

Bermingham, N., Costan, G., Blaise, C. and Patenaude, L. (1996) Use of micro-scale aquatic toxicity tests in ecolabelling guidelines for general purpose cleaners, in M. Richardson (ed.), *Environmental Xenobiotics*, Taylor & Francis Books Ltd, London, England, pp. 195-212.

Bervoets, L. and Blust, R. (2003) Metal concentrations in water, sediment and gudgeon (*Gobio gobio*) from a pollution gradient: relationship with fish condition factor, *Environmental Pollution* **126** (1), 9-19.

Bervoets, L., Baillieul, M., Blust, R. and Verheyen, R. (1996) Evaluation of effluent toxicity and ambient toxicity in a polluted lowland river, *Environmental Pollution* **91** (3), 333-341.

Bervoets, L., De Bruyn, L., Van Ginneken, L. and Blust, R. (2003) Accumulation of ^{137}Cs by larvae of the midge *Chironomus riparius* from sediment: effect of potassium, *Environmental Toxicology and Chemistry* **22** (7), 1589-1596.

Bervoets, L., Meregalli, G., De Cooman, W., Goddeeris, B. and Blust, R. (2004) Caged midge larvae (*Chironomus riparius*) for the assessment of metal bioaccumulation from sediments *in situ*, *Environmental Toxicology and Chemistry* **23** (2), 443-454.

Besser, J.M., Kubitz, J.A., Ingersoll, C.G., Braselton, W.E. and Giesy, J.P. (1995) Influences on copper bioaccumulation, growth, and survival of the midge, *Chironomus tentans*, in metal contaminated sediments, *Journal of Aquatic Ecosystem Health* **4** (3), 157-168.

Besser, J.M., Ingersoll, C.G. and Giesy, J.P. (1996) Effects of spatial and temporal variation of acid-volatile sulfide on the bioavailability of copper and zinc in freshwater sediments, *Environmental Toxicology and Chemistry* **15** (3), 286-293.

Besser, J.M., Ingersoll, C.G., Leonard, E.N. and Mount, D.R. (1998) Effect of zeolite on toxicity of ammonia in freshwater sediments: implications for toxicity identification evaluation procedures, *Environmental Toxicology and Chemistry* **17** (11), 2310-2317.

Besser, J.M., Brumbaugh, W.G., May, T.W. and Ingersoll, C.G. (2003) Effects of organic amendments on the toxicity and bioavailability of cadmium and copper in spiked formulated sediments, *Environmental Toxicology and Chemistry* **22** (4), 805-815.

Bettinetti, R., Giarei, C. and Provini, A. (2003) Chemical analysis and sediment toxicity bioassays to assess the contamination of the River Lambro (Northern Italy), *Archives of Environmental Contamination and Toxicology* **45** (1), 72-78.

Bierkens, J., Maes, J. and Plaetse, F.V. (1998) Dose-dependent induction of heat shock protein 70 synthesis in *Raphidocelis subcapitata* following exposure to different classes of environmental pollutants, *Environmental Pollution* **101** (1), 91-97.

Bitton, G. and Koopman, B. (1992) Bacterial and enzymatic bioassays for toxicity testing in the environment, *Reviews of Environmental Contamination and Toxicology* **125**, 1-22.

Bitton, G. and Morel, J.L. (1998) Microbial enzyme assays for the detection of heavy metal toxicity, in P.G. Wells, K. Lee and C. Blaise (eds.), *Microscale Testing in Aquatic Toxicology: Advances, Techniques, and Practice*, CRC Press, Boca Raton, FL, pp. 143-152.

Bitton, G., Koopman, B. and Agami, O. (1992) MetPAD™: a bioassay for rapid assessment of heavy metal toxicity in wastewater, *Water Environment Research* **64** (6), 834-836.

Bitton, G., Jung, K. and Koopman, B. (1994) Evaluation of a microplate assay specific for heavy metal toxicity, *Archives of Environmental Contamination and Toxicology* **27** (1), 25-28.

Blaise, C. (1993) Practical laboratory applications with micro-algae for hazard assessment of aquatic contaminants, in M. Richardson (ed.), *Ecotoxicology Monitoring*, VCH Publishers, Weinheim, Germany, pp. 83-107.

Blaise, C. (2002) Use of microscopic algae in toxicity testing, in G. Bitton (ed.), *Encyclopedia of Environmental Microbiology*, John Wiley & Sons Inc., New York, pp. 3219-3230.

Blaise, C. (2003) Canadian application of bioassays for environmental management: a review, in M. Munawar (ed.), *Sediment Quality Assessment and Management: Insight and Progress*, Aquatic Ecosystem Health and Management Society, Canada, pp. 39-58.

Blaise, C. and Costan, G. (1987) La toxicité létale aiguë des effluents industriels au Québec vis-à-vis de la truite arc-en-ciel, *Water Pollution Research Journal of Canada* **22** (3), 385-402.

Blaise, C. and Kusui, T. (1997) Acute toxicity assessment of industrial effluents with a microplate-based *Hydra attenuata* assay, *Environmental Toxicology and Water Quality* **12** (1), 53-60.

Blaise, C. and Ménard, L. (1998) A micro-algal solid-phase test to assess the toxic potential of freshwater sediments, *Water Quality Research Journal of Canada* **33** (1), 133-151.

Blaise, C., Bermingham, N. and Van Collie, R. (1985) The integrated ecotoxicological approach to assessment of ecotoxicity, *Water Quality Bulletin* **10** (1), 3-10.

Blaise, C., Van Coillie, R., Bermingham, N. and Coulombe, G. (1987) Comparaison des réponses toxiques de trois indicateurs biologiques (bactéries, algues, poissons) exposés à des effluents de fabriques de pâtes et papiers, *Revue Internationale des Sciences de l'Eau* **3** (1), 9-17.

Blaise, C., Sergy, G., Wells, P., Bermingham, N. and Van Coillie, R. (1988) Biological Testing-Development, Application, and Trends in Canadian Environmental Protection Laboratories, *Toxicity Assessment* **3**, 385-406.

Blaise, C., Forghani, R., Legault, R., Guzzo, J. and Dubow, M.S. (1994) A bacterial toxicity assay performed with microplates, microluminometry and Microtox reagent, *Biotechniques* **16** (5), 932-937.

Blaise, C., Wells, P.G. and Lee, K. (1998a) Microscale testing in aquatic toxicology: introduction, historical perspective, and context, in P.G. Wells, K. Lee and C. Blaise (eds.), *Microscale Testing in Aquatic Toxicology: Advances, Techniques, and Practice*, CRC Press, Boca Raton, FL, pp. 1-9.

Blaise, C., Férard, J.-F. and Vasseur, P. (1998b) Microplate toxicity tests with microalgae: a review, in P.G. Wells, K. Lee and C. Blaise (eds.), *Microscale Testing in Aquatic Toxicology: Advances, Techniques, and Practice*, CRC Press, Boca Raton, FL, pp. 269-288.

Blaise, C., Gagné, F. and Bombardier, M. (2000) Recent developments in microbiotesting and early millennium prospects, *Water, Air, and Soil Pollution* **123** (1-4), 11-23.

Blaise, C, Gagné, F., Chèvre, N., Harwood, M., Lee, K., Lappalainen, J, Chial, B., Persoone, G. and Doe, K. (2004) Toxicity assessment of oil-contaminated freshwater sediments, *Environmental Toxicology* **19**, 329-335.

Bleckmann, C.A., Rabe, B., Edgmon, S.J. and Fillingame, D. (1995) Aquatic toxicity variability for fresh- and saltwater species in refinery wastewater effluent, *Environmental Toxicology and Chemistry* **14** (7), 1219-1223.

Blinova, I. (2000) Comparison of the sensitivity of aquatic test species for toxicity evaluation of various environmental samples, in G. Persoone, C. Janssen and W.M. De Coen (eds.), *New Microbiotests for Routine Toxicity Screening and Biomonitoring*, Kluwer Academic/Plenum Publishers, New York, pp. 217-220.

Boluda, R., Quintanilla, J.F., Bonilla, J.A., Saez, E. and Gamon, M. (2002) Application of the Microtox test and pollution indices to the study of water toxicity in the Albufera Natural Park (Valencia, Spain), *Chemosphere* **46** (2), 355-369.

Bombardier, M. and Blaise, C. (2000) Comparative study of the sediment-toxicity index, benthic community metrics and contaminant concentrations, *Water Quality Research Journal of Canada* **35** (4), 753-780.

Bonnemoy, F., Lavédrine, B. and Boulkamh, A. (2004) Influence of UV irradiation on the toxicity of phenylurea herbicides using Microtox® test, *Chemosphere* **54** (8), 1183-1187.

Bonnet, C. (2000) Développement de bioessais sur sédiments et applications à l'étude, en laboratoire, de la toxicité de sédiments dulçaquicoles contaminés, UFR Sciences Fondamentales et Appliquées, Université de METZ, Metz, France, 326 pages.

Borgmann, U. and Munawar, M. (1989) A new standardized sediment bioassay protocol using the amphipod *Hyalella azteca* (Saussure), in M. Munawar, G. Dixon, C.I. Mayfield, T. Reynoldson and M.H. Sadar (eds.), *Environmental Bioassay Techniques and their Application: Proceedings of the 1st International Conference held in Lancaster, England, 11-14 July 1988*, Kluwer Academic Publishers, Dordrecht, Netherlands, pp. 425-531.

Botsford, J.L. (1998) A simple assay for toxic chemicals using a bacterial indicator, *World Journal of Microbiology and Biotechnology* **14** (3), 369-376.

Boulanger, B. and Nikolaidis, N.P. (2003) Mobility and aquatic toxicity of copper in an urban watershed, *Journal of the American Water Resources Association* **39** (2), 325-336.

Bradfield, A.D., Flexner, N.M. and Webster, D.A. (1993) *Water quality, organic chemistry of sediment, and biological conditions of streams near an abandoned wood-preserving plant site at Jackson, Tennessee*, Water-resources investigations report; 93-4148, USGS, Earth Science Information Center, Denver, CO, pp. 1-50.

Brent, R.N. and Herricks, E.E. (1998) Postexposure effects of brief cadmium, zinc, and phenol exposures on freshwater organisms, *Environmental Toxicology and Chemistry* **17** (10), 2091-2099.

Bridges, T.S., Wright, R.B., Gray, B.R., Gibson, A.B. and Dillon, T.M. (1996) Chronic toxicity of Great Lakes sediments to *Daphnia magna*: elutriate effects on survival, reproduction and population growth, *Ecotoxicology* **5**, 83-102.

Brix, K.V., DeForest, D.K. and Adams, W.J. (2001a) Assessing acute and chronic copper risks to freshwater aquatic life using species sensitivity distributions for different taxonomic groups, *Environmental Toxicology and Chemistry* **20** (8), 1846-1856.

Brix, K.V., Henderson, D.G., Adams, W.J., Reash, R.J., Carlton, R.G. and McIntyre, D.O. (2001b) Acute toxicity of sodium selenate to two daphnids and three amphipods, *Environmental Toxicology* **16** (2), 142-150.

Brix, K.V., Volosin, J.S., Adams, W.J., Reash, R.J., Carlton, R.G. and McIntyre, D.O. (2001c) Effects of sulfate on the acute toxicity of selenate to freshwater organisms, *Environmental Toxicology and Chemistry* **20** (5), 1037-1045.

Broomhall, S. (2002) The effects of endosulfan and variable water temperature on survivorship and subsequent vulnerability to predation in *Litoria citropa* tadpoles, *Aquatic Toxicology* **61** (3-4), 243-250.

Brorson, T., Björklund, I., Svenstam, G. and Lantz, R. (1994) Comparison of two strategies for assessing ecotoxicological aspects of complex wastewater from a chemical-pharmaceutical plant, *Environmental Toxicology and Chemistry* **13** (4), 543-552.

Brouwer, H., Murphy T. and McArdle, L. (1990) A sediment contact assay with *Photobacterium phosphoreum*, *Environmental Toxicology and Chemistry* **9**, 1353-1358.

Buchwalter, D.B., Jenkins, J.J. and Curtis, L.R. (2003) Temperature influences on water permeability and chlorpyrifos uptake in aquatic insects with differing respiratory strategies, *Environmental Toxicology and Chemistry* **22** (11), 2806-2812.

Bulich, A.A. and Bailey, G. (1995) Environmental toxicity assessment using luminescent bacteria, in M. Richardson (ed.), *Environmental Toxicology Assessment*, Taylor & Francis Ltd., London, England, pp. 29-40.

Bulich, A.A., Huynh, H. and Ulitzur, S. (1996) The use of luminescent bacteria for measuring chronic toxicity, in G.K. Ostrander (ed.), *Techniques in Aquatic Toxicology*, CRC Press, Boca Raton, FL, pp. 3-12.

Burba, A. (1999) The design of an experimental system of estimation methods for effects of heavy metals and their mixtures on *Daphnia magna*, *Acta Zoologica Lituanica, Hydrobiologia* **9** (2), 21-29.

Burton Jr, G.A., Nelson, M.K. and Ingersoll, C.G. (1992) Freshwater benthic toxicity tests, in G.A. Burton Jr. (ed.), *Sediment Toxicity Assessment*, Lewis Publishers, Boca Raton, FL, pp. 213-240.

Burton Jr, G.A., Norberg-King, T.J., Ingersoll, C.G., Benoit, D.A., Ankley, G.T., Winger, P.V., Kubitz, J.A., Lazorchak, J.M., Smith, M.E., Greer, E., Dwyer, F.J., Call, D.J., Day, K.E., Kennedy, P. and Stinson, M. (1996) Interlaboratory study of precision: *Hyallela azteca* and *Chironomus tentans* freshwater sediment toxicity assays, *Environmental Toxicology and Chemistry* **15** (8), 1335-1343.

Burton Jr, G.A., Baudo, R., Beltrami, M. and Rowland, C. (2001) Assessing sediment contamination using six toxicity assays, *Journal of Limnology* **60** (2), 263-267.

Bury, N.R., Shaw, J., Glover, C. and Hogstrand, C. (2002) Derivation of a toxicity-based model to predict how water chemistry influences silver toxicity to invertebrates, *Comparative Biochemistry and Physiology, Part C* **133** (1-2), 259-270.

Cairns Jr, J. and Pratt, J.R. (1989) The scientific basis of bioassays, in M. Munawar, G. Dixon, C.I. Mayfield, T. Reynoldson and M.H. Sadar (eds.), *Environmental Bioassay Techniques and their Application: Proceedings of the 1st International Conference held in Lancaster, England, 11-14 July 1988*, Kluwer Academic Publishers, Dordrecht, Netherlands, pp. 5-20.

Call, D.J., Liber, K., Whiteman, F.W., Dawson, T.D. and Brooke, L.T. (1999) Observations on the 10-day *Chironomus tentans* survival and growth bioassay in evaluating Great Lakes sediments, *Journal of the Great Lakes Research* **25**, 171-178.

Call, D.J., Markee, T.P., Geiger, D.L., Brooke, L.T., VandeVenter, F.A., Cox, D.A., Genisot, K.I., Robillard, K.A., Gorsuch, J.W., Parkerton, T.F., Reiley, M.C., Ankley, G.T. and Mount, D.R. (2001) An assessment of the toxicity of phthalate esters to freshwater benthos. 1. Aqueous exposures, *Environmental Toxicology and Chemistry* **20** (8), 1798-1804.

Calleja, M.C. and Persoone, G. (1993) The influence of solvents on the acute toxicity of some lipophilic chemicals to aquatic invertebrates, *Chemosphere* **26** (11), 2007-2022.

Calow, P. (1989) The choice and implementation of environmental bioassays, in M. Munawar, G. Dixon, C.I. Mayfield, T. Reynoldson and M.H. Sadar (eds.), *Environmental Bioassay Techniques and their Application: Proceedings of the 1st International Conference held in Lancaster, England, 11-14 July 1988*, Kluwer Academic Publishers, Dordrecht, Netherlands, pp. 61-64.

Campbell, D.L., Lawton, L.A., Beattie, K.A. and Codd, G.A. (1994) Comparative assessment of the specificity of the brine shrimp and Microtox assays to hepatotoxic (microcystin-LR-containing) cyanobacteria, *Environmental Toxicology and Water Quality* **9** (1), 71-77.

Cancilla, D.A., Holtkamp, A., Matassa, L. and Fang, X. (1997) Isolation and characterization of Microtox®-active components from aircraft de-icing/anti-icing fluids, *Environmental Toxicology and Chemistry* **16**(3), 430-434.

Canfield, T.J., Brunson, E.L., Dwyer, F.J., Ingersoll, C.G. and Kemble, N.E. (1998) Assessing sediments from Upper Mississippi River navigational pools using a benthic invertebrate community evaluation and the sediment quality triad approach, *Archives of Environmental Contamination and Toxicology* **35** (2), 202-212.

Cangiano, T., Dellagreca, M., Fiorentino, A., Isidori, M., Monaco, P. and Zarrelli, A. (2002) Effect of ent-labdane diterpenes from *Potamogetonaceae* on *Selenastrum capricornutum* and other aquatic organisms, *Journal of Chemical Ecology* **28** (6), 1091-1102.

Canivet, V. and Gibert, J. (2002) Sensitivity of epigean and hypogean freshwater macroinvertebrates to complex mixtures. Part I: Laboratory experiments, *Chemosphere* **46** (7), 999-1009.

CANMET (1996) Comparison of results from alternative acute toxicity tests with rainbow trout for selected mine effluents, *Aquatic Effects Technology Evaluation (AETE) Program*, Project 1.1.4, Canada Centre for Mineral and Energy Technology (CANMET), Mining Association of Canada (MAC), Ottawa, Ontario, pp. 1-228.

CANMET (1997a) Review of methods for sublethal aquatic toxicity tests relevant to the Canadian metal-mining industry, *Aquatic Effects Technology Evaluation (AETE) Program*, Project 1.2.1, Canada Centre for Mineral and Energy Technology (CANMET), Mining Association of Canada (MAC), Ottawa, Ontario, pp. 1-132.

CANMET (1997b) Laboratory screening of sublethal toxicity tests for selected mine effluents, *Aquatic Effects Technology Evaluation (AETE) Program*, Project 1.2.2, Canada Center for Mineral and Energy Technology (CANMET), Mining Association of Canada (MAC), Ottawa, Ontario, pp. 1-69.

CANMET (1997c) Toxicity assessment of highly mineralized waters from potential mine sites, *Aquatic Effects Technology Evaluation (AETE) Program*, Project 1.2.4, Canada Centre for Mineral and Energy Technology (CANMET), Mining Association of Canada (MAC), Ottawa, Ontario, 38 pp.

CANMET (1998) Toxicity assessment of mining effluents using up-stream or reference site waters and test organism acclimation techniques, *Aquatic Effects Technology Evaluation (AETE) Program*, Project 4.1.2a, Canada Centre for Mineral and Energy Technology (CANMET), Mining Association of Canada (MAC), Ottawa, Ontario, 81 pp.

CANMET (1999) Technical evaluation of determining mining related impacts utilizing benthos macroinvertebrate fitness parameters, *Aquatic Effects Technology Evaluation (AETE) Program*, Project 2.1.5, Canada Centre for Mineral and Energy Technology (CANMET), Mining Association of Canada (MAC), Ottawa, Ontario, 81 pp.

Canna-Michaelidou, S., Nicolaou, A.S., Neopfytou, E. and Christodoulidou, M., (2000) The use of a battery of microbiotests as a tool for integrated pollution control evaluation and perspectives in Cyprus, in G. Persoone, C. Janssen and W.M. De Coen (eds.), *New Microbiotests for Routine Toxicity Screening and Biomonitoring,* Kluwer Academic / Plenum Publishers, New York, pp. 39-48.

Carter, J.A., Mroz, R.E., Tay, K.L. and Doe, K.G. (1998) An evaluation of the use of soil and sediment bioassays in the assessment of three contaminated sites in Atlantic Canada, *Water Quality Research Journal of Canada* **33** (2), 295-317.

Castaño, A., Bols, N.C., Braunbeck, T., Dierickx, P.J., Halder, M., Isomaa, B., Kawahara, K., Lee, L.E.J., Mothersill, C., Pärt, P., Repetto, G., Sintes, J.R., Rufli, H., Smith, R., Wood, C. and Segner, H. (2003) The use of fish cells in ecotoxicology. The report and recommendations of ECVAM Workshop 47, *ATLA (Alternatives To Laboratory Animals)* **31** (3), 317-351.

Castillo, G.C., Vila, I.C. and Neild, E. (2000) Ecotoxicity assessment of metals and wastewater using multitrophic assays, *Environmental Toxicology* **15** (5), 370-375.

Cataldo, D., Colombo, J.C., Boltovskoy, D., Bilos, C. and Landoni, P. (2001) Environmental toxicity assessment in the Paraná river delta (Argentina): simultaneous evaluation of selected pollutants and mortality rates of *Corbicula fluminea* (Bivalvia) early juveniles, *Environmental Pollution* **112** (3), 379-389.

Cerejeira, M.J., Pereira, T. and Silva-Fernandes, A. (1998) Use of new microbiotests with *Daphnia magna* and *Selenastrum capricornutum* immobilized forms, *Chemosphere* **37** (14-15), 2949-2955.

Chan, Y.K., Wong, C.K., Hsieh, D.P.H., Ng, S.P., Lau, T.K. and Wong, P.K. (2003) Application of a toxicity identification evaluation for a sample of effluent discharged from a dyeing factory in Hong Kong, *Environmental Toxicology* **18** (5), 312-316.

Chapman, P.M. (1998) Death by mud: amphipod sediment toxicity tests, in P.G. Wells, K. Lee and C. Blaise (eds.), *Microscale Testing in Aquatic Toxicology: Advances, Techniques, and Practice*, CRC Press, Boca Raton, FL, pp. 451-463.

Chapman, K.K., Benton, M.J., Brinkhurst, R.O. and Scheuerman, P.R. (1999) Use of the aquatic oligochaetes *Lumbriculus variegatus* and *Tubifex tubifex* for assessing the toxicity of copper and cadmium in a spiked-artificial-sediment toxicity test, *Environmental Toxicology* **14** (2), 271-278.

Charles, A.L., Markich, S.J., Stauber, J.L. and De Filippis, L.F. (2002) The effect of water hardness on the toxicity of uranium to a tropical freshwater alga (*Chlorella* sp.), *Aquatic Toxicology* **60** (1-2), 61-73.

Cheam, V., Reynoldson, T., Garbai, G., Rajkumar, J. and Milani, D. (2000) Local impacts of coal mines and power plants across Canada. II. Metals, organics and toxicity in sediments, *Water Quality Research Journal of Canada* **35** (4), 609-631.

Chen, C.-Y. and Lin, K.-C. (1997) Optimization and performance evaluation of the continuous algal toxicity test, *Environmental Toxicology and Chemistry* **16** (7), 1337-1344.

Chen, C.-Y., Huang, J.-B. and Chen, S.-D. (1997) Assessment of the microbial toxicity test and its application for industrial wastewaters, *Water Science and Technology* **36** (12), 375-382.

Chen, C.-Y., Chen, J.-N. and Chen, S.-D. (1999) Toxicity assessment of industrial wastewater by microbial testing method, *Water Science and Technology* **39** (10-11), 139-143.

Chial, B.Z. and Persoone, G. (2002) Cyst-based toxicity tests XIII - Development of a short chronic sediment toxicity test with the ostracod crustacean *Heterocypris incongruens*: Methodology and precision, *Environmental Toxicology* **17** (6), 528-532.

Chial, B.Z., Persoone, G. and Blaise, C. (2003) Cyst-based toxicity tests. XVIII. Application of ostracodtoxkit microbiotest in a bioremediation project of oil-contaminated sediments: Sensitivity comparison with *Hyalella azteca* solid-phase assay, *Environmental Toxicology* **18** (5), 279-283.

Choi, K. and Meier, P.G. (2001) Toxicity evaluation of metal plating wastewater employing the Microtox® assay: a comparison with cladocerans and fish, *Environmental Toxicology* **16** (2), 136-141.

Choi, K., Sweet, L.I., Meier, P.G. and Kim, P.G. (2004) Aquatic toxicity of four alkylphenols (3-tert-butylphenol, 2-isopropylphenol, 3-isopropylphenol, and 4-iso-propylphenol) and their binary mixtures to microbes, invertebrates and fish, *Environmental Toxicology* **19**, 45-50.

Chu, S., He, Y. and Xu, X. (1997) Determination of acute toxicity of polychlorinated biphenyls to *Photobacterium phosphoreum*, *Bulletin of Environmental Contamination and Toxicology* **58** (2), 263-267.

Cieniawski, S. and Collier, D. (2003) *Post-Remediation Sediment Sampling on the Raisin River Near Monroe, Michigan*, U.S. Environmental Protection Agency, Great Lakes National Programm Office, Chicago, Illinois, 52 pp.

Clément, B. (2000) The use of microbiotests for assessing the influence of the dilution medium quality on the acute toxicity of chemicals and effluents, in G. Persoone, C. Janssen and W.M. De Coen (eds.), *New Microbiotests for Routine Toxicity Screening and Biomonitoring*, Kluwer Academic/Plenum Publishers, New York, pp. 221-228.

Clément, B., Persoone, G., Janssen, C. and Le Dû-Delepierre, A. (1996) Estimation of the hazard of landfills through toxicity testing of leachates - I. Determination of leachate toxicity with a battery of acute tests, *Chemosphere* **33** (11), 2303-2320.

Cleuvers, M. and Ratte, H.T. (2002a) Phytotoxicity of coloured substances: is *Lemna* duckweed an alternative to the algal growth inhibition test?, *Chemosphere* **49** (1), 9-15.

Cleuvers, M. and Ratte, H.T. (2002b) The importance of light intensity in algal tests with coloured substances, *Water Research* **36** (9), 2173-2178.

Cleuvers, M. and Weyers, A. (2003) Algal growth inhibition test: does shading of coloured substances really matter?, *Water Research* **37** (11), 2718-2722.

Codina, J.C., Pérez-García, A. and de Vicente, A. (1994) Detection of heavy metal toxicity and genotoxicity in wastewaters by microbial assay, *Water Science and Technology* **30** (10), 145-151.

Coleman, R.N. and Qureshi, A.A. (1985) Microtox and *Spirillum volutans* tests for assessing toxicity of environmental samples, *Bulletin of Environmental Contamination and Toxicology* **35** (4), 443-451.

Collier, D. and Cieniawski, S. (2003) *Survey of sediment contamination in the Chicago River, Chicago, Illinois*, U.S. Environmental Protection Agency, Great Lakes National Program Office, Chicago, IL, (2003-12-30), 46 pp.; http://www.epa.gov/glnpo/sediment/ChgoRvr/chgorvrpt.pdf.

Connon, R., Printes, L.B., Dewhurst, R.E., Crane, M. and Callaghan, A. (2000) *Groundwater Pollution: development of biomarkers for the assessment of sublethal toxicity*, URGENT Annual Meeting 2000 Proceedings of the NERC URGENT Thematic Programme, Cardiff University, Wales, UK (2003-12-22); http://urgent.nerc.ac.uk/Meetings/2000/2000Proc/water/connon.htm.

Conrad, A.U., Fleming, R.J. and Crane, M. (1999) Laboratory and field response of *Chironomus riparius* to a pyrethroid insecticide, *Water Research* **33**(7), 1603-1610.

Cooman, K., Gajardo, M., Nieto, J., Bornhardt, C. and Vidal, G. (2003) Tannery wastewater characterization and toxicity effects on *Daphnia* spp., *Environmental Toxicology* **18** (1), 45-51.

Cortes, G., Mendoza, A. and Muñoz, D. (1996) Toxicity evaluation using bioassays in rural developing district 063 Hidalgo, Mexico, *Environmental Toxicology and Water Quality* **11** (2), 137-143.

Côté, C., Blaise, C., Michaud, J.-R., Ménard, L., Trottier, S., Gagné, F. and Lifshitz, R. (1998a) Comparisons between microscale and whole-sediment assays for freshwater sediment toxicity assessment, *Environmental Toxicology and Water Quality* **13** (1), 93-110.

Côté, C., Blaise, C., Schroeder, J., Douville, M. and Michaud, J.-R. (1998b) Investigating the adequacy of selected micro-scale bioassays to predict the toxic potential of freshwater sediments through a tier process, *Water Quality Research Journal of Canada* **33** (2), 253-277.

Côté, C., Douville, M. and Michaud, J.-R. (1999) *Eaux usées industrielles : évaluation de micro-bioessais pour la surveillance et l'identification de la toxicité des effluents de l'industrie papetière*, Saint-Laurent Vision 2000, Environnement Canada, Québec.

Couture, P., Blaise, C., Cluis, D. and Bastien, C. (1989) Zirconium toxicity assessment using bacteria, algae and fish assay, *Water, Air, and Soil Pollution* **47** (1-2), 87-100.

Cowgill, U.M. (1986) Why round-robin testing with zooplankton often fails to provide acceptable results, in T.M. Poston and R. Purdy (eds.), *Aquatic Toxicology and Environmental Fate: 9th Volume, ASTM STP 921*, American Society for Testing and Materials, Philadelphia, PA, pp. 349-356.

Coya, B., Marañón, E. and Sastre, H. (1996) Evaluation of the ecotoxicity of industrial wastes by microtox bioassay, *Toxicology Letters* **88** (Supplement 1), 79-79.

Coz, A., Andrés, A. and Irabien, A. (2004) Ecotoxicity assessment of stabilized/solidified foundry sludge, *Environmental Science and Technology* **38**, 1897-1900.

Crane, M., Delaney, P., Mainstone, C. and Clarke, S. (1995) Measurement by *in situ* bioassay of water quality in an agricultural catchment, *Water Research* **29** (11), 2441-2448.

Crane, M., Attwood, C., Sheahan, D. and Morris, S. (1999) Toxicity and bioavailability of the organophosphorus insecticide pirimiphos methyl to the freshwater amphipod *Gammarus pulex* L. in laboratory and mesocosm systems, *Environmental Toxicology and Chemistry* **18** (7), 1456-1461.

Czerniawska-Kusza, I. and Ebis, M. (2000) Toxicity of waste dump leachates and sugar factory effluents and their impact on groundwater and surface water quality in the Opole Province in Poland, in G. Persoone, C. Janssen and W.M. De Coen (eds.), *New Microbiotests for Routine Toxicity Screening and Biomonitoring*, Kluwer Academic/Plenum Publishers, New York, pp. 319-322.

Daniels, S.A., Munawar, M. and Mayfield, C.I. (1989) An improved elutriation technique for the bioassessment of sediment contaminants, in M. Munawar, G. Dixon, C.I. Mayfield, T. Reynoldson and M.H. Sadar (eds.), *Environmental Bioassay Techniques and their Application: Proceedings of the 1st International Conference held in Lancaster, England, 11-14 July 1988*, Kluwer Academic Publishers, Dordrecht, Netherlands, pp. 619-631.

Dauble, D.D., Fallon, W.E., Gray, R.H. and Bean, R.M. (1982) Effects of coal liquid water-soluble fractions on growth and survival of four aquatic organisms, *Archives of Environmental Contamination and Toxicology* **11** (5), 553-560.

Dave, G. and Dennegard, B. (1994) Sediment toxicity and heavy metals in the Kattegat and Skaggerak, *Journal of Aquatic Ecosystem Health* **3** (3), 207-219.

Day, K.E., Holtze, K.E., Metcalfe-Smith, J.L., Bishop, C.T. and Dutka, B.J. (1993) Toxicity of leachate from automobile tires to aquatic biota, *Chemosphere* **27** (4), 665-675.

Day, K.E., Dutka, B.J., Kwan, K.K., Batista, N., Reynoldson, T.B. and Metcalfe-Smith, J.L. (1995a) Correlations between solid-phase microbial screening assays, whole-sediment toxicity tests with macroinvertebrates and *in situ* benthic community structure, *Journal of the Great Lakes Research* **21** (2), 192-206.

Day, K.E., Kirby, R.S. and Reynoldson, T.B. (1995b) The effect of manipulations of freshwater sediments on responses of benthic invertebrates in whole-sediment toxicity tests, *Environmental Toxicology and Chemistry* **14** (8), 1333-1343.

Day, K.E., Maguire, R.J., Milani, D. and Batchelor, S.P. (1998) Toxicity of tributyltin to four species of freshwater benthic invertebrates using spiked sediment bioassays, *Water Quality Research Journal of Canada* **33** (1), 111-132.

De Coen, W.M. and Janssen, C.R. (1997) The use of biomarkers in *Daphnia magna* toxicity testing II. Digestive enzyme activity in *Daphnia magna* exposed to sublethal concentrations of cadmium, chromium and mercury, *Chemosphere* **35** (5), 1053-1067.

de Haas, E.M., Reuvers, B., Moermond, C.T.A., Koelmans, A.A. and Kraak, M.H.S. (2002) Responses of benthic invertebrates to combined toxicant and food input in floodplain lake sediments, *Environmental Toxicology and Chemistry* **21** (10), 2165-2171.

de Haas, E.M., Paumen, M.L., Koelman, A.A. and Kraak, M.H.S. (2004) Combined effects of copper and food on the midge *Chironomus riparius* in whole-sediment bioassays, *Environmental Pollution* **127** (1), 99-107.

de Jonge, J., Brils, J.M., Hendriks, A.J. and Ma, C. (1999) Ecological and ecotoxicological surveys of moderately contaminated floodplain ecosystems in the Netherlands, *Aquatic Ecosystem Health and Management* **2** (1), 9-18.

de Vlaming, V. and Norberg-King, T.J. (1999) *A review of single species toxicity tests: are the tests reliable predictors of aquatic ecosystem community reponses?*, EPA 600/R-97/114, Office of Research and Development, U.S. Environmental Protection Agency, Duluth, MN.

DellaGreca, M., Fiorentino, A., Isidori, M., Monaco, P., Temussi, F. and Zarrelli, A. (2001) Antialgal furano-diterpenes from *Potamogeton natans* L., *Phytochemistry* **58** (2), 299-304.

DellaGreca, M., Fiorentino, A., Isidori, M., Lavorgna, M., Monaco, P., Previtera, L. and Zarrelli, A. (2002) Phenanthrenoids from the wetland *Juncus acutus*, *Phytochemistry* **60** (6), 633-638.

DellaGreca, M., Fiorentino, A., Isidori, M., Lavorgna, M., Previtera, L., Rubino, M. and Temussi, F. (2004) Toxicity of prednisolone, dexamethasone and their photochemical derivatives on aquatic organisms, *Chemosphere* **54** (5), 629-637.

den Besten, P.J. and Tuk, C.W. (2000) Relation between responses in the neutral red retention test and the comet assay and life history parameters of *Daphnia magna*, *Marine Environment Research* **50** (1-5), 513-516.

den Besten, P.J., Naber, A., Grootelaar, E.M.M. and van de Guchte, C. (2003) *In situ* bioassays with *Chironomus riparius*: laboratory-field comparisons of sediment toxicity and effects during wintering, *Aquatic Ecosystem Health and management* **6** (2), 217 - 228.

Deniseger, J. and Kwong, Y.T.J. (1996) Risk Assessment of Copper-Contaminated Sediments in the Tsolum River Near Courtenay, British Columbia, *Water Quality Research Journal of Canada* **31** (4), 725-740.

Denizeau, F. (1998) The use of fish cells in the toxicological evaluation of environmental contaminants, in P.G. Wells, K. Lee and C. Blaise (eds.), *Microscale Testing in Aquatic Toxicology: Advances, Techniques, and Practice*, CRC Press, Boca Raton, FL, pp. 113-128.

Dewhurst, R.E., Connon, R., Crane, M., Callaghan, A. and Mather, J.D. (2001) *URGENT in Hounslow and Heathrow. The application of acute and sub-lethal ecotoxicity tests to groundwater quality assessment*, URGENT Annual Meeting 2000 Proceedings of the NERC URGENT Thematic Programme, Cardiff University, Wales, UK (2004-01-02); http://urgent.nerc.ac.uk/Meetings/2001/Abstracts/mather.htm.

Dias, N. and Lima, N. (2002) A comparative study using a fluorescence-based and a direct-count assay to determine cytotoxicity in *Tetrahymena pyriformis*, *Research in Microbiology* **153** (5), 313-322.

Diaz-Baez, M.C. and Roldan, F. (1996) Evaluation of the agar plate method for rapid toxicity assessment with some heavy metals and environmental samples, *Environmental Toxicology and Water Quality* **11** (3), 259-263.

Diaz-Baez, M.C., Sanchez, W.A., Dutka, B.J., Ronco, A., Castillo, G., Pica-Granados, Y., Castillo, L.E., Ridal, J., Arkhipchuk, V. and Srivastava, R.C. (2002) Overview of results from the WaterTox intercalibration and environmental testing phase II program: part 2, ecotoxicological evaluation of drinking water supplies, *Environmental Toxicology* **17** (3), 241-249.

Dieter, C.D., Hamilton, S.J., Duffy, W.G. and Flake, L.D. (1994) Evaluation of the Microtox test to detect phorate contamination in wetlands, *Journal of Freshwater Ecology* **9** (4), 271-280.

Dive, D., Blaise, C., Robert, S., Le Du, A., Bermingham, N., Cardin, R., Kwan, A., Legault, R., Mac Carthy, L., Moul, D. and Veilleux, L. (1990) Canadian workshop on the *Colpidium campylum* ciliate protozoan growth inhibition test, *Zeitschrift für angewandte Zoologie* **76** (1), 49-63.

Dive, D., Blaise, C. and Le Du, A. (1991) Standard protocol proposal for undertaking the *Colpidium campylum* ciliate protozoan growth inhibition test, *Zeitschrift für angewandte Zoologie* **78** (1), 79-90.

Dizer, H., Wittekindt, E., Fischer, B. and Hansen, P.-D. (2002) The cytotoxic and genotoxic potential of surface water and wastewater effluents as determined by bioluminescence, umu-assays and selected biomarkers, *Chemosphere* **46** (2), 225-233.

Dmitruk, U. and Dojlido, J. (2000) Application of Toxkit microbiotests for toxicity evaluation of river waters and waste waters in the region of Warsaw in Poland, in G. Persoone, C. Janssen and W.M. De Coen (eds.), *New Microbiotests for Routine Toxicity Screening and Biomonitoring*, Kluwer Academic/Plenum Publishers, New York, pp. 323-325.

Dodard, S.G., Renoux, A.Y., Hawari, J., Ampleman, G., Thiboutot, S. and Sunahara, G.I. (1999) Ecotoxicity characterization of dinitrotoluenes and some of their reduced metabolites, *Chemosphere* **38** (9), 2071-2079.

Dodson, S.I., Merritt, C.M., Shannahan, J.-P. and Shults, C.M. (1999) Low exposure concentrations of atrazine increase male production in *Daphnia pulicaria*, *Environmental Toxicology and Chemistry* **18** (7), 1568-1573.

Doherty, F.G. (2001) A review of the Microtox toxicity test system for assessing the toxicity of sediments and soils, *Water Quality Research Journal of Canada* **36** (3), 475-518.

Doherty, F.G., Qureshi, A.A. and Razza, J.B. (1999) Comparison of the *Ceriodaphnia dubia* and Microtox® inhibition tests for toxicity assessment of industrial and municipal wastewaters, *Environmental Toxicology* **14** (4), 375-382.

Doi, J. and Grothe, D.R. (1989) Use of fractionation and chemical analysis schemes for plant effluent toxicity evaluations, in G.W. Suter II and M.A. Lewis (eds.), *Aquatic Toxicology and Environmental Fate: Eleventh Volume, ASTM STP 1007*, American Society for Testing and Materials, Philadelphia, PA, pp. 204-215.

Dombroski, E.C., Smiley, K.L., Johnson, C.I., Florence, L.Z. and Dieken, F.P. (1993) A comparison of bioassay results from untreated CTMP effluent, *Canadian Technical Report of Fisheries and Aquatic Sciences* **1942**, 96-104.

Draper III, A.C. and Brewer, W.S. (1979) Measurement of the aquatic toxicity of volatile nitrosamines, *Journal of Toxicology and Environmental Health* **5** (6), 985-993.

Dunbar, A.M., Lazorchak, J.M. and Waller, W.T. (1983) Acute and chronic toxicity of sodium selenate to *Daphnia magna* Straus, *Environmental Toxicology and Chemistry* **2** (2), 239-244.

Dutka, B.J., Tuominen, T., Churchland, L. and Kwan, K.K. (1989) Fraser river sediments and waters evaluated by the battery of screening tests technique, in M. Munawar, G. Dixon, C.I. Mayfield, T. Reynoldson and M.H. Sadar (eds.), *Environmental Bioassay Techniques and their Application: Proceedings of the 1st International Conference held in Lancaster, England, 11-14 July 1988*, Kluwer Academic Publishers, Dordrecht, Netherlands, pp. 301-315.

Dutka, B.J., McInnis, R., Jurkovic, A., Liu, D. and Castillo, G. (1996) Water and sediment ecotoxicity studies in Temuco and Rapel River Basin, Chile, *Environmental Toxicology and Water Quality* **11** (3), 237-247.

Dyatlov, S. (2000) Comparison of Ukrainian standard methods and new microbiotests for water toxicity assessment, in G. Persoone, C. Janssen and W.M. De Coen (eds.), *New Microbiotests for Routine Toxicity Screening and Biomonitoring*, Kluwer Academic/Plenum Publishers, New York, pp. 229-232.

Eggen, R.I.L., Behra, R., Burkhardt-Holm, P., Escher, B.I. and Schweigert, N. (2004) Challenges in ecotoxicology, *Environmental Science and Technology*, February 1, 2004, pp. 59A-64A.

Eleftheriadis, K., Angelaki, A., Kungolos, A., Nalbandian, L. and Sakellaropoulos, G.P. (2000) Assessing the impact of atmospheric wet and dry deposition using chemical and toxicological analysis, in G. Persoone, C. Janssen and W.M. De Coen (eds.), *New Microbiotests for Routine Toxicity Screening and Biomonitoring*, Kluwer Academic/Plenum Publishers, New York, pp. 469-473.

Environment Canada (1990) Guidance document on control of toxicity test precision using reference toxicants, Report EPS 1/RM/12, Environment Canada, Ottawa, 85 pp.

Environment Canada (1994) Guidance document on collection and preparation of sediment for physicochemical characterization and biological testing, Report EPS 1/RM/29, Environment Canada, Ottawa, 144 pp.

Environment Canada (1995) Guidance document on measurement of toxicity test precision using control sediments spiked with a reference toxicant, Report EPS 1/RM/30, Environment Canada, Ottawa, 56 pp.

Environment Canada (1999) Guidance document on application and interpretation of single-species tests in environmental toxicology, Report EPS 1/RM/34, Environment Canada, Ottawa, 203 pp.

Environment Canada (2004a) Guidance document for testing the pathogenicity and toxicity of new microbial substances to aquatic and terrestrial organisms, Report EPS 1/RM/44, Environment Canada, Ottawa, 171 pp.

Environment Canada (2004b) Guidance document on statistical methods to determine endpoints to toxicity tests, Report EPS 1/RM/46, Environment Canada, Ottawa, 265 pp.

Erten-Unal, M., Wixson, B.G., Gale, N. and Pitt, J.L. (1998) Evaluation of toxicity, bioavailability and speciation of lead, zinc and cadmium in mine/mill wastewaters, *Chemical Speciation and Bioavailability* **10** (2), 37-46.

Evandri, M.G., Costa, L.G. and Bolle, P. (2003) Evaluation of brominated diphenyl ether-99 toxicity with *Raphidocelis subcapitata* and *Daphnia magna*, *Environmental Toxicology and Chemistry* **22** (9), 2167-2172.

Farrell, A.P., Kennedy, C.J., Wood, A., Johnston, B.D. and Bennett, W.R. (1998) Acute toxicity of a didecyldimethylammonium chloride-based wood preservative, bardac 2280, to aquatic species, *Environmental Toxicology and Chemistry* **17** (8), 1552-1557.

Farrell, A.P., Kennedy, C., Cheng, W. and Lemke, M.A. (2001) Acute toxicity of monochloramine to juvenile chinook salmon (*Oncorhynchus tshawytscha* Walbaum) and *Ceriodaphnia dubia*, *Water Quality Research Journal of Canada* **36** (1), 133-149.

Fearnside, D. and Hiley, P.D. (1993) The role of Microtox® in the detection and control of toxic trade effluents and spillages, in M. Richardson (ed.), *Ecotoxicology Monitoring*, VCH Publishers, Weinheim, Germany, pp. 319-332.

Feng, Q., Boone, A.N. and Vijayan, M.M. (2003) Copper impact on heat shock protein 70 expression and apoptosis in rainbow trout hepatocytes, *Comparative Biochemistry and Physiology, Part C* **135** (3), 345-355.

Fent, K. (2001) Fish cell lines as versatile tools in ecotoxicology: assessment of cytotoxicity, cytochrome P4501A induction potential and estrogenic activity of chemicals and environmental samples, *Toxicology In Vitro* **15** (4-5), 477-488.

Fentem, J. and Balls, M. (1993) Replacement of fish in ecotoxicology testing: use of bacteria, other lower organisms and fish cells *in vitro*, in M. Richardson (ed.), *Ecotoxicology Monitoring*, VCH Publishers, Weinheim, Germany, pp. 71-81.

Fernández-Alba, A.R., Hernando, M.D., Piedra, L. and Chisti, Y. (2002) Toxicity evaluation of single and mixed antifouling biocides measured with acute toxicity bioassays, *Analytica Chimica Acta* **456** (2), 303-312.

Fernández-Casalderrey, A., Ferrando, M.D. and Andreu-Moliner, E. (1993) Chronic toxicity of methylparathion to the rotifer *Brachionus calyciflorus* fed on *Nannochloris oculata* and *Chlorella pyrenoidosa*, *Hydrobiologia* **255/256**, 41-49.

Fernández-Sempere, J., Barrueso-Martínez, M.L., Font-Montesinos, R. and Sabater-Lillo, M.C. (1997) Characterization of tannery wastes. Comparison of three leachability tests, *Journal of Hazardous Materials* **54** (1-2), 31-45.

Ferrari, B., Radetski, C.M., Veber, A.-M. and Férard, J.-F. (1999) Ecotoxicological assessment of solid wastes: a combined liquid- and solid-phase testing approach using a battery of bioassays and biomarkers, *Environmental Toxicology and Chemistry* **18** (6), 1195-1202.

Fialkowski, W., Klonowska-Olejnik, M., Smith, B.D. and Rainbow, P.S. (2003) Mayfly larvae (*Baetis rhodani* and *B. vernus*) as biomonitors of trace metal pollution in streams of a catchment draining a zinc and lead mining area of Upper Silesia, Poland, *Environmental Pollution* **121** (2), 253-267.

Finger, S.E., Little, E.F., Henry, M.G., Fairchild, J.F. and Boyle, T.P. (1985) Comparison of laboratory and field assessment of fluorene - Part I: Effects of fluorene on the survival, growth, reproduction, and behavior of aquatic organisms in laboratory tests, in T.P. Boyle (ed.), *Validation and Predictability of Laboratory Methods for Assessing the Fate and Effects of Contaminants in Aquatic Ecosystems, ASTM STP 865*, American Society for Testing and Materials, Philadelphia, PA, pp. 120-133.

Fisher, D.J., Hersh, C.M., Paulson, R.L., Burton, D.T. and Hall Jr., L.W. (1989) Acute toxicity of industrial and municipal effluents in the state of Maryland, USA: results from one year of toxicity testing, in M. Munawar, G. Dixon, C.I. Mayfield, T. Reynoldson and M.H. Sadar (eds.), *Environmental Bioassay Techniques and their Application: Proceedings of the 1st International Conference held in Lancaster, England, 11-14 July 1988*, Kluwer Academic Publishers, Dordrecht, Netherlands, pp. 641-648.

Fisher, D.J., Knott, M.H., Turley, B.S., Yonkos, L.T. and Ziegler, G.P. (1998) Acute and chronic toxicity of industrial and municipal effluents in Maryland, U.S., *Water Environment Research* **10** (1), 101-107.

Fisher, S.W. (1986) Effects of temperature on the acute toxicity of PCP in the midge *Chironomus riparius* Meigen, *Bulletin of Environmental Contamination and Toxicology* **36** (5), 744-748.

Fisher, S.W. and Wadleigh, R.W. (1986) Effects of pH on the acute toxicity and uptake of [^{14}C]pentachlorophenol in the midge, *Chironomus riparius*, *Ecotoxicology and Environmental Safety* **11** (1), 1-8.

Fleming, R.J., Holmes, D. and Nixon, S.J. (1998) Toxicity of permethrin to *Chironomus riparius* in artificial and natural sediments, *Environmental Toxicology and Chemistry* **17** (7), 1332-1337.

Fliedner, A. (1997) Ecotoxicity of poorly water-soluble substances, *Chemosphere* **35** (1-2), 295-305.

Fochtman, P., Raszka, A. and Nierzedska, E. (2000) The use of conventional bioassays, microbiotests, and some rapid methods in the selection of an optimal test battery for the assessment of pesticides toxicity, *Environmental Toxicology* **15** (5), 376-384.

Font, R., Gomis, V., Fernandez, J. and Sabater, M.C. (1998) Physico-chemical characterization and leaching of tannery wastes, *Waste Management and Research* **16** (2), 139-149.

Forget, G., Gagnon, P., Sanchez, W.A. and Dutka, B.J. (2000) Overview of methods and results of the eight country International Development Research Centre (IDRC) WaterTox project, *Environmental Toxicology* **15** (4), 264-276.

Franklin, N.M., Stauber, J.L., Markich, S.J. and Lim, R.P. (2000) pH-dependent toxicity of copper and uranium to a tropical freshwater alga (*Chlorella* sp.), *Aquatic Toxicology* **48** (2-3), 275-289.

Franklin, N.M., Stauber, J.L., Apte, S.C. and Lim, R.P. (2002) Effect of initial cell density on the bioavailability and toxicity of copper in microalgal bioassays, *Environmental Toxicology and Chemistry* **21** (4), 742-751.

Froehner, K., Backhaus, T. and Grimme, L.H. (2000) Bioassays with *Vibrio fischeri* for the assessment of delayed toxicity, *Chemosphere* **40** (8), 821-828.

Froehner, K., Meyer, W. and Grimme, L.H. (2002) Time-dependent toxicity in the long-term inhibition assay with *Vibrio fischeri*, *Chemosphere* **46** (7), 987-997.

Fu, L.-J., Staples, R.E. and Stahl Jr, R.G. (1991) Application of the *Hydra attenuata* assay for identifying developmental hazards among natural waters and wastewaters, *Ecotoxicology and Environmental Safety* **22** (3), 309-319.

Fu, L.-J., Staples, C.A. and Stahl Jr, R.G. (1994) Assessing acute toxicities of pre- and post-treatment industrial wastewaters with *Hydra attenuata*: a comparative study of acute toxicity with the fathead minnow, *Pimephales promelas*, *Environmental Toxicology and Chemistry* **13** (4), 563-569.

Fuchsman, P.C., Barber, T.R. and Sheehan, P.J. (1998) Sediment toxicity evaluation for hexachlorobenzene: spiked sediment tests with *Leptocheirus plumulosus*, *Hyalella azteca*, and *Chironomus tentans*, *Archives of Environmental Contamination and Toxicology* **35** (4), 573-579.

Gabrielson, J., Kühn, I., Colque-Navarro, P., Hart, M., Iversen, A., McKenzie, D. and Möllby, R. (2003) Microplate-based microbial assay for risk assessment and (eco)toxic fingerprinting of chemicals, *Analytica Chimica Acta* 485, 121-130.

Gagné, F. and Blaise, C. (1993) Hepatic metallothionein level and mixed function oxidase activity in fingerling rainbow trout (*Oncorhynchus mykiss*) after acute exposure to pulp and paper mill effluents, *Water Research* 27 (11), 1669-1682.

Gagné, F. and Blaise, C. (1997) Evaluation of industrial wastewater quality with a chemiluminescent peroxidase activity assay, *Environmental Toxicology and Water Quality* 12 (4), 315-320.

Gagné, F. and Blaise, C. (1998a) Toxicological evaluation of municipal wastewaters to rainbow trout hepatocytes, *Toxicology Letters* 95 (Supplement 1), 194-194.

Gagné, F. and Blaise, C. (1998b) Estrogenic properties of municipal and industrial wastewaters evaluated with a rapid and sensitive chemoluminescent in situ hybridization assay (CISH) in rainbow trout hepatocytes, *Aquatic Toxicology* 44 (1), 83-91.

Gagné, F. and Blaise, C. (1998c) Differences in the measurement of cytotoxicity of complex mixtures with rainbow trout hepatocytes and fibroblasts, *Chemosphere* 37 (4), 753-769.

Gagné, F. and Blaise, C. (1999) Toxicological effects of municipal wastewaters to rainbow trout hepatocytes, *Bulletin of Environmental Contamination and Toxicology* 63 (4), 503-510.

Gagné, F. and Blaise, C. (2001) Acute cytotoxicity assessment of liquid samples using rainbow trout (*Oncorhynchus mykiss*) hepatocytes, *Environmental Toxicology* 16 (1), 104-109.

Gagné, F., Blaise, C., van Aggelen, G., Boivin, P., Martel, P., Chong-Kit, R., Jonczyk, E., Marion, M., Kennedy, S.W., Legault, R. and Goudreault, J. (1999a) Intercalibration study in the evaluation of toxicity with rainbow trout hepatocytes, *Environmental Toxicology* 14 (4), 429-437.

Gagné, F., Pardos, M., Blaise, C., Turcotte, P., Quémerais, B. and Fouquet, A. (1999b) Toxicity evaluation of organic sediment extracts resolved by size exclusion chromatography using rainbow trout hepatocytes, *Chemosphere* 39 (9), 1545-1570.

Gagné, F., Ridal, J., Blaise, C. and Brownlee, B. (1999c) Toxicological effects of geosmin and 2-methylisoborneol on rainbow trout hepatocytes, *Bulletin of Environmental Contamination and Toxicology* 63 (2), 174-180.

Gale, S.A., Smith, S.V., Lim, R.P., Jeffree, R.A. and Petocz, P. (2003) Insights into the mechanisms of copper tolerance of a population of black-banded rainbowfish (*Melanotaenia nigrans*) (Richardson) exposed to mine leachate, using 64/67Cu, *Aquatic Toxicology* 62 (2), 135-153.

Garay, V., Roman, G. and Isnard, P. (2000) Evaluation of PNEC values: extrapolation from Microtox®, algae, daphnid, and fish data to HC5, *Chemosphere* 40 (3), 267-273.

Gasith, A., Jop, K.M., Dickson, K.L., Parkerton, T.F. and Kaczmarek, S.A. (1988) Protocol for the identification of toxic fractions in industrial wastewater effluents, in W.J. Adams, G.A. Chapman and W.G. Landis (eds.), *Aquatic Toxicology and Hazard Assessment: 10th volume, ASTM STP 971*, American Society for Testing and Materials, Philadelphia, PA, pp. 204-215.

Geis, S.W., Fleming, K.L., Korthals, E.T., Searle, G., Reynolds, L. and Karner, D.A. (2000) Modifications to the algal growth inhibition test for use a regulatory assay, *Environmental Toxicology and Chemistry* 19 (1), 36-41.

Geis, S.W., Fleming, K.L., Mager, A. and Reynolds, L. (2003) Modifications to the fathead minnow (*Pimephales promelas*) chronic test method to remove mortality due to pathogenic organisms, *Environmental Toxicology and Chemistry* 22 (10), 2400-2404.

Gensemer, R.W., Naddy, R.B., Stubblefield, W.A., Hockett, J.R., Santore, R. and Paquin, P. (2002) Evaluating the role of ion composition on the toxicity of copper to *Ceriodaphnia dubia* in very hard waters, *Comparative Biochemistry and Physiology, Part C* 133 (1-2), 87-97.

George, D.B., Berk, S.G., Adams, V.D., Ting, R.S., Roberts, R.O., Parks, L.H. and Lott, R.C. (1995) Toxicity of alum sludge extracts to a freshwater alga, protozoan, fish, and marine bacterium, *Archives of Environmental Contamination and Toxicology* 29 (2), 149-158.

Gerhardt, A., Janssens de Bisthoven, L., Mo, Z., Wang, C., Yang, M. and Wang, Z. (2002a) Short-term responses of *Oryzias latipes* (Pisces: Adrianichthyidae) and *Macrobrachium nipponense* (Crustacea: Palaemonidae) to municipal and pharmaceutical waste water in Beijing, China: survival, behaviour, biochemical biomarkers, *Chemosphere* 47 (1), 35-47.

Gerhardt, A., Schmidt, S. and Höss, S. (2002b) Measurement of movement patterns of *Caenorhabditis elegans* (Nematoda) with the Multispecies Freshwater Biomonitor® (MFB) - a potential new method to study a behavioral toxicity parameter of nematodes in sediments, *Environmental Pollution* 120 (3), 513-516.

Ghosal, T.K. and Kaviraj, A. (2002) Combined effects of cadmium and composted manure to aquatic organisms, *Chemosphere* 46(7), 1099-1105.

Giesy, J.P. and Hoke, R.A. (1989) Freshwater sediment toxicity bioassessment: rationale for species selection and test design, *Journal of the Great Lakes Research* **15** (4), 539-569.

Giesy, J.P., Graney, R.L., Newsted, J.L., Rosiu, C.J., Benda, A., Kreis, J.R.G. and Horvath, F.J. (1988) Comparison of three sediment bioassay methods using Detroit River sediments, *Environmental Toxicology and Chemistry* **7**, 483-498.

Giesy, J.P., Rosiu, C.J., Graney, R.L. and Henry, M.G. (1990) Benthic invertebrate bioassays with toxic sediment and pore water, *Environmental Toxicology and Chemistry* **9** (2), 233-248.

Gilli, G. and Meineri, V. (2000) Assessment of the toxicity and genotoxicity of wastewaters treated in a municipal plant, in G. Persoone, C. Janssen and W.M. De Coen (eds.), *New Microbiotests for Routine Toxicity Screening and Biomonitoring*, Kluwer Academic/Plenum Publishers, New York, pp. 327-338.

Gillis, P.L., Diener, L.C., Reynoldson, T.B. and Dixon, D.G. (2002) Cadmium-induced production of a metallothioneinlike protein in *Tubifex tubifex* (oligochaeta) and *Chironomus riparius* (diptera): correlation with reproduction and growth, *Environmental Toxicology and Chemistry* **21** (9), 1836-1844.

Gilron, G.L. and Lynn, D.H. (1998) Ciliated protozoa as test organisms in toxicity assessments, in P.G. Wells, K. Lee and C. Blaise (eds.), *Microscale Testing in Aquatic Toxicology: Advances, Techniques, and Practice*, CRC Press, Boca Raton, FL, pp. 323-336.

Gilron, G.L., Gransden, S.G., Lynn, D.H., Broadfoot, J. and Scroggins, R. (1999) A behavioral toxicity test using the ciliated protozoan *Tetrahymena thermophila*. I. Method description, *Environmental Toxicology and Chemistry* **18** (8), 1813-1816.

Girling, A.E., Pascoe, D., Janssen, C.R., Peither, A., Wenzel, A., Schäfer, H., Neumeier, B., Mitchell, G.C., Taylor, E.J., Maund, S.J., Lay, J.P., Jüttner, I., Crossland, N.O., Stephenson, R.R. and Persoone, G. (2000) Development of methods for evaluating toxicity to freshwater ecosystems, *Ecotoxicology and Environmental Safety* **45** (2), 148-176.

Gonzalez, A.M. (1996) A laboratory formulated sediment incorporating synthetic acid-volatile sulfide, *Environmental Toxicology and Chemistry* **15** (12), 2209-2220.

Gorbi, G., Corradi, M.G., Invidia, M., Rivara, L. and Bassi, M. (2002) Is Cr(VI) toxicity to *Daphnia magna* modified by food availability or algal exudates? The hypothesis of a specific chromium/algae/exudates interaction, *Water Research* **36** (8), 1917-1926.

Graff, L., Isnard, P., Cellier, P., Bastide, J., Cambon, J.-P., Narbonne, J.-F., Budzinski, H. and Vasseur, P. (2003) Toxicity of chemicals to microalgae in river and in standard waters, *Environmental Toxicology and Chemistry* **22** (6), 1368-1379.

Gray, N.F. and O'Neill, C. (1997) Acid mine-drainage toxicity testing, *Environmental Geochemistry and Health* **19** (4), 165-171.

Gregor, D.J. and Munawar, M. (1989) Assessing toxicity of Lake Diefenbaker (Saskatchewan, Canada) sediments using algal and nematode bioassays, in M. Munawar, G. Dixon, C.I. Mayfield, T. Reynoldson and M.H. Sadar (eds.), *Environmental Bioassay Techniques and their Application: Proceedings of the 1st International Conference held in Lancaster, England, 11-14 July 1988*, Kluwer Academic Publishers, Dordrecht, Netherlands, pp. 291-300.

Grothe, D.R., Dickson, K.L. and Reed-Judkins, D.K. (eds.) (1996) *Whole effluent toxicity testing: an evaluation of methods and prediction of receiving system impacts*, Proceedings from a SETAC - sponsored Pellston Workshop, Society of Environmental Toxicology and Chemistry, Pensacola, FL, 346 pp.

Guéguen, C., Koukal, B., Dominik, J. and Pardos, M. (2003) Competition between alga (*Pseudokirchneriella subcapitata*), humic substances and EDTA for Cd and Zn control in the algal assay procedure (AAP) medium, *Chemosphere* **53** (8), 927-934.

Guerra, R. (2001) Ecotoxicological and chemical evaluation of phenolic compounds in industrial effluents, *Chemosphere* **44** (8), 1737-1747.

Guerrero, N.R.V., Taylor, M.G., Wider, E.A. and Simkiss, K. (2003) Influence of particle characteristics and organic matter content on the bioavailability and bioaccumulation of pyrene by clams, *Environmental Pollution* **121** (1), 115-122.

Guilhermino, L., Diamantino, T.C., Ribeiro, R., Gonçalves, F. and Soares, A.M. (1997) Suitability of test media containing EDTA for the evaluation of acute metal toxicity to *Daphnia magna* Straus, *Ecotoxicology and Environmental Safety* **38** (3), 292-295.

Guilhermino, L., Lacerda, M.N., Nogueira, A.J.A. and Soares, A.M.V.M. (2000) *In vitro* and *in vivo* inhibition of *Daphnia magna* acetylcholinesterase by surfactant agents: possible implications for contamination biomonitoring, *The Science of The Total Environment* **247** (2-3), 137-141.

Gunn, A.M., Hunt, D.T.E. and Winnard, D.A. (1989) The effect of heavy metal speciation in sediment on bioavailability to tubificid worms, in M. Munawar, G. Dixon, C.I. Mayfield, T. Reynoldson and M.H. Sadar (eds.), *Environmental Bioassay Techniques and their Application: Proceedings of the 1st International Conference held in Lancaster, England, 11-14 July 1988*, Kluwer Academic Publishers, Dordrecht, Netherlands, pp. 487-496.

Gustavson, K.E., Svenson, A. and Harkin, J.M. (1998) Comparison of toxicities and mechanism of action of *n*-alkanols in the submitochondrial particle and the *Vibrio fischeri* bioluminescence (Microtox®) bioassay, *Environmental Toxicology and Chemistry* **17** (10), 1917-1921.

Gustavson, K.E., Sonsthagen, S.A., Crunkilton, R.A. and Harkin, J.M. (2000) Groundwater toxicity assessment using bioassay, chemical, and toxicity identification evaluation analyses, *Environmental Toxicology* **15** (5), 421-430.

Guzzella, L. (1998) Comparison of test procedures for sediment toxicity evaluation with *Vibrio fischeri* bacteria, *Chemosphere* **37** (14-15), 2895-2909.

Guzzella, L. and Mingazzini, M. (1994) Biological assaying of organic compounds in surface waters, *Water Science and Technology* **30** (10), 113-124.

Guzzella, L., Bartone, C., Ross, P., Tartari, G. and Muntau, H. (1996) Toxicity identification evaluation of Lake Orta (Northern Italy) sediments using the Microtox system, *Ecotoxicology and Environmental Safety* **35** (3), 231-235.

Haller, W.T. and Stocker, R.K. (2003) Toxicity of 19 adjuvants to juvenile *Lepomis macrochirus* (bluegill sunfish), *Environmental Toxicology and Chemistry* **22**(3), 615-619.

Hamm, J.T., Wilson, B.W. and Hinton, D.E. (2001) Increasing uptake and bioactivation with development positively modulate diazinon toxicity in early life stage medaka (*Oryzias latipes*), *Toxicological Sciences* **61** (2), 304-313.

Hankenson, K. and Schaeffer, D.J. (1991) Microtox assay of trinitrotoluene, diaminonitrotoluene, and dinitromethylaniline mixtures, *Bulletin of Environmental Contamination and Toxicology* **46**(4), 550-553.

Hansen, P.D. (1993) Regulatory significance of toxicological monitoring by and summarizing effect parameters, in M. Richardson (ed.), *Ecotoxicology Monitoring*, VCH Publishers, Weinheim, Germany, pp. 273-286.

Hao, O.J., Shin, C.-J., Lin, C.-F., Jeng, F.-T. and Chen, Z.-C. (1996) Use of microtox tests for screening industrial wastewater toxicity, *Water Science and Technology* **34** (10), 43-50.

Harkey, G.A., Landrum, P.F. and Klaine, S.J. (1994) Comparison of whole-sediment, elutriate and pore-water exposures for use in assessing sediment-associated organic contaminants in bioassays, *Environmental Toxicology and Chemistry* **13** (8), 1315-1329.

Harrahy, E.A. and Clements, W.H. (1997) Toxocity and bioaccumulation of a mixture of heavy metals in *Chironomus tentans* (Diptera: Chironomidae) in synthetic sediment, *Environmental Toxicology and Chemistry* **16** (2), 317-327.

Hartgers, E.M., Aalderink, G.H.R., Van den Brink, P.J., Gylstra, R., Wiegman, J.W.F. and Brock, T.C.M. (1998) Ecotoxicological threshold levels of a mixture of herbicides (atrazine, diuron and metolachlor) in freshwater microcosms, *Aquatic Ecology* **32** (2), 135-152.

Hatch, A.C. and Burton Jr, G.A. (1999) Sediment toxicity and stormwater runoff in a contaminated receiving system: consideration of different bioassays in the laboratory and field, *Chemosphere* **39** (6), 1001-1017.

Hauser, B., Schrader, G. and Bahadir, M. (1997) Comparison of acute toxicity and genotoxic concentrations of single compounds and waste elutriates using the Microtox/Mutatox test system, *Ecotoxicology and Environmental Safety* **38** (3), 227-231.

Havas, M. and Likens, G.E. (1985) Toxicity of aluminum and hydrogen ions to *Daphnia catawba*, *Holopedium gibberum*, *Chaoborus punctipennis*, and *Chironomus anthrocinus* from Mirror Lake, New Hampshire, *Canadian Journal of Zoology* **63** (5), 1114-1119.

Hayes, K.R., Douglas, W.S. and Fischer, J. (1996) Inter- and intra-laboratory testing of the *Daphnia magna* IQ toxicity test, *Bulletin of Environmental Contamination and Toxicology* **57** (4), 660-666.

Heida, H. and van der Oost, R. (1996) Sediment pore water toxicity testing, *Water Science and Technology* **34** (7-8), 109-116.

Heijerick, D.G., Janssen, C.R., Karlèn, C., Wallinder, I.O. and Leygraf, C. (2002) Bioavailability of zinc in runoff water from roofing materials, *Chemosphere* **47** (10), 1073-1080.

Heinis, F., Brils, J.M., Klapwijk, S.P. and De Poorter, L.R.M. (2000) From microbiotest to decision support system: an assessment framework for surface water toxicity, in G. Persoone, C. Janssen and W.M. De Coen (eds.), *New Microbiotests for Routine Toxicity Screening and Biomonitoring*, Kluwer Academic / Plenum Publishers, New York, pp. 65-72.

Helma, C., Eckl, P., Gottmann, E., Kassie, F., Rodinger, W., Steinkellner, H., Windpassinger, C., Schulte-Hermann, R. and Knasmüller, S. (1998) Genotoxic and ecotoxic effects of groundwaters and their relation to routinely measured chemical parameters, *Environmental Science and Technology* **32** (12), 1799-1805.

Herbrandson, C., Bradbury, S.P. and Swackhamer, D.L. (2003a) Influence of suspended solids on acute toxicity of carbofuran to *Daphnia magna*: I. Interactive effects, *Aquatic Toxicology* **63** (4), 333-342.

Herbrandson, C., Bradbury, S.P. and Swackhamer, D.L. (2003b) Influence of suspended solids on acute toxicity of carbofuran to *Daphnia magna*: II. An evaluation of potential interactive mechanisms, *Aquatic Toxicology* **63**(4), 343-355.

Heugens, E.H., Jager, T., Creyghton, R., Kraak, M.H., Hendriks, A.J., Van Straalen, N.M. and Admiraal, W. (2003) Temperature-dependent effects of cadmium on *Daphnia magna*: accumulation versus sensitivity, *Environmental Science and Technology* **37** (10), 2145-2151.

Hickey, C.W., Blaise, C. and Costan, G. (1991) Microtesting appraisal of ATP and cell recovery toxicity end points after acute exposure of *Selenastrum capricornutum* to selected chemicals, *Environmental Toxicology and Water Quality* **6**, 383-403.

Hill, L. and Jooste, S. (1999) The effects of contaminated sediments of the Blesbok Spruit near Witbank on water quality and the toxicity thereof to *Daphnia pulex*, *Water Science and Technology* **39** (10-11), 173-176.

Hoffmann, C. and Christofi, N. (2001) Testing the toxicity of influents to activated sludge plants with the *Vibrio fischeri* bioassay utilising a sludge matrix, *Environmental Toxicology* **16** (5), 422-427.

Hoke, R.A., Giesy, J.P., Zabik, M. and Ungers, M. (1993) Toxicity of sediments and sediment pore waters from the Grand Calumet River - Indiana Harbor, Indiana area of concern, *Ecotoxicology and Environmental Safety* **26** (1), 86-112.

Holdway, D.A., Lok, K. and Semaan, M. (2001) The acute and chronic toxicity of cadmium and zinc to two *Hydra* species, *Environmental Toxicology* **16** (6), 557-565.

Hong, L.C.D., Becker-van Slooten, K., Sauvain, J.-J., Minh, T.L. and Tarradellas, J. (2000) Toxicity of sediments from the Ho Chi Minh City canals and Saigon River, Viet Nam, *Environmental Toxicology* **15** (5), 469-475.

Huggett, D.B., Brooks, B.W., Peterson, B., Foran, C.M. and Schlenk, D. (2002) Toxicity of select beta adrenergic receptor-blocking pharmaceuticals (B-blockers) on aquatic organisms, *Archives of Environmental Contamination and Toxicology* **43** (2), 229-235.

Huuskonen, S.E., Ristola, T.E., Tuvikene, A., Hahn, M.E., Kukkonen, J.V.K. and Lindström-Seppä, P. (1998) Comparison of two bioassays, a fish liver cell line (PLHC-1) and a midge (*Chironomus riparius*), in monitoring freshwater sediments, *Aquatic Toxicology* **44** (1-2), 47-67.

Hyötyläinen, T. and Oikari, A. (1999) The toxicity and concentrations of PAHs in creosote-contaminated lake sediment, *Chemosphere* **38** (5), 1135-1144.

IGETG (Inter-Governmental Ecotoxicological Testing Group) (2004) The evolution of toxicological testing in Canada, Environment Canada, Environmental Technology Centre Report, January 2004, Ottawa, Ontario, K1A 0H3, 19 pp.

Ingersoll, C.G., Ankley, G.T., Benoit, D.A., Brunson, E.L., Burton, G.A., Dwyer, F.J., Hoke, R.A., Landrum, P.F., Norberg-King, T.J. and Winger, P.V. (1995) Toxicity and bioaccumulation of sediment-associated contaminants using freshwater invertebrates: a review of methods and applications, *Environmental Toxicology and Chemistry* **14** (11), 1885-1894.

Ingersoll, C., Besser, J. and Dwyer, J. (1997) *Development and application of methods for assessing the bioavailability of contaminants associated with sediments: I. Toxicity and the sediment quality triad*, Proceedings of the U.S. Geological Survey (USGS) Sediment Workshop, U.S. Geological Survey, Columbia, Missouri (2003-12-22); http://water.usgs.gov/osw/techniques/workshop/ingersoll.html.

Ingersoll, C.G., MacDonald, D.D., Wang, N., Crane, J.L., Field, L.J., Haverland, P.S., Kemble, N.E., Lindskoog, R.A., Severn, C. and Smorong, D.E. (2000) *Prediction of sediment toxicity using consensus-based freshwater sediment quality guidelines*, EPA 905/R-00/007, U.S. Enviromnental Protection Agency, Great Lakes National Program Office, Chicago, IL (2004-02-25); http://www.cerc.usgs.gov/pubs/center/pdfdocs/91126.pdf.

Ingersoll, C.G., MacDonald, D.D., Brumbaugh, W.G., Johnson, B.T., Kemble, N.E., Kunz, J.L., May, T.W., Wang, N., Smith, J.R., Sparks, D.W. and Ireland, D.S. (2002) Toxicity assessment of sediments from the Grand Calumet River and Indiana Harbor Canal in Northwest Indiana, USA., *Archives of Environmental Contamination and Toxicology* **43** (2), 156-167.

Isidori, M. (2000) Toxicity monitoring of waste waters from tanneries with microbiotests, in G. Persoone, C. Janssen and W.M. De Coen (eds.), *New Microbiotests for Routine Toxicity Screening and Biomonitoring*, Kluwer Academic / Plenum Publishers, New York, pp. 339-345.

Isidori, M., Parrella, A., Piazza, C.M.L. and Strada, R. (2000) Toxicity screening of surface waters in southern Italy with Toxkit microbiotests, in G. Persoone, C. Janssen and W.M. De Coen (eds.), *New Microbiotests for Routine Toxicity Screening and Biomonitoring*, Kluwer Academic/Plenum Publishers, New York, pp. 289-293.

Isnard, P., Flammarion, P., Roman, G., Babut, M., Bastien, P., Bintein, S., Esserméant, L., Férard, J.F., Gallotti-Schmitt, S., Saouter, E., Saroli, M., Thiébaud, H., Tomassone, R. and Vindimian, E. (2001) Statitical analysis of regulatory ecotoxicity tests, *Chemosphere* **45**, 659-669.

ISO (2003) Water quality - Freshwater algal growth inhibition test with unicellular green algae, (ISO/FDIS 8692:2004), International Standard (under development). Water quality - Freshwater algal growth inhibition test with unicellular green algae.

Isomaa, B., Lilius, H., Sandbacka, M. and Holmström, T. (1995) The use of freshly isolated rainbow trout hepatocytes and gill epithelial cells in toxicity testing, *Toxicology Letters* **78** (1), 42-42.

Jackson, M., Milne, J., Johnston, H. and Dermott, R. (1995) Assays of Hamilton Harbour sediments using *Diporeia hoyi* (Amphipoda) and *Chironomus plumosus* (Diptera), *Canadian Technical Report of Fisheries and Aquatic Sciences* **2039**, 1-21.

Jacobs, M.W., Delfino, J.J. and Bitton, G. (1992) The toxicity of sulfur to Microtox® from acetonitrile extracts of contaminated sediments, *Environmental Toxicology and Chemistry* **11** (8), 1137-1143.

Jak, R.G., Maas, J.L. and Scholten, M.C.T. (1998) Ecotoxicity of 3,4-dichloroaniline in enclosed freshwater plankton communities at different nutrient levels, *Ecotoxicology* **7** (1), 49-60.

Janati-Idrissi, M., Guerbet, M. and Jouany, J.M. (2001) Effect of cadmium on reproduction of daphnids in a small aquatic microcosm, *Environmental Toxicology* **16**(4), 361-364.

Janssen, C.R., Vangheluwe, M. and Van Sprang, P. (2000) A brief review and critical evaluation of the status of microbiotests, in G. Persoone, C. Janssen and W. M. De Coen (eds.), *New Microbiotests for Routine Toxicity Screening and Biomonitoring*, Kluwer Academic / Plenum Publishers, New York, pp. 27-37.

Jaworska, J.S., Schowanek, D. and Feijtel, T.C. (1999) Environmental risk assessment for trisodium [S,S]-ethylene diamine disuccinate, a biodegradable chelator used in detergent applications, *Chemosphere* **38** (15), 3597-3625.

Jenner, H.A. and Janssen-Mommen, J.P.M. (1989) Phytomonitoring of pulverized fuel ash leachates by the duckweed *Lemna minor*, in M. Munawar, G. Dixon, C.I. Mayfield, T. Reynoldson and M.H. Sadar (eds.), *Environmental Bioassay Techniques and their Application: Proceedings of the 1st International Conference held in Lancaster, England, 11-14 July 1988*, Kluwer Academic Publishers, Dordrecht, Netherlands, pp. 361-366.

Johnson, B.T., Petty, J.D., Huckins, J.N., Lee, K. and Gauthier, J. (2004) Hazard assessment of a simulated oil spill on intertidal areas of the St-Lawrence River with SPMP-TOX, *Environmental Toxicology* **19**, 329-335.

Johnson, I. (2000) Criteria-based procedure for selecting test methods for effluent testing and its application to Toxkit microbiotests, in G. Persoone, C. Janssen and W.M. De Coen (eds.), *New Microbiotests for Routine Toxicity Screening and Biomonitoring*, Kluwer Academic / Plenum Publishers, New York, pp. 73-94.

Johnson, I. and Delaney, P. (1998) Development of a 7-day *Daphnia magna* growth test using image analysis, *Bulletin of Environmental Contamination and Toxicology* **61** (3), 355-362.

Johnson, I., Butler, R., Milne, R. and Redshaw, C.J. (1993) The role of Microtox® in the monitoring and control of effluents, in M. Richardson (ed.), *Ecotoxicology Monitoring*, VCH Publishers, Weinheim, Germany, pp. 309-317.

Jooste, S. and Thirion, C. (1999) An ecological risk assessment for a South African acid mine drainage, *Water Science and Technology* **39** (10-11), 297-303.

Jop, K.M., Foster, R.B. and Askew, A.M. (1991) Factors affecting toxicity identification evaluation: the role of source water use in industrial processes, in M.A. Mayes and M.G. Barron (eds.), *Aquatic Toxicology and Risk Assessment: Fourteenth Volume, ASTM STP 1124*, American Society for Testing and Materials, Philadelphia, PA, pp. 84-93.

Jop, K.M., Askew, A.M., Terrio, K.F. and Simoes, A.T. (1992) Application of the short-term chronic test with *Ceriodaphnia dubia* in identifying sources of toxicity in industrial wastewaters, *Bulletin of Environmental Contamination and Toxicology* **49** (5), 765-771.

Jos, A., Repetto, G., Rios, J.C., Hazen, M.J., Molero, M.L., del Peso, A., Salguero, M., Fernández-Freire, P., Pérez-Martín, M. and Cameán, A. (2003) Ecotoxicological evaluation of carbamazepine using six different model systems with eighteen endpoints, *Toxicology in Vitro* **17** (5-6), 525-532.

Joutti, A., Schultz, E., Tuukkanen, E. and Vaajasaari, K. (2000) Industrial waste leachates toxicity detection with microbiotests and biochemical tests, in G. Persoone, C. Janssen and W.M. De Coen (eds.), *New Microbiotests for Routine Toxicity Screening and Biomonitoring*, Kluwer Academic/Plenum Publishers, New York, pp. 347-355.

Jung, K. and Bitton, G. (1997) Use of Ceriofast™ for monitoring the toxicity of industrial effluents: comparison with the 48-h acute *Ceriodaphnia* toxicity test and Microtox®, *Environmental Toxicology and Chemistry* **16** (11), 2264-2267.

Junghans, M., Backhaus T., Faust M., Scholze M., Grimme L.H. (2003) Predictability of combined effects of eight chloroacetanilide herbicides on algal reproduction, *Pest Management Science* **59**, 1101-1110.

Juvonen, R., Martikainen, E., Schultz, E., Joutti, A., Ahtiainen, J. and Lehtokari, M. (2000) A battery of toxicity tests as indicators of decontamination in composting oily waste, *Ecotoxicology and Environmental Safety* **47** (2), 156-166.

Kahru, A., Kurvet, M. and Külm, I. (1996) Toxicity of phenolic wastewater to luminescent bacteria *Photobacterium phosphoreum* and activated sludges, *Water Science and Technology* **33**(6), 139-146.

Kahru, A., Põllumaa, L., Reiman, R. and Rätsep, A. (1999) Predicting the toxicity of oil-shale industry wastewater by its phenolic composition, *ATLA (Alternatives To Laboratory Animals)* **27** (3), 359-366.

Kahru, A., Põllumaa, L., Reiman, R. and Rätsep, A. (2000) Microbiotests for the evaluation of the pollution from the oil shale industry, in G. Persoone, C. Janssen and W.M. De Coen (eds.), *New Microbiotests for Routine Toxicity Screening and Biomonitoring* Kluwer Academic/Plenum Publishers, New York, pp. 357-365.

Kaiser, K.L.E. and McKinnon, M.B. (1993) Qualitative and quantitative relationships of Microtox data with acute and subchronic toxicity data for other aquatic species, *Canadian Technical Report of Fisheries and Aquatic Sciences* **1942**, 1-24.

Kamaya, Y., Kurogi, Y. and Suzuki, K. (2003) Acute toxicity of fatty acids to the freshwater green alga *Selenastrum capricornutum*, *Environmental Toxicology* **18** (5), 289-294.

Karen, D.J., Ownby, D.R., Forsythe, B.L., Bills, T.P., La Point, T.W., Cobb, G.B. and Klaine, S.J. (1999) Influence of water quality on silver toxicity to rainbow trout (*Onchorhynchus mykiss*), fathead minnow (*Pimephales promelas*), and water fleas (*Daphnia magna*), *Environmental Toxicology and Chemistry* **18** (1), 63-70.

Keller, A.E., Ruessler, D.S. and Chaffee, C.M. (1998) Testing the toxicity of sediments contaminated with diesel fuel using glochidia and juvenile mussels (Bivalvia, Unionidae), *Aquatic Ecosystem Health and Management* **1** (1), 37-47.

Kemble, N.E., Brumbaugh, W.G., Brunson, E.L., Dwyer, F.J., Ingersoll, C.G., Monda, D.P. and Woodward, D.F. (1994) Toxicity of metal-contaminated sediments from the Upper Clark Fork River, Montana, to aquatic invertebrates and fish in laboratory exposures, *Environmental Toxicology and Chemistry* **13**, 1985-1997.

Kemble, N.E., Brunson, E.L., Canfield, T.J., Dwyer, F.J. and Ingersoll, C.G. (1998) Assessing sediment toxicity from navigational pools of the Upper Mississippi River using a 28-D *Hyalella azteca* test., *Archives of Environmental Contamination and Toxicology* **35** (2), 181-190.

Kemble, N.E., Dwyer, F.J., Ingersoll, C.G., Dawson, T.D. and Norberg-King, T.J. (1999) Tolerance of freshwater test organisms to formulated sediments for use as control materials in whole-sediment toxicity tests, *Environmental Toxicology and Chemistry* **18** (2), 222-230.

Kemble, N.E., Ingersoll, C.G. and Kunz, J.L. (2002) Toxicity assessment of sediment samples collected from North Carolina streams, U.S. Geological Survey, Columbia, Missouri, Columbia Environmental Research Center, Columbia, MO, Final Report CERC-8335-FY03-20-01, 69 pages.

Koivisto, S. and Ketola, M. (1995) Effects of copper on life-history traits of *Daphnia pulex* and *Bosmina longirostris*, *Aquatic Toxicology* **32** (2-3), 255-269.

Kondo, S., Fujiwara, M., Ohba, M. and Ishii, T. (1995) Comparative larvicidal activities of the four *Bacillus thuringiensis* serovars against a chironomid midge, *Paratanytarsus grimmii* (Diptera: Chironomidae), *Microbiological Research* **150** (4), 425-428.

Kosmala, A., Charvet, S., Roger, M.-C. and Faessel, B. (1999) Impact assessment of a wastewater treatment plant effluent using instream invertebrates and the *Ceriodaphnia dubia* chronic toxicity test, *Water Research* **33** (1), 266-278.

Koukal, B., Guéguen, C., Pardos, M. and Dominik, J. (2003) Influence of humic substances on the toxic effects of cadmium and zinc to the green alga *Pseudokirchneriella subcapitata*, *Chemosphere* **53** (8), 953-961.

Kovacs, T., Gibbons, J.S., Naish, V. and Voss, R. (2002) Complying with effluent toxicity regulation in Canada, *Water Quality Research Journal of Canada* **37** (4), 671-679.

Kross, B.C. and Cherryholmes, K. (1993) Toxicity screening of sanitary landfill leachates: a comparative evaluation with Microtox® analyses, chemical, and other toxicity screening methods, in M. Richardson (ed.), *Ecotoxicology Monitoring*, VCH Publishers, Weinheim, Germany, pp. 225-249.

Kszos, L.A., Morris, G.W. and Konetsky, B.K. (2004) Source of toxicity in storm water: zinc from commonly used paint, *Environmental Toxicology and Chemistry* **23** (1), 12-16.

Kubitz, J.A., Besser, J.M. and Giesy, J.P. (1996) A two-step experimental design for a sediment bioassay using growth of the amphipod *Hyallela azteca* for the test end point, *Environmental Toxicology and Chemistry* **15** (10), 1783-1792.

Kuhne, W.W., Caldwell, C.A., Gould, W.R., Fresquez, P.R. and Finger, S.E. (2002) Effects of depleted uranium on the health and survival of *Ceriodaphnia dubia* and *Hyallela azteca*, *Environmental Toxicology and Chemistry* **21** (10), 2198-2203.

Kungolos, A., Samaras, P., Kimeroglu, V., Dabou, X. and Sakellaropoulos, G.P. (1998) Water quality and toxicity assessment in Koronia Lake, Greece, *Fresenius Environmental Bulletin* **7** (7A-8A, Sp.), 615-622.

Kusui, T. and Blaise, C. (1999) Ecotoxicological assessment of japanese industrial effluents using a battery of small-scale toxicity tests, in S.S. Rao (ed.), *Impact Assessment of Hazardous Aquatic Contaminants: Concepts and Approaches*, Lewis Publishers, Boca Raton, Florida, pp. 161-181.

Kwan, K.K. (1995) Direct sediment toxicity testing procedure using sediment-chromotest kit, *Environmental Toxicology and Water Quality* **10**, 193-196.

Kwan, K.K. and Dutka, B.J. (1992) Evaluation of Toxi-Chromotest direct sediment toxicity testing procedure and Microtox solid-phase testing procedure, *Bulletin of Environmental Contamination and Toxicology* **49** (5), 656-662.

Kwan, K.K. and Dutka, B.J. (1995) Comparative assessment of two solid-phase toxicity bioassays: the direct sediment toxicity testing procedure (DSTTP) and the Microtox® solid-phase test (SPT), *Bulletin of Environmental Contamination and Toxicology* **55** (3), 338-346.

La Point, T.W., Cobb, G.P., Klaine, S.J., Bills, T., Forsythe, B., Jeffers, R., Waldrop, V.C. and Wenholz, M. (1996) Water quality components affecting silver toxicity in *Daphnia magna* and *Pimephales promelas*, in A.W. Andren and T.W. Bober (eds.), *Proceedings of the Fourth International Conference on Transport, Fate and Effects of Silver in the Environment*, International Argentum Conference, Madison, Wisconsin, USA, pp. 121-124.

Lacaze, J.C., Chesterikoff, A. and Garban, B. (1989) Bioévaluation de la pollution des sédiments de la Seine (région parisienne) par l'emploi d'un bioessai basé sur la croissance à court terme de la microalgue *Selenastrum capricornutum* Printz, *Revue des Sciences de l'Eau* **2**, 405-427.

Lacey, R., Watzin, M.C. and McIntosh, A.W. (1999) Sediment organic matter content as a confounding factor in toxicity tests with *Chironomus tentans*, *Environmental Toxicology and Chemistry* **18** (2), 231-236.

Lahr, J. (1998) An ecological assessment of the hazard of eight insecticides used in Desert Locust control, to invertebrates in temporary ponds in the Sahel, *Aquatic Ecology* **32** (2), 153-162.

Lahr, J., Maas-Diepeveen, J.L., Stuijfzand, S.C., Leonards, P.E.G., Drüke, J.M., Lücker, S., Espeldoorn, A., Kerkum, L.C.M., van Stee, L.L.P. and Hendriks, A.J., (2003) Responses in sediment bioassays used in the Netherlands: can observed toxicity be explained by routinely monitored priority pollutants?, *Water Research* **37** (8), 1691-1710.

Lambolez, L., Vasseur, P., Férard, J.-F. and Gisbert, T. (1994) The environmental risks of industrial waste disposal: an experimental approach including acute and chronic toxicity studies, *Ecotoxicology and Environmental Safety* **28** (3), 317-328.

Lamy-Enrici, M.-H., Dondeyne, A. and Thybaud, E. (2003) Influence of the organic matter on the bioavailability of phenanthrene for benthic organisms, *Aquatic Ecosystem Health and Management* **6** (4), 391-396.

Landrum, P.F., Leppänen, M.T., Robinson, S.D., Gossiaux, D.C., Burton, G.A., Greenberg, M., Kukkonen, J.V.K., Eadie, B.J. and Lansing, M.B. (2004) Comparing behavioral and chronic endpoints to evaluate the response of *Lumbriculus variegatus* to 3,4,3',4'-tetrachlorobiphenyl sediment exposures, *Environmental Toxicology and Chemistry* **23** (1), 187-194.

Länge, R. and Dietrich, D. (2002) Environmental risk assessment of pharmaceutical drug substances - conceptual considerations, *Toxicology Letters* **131** (1-2), 97-104.

Lappalainen, J., Juvonen, R., Vaajasaari, K. and Karp, M. (1999) A new flash method for measuring the toxicity of solid and colored samples, *Chemosphere* **38** (5), 1069-1083.

Larsen, J., Schultz, T.W., Rasmussen, L., Hooftman, R. and Pauli, W. (1997) Progress in an ecotoxicological standard protocol with protozoa: results from a pilot ring test with *Tetrahymena pyriformis*, *Chemosphere* **35** (5), 1023-1041.

Lasier, P.J., Winger, P.V. and Reinert, R.E. (1997) Toxicity of alkalinity to *Hyalella azteca*, *Bulletin of Environmental Contamination and Toxicology* **59** (5), 807-814.

Lasier, P.J., Winger, P.V. and Bogenrieder, K.J. (2000) Toxicity of manganese to *Ceriodaphnia dubia* and *Hyalella azteca*, *Archives of Environmental Contamination and Toxicology* **38** (3), 298-304.

Latif, M. and Zach, A. (2000) Toxicity studies of treated residual wastes in Austria using different types of conventional assays and cost-effective microbiotests, in G. Persoone, C. Janssen and W.M. De Coen (eds.), *New Microbiotests for Routine Toxicity Screening and Biomonitoring*, Kluwer Academic/Plenum Publishers, New York, pp. 367-383.

Latif, M. and Licek, E. (2004) Toxicity assessment of wastewaters, river waters, and sediments in Austria using cost-effective microbiotests, *Environmental Toxicology* **19**, 302-309.

Lauten, K.P. (1993) Sediment toxicity assessment - North Saskatchewan River, *Canadian Technical Report of Fisheries and Aquatic Sciences* **1942**, 360-367.

Leal, H.E., Rocha, H.A. and Lema, J.M. (1997) Acute toxicity of hardboard mill effluents to different bioindicators, *Environmental Toxicology and Water Quality* **12** (1), 39-42.

LeBlond, J.B. and Duffy, L.K. (2001) Toxicity assessment of total dissolved solids in effluent of Alaskan mines using 22-h chronic Microtox® and *Selenastrum capricornatum* assays, *The Science of The Total Environment* **271** (1-3), 49-59.

Lechelt, M., Blohm, W., Kirschneit, B., Pfeiffer, M., Gresens, E., Liley, J., Holz, R., Lüring, C. and Moldaenke, C. (2000) Monitoring of surface water by ultrasensitive *Daphnia* toximeter, *Environmental Toxicology* **15** (5), 390-400.

Lee, J.-S., Lee, B.-G., Luoma, S.N. and Yoo, H. (2004) Importance of equilibration time in the partitioning and toxicity of zinc in spiked sediment bioassays, *Environmental Toxicology and Chemistry* **23** (1), 65-71.

Lewis, M.A. (1995) Use of freshwater plants for phytotoxicity testing: a review, *Environmental Pollution* **87** (3), 319-336.

Leynen, M., Duvivier, L., Girboux, P. and Ollevier, F. (1998) Toxicity of ozone to fish larvae and *Daphnia magna*, *Ecotoxicology and Environmental Safety* **41** (2), 176-179.

Liao, C.M. and Lin, M.C. (2001) Acute toxicity modeling of rainbow trout and silver sea bream exposed to waterborne metals, *Environmental Toxicology* **16** (4), 349-360.

Liao, C.-M., Chen, B.-C., Singh, S., Li, M.-C., Liu, C.-W. and Han, B.-C. (2003) Acute toxicity and bioaccumulation of arsenic in tilapia (*Oreochromis mossambicus*) from a blackfoot disease area in Taiwan, *Environmental Toxicology* **18** (4), 252-259.

Liu, D.H.W., Bailey, H.C. and Pearson, J.G. (1983) Toxicity of a complex munitions wastewater to aquatic organisms, in W.E. Bishop, R.D. Cardwell and B.B. Heidolph (eds.), *Aquatic Toxicology and Hazard Assessment: Sixth Symposium, ASTM STP 802*, American Society for Testing and Materials, Philadelphia, PA, pp. 135-150.

Liu, M.C., Chen, C.M., Cheng, H.Y., Chen, H.Y., Su, Y.C. and Hung, T.Y. (2002) Toxicity of different industrial effluents in Taiwan: a comparison of the sensitivity of *Daphnia similis* and Microtox®, *Environmental Toxicology* **17** (2), 93-97.

Lockhart, W.L., Billeck, B.N. and Baron, C.L. (1989) Bioassays with a floating aquatic plant (*Lemna minor*) for effects of sprayed and dissolved glyphosate, in M. Munawar, G. Dixon, C.I. Mayfield, T. Reynoldson and M.H. Sadar (eds.), *Environmental Bioassay Techniques and their Application: Proceedings of the 1st International Conference held in Lancaster, England, 11-14 July 1988*, Kluwer Academic Publishers, Dordrecht, Netherlands, pp. 353-359.

Long, E.R., MacDonald, D.D., Cubbage, J.C. and Ingersoll, C.G. (1998) Predicting the toxicity of sediment associated trace metals with simultaneously extracted trace metal: acid-volatile sulfide concentrations and dry weight-normalized concentrations: a critical comparison, *Environmental Toxicology and Chemistry* **17** (5), 972-974.

Long, K.E., Van Genderen, E.J. and Klaine, S.J. (2004) The effects of low hardness and pH on copper toxicity to *Daphnia magna*, *Environmental Toxicology and Chemistry* **23** (1), 72-75.

Lopes, I., Gonçalves, F., Soares, A.M.V.M. and Ribeiro, R. (1999) Discriminating the ecotoxicity due to metals and to low pH in acid mine drainage, *Ecotoxicology and Environmental Safety* **44** (2), 207-214.

Lotufo, G.R. (1998) Lethal and sublethal toxicity of sediment-associated fluoranthene to benthic copepods: application of the critical-body-residue approach, *Aquatic Toxicology* **44** (1-2), 17-30.

Lucivjanská, V., Lucivjanská, M. and Cízek, V. (2000) Sensitivity comparison of the ISO *Daphnia* and algal test procedures with Toxkit microbiotests, in G. Persoone, C. Janssen and W.M. De Coen (eds.), *New Microbiotests for Routine Toxicity Screening and Biomonitoring*, Kluwer Academic/Plenum Publishers, New York, pp. 243-246.

Ma, M., Zhu, W., Wang, Z. and Witkamp, G.J. (2003) Accumulation, assimilation and growth inhibition of copper on freshwater alga (*Scenedesmus subspicatus* 86.81 SAG) in the presence of EDTA and fulvic acid, *Aquatic Toxicology* **63** (3), 221-228.

MacDonald, D.D. and Ingersoll, C.G. (2002a) A guidance manual to support the assessment of contaminated sediments in freshwater ecosystems. Volume II - Design and implementation of sediment quality investigations, *EPA-905-B02-001-B*, U.S. Environmental Protection Agency, Great Lakes National Program Office, Chicago, IL, 136 pp.

MacDonald, D.D. and Ingersoll, C.G. (2002b) A guidance manual to support the assessment of contaminated sediments in freshwater ecosystems. Volume III - Interpretation of the results of sediment quality investigations, *EPA-905-B02-001-C*, U.S. Environmental Protection Agency, Great Lakes National Program Office, Chicago, Il, 232 pp.

MacGregor, D.J. and Wells, P.G. (1984) The role of ecotoxicological testing of effluents and chemicals in the Environmental Protection Service, A working paper for E.P.S., Environment Canada, Ottawa, Ontario, November 1984, 56 pp.

Mackay, D.W., Holmes, P.J. and Redshaw, C.J. (1989) The application of bioassay techniques to water pollution problems - The United Kingdom experience, in M. Munawar, G. Dixon, C.I. Mayfield, T. Reynoldson and M.H. Sadar (eds.), *Environmental Bioassay Techniques and their Application: Proceedings of the 1st International Conference held in Lancaster, England, 11-14 July 1988*, Kluwer Academic Publishers, Dordrecht, Netherlands, pp. 77-86.

Madoni, P. (2000) The acute toxicity of nickel to freshwater ciliates, *Environmental Pollution* **109** (1), 53-59.

Mäenpää, K.A., Sormunen, A.J. and Kukkonen, J.V.K. (2003) Bioaccumulation and toxicity of sediment associated herbicides (ioxynil, pendimethalin, and bentazone) in *Lumbriculus variegatus* (Oligochaeta) and *Chironomus riparius* (Insecta), *Ecotoxicology and Environmental Safety* **56** (3), 398-410.

Maier, K.J. and Knight, A.W. (1993) Comparative acute toxicity and bioconcentration of selenium by the midge *Chironomus decorus* exposed to selenate, selenite, and seleno-DL-methionine, *Archives of Environmental Contamination and Toxicology* **25** (3), 365-370.

Malá, J., Maršálková, E. and Rovnaníková, P. (2000) Toxicity testing of solidified waste leachates with microbiotests, in G. Persoone, C. Janssen and W.M. De Coen (eds.), *New Microbiotests for Routine Toxicity Screening and Biomonitoring*, Kluwer Academic/Plenum Publishers, New York, pp. 385-390.

Maltby, L. and Calow, P. (1989) The application of bioassays in the resolution of environmental problems; past, present and future, in M. Munawar, G. Dixon, C.I. Mayfield, T. Reynoldson and M.H. Sadar (eds.), *Environmental Bioassay Techniques and their Application: Proceedings of the 1st International Conference held in Lancaster, England, 11-14 July 1988*, Kluwer Academic Publishers, Dordrecht, Netherlands, pp. 65-76.

Manasherob, R., Ben-Dov, E., Zaritsky, A. and Barak, Z. (1994) Protozoan-enhanced toxicity of *Bacillus thuringiensis* var. israelensis - Endotoxin against *Aedes aegypti* larvae, *Journal of Invertebrate Pathology* **63** (3), 244-248.

Mandal, R., Hassan, N.M., Murimboh, J., Chakrabarti, C.L., Back, M.H., Rahayu, U. and Lean, D.R.S. (2002) Chemical speciation and toxicity of nickel species in natural waters from the Sudbury area (Canada), *Environmental Science and Technology* **36** (7), 1477-1484.

Manusadžianas, L., Balkelyte, L., Sadauskas, K. and Stoškus, L. (2000) Microbiotests for the toxicity assessment of various types of water samples, in G. Persoone, C. Janssen and W.M. De Coen (eds.), *New Microbiotests for Routine Toxicity Screening and Biomonitoring*, Kluwer Academic/Plenum Publishers, New York, pp. 391-399.

Manusadžianas, L., Balkelyte, L., Sadauskas, K., Blinova, I., Põllumaa, L. and Kahru, A. (2003) Ecotoxicological study of Lithuanian and Estonian wastewaters: selection of the biotests, and correspondence between toxicity and chemical-based indices, *Aquatic Toxicology* **63** (1), 27-41.

Maršálek, B. and Bláha, L. (2000) Microbiotests for cyanobacterial toxins screening, in G. Persoone, C. Janssen and W.M. De Coen (eds.), *New Microbiotests for Routine Toxicity Screening and Biomonitoring*, Kluwer Academic/Plenum Publishers, New York, pp. 519-525.

Maršálek, B. and Rojíčková-Padrtová, R. (2000) Selection of a battery of microbiotests for various purposes - the Czech experience, in G. Persoone, C. Janssen and W.M. De Coen (eds.), *New Microbiotests for Routine Toxicity Screening and Biomonitoring*, Kluwer Academic/Plenum Publishers, New York, pp. 95-101.

Marsalek, J., Rochfort, Q., Brownlee, B., Mayer, T. and Servos, M. (1999) An exploratory study of urban runoff toxicity, *Water Science and Technology* **39** (12), 33-39.

Martinez-Madrid, M., Rodriguez, P. and Perez-Iglesias, J.I. (1999) Sediment toxicity bioassays for assessment of contaminated sites in the Nervion River (Northern Spain). I. Three-brood sediment chronic bioassay of *Daphnia magna* Straus, *Ecotoxicology* **8**, 97-109.

Marvin, C.H., Howell, E.T., Kolic, T.M. and Reiner, E.J. (2002) Polychlorinated dibenzo-p-dioxins and dibenzofurans and dioxinlike polychlorianted biphenyls in sediments and mussels at three sites in the lower Great Lakes, North America, *Environmental Toxicology and Chemistry* **21** (9), 1908–1921.

Maycock, D.S., Prenner, M.M., Kheir, R., Morris, S., Callaghan, A., Whitehouse, P., Morritt, D. and Crane, M. (2003) Incorporation of *in situ* and biomarker assays in higher-tier assessment of the aquatic toxicity of insecticides, *Water Research* **37** (17), 4180-4190.

McCarthy, L.H., Williams, T.G., Stephens, G.R., Peddle, J., Robertson, K. and Gregor, D.J. (1997) Baseline studies in the Slave River, NWT, 1990-1994: Part I. Evaluation of the chemical quality of water and suspended sediment from the Slave River (NWT), *The Science of The Total Environment* **197** (1-3), 21-53.

McDaniel, M. and Snell, T.W. (1999) Probability distributions of toxicant sensitivity for freshwater rotifer species, *Environmental Toxicology* **14** (3), 361-366.

McDonald, S.F., Hamilton, S.J., Buhl, K.J. and Heisinger, J.F. (1996) Acute toxicity of fire control chemicals to *Daphnia magna* (Straus) and *Selenastrum capricornutum* (Printz), *Ecotoxicology and Environmental Safety* **33** (1), 62-72.

McKnight, D.M., Feder, G.L. and Stiles, E.A. (1981) Toxicity of volcanic-ash leachate to a blue-green alga. Results of a preliminary bioassay experiment, *Environmental Science and Technology* **15** (3), 362-364.

Miana, P., Scotto, S., Perin, G. and Argese, E. (1993) Sensitivity of *Selenastrum capricornutum*, *Daphnia magna* and submitochondrial particles to tributyltin, *Environmental Technology (Letters) ETLEDB* **14** (2), 175-181.

Michel, K., Brinkmann, C., Hahn, S., Dott, W. and Eisentraeger, A. (2004) Acute toxicity investigations of ester-based lubricants by using biotests with algae and bacteria, *Environmental Toxicology* **19**, 445-448.

Michniewicz, M., Nalecz-Jawecki, G., Stufka-Olczyk, J. and Sawicki, J. (2000) Comparison of chemical composition and toxicity of wastewaters from pulp industry, in G. Persoone, C. Janssen and W.M. De Coen (eds.), *New Microbiotests for Routine Toxicity Screening and Biomonitoring*, Kluwer Academic/Plenum Publishers, New York, pp. 401-411.

Middaugh, D.P., Beckham, N., Fournie, J.W. and Deardorff, T.L. (1997) Evaluation of bleached kraft mill process water using Microtox®, *Ceriodaphnia dubia*, and *Menidia beryllina* toxicity tests, *Archives of Environmental Contamination and Toxicology* **32** (4), 367-375.

Milam, C.D. and Farris, J.L. (1998) Risk identification associated with iron-dominated mine discharges and their effect upon freshwater bivalves, *Environmental Toxicology and Chemistry* **17** (8), 1611-1619.

Milam, C.D., Farris, J.L. and Wilhide, J.D. (2000) Evaluating mosquito control pesticides for effect on target and nontarget organisms, *Archives of Environmental Contamination and Toxicology* **39** (3), 324-328.

Milani, D., Reynoldson, T.B., Borgmann, U. and Kolasa, J. (2003) The relative sensitivity of four benthic invertebrates to metals in spiked-sediment exposures and application to contaminated field sediment, *Environmental Toxicology and Chemistry* **22** (4), 845-854.

Milner, R.J., Lim, R.P. and Hunter, D.M. (2002) Risks to the aquatic ecosystem from the application of *Metarhizium anisopliae* for locust control in Australia, *Pest Management Science* **58** (7), 718-723.

Mkandawire, M., Lyubun, Y.V., Kosterin, P.V. and Dudel, E.G. (2004) Toxicity of arsenic species to *Lemna gibba* L. and the influence of phosphate on arsenic bioavailability, *Environmental Toxicology* **19** (1), 26-34.

Monarca, S., Feretti, D., Collivignarelli, C., Guzzella, L., Zerbini, I., Bertanza, G. and Pedrazzani, R. (2000) The influence of different disinfectants on mutagenicity and toxicity of urban wastewater, *Water Research* **34** (17), 4261-4269.

Monda, D.P., Galat, D.L., Finger, S.E. and Kaiser, M.S. (1995) Acute toxicity of ammonia (NH3-N) in sewage effluent to *Chironomus riparius*: II. Using a generalized linear model, *Archives of Environmental Contamination and Toxicology* **28** (3), 385-390.

Monkiédjé, A., Njiné, T., Tamatcho, B. and Démanou, J. (2000) Assessment of the acute toxic effects of the fungicide Ridomil plus 72 on aquatic organisms and soil micro-organisms, *Environmental Toxicology* **15** (1), 65-70.

Moran, T. and Chiles, C. (1993) Multi-species toxicity assessment of sediments from the St-Clair River using *Hyalella azteca*, *Daphnia magna* and Microtox (*Photobacterium phosphoreum*) as test organisms, *Canadian Technical Report of Fisheries and Aquatic Sciences* **1942**, 447-456.

Moreno-Garrido, I., Lubián, L.M. and Soares, A.M.V.M. (2000) Influence of cellular density on determination of EC(50) in microalgal growth inhibition tests, *Ecotoxicology and Environmental Safety* **47** (2), 112-116.

Mowat, F.S. and Bundy, J.G. (2002) Experimental and mathematical/computational assessment of the acute toxicity of chemical mixtures from the Microtox® assay, *Advances in Environmental Research* **6** (4), 547-558.

Mueller, D.C., Bonner, J.S., McDonald, S.J., Autenrieth, R.L., Donnelly, K.C., Lee, K., Doe, K. and Anderson, J. (2003) The use of toxicity bioassays to monitor the recovery of oiled wetland sediments, *Environmental Toxicology and Chemistry* **22** (9), 1945-1955.

Mummert, A.K., Neves, R.J., Newcomb, T.J. and Cherry, D.S. (2003) Sensitivity of juvenile freshwater mussels (*Lampsilis fasciola*, *Villora iris*) to total and un-ionized ammonia, *Environmental Toxicology and Chemistry* **22** (11), 2545-2553.

Munawar, M., Dermott, R., McCarthy, L.H., Munawar, S.F. and van Stam, H.A. (1999) A comparative bioassessment of sediment toxicity in lentic and lotic ecosystems of the North American Great Lakes, *Aquatic Ecosystem Health and Management* **2** (4), 367-378.

Munawar, M., Munawar, I.F., Sergeant, D. and Wenghofer, C. (2000) A preliminary bioassessment of Lake Baikal sediment toxicity in the vicinity of a pulp and paper mill, *Aquatic Ecosystem Health and Management* **3** (2), 249-257.

Munawar, M., Munawar, I.F., Burley, M., Carou, S. and Niblock, H. (2003) Multi-trophic bioassessment of stressed "Areas of Concern" of the Lake Erie watershed, in M. Munawar (ed.), *Sediment Quality Assessment and Management: Insight and Progress*, Aquatic Ecosystem Health and Management Society, Canada, pp. 169-192.

Muyssen, B.T. and Janssen, C.R. (2001) Zinc acclimation and its effect on the zinc tolerance of *Raphidocelis subcapitata* and *Chlorella vulgaris* in laboratory experiments, *Chemosphere* **45** (4-5), 507-514.

Muyssen, B.T. and Janssen, C.R. (2002) Tolerance and acclimation to zinc of *Ceriodaphnia dubia*, *Environmental Pollution* **117**(2), 301-306.

Naddy, R.B. and Klaine, S.J. (2001) Effect of pulse frequency and interval on the toxicity of chlorpyrifos to *Daphnia magna*, *Chemosphere* **45** (4-5), 497-506.

Naddy, R.B., Stern, G.R. and Gensemer, R.W. (2003) Effect of culture water hardness on the sensitivity of *Ceriodaphnia dubia* to copper toxicity, *Environmental Toxicology and Chemistry* **22** (6), 1269-1271.

Naimo, T.J., Cope, W.G. and Bartsch, M.R. (2000) Sediment-contact and survival of fingernail clams: implications for conducting short-term laboratory tests, *Environmental Toxicology* **15** (1), 23-27.

Nalecz-Jawecki, G. (2004) Spirotox – *Spirostomum ambiguum* acute toxicity test – 10 years of experience, *Environmental Toxicology* **19**, 359-364.

Nalecz-Jawecki, G. and Sawicki, J. (2002) A comparison of sensitivity of spirotox biotest with standard toxicity tests, *Archives of Environmental Contamination and Toxicology* **42** (4), 389-395.

Naudin, S., Pardos, M. and Quiniou, F. (1995) Toxicité des sédiments du bassin versant du Stang Alar (Brest) déterminée par une batterie de bio-essais, Cemagref, France, *La revue Ingénieries - EAT no spécial Rade de Brest*, 67-74.

Naylor, C. and Howcroft, J. (1997) Sediment bioassays with *Chironomus riparius*: understanding the influence of experimental design on test sensitivity, *Chemosphere* **35** (8), 1831-1845.

Nebeker, A.V., Cairns, M.A., Gakstatter, J.H., Malueg, K.W., Schuytema, G.S. and Krawczyk, D.F. (1984) Biological methods for determining toxicity of contaminated freshwater sediments to invertebrates, *Environmental Toxicology and Chemistry* **3** (4), 617-630.

Nebeker, A.V., Onjukka, S.T. and Cairns, M.A. (1988) Chronic effects of contaminated sediment on *Daphnia magna* and *Chironomus tentans*, *Bulletin of Environmental Contamination and Toxicology* **41**, 574-581.

Newton, T.J., Allran, J.W., O'Donnell, J.A., Bartsch, M.R. and Richardson, W.B. (2003) Effects of ammonia on juvenile unionid mussels (*Lampsilis cardium*) in laboratory sediment toxicity tests, *Environmental Toxicology and Chemistry* **22** (11), 2554-2560.

Nicolau, A., Dias, N., Mota, M. and Lima, N. (2001) Trends in the use of protozoa in the assessment of wastewater treatment, *Research in Microbiology* **152** (7), 621-630.

Nipper, M.G. (1998) The development and application of sediment toxicity tests for regulatory purposes, in P.G. Wells, K. Lee and C. Blaise (eds.), *Microscale Testing in Aquatic Toxicology: Advances, Techniques, and Practice*, CRC Press, Boca Raton, FL, pp. 631-643.

Nyström, B. and Blanck, H. (1998) Effects of the sulfonylurea herbicide metsulfuron methyl on growth and macromolecular synthesis in the green alga *Selenastrum capricornutum*, *Aquatic Toxicology* **43** (1), 25-39.

Oanh, N.T.K. (1996) A comparative study of effluent toxicity for three chlorine-bleached pulp and paper mills in Southeast Asia, *Resources, Conservation and Recycling* **18** (1-4), 87-105.

Oanh, N.T.K. and Bengtsson, B.-E. (1995) Toxicity to Microtox, micro-algae and duckweed of effluents from the Bai Bang paper company (BAPACO), a Vietnamese bleached kraft pulp and paper mill, *Environmental Pollution* **90** (3), 391-399.

OECD (1987) The use of biological tests for water pollution assessment and control, Environment Monograph No. 11, 70 pp.

OECD (2001a) Proposal for a new guideline 218: Sediment-water Chironomid toxicity test using spiked sediment, OECD Guidelines for the Testing of Chemicals, Organisation for Economic Co-operation and Development (OECD), Washington, DC (2003-12-23); http://www.oecd.org/dataoecd/40/3/2739721.pdf.

OECD (2001b) Proposal for a new guideline 219: Sediment-water Chironomid toxicity test using spiked water, OECD Guidelines for the Testing of Chemicals, Organisation for Economic Co-operation and Development (OECD), Washington, DC (2004-02-25); http://www.oecd.org/dataoecd/40/45/2739742.pdf.

OECD (2002a) Proposal for updating guideline 201: Freshwater alga and cyanobacteria, growth inhibition test, OECD Guidelines for the Testing of Chemicals, Organisation for Economic Co-operation and Development (OECD), Washington, DC (2004-02-25); http://www.oecd.org/dataoecd/58/60/1946914.pdf.

OECD (2002b) Revised proposal for a new guideline 221: Lemna sp. growth inhibition test, OECD Guidelines for the Testing of Chemicals, Organisation for Economic Co-operation and Development (OECD), Washington, DC (2004-02-25); http://www.oecd.org/dataoecd/16/51/1948054.pdf.

O'Farrell, I., Lombardo, R.J., de Tezanos Pinto, P. and Loez, C. (2002) The assessment of water quality in the Lower Lujan River (Buenos Aires, Argentina): phytoplankton and algal bioassays, *Environmental Pollution* **120** (2), 207-218.

Okamura, H., Luo, R., Aoyama, I. and Liu, D. (1996) Ecotoxicity assessment of the aquatic environment around Lake Kojima, Japan, *Environmental Toxicology and Water Quality* **11** (3), 213-221.

Okamura, H., Piao, M., Aoyama, I., Sudo, M., Okubo, T. and Nakamura, M. (2002) Algal growth inhibition by river water pollutants in the agricultural area around Lake Biwa, Japan, *Environmental Pollution* **117** (3), 411-419.

Oladimeji, A.A. and Offem, B.O. (1989) Toxicity of lead to *Clarias lazera*, *Oreochromis niloticus*, *Chironomus tentans* and *Benacus* sp., *Water, Air, and Soil Pollution* **44** (3-4), 191-201.

Onorati, F., Pellegrini, D. and Ausili, A. (1998) Sediment toxicity assessment with *Photobacterium phosphoreum*: a preliminary evaluation of natural matrix effect, *Fresenius Environmental Bulletin* **7** (Special), 596-604.

Oris, J.T., Winner, R.W. and Moore, M.V. (1991) A four day survival and reproduction toxicity test for *Ceriodaphnia dubia*, *Environmental Toxicology and Chemistry* **10** (2), 217-224.

Ortego, L.S. and Benson, W.H. (1992) Effects of dissolved humic material on the toxicity of selected pyrethroid insecticides, *Environmental Toxicology and Chemistry* **11** (2), 261-265.

OSPAR (2000) Briefing document on the work of DYNAMEC and the DYNAMEC mechanism for the selection and prioritisation of hazardous substances. OSPAR Commission PRAM 2000. Summary Record (PRAM 00/12/1, Annex 5).

Pablos, V., Fernández, C., Valdovinos, C., Castaño, A., Muñoz, M.J. and Tarazona, J.V. (1996) Use of ecotoxicity tests as biological detectors of toxic chemicals in the environmental analysis of complex sewages, *Toxicology Letters* **88** (Supplement 1), 82-82.

Painter, H.A. (1993) A review of tests for inhibition of bacteria (especially those agreed internationally), in M. Richardson (ed.), *Ecotoxicology Monitoring*, VCH Publishers, Weinheim, Germany, pp. 17-36.

Paixão, S.M. and Anselmo, A.M. (2002) Effect of olive mill wastewaters on the oxygen consumption by activated sludge microorganisms: an acute toxicity test method, *Journal of Applied Toxicology* **22** (3), 173-176.

Paixão, S.M., Mendonça, E., Picado, A. and Anselmo, A.M. (1999) Acute toxicity evaluation of olive oil mill wastewaters: A comparative study of three aquatic organisms, *Environmental Toxicology* **14** (2), 263-269.

Pape-Lindstrom, P.A. and Lydy, M.J. (1997) Synergistic toxicity of atrazine and organophosphate insecticides contravenes the response addition mixture model, *Environmental Toxicology and Chemistry* **16** (11), 2415-2420.

Pardos, M., Benninghoff, C., Guéguen, C., Thomas, R., Dobrowolski, J. and Dominik, J. (1999a) Acute toxicity assessment of Polish (waste)water with a microplate-based *Hydra attenuata* assay: a comparison with the Microtox® test, *The Science of The Total Environment* **243-244**, 141-148.

Pardos, M., Benninghoff, C., Thomas, R.L. and Khim-Heang, S. (1999b) Confirmation of elemental sulfur toxicity in the Microtox® assay during organic extracts assessment of freshwater sediments, *Environmental Toxicology and Chemistry* **18** (2), 188-193.

Parrott, J.L. and Sprague, J.B. (1993) Patterns in toxicity of sublethal mixtures of metals and organic chemicals determined by Microtox® and by DNA, RNA, and protein content of fathead minnows (*Pimephales promelas*), *Canadian Journal of Fisheries and Aquatic Sciences* **50** (10), 2245-2253.

Parrott, J.L., Wood, C.S., Boutot, P. and Dunn, S. (2003) Changes in growth and secondary sex characteristics of fathead minnows exposed to bleached sulfite mill effluent, *Environmental Toxicology and Chemistry* **22** (12), 2908-2915.

Pascoe, D., Wenzel, A., Janssen, C.R., Girling, A.E., Jüttner, I., Fliedner, A., Blockwell, S.J., Maund, S.J., Taylor, E.J., Diedrich, M., Persoone, G., Verhelst, P., Stephenson, R.R., Crossland, N.O., Mitchell, G.C., Pearson, N., Tattersfield, L., Lay, J.P., Peither, A., Neumeier, B. and Velletti, A.R. (2000) The development of toxicity tests for freshwater pollutants and their validation in stream and pond mesocosms, *Water Research* **34** (8), 2323-2329.

Pasteris, A., Vecchi, M., Reynoldson, T.B. and Bonomi, G. (2003) Toxicity of copper-spiked sediments to *Tubifex tubifex* (Oligochaeta, Tubificidae): a comparison of the 28-day reproductive bioassay with a 6-month cohort experiment, *Aquatic Toxicology* **65** (3), 253-265.

Pastorok, R.A., Peek, D.C., Sampson, J.R. and Jacobson, M.A. (1994) Ecological risk assessment for river sediments contaminated by creosote, *Environmental Toxicology and Chemistry* **13** (12), 1929-1941.

Pauli, W. and Berger, S. (1997) Toxicological comparisons of *Tetrahymena* species, end points and growth media: supplementary investigations to the pilot ring test, *Chemosphere* **35** (5), 1043-1052.

Peck, M.R., Klessa, D.A. and Baird, D.J. (2002) A tropical sediment toxicity test using the dipteran *Chironomus crassiforceps* to test metal bioavailability with sediment pH change in tropical acid-sulfate sediments, *Environmental Toxicology and Chemistry* **21** (4), 720-728.

Peeters, E.T.H.M., Dewitte, A., Koelmans, A.A., van der Velden, J.A. and den Besten, P.J. (2001) Evaluation of bioassays versus contaminant concentrations in explaining the macroinvertebrate community structure in the Rhine-Meuse delta, the Netherlands, *Environmental Toxicology and Chemistry* **20** (12), 2883-2891.

Pereira, A.M.M., Soares, A.M.V.M., Gonçalves, F. and Ribeiro, R. (1999) Test chambers and test procedures for *in situ* toxicity testing with zooplankton, *Environmental Toxicology and Chemistry* **18** (9), 1956-1964.

Pereira, A.M.M., Soares, A.M.V.M., Gonçalves, F. and Ribeiro, R. (2000) Water-column, sediment, and *in situ* chronic bioassays with cladocerans, *Ecotoxicology and Environmental Safety* **47** (1), 27-38.

Pérez, S., Farré, M., García, M.J. and Barceló, D. (2001) Occurrence of polycyclic aromatic hydrocarbons in sewage sludge and their contribution to its toxicity in the toxalert 100 bioassay, *Chemosphere* **45** (6-7), 705-712.

Perkins Jr, E.J. and Schlenk, D. (2000) *In vivo* acetylcholinesterase inhibition, metabolism, and toxicokinetics of aldicarb in channel catfish: role of biotransformation in acute toxicity, *Toxicological Sciences* **53** (2), 308-315.

Persoone, G. (1998) Development and first validation of a "stock-culture free" algal microbiotest: the Algaltoxkit, in P.G. Wells, K. Lee and C. Blaise (eds.), *Microscale Testing in Aquatic Toxicology: Advances, Techniques, and Practice*, CRC Press, Boca Raton, FL, pp. 311-320.

Persoone, G. and Vangheluwe, M.L. (2000) Toxicity determination of the sediments of the river Seine in France by application of a battery of microbiotests, in G. Persoone, C. Janssen and W.M. De Coen (eds.), *New Microbiotests for Routine Toxicity Screening and Biomonitoring*, Kluwer Academic / Plenum Publishers, New York, pp. 427-439.

Persoone, G., Van de Vel, A., Van Steertegem, M. and De Nayer, B. (1989) Predictive value of laboratory tests with aquatic invertebrates: influence of experimental conditions, *Aquatic Toxicology* **14** (2), 149-167.
Persoone, G., Blaise, C., Snell, T., Janssen, C. and Van Steertegem, M. (1993) Cyst-based toxicity tests: II. - Report on an international intercalibration exercise with three cost-effective Toxkits, *Zeitschrift für Angewandte Zoologie* **79** (1), 17-36.
Persoone, G. Janssen, C. and De Coen, W. (2000) New microbiotests for routine toxicity screening and biomonitoring, Kluwer Academic / Plenum Publishers, New York, 550 pp.
Persoone, G., Marsalek, B., Blinova, I., Törökné, A., Zarina, D., Manusadžianas, L., Nalecz-Jawecki, G., Tofan, L., Stepanova, N., Tothova, L. and Kolar, B. (2003) A practical and user-friendly toxicity classification system with microbiotests for natural waters and wastewaters, *Environmental Toxicology* **18** (6), 395-402.
Péry, A.R.R., Ducrot, V., Mons, R. and Garric, J. (2003) Modelling toxicity and mode of action of chemicals to analyse growth and emergence tests with the midge *Chironomus riparius*, *Aquatic Toxicology* **65** (3), 281-292.
Pesonen, M. and Andersson, T.B. (1997) Fish primary hepatocyte culture; an important model for xenobiotic metabolism and toxicity studies, *Aquatic Toxicology* **37** (2-3), 253-267.
Petänen, T., Lyytikäinen, M., Lappalainen, J., Romantschuk, M. and Kukkonen, J.V. K. (2003) Assessing sediment toxicity and arsenite concentration with bacterial and traditional methods, *Environmental Pollution* **122** (3), 407-415.
Peter, S., Siersdorfer, C., Kaltwasser, H. and Geiger, M. (1995) Toxicity estimation of treated coke plant wastewater using the luminescent bacteria assay and the algal growth inhibition test, *Environmental Toxicology and Water Quality* **10**, 179-184.
Pettersson, A., Adamsson, M. and Dave, G. (2000) Toxicity and detoxification of Swedish detergents and softener products, *Chemosphere* **41** (10), 1611-1620.
Phipps, J.L., Mattson, V.R. and Ankley, G.T. (1995) Relative sensitivity of three freshwater benthic macroinvertebrates to ten contaminants, *Archives of Environmental Contamination and Toxicology* **28** (3), 281-286.
Pica-Granados, Y., Trujillo, G.D. and Hernández, H.S. (2000) Bioassay standardization for water quality monitoring in Mexico, *Environmental Toxicology* **15** (4), 322-330.
Pintar, A., Besson, M., Gallezot, P., Gibert, J.J. and Martin, D. (2004) Toxicity to *Daphnia magna* and *Vibrio fischeri* of Kraft bleach plant effluents treated by catalytic wet-air oxidation, *Water Research* **38** (2), 289-300.
Preston, B.L., Snell, T.W., Fields, D.M. and Weissburg, M.J. (2001) The effects of fluid motion on toxicant sensitivity of the rotifer *Brachionus calyciflorus*, *Aquatic Toxicology* **52** (2), 117-131.
Priha, M.H. (1996) Ecotoxicological impacts of pulp mill effluents in Finland, in M.R. Servos, K.R. Munkittrick, J.H. Carey and G.J. Van Der Kraak (eds.), *Environmental Fate and Effects of Pulp and Paper Mill Effluents*, St- Lucie Press, Delray Beach, FL, pp. 637-650.
Radetski, C.M, Férard, J.F. and Blaise, C. (1995) A semi-static microplate-based phytotoxicity test, *Environmental Toxicology and Chemistry* **14**, 299-302.
Ramirez, N.E., Vargas, M.C. and Sanchez, F.N. (1996) Use of the sediment Chromotest for monitoring simulated hydrocarbon biodegradation processes, *Environmental Toxicology and Water Quality* **11**, 223-230.
Ramos, E.U., Vermeer, C., Vaes, W.H.J. and Hermens, J.L.M. (1998) Acute toxicity of polar narcotics to three aquatic species (*Daphnia magna*, *Poecilia reticulata* and *Lymnaea stagnalis*) and its relation to hydrophobicity, *Chemosphere* **37** (4), 633-650.
Rao, S.S., Burnison, B.K., Rokosh, D.A. and Taylor, C.M. (1994) Mutagenicity and toxicity assessment of pulp mill effluent, *Chemosphere* **28** (10), 1859-1870.
Rediske, R., Thompson, C., Schelske, C., Gabrosek, J., Nalepa, T. and Peaslee, G. (2002) *Preliminary investigation of the extent of sediment contamination in Muskegon Lake*, U.S. Environmental Protection Agency, Great Lakes National Program Office, Chicago, IL, #GL-97520701, (2003-12-30), 112 pp.; http://www.epa.gov/glnpo/sediment/muskegon/MuskRpt8.pdf.
Redondo, M.J., López-Jaramillo, L., Ruiz, M.J. and Font, G. (1996) Toxicity assessment using the microtox test and determination of pesticides in soil and water samples by chromatographic techniques, *Toxicology Letters* **88** (Supplement 1), 30-30.
Reinhold-Dudok van Heel, H.C. and den Besten, P.J. (1999) The relation between macroinvertebrate assemblages in the Rhine–Meuse delta (The Netherlands) and sediment quality, *Aquatic Ecosystem Health and Management* **2** (1), 19-38.

Ren, S. and Frymier, P.D. (2003) Use of multidimensional scaling in the selection of wastewater toxicity test battery components, *Water Research* **37** (7), 1655-1661.

Repetto, G., del Peso, A. and Repetto, M. (2000) Alternative ecotoxicological methods for the evaluation, control and monitoring of environmental pollutants, *Ecotoxicology and Environmental Restoration* **3** (1), 47-51.

Repetto, G., Jos, A., Hazen, M.J., Molero, M.L., del Peso, A., Salguero, M., del Castillo, P.D., Rodríguez-Vicente, M.C. and Repetto, M. (2001) A test battery for the ecotoxicological evaluation of pentachlorophenol, *Toxicology In Vitro* **15** (4-5), 503-509.

Reynoldson, T.B., Day, K.E., Clarke, C. and Milani, D. (1994) Effect of indigenous animals on chronic end points in freshwater sediment toxicity tests, *Environmental Toxicology and Chemistry* **13** (6), 973-977.

Ribeiro, I.C., Veríssimo, I., Moniz, L., Cardoso, H., Sousa, M.J., Soares, A.M. and Leão, C. (2000) Yeasts as a model for assessing the toxicity of the fungicides penconazol, cymoxanil and dichlofluanid, *Chemosphere* **41** (10), 1637-1642.

Ribo, J.M. (1997) Interlaboratory comparison studies of the luminescent bacteria toxicity bioassay, *Environmental Toxicology and Water Quality* **12** (4), 283-294.

Ribo, J.M., Yang, J.E. and Huang, P.M. (1989) Luminescent bacteria toxicity assay in the study of mercury speciation, in M. Munawar, G. Dixon, C.I. Mayfield, T. Reynoldson and M.H. Sadar (eds.), *Environmental Bioassay Techniques and their Application: Proceedings of the 1st International Conference held in Lancaster, England, 11-14 July 1988*, Kluwer Academic Publishers, Dordrecht, Netherlands, pp. 155-162.

Richardson, J.S., Hall, K.J., Kiffney, P.M., Smith, J.A. and Keen, P. (1998) *Ecological impacts of contaminants in an urban watershed*, DOE FRAP 1998-25, Environment Canada, Environmental Conservation Branch, Aquatic and Atmospheric Sciences Division, Vancouver, BC, 22 pp.

Riisberg, M., Bratlie, E. and Stenersen, J. (1996) Comparison of the response of bacterial luminescence and mitochondrial respiration to the effluent of an oil refinery, *Environmental Toxicology and Chemistry* **15** (4), 501-502.

Rippon, G.D. and Riley, S.J. (19960 Environmental impact assessment of tailings dispersal from a uranium mine using toxicity testing protocols, *Water Resources Bulletin* **32** (6), 1167-1175.

Rissanen, E., Krumschnabel, G. and Nikinmaa, M. (2003) Dehydroabietic acid, a major component of wood industry effluents, interferes with cellular energetics in rainbow trout hepatocytes, *Aquatic Toxicology* **62** (1), 45-53.

Ristola, T., Pellinen, J., Leppänen, M. and Kukkonen, J. (1996) Characterization of Lake Ladoga sediments. I. Toxicity to *Chironomus riparius* and *Daphnia magna*, *Chemosphere* **32** (8), 1165-1178.

Ristola, T., Parker, D. and Kukkonen, J.V.K. (2001) Life-cycle effects of sediment-associated 2,4,5-trichlorophenol on two groups of the midge *Chironomus riparius* with different exposure histories, *Environmental Toxicology and Chemistry* **20** (8), 1772-1777.

Roberts, R.O. and Berk, S.G. (1993) Effect of copper, herbicides, and a mixed effluent on chemoattraction of *Tetrahymena pyriformis*, *Environmental Toxicology and Water Quality* **8** (1), 73-85.

Robinson, P.W. and Scott, R.R. (1995) The toxicity of cyromazine to *Chironomus zealandicus* (Chironomidae) and *Deleatidium* sp. (Leptophlebiidae), *Pesticide Science* **44** (3), 283-292.

Rodgers, D.W., Schröder, J. and Sheehan, L.V. (1996) Comparison of *Daphnia magna*, rainbow trout and bacterial-based toxicity tests of Ontario Hydro aquatic effluents, *Water, Air, and Soil Pollution* **90** (1-2), 105-112.

Roghair, C.J., Buijze, A. and Schoon, H.N.P. (1992) Ecotoxicological risk evaluation of the cationic fabric softener DTDMAC. I. Ecotoxicological effects, *Chemosphere* **24** (5), 599-609.

Ronco, A.E., Castillo, G. and Díaz-Baez, M.C. (2000) Development and application of microbioassays for routine testing and biomonitoring in Argentina, Chile and Colombia, in G. Persoone, C. Janssen and W.M. De Coen (eds.), *New Microbiotests for Routine Toxicity Screening and Biomonitoring*, Kluwer Academic / Plenum Publishers, New York, pp. 49-61.

Ronco, A., Gagnon, P., Díaz-Baez, M.C., Arkhipchuk, V., Castillo, G., Castillo, L.E., Dutka, B.J., Pica-Granados, Y., Ridal, J., Srivastava, R.C. and Sanchez, A. (2002) Overview of results from the WaterTox intercalibration and environmental testing phase II program: Part 1, statistical analysis of blind sample testing, *Environmental Toxicology* **17** (3), 232-240.

Roseth, S., Edvardsson, T., Botten, T.M., Fuglestad, J., Fonnum, F. and Stenersen, J. (1996) Comparison of acute toxicity of process chemicals used in the oil refinery industry, tested with the diatom *Chaetoceros gracilis*, the flagellate *Isochrysis galbana*, and the zebra fish, *Brachydanio rerio*, *Environmental Toxicology and Chemistry* **15** (7), 1211-1217.

Ross, P. (1993) The use of bacterial luminescence systems in aquatic toxicity testing, in M. Richardson (ed.), *Ecotoxicology Monitoring*, VCH Publishers, Weinheim, Germany, pp. 185-195.

Ross, P. (1998) Role of microbiotests in contaminated sediment assessment batteries, in P.G. Wells, K. Lee and C. Blaise (eds.), *Microscale Testing in Aquatic Toxicology: Advances, Techniques, and Practice*, CRC Press, Boca Raton, FL, pp. 549-556.

Ross, P. and Leitman, P.A. (1995) Solid phase testing of aquatic sediments using *Vibrio fischeri*: test design and data interpretation, in M. Richardson (ed.), *Environmental Toxicology Assessment*, Taylor & Francis Ltd., London, England, pp. 65-76.

Ross, P.E., Burton Jr, G.A., Crecelius, E.A., Filkins, J.C., Giesy, J.P., Ingersoll, C.G., Landrum, P.F., Mac, M.J., Murphy, T.J., Rathbun, J.E., Smith, V.E., Tatem, H.E. and Taylor, R.W. (1992) Assessment of sediment contamination at Great Lakes areas of concern: the ARCS Program Toxicity-Chemistry Work Group strategy, *Journal of Aquatic Ecosystem Health* **1** (3), 193-200.

Ross, P., Burton Jr, G.A., Greene, M., Ho, K., Meier, P.G., Sweet, L.I., Auwarter, A., Bispo, A., Doe, K., Erstfeld, K., Goudey, S., Goyvaerts, M., Henderson, D.G., Jourdain, M., Lenon, M., Pandard, P., Qureshi, A., Rowland, C., Schipper, C., Schreurs, W., Trottier, S. and Van Aggelen, G. (1999) Interlaboratory precision study of a whole sediment toxicity test with the bioluminescent bacterium *Vibrio fischeri*, *Environmental Toxicology* **14** (3), 339-345.

Rossi, D. and Beltrami, M. (1998) Sediment ecological risk assessment: *in situ* and laboratory toxicity testing of Lake Orta sediments, *Chemosphere* **37** (14-15), 2885-2894.

Rousch, J.M., Simmons, T.W., Kerans, B.L. and Smith, B.P. (1997) Relative acute effects of low pH and high iron on the hatching and survival of the water mite (*Arrenurus manubriator*) and the aquatic insect (*Chironomus riparius*), *Environmental Toxicology and Chemistry* **16** (10), 2144-2150.

Ruck, J.G., Martin, M. and Mabon, M. (2000) Evaluation of Toxkits as methods for monitoring water quality in New Zealand, in G. Persoone, C. Janssen and W.M. De Coen (eds.), *New Microbiotests for Routine Toxicity Screening and Biomonitoring*, Kluwer Academic / Plenum Publishers, New York, pp. 103-119.

Rue, W.J., Fava, J.A. and Grothe, D.R. (1988) A review of inter- and intralaboratory effluent toxicity test method variablility, in M.S. Adams, G.A. Chapman and W.G. Landis (eds.), *Aquatic Toxicology and Hazard Assessment: 10th Volume, ASTM STP 971*, American Society for Testing and Materials, Philadelphia, PA, pp. 190-203.

Rutherford, L.A., Matthews, S.L., Doe, K.G. and Julien, G.R.J. (2000) Aquatic toxicity and environmental impact of leachate discharges from a municipal landfill, *Water Quality Research Journal of Canada* **35** (1), 39-57.

Sabaliunas, D., Lazutka, J. and Sabaliuniené, I. (2000) Acute toxicity and genotoxicity of aquatic hydrophobic pollutants sampled with semipermeable membrane devices, *Environmental Pollution* **109** (2), 251-265.

Sae-Ma, B., Meier, P.G. and Landrum, P.F. (1998) Effect of extended storage time on the toxicity of sediment-associated cadmium on midge larvae (*Chironomus tentans*), *Ecotoxicology* **7** (3), 133-139.

Sakai, M. (2001) Chronic toxicity tests with *Daphnia magna* for examination of river water quality, *Journal of Environmental Science and Health, Part B* **36** (1), 67-74.

Sakai, M. (2002a) Use of chronic tests with *Daphnia magna* for examination of diluted river water, *Ecotoxicology and Environmental Safety* **53** (3), 376-381.

Sakai, M. (2002b) Determination of pesticides and chronic test with *Daphnia magna* for rainwater samples, *Journal of Environmental Science and Health, Part B*, **37**(3), 247-254.

Salizzato, M., Pavoni, B., Ghirardini, A.V. and Ghetti, P.F. (1998) Sediment toxicity measured using *Vibrio fischeri* as related to the concentrations of organic (PCBs, PAHs) and inorganic (metals, sulfur) pollutants, *Chemosphere* **36** (14), 2949-2968.

Samaras, P., Sakellaropoulos, G.P., Kungolos, A. and Dermissi, S. (1998) Toxicity assessment assays in Greece, *Fresenius Environmental Bulletin* **7** (7-8, Special), 623-630.

Samel, A., Ziegenfuss, M., Goulden, C.E., Banks, S. and Baer, K.N. (1999) Culturing and bioassay testing of *Daphnia magna* using Elendt M4, Elendt M7, and COMBO media, *Ecotoxicology and Environmental Safety* **43** (1), 103-110.

Sánchez-Mata, J.D., Fernández, V., Chordi, A. and Tejedor, C. (2001) Toxicity and mutagenecity of urban wastewater treated with different purifying processes, *Aquatic Ecosystem Health and Management* **4** (1), 61-72.

Sandbacka, M., Pärt, P. and Isomaa, B. (1999) Gill epithelial cells as tools for toxicity screening - comparison between primary cultures, cells in suspension and epithelia on filters, *Aquatic Toxicology* **46** (1), 23-32.

Santiago, S., Thomas, R.L., Larbaigt, G., Rossel, D., Echeverria, M.A., Tarradellas, J., Loizeau, J.L., McCarthy, L., Mayfield, C.I. and Corvi, C. (1993) Comparative ecotoxicity of suspended sediment in the lower Rhone River using algal fractionation, Microtox and *Daphia magna* bioassays, *Hydrobiologia* **252** (3), 231-244.

Santiago, S., van Slooten, K.B., Chèvre, N., Pardos, M., Benninghoff, C., Dumas, M., Thybaud, E. and Garrivier, F. (2002) *Guide pour l'Utilisation des Tests Ecotoxicologiques avec les Daphnies, les Bactéries Luminescentes et les Algues Vertes, Appliqués aux Echantillons de l'Environnement*, Soluval Institut Forel, Genève, 56 pp.

Sarma, S.S.S., Nandini, S. and Flores, J.L.G. (2001) Effect of methyl parathion on the population growth of the rotifer *Brachionus patulus* (O.F. Muller) under different algal food (*Chlorella vulgaris*) densities, *Ecotoxicology and Environmental Safety* **48** (2), 190-195.

Sauvant, M.P., Pépin, D. and Piccinni, E. (1999) *Tetrahymena pyriformis*: a tool for toxicological studies. A review, *Chemosphere* **38** (7), 1631-1669.

Sauvant, M.P., Pépin, D., Bohatier, J., Grolière, C.A. and Veyre, A. (1994) Comparative study of two in vitro models ($_L$-929 fibroblasts and *Tetrahymena pyriformis* GL) for the cytotoxicological evaluation of packaged water, *The Science of The Total Environment* **156** (2), 159-167.

Schramm, K.-W., Kaune, A., Beck, B., Thumm, W., Behechti, A., Kettrup, A. and Nickolova, P. (1996) Acute toxicities of five nitromusk compounds in *Daphnia*, algae and photoluminescent bacteria, *Water Research* **30** (10), 2247-2250.

Schultz, T.W., Sinks, G.D. and Bearden, A.P. (1998) QSAR in aquatic toxicology: a mechanism of action approach comparing toxic potency to *Pimephales promelas*, *Tetrahymena pyriformis*, and *Vibrio fischeri*, in J. Devillers (ed.), *Comparative QSAR*, Taylor & Francis, New York, pp. 51-109.

Schulz, R., Peall, S.K., Dabrowski, J.M. and Reinecke, A.J. (2001) Spray deposition of two insecticides into surface waters in a South African orchard area, *Journal of Environmental Quality* **30** (3), 814-822.

Schultz, T.W., Seward-Nagel, J., Foster, K.A. and Tucker, V.A. (2004) Population growth impairment of aliphatic alcohols to *Tetrahymena*, *Environmental Toxicology* **19**, 1-10.

Schweigert, N., Eggen, R.I., Escher, B.I., Burkhardt-Holm, P. and Behra, R. (2002) Ecotoxicological assessment of surface waters: a modular approach integrating in vitro methods, *ALTEX (Alternatives to Animal Experiments)* **19** (Suppl 1), 30-37.

Scroggins, R., van Aggelen, G. and Schroeder, J. (2002a, Monitoring sublethal toxicity in effluent under the metal mining EEM Program, *Water Quality Research Journal of Canada* **37** (1), 279-294.

Scroggins, R.P., Miller, J.A., Borgmann, A.I. and Sprague, J.B. (2002b) Sublethal toxicity findings by the pulp and paper industry for cycles 1 and 2 of the environmental effects monitoring program, *Water Quality Research Journal of Canada* **37** (1), 21-48.

Seco, J.I., Fernández-Pereira, C. and Vale, J. (2003) A study of the leachate toxicity of metal-containing solid wastes using *Daphnia magna*, *Ecotoxicology and Environmental Safety* **56** (3), 339-350.

Sekkat, N., Guerbet, M. and Jouany, J.-M. (2001) Étude comparative de huit bioessais à court terme pour l'évaluation de la toxicité de lixiviats de déchets urbains et industriels, *Revue des Sciences de l'Eau* **14** (1), 63-72.

Sepúlveda, M.S., Quinn, B.P., Denslow, N.D., Holm, S.E. and Gross, T.S. (2003) Effects of pulp and paper mill effluent on reproductive success of largemouth bass, *Environmental Toxicology and Chemistry* **22** (1), 205-213.

Sergy, G. (1987) Recommendations on aquatic biological tests and procedures for environmental protection, Conservation and Protection, Department of Environment Manuscript Report, Environment Canada, Ottawa, Ontario, 102 pp.

Seymour, D.T., Verbeek, A.G., Hrudey, S.E. and Fedorak, P.M. (1997) Acute toxicity and aqueous solubility of some condensed thiophenes and their microbial metabolites, *Environmental Toxicology and Chemistry* **16** (4), 658-665.

Sherrard, R.M., Murray-Gulde, C.L., Rodgers Jr, J.H. and Shah, Y.T. (2003) Comparative toxicity of chlorothalonil: *Ceriodaphnia dubia* and *Pimephales promelas*, *Ecotoxicology and Environmental Safety* **56** (3), 327-333.

Sherry, J.P., Scott, B.F., Nagy, V. and Dutka, B.J. (1994) Investigation of the sublethal effects of some petroleum refinery effluents, *Journal of Aquatic Ecosystem Health* **3** (2), 129-137.

Sherry, J.P, Scott, B.F and Dutka, B. (1997) Use of various acute, sublethal and early life-stage tests to evaluate the toxicity of refinery effluents, *Environmental Toxicology and Chemistry* **16** (11), 2249-2257.

Sibley, P.K., Ankley, G.T., Cotter, A.M. and Leonard, E.N. (1996) Predicting chronic toxicity of sediments spiked with zinc: an evaluation of the acid-volatile sulfide model using a life-cycle test with the midge *Chironomus tentans*, *Environmental Toxicology and Chemistry* **15** (12), 2102-2112.

Sibley, P.K., Benoit, D.A. and Ankley, G.T. (1997a) The significance of growth in *Chironomus tentans* sediment toxicity tests: relationship to reproduction and demographic endpoints, *Environmental Toxicology and Chemistry* **16** (2), 336-345.

Sibley, P.K., Legler, J., Dixon, D.G. and Barton, D.R. (1997b) Environmental health assessment of the benthic habitat adjacent to a pulp mill discharge. I. Acute and chronic toxicity of sediments to benthic macroinvertebrates, *Archives of Environmental Contamination and Toxicology* **32** (3), 274-284.

Sibley, P.K., Monson, P.D. and Ankley, G.T. (1997c) The effect of gut contents on dry weight estimates of *Chironomus tentans* larvae: implications for interpreting toxicity in freshwater sediment toxicity tests, *Environmental Toxicology and Chemistry* **16** (8), 1721-1726.

Sildanchandra, W. and Crane, M. (2000) Influence of sexual dimorphism in *Chironomus riparius* Meigen on toxic effects of cadmium, *Environmental Toxicology and Chemistry* **19** (9), 2309-2313.

Sloterdijk, H., Champoux, L., Jarry, V., Couillard, Y. and Ross, P. (1989) Bioassay responses of microorganisms to sediment elutriates from the St. Lawrence River (Lake St. Louis), in M. Munawar, G. Dixon, C.I. Mayfield, T. Reynoldson and M.H. Sadar (eds.), *Environmental Bioassay Techniques and their Application: Proceedings of the 1st International Conference held in Lancaster, England, 11-14 July 1988*, Kluwer Academic Publishers, Dordrecht, Netherlands, pp. 317-335.

Snell, T.W. (2000) The distribution of endpoint chronic value, for freshwater rotifer, in G. Persoone, C. Janssen and W.M. De Coen (eds.), *New Microbiotests for Routine Toxicity Screening and Biomonitoring*, Kluwer Academic/Plenum Publishers, New York, pp. 185-190.

Snell, T.W. and Persoone, G. (1989) Acute toxicity bioassays using rotifers. II. A freshwater test with *Brachionus rubens*, *Aquatic Toxicology* **14** (1), 81-91.

Snell, T.W. and Janssen, C.R. (1995) Rotifers in ecotoxicology: a review, *Hydrobiologia* **313/314**, 231-247.

Snell, T.W. and Janssen, C.R. (1998) Microscale toxicity testing with rotifers, in P.G. Wells, K. Lee and C. Blaise (eds.), *Microscale Testing in Aquatic Toxicology: Advances, Techniques, and Practice*, CRC Press, Boca Raton, FL, pp. 409-422.

Sosak-Swiderska, B. and Tyrawska, D. (1996) The role of algae in ecotoxicological tests, in M. Richardson (ed.), *Environmental Xenobiotics*, Taylor & Francis Books Ltd, London, England, pp. 179-193.

Sponza, D.T. (2001) Toxicity studies of tobacco wastewater, *Aquatic Ecosystem Health and Management* **4**(4), 479-492.

Staples, C.A. and Davis, J.W. (2002) An examination of the physical properties, fate, ecotoxicity and potential environmental risks for a series of propylene glycol ethers, *Chemosphere* **49** (1), 61-73.

Stepanova, N.J., Petrov, A.M., Gabaydullin, A.G. and Shagidullin, R.R. (2000) Toxicity of snow cover for the assessment of air pollution as determined with microbiotests, in G. Persoone, C. Janssen and W.M. De Coen (eds.), *New Microbiotests for Routine Toxicity Screening and Biomonitoring*, Kluwer Academic/Plenum Publishers, New York, pp. 475-478.

Stephenson, G.L., Kaushik, N.K. and Solomon, K.R. (1991) Acute toxicity of pure pentachlorophenol and a technical formulation to three species of *Daphnia*, *Archives of Environmental Contamination and Toxicology* **20** (1), 73-80.

Stewart, K.M. and Thompson, R.S. (1995) Fluoranthene as a model toxicant in sediment studies with *Chironomus riparius*, *Journal of Aquatic Ecosystem Health* **4** (4), 231-238.

St-Laurent, D. and Blaise, C. (1995) Validation of a microplate-based algal lethality test developed with the help of flow cytometry, in M. Richardson (ed.), *Environmental Toxicology Assessment*, Taylor & Francis Ltd., London, England, pp. 137-155.

Stratton, G.W. (1987) The effects of pesticides and heavy metals towards phototrophic microorganisms. In: E. Hodgson, Editor, Reviews in Environmental Toxicology vol. 3, Elsevier, NY (1987), pp. 71-147.

Stuhlfauth, T. (1995) Ecotoxicological monitoring of industrial effluents, in M. Richardson (ed.), *Environmental Toxicology Assessment*, Taylor & Francis Ltd., London, England, pp. 187-198.

Stuijfzand, S.C., Drenth, A., Helms, M. and Kraak, M.H. (1998) Bioassays using the midge *Chironomus riparius* and the zebra mussel *Dreissena polymorpha* for evaluation of river water quality, *Archives of Environmental Contamination and Toxicology* **34** (4), 357-363.

Sturm, A. and Hansen, P. (1999) Altered cholinesterase and monooxygenase levels in *Daphnia magna* and *Chironomus riparius* exposed to environmental pollutants, *Ecotoxicology and Environmental Safety* **42** (1), 9-15.

Suedel, B.C. and Rodgers Jr, J.H. (1994) Development of formulated reference sediments for freshwater and estuarine sediment testing, *Environmental Toxicology and Chemistry* **13** (7), 1163-1175.

Suedel, B.C. and Rodgers Jr, J.H. (1996) Toxicity of fluoranthene to *Daphnia magna, Hyalella azteca, Chironomus tentans*, and *Stylaria lacustris* in water-only and whole sediment exposures, *Bulletin of Environmental Contamination and Toxicology* **57** (1), 132-138.

Suedel, B.C., Deaver, E. and Rodgers Jr, J.H. (1996) Experimental factors that may affect toxicity of aqueous and sediment-bound copper to freshwater organisms, *Archives of Environmental Contamination and Toxicology* **30** (1), 40-46.

Suedel, B.C., Rodgers Jr, J.H. and Deaver, E. (1997) Experimental factors that may affect toxicity of cadmium to freshwater organisms, *Archives of Environmental Contamination and Toxicology* **33** (2), 188-193.

Suter, G.W. II. (1993) Ecological risk assessment, Lewis Publishers, Boca Raton, Fl., U.S.A., 538 pp.

Svenson, A. and Zhang, L. (1995) Acute aquatic toxicity of protolyzing substances studied as the Microtox effect, *Ecotoxicology and Environmental Safety* **30** (3), 283-288.

Svenson, A., Linlin, Z. and Kaj, L. (1992) Primary chemical and physical characterization of acute toxic components in wastewaters, *Ecotoxicology and Environmental Safety* **24** (2), 234-242.

Svenson, A., Edsholt, E., Ricking, M., Remberger, M. and Röttorp, J. (1996) Sediment contaminants and Microtox toxicity tested in a direct contact exposure test, *Environmental Toxicology and Water Quality* **11** (4), 293-300.

Svenson, A., Sandén, B., Dalhammar, G., Remberger, M. and Kaj, L. (2000) Toxicity identification and evaluation of nitrification inhibitors in wastewaters, *Environmental Toxicology* **15** (5), 527-532.

Sweet, L.I., Travers, D.F. and Meier, P.G. (1997) Short Communication-chronic toxicity evaluation of wastewater treatment plant effluents with bioluminescent bacteria: a comparison with invertebrates and fish, *Environmental Toxicology and Chemistry* **16** (10), 2187-2189.

Tarczynska, M., Nalecz-Jawecki, G., Brzychcy, B., Zalewski, M. and Sawicki, J. (2000) The toxicity of cyanobacterial blooms as determined by microbiotests and mouse assays, in G. Persoone, C. Janssen and W.M. De Coen (eds.), *New Microbiotests for Routine Toxicity Screening and Biomonitoring*, Kluwer Academic/Plenum Publishers, New York, pp. 527-532.

Tarkpea, M. and Hansson, M. (1989) Comparison between two Microtox test procedures, *Ecotoxicology and Environmental Safety* **18** (2), 204-210.

Taylor, L.N., Wood, C.M. and McDonald, D.G. (2003) An evaluation of sodium loss and gill metal binding properties in rainbow trout and yellow perch to explain species differences in copper tolerance, *Environmental Toxicology and Chemistry* **22** (9), 2159-2166.

Tchounwou, P.B. and Reed, L. (1999) Assessment of lead toxicity to the marine bacterium, *Vibrio fischeri*, and to a heterogeneous population of microorganisms derived from the Pearl River in Jackson, Mississippi, USA, *Review of Environmental Health* **14** (2), 51-61.

Tellez, M.R., Dayan, F.E., Schrader, K.K., Wedge, D.E. and Duke, S.O. (2000) Composition and some biological activities of the essential oil of *Callicarpa americana* (L.), *Journal of Agricultural and Food Chemistry* **48** (7), 3008-3012.

Tellez, M.R., Khan, I.A., Kobaisy, M., Schrader, K.K., Dayan, F.E. and Osbrink, W. (2002) Composition of the essential oil of *Lepidium meyenii* (Walp), *Phytochemistry* **61** (2), 149-155.

Terzaghi, C., Buffagni, M., Cantelli, D., Bonfanti, P. and Camatini, M. (1998) Physical-chemical and ecotoxicological evaluation of water based drilling fluids used in Italian off-shore, *Chemosphere* **37** (14-15), 2859-2871.

Tessier, L., Unfer, S., Férard, J.F., Loiseau, C., Richard, E. and Brumas, V. (1999) Potential of acoustic wave microsensors for aquatic ecotoxicity assessment based on microplates. *Sensors and Actuators* **B 59**: 177-179.

Tetreault, G.R., McMaster, M.E., Dixon, D.G. and Parrott, J.L. (2003) Physiological and biochemical responses of Ontario slimy sculpin (*Cottus cognatus*) to sediment from the Athabasca oil sands area, *Water Quality Research Journal of Canada* **38** (2), 361-377.

Thellen, C., Blaise, C., Roy, Y. and Hickey, C. (1989) Round Robin testing with the *Selenastrum capricornutum* microplate toxicity assay, in M. Munawar, G. Dixon, C.I. Mayfield, T. Reynoldson and M.H. Sadar (eds.), *Environmental Bioassay Techniques and their Application: Proceedings of the 1st International Conference held in Lancaster, England, 11-14 July 1988*, Kluwer Academic Publishers, Dordrecht, Netherlands, pp. 259-268.

Thomulka, K.W., Schroeder, J.A. and Lange, J.H. (1997) Use of *Vibrio harveyi* in an aquatic bioluminescent toxicity test to assess the effects of metal toxicity: Treatment of sand and water-buffer, with and without EDTA, *Environmental Toxicology and Water Quality* **12** (4), 343-348.
Tietge, J.E., Hockett, J.R. and Evans, J.M. (1997) Major ion toxicity of six produced waters to three freshwater species: application of ion toxicity models and TIE procedures, *Environmental Toxicology and Chemistry* **16** (10), 2002-2008.
Tišler, T. and Zagorc-Koncan, J. (1997) Comparative assessment of toxicity of phenol, formalhehyde, and industrial wastewater to aquatic organisms, *Water, Air, and Soil Pollution* **97** (3-4), 315-322.
Tišler, T. and Zagorc-Koncan, J. (1999) Toxicity evaluation of wastewater from the pharmaceutical industry to aquatic organisms, *Water Science and Technology* **39** (10-11), 71-76.
Törökné, A.K. (2000) The potential of the Thamnotoxkit microbiotest for routine detection of cyanobacterial toxins, in G. Persoone, C. Janssen and W.M. De Coen (eds.), *New Microbiotests for Routine Toxicity Screening and Biomonitoring*, Kluwer Academic / Plenum Publishers, New York, pp. 533-539.
Törökné, A., Oláh, B., Reskóné, M., Báskay, I. and Bérciné, J. (2000) Utilization of microbiotests to assess the contamination of water-bases, *Central European Journal of Public Health* **8** (8), 97-99.
Tran, D., Ciret, P., Ciutat, A., Durrieu, G. and Massabuau, J.-C. (2003) Estimation of potential and limits of bivalve closure response to detect contaminants: application to cadmium, *Environmental Toxicology and Chemistry* **22** (4), 914-920.
Traunspurger, W., Schäfer, H. and Remde, A. (1996) Comparative investigation on the effect of a herbicide on aquatic organisms in single species tests and aquatic microcosms, *Chemosphere* **33** (6), 1129-1141.
Trottier, S., Blaise, C., Kusui, T. and Johnson, E.M. (1997) Acute toxicity assessment of aqueous samples using a microplate-based *Hydra attenuata* assay, *Environmental Toxicology and Water Quality* **12** (3), 265-271.
Tsui, M.T.K. and Chu, L.M. (2003) Aquatic toxicity of glyphosate-based formulations: comparison between different organisms and the effects of environmental factors, *Chemosphere* **52** (7), 1189-1197.
Twagilimana, L., Bohatier, J., Grolière, C.A., Bonnemoy, F. and Sargos, D. (1998) A new low-cost microbiotest with the Protozoan *Spirostomum teres*: culture conditions and assessment of sensitivity of the ciliate to 14 pure chemicals, *Ecotoxicology and Environmental Safety* **41** (3), 231-244.
Ulitzur, S., Lahav, T. and Ulitzur, N. (2002) A novel and sensitive test for rapid determination of water toxicity, *Environmental Toxicology* **17** (3), 291-296.
U.S. EPA (2002a) Methods for measuring the acute toxicity of effluents and receiving waters to freshwater and marine organisms, EPA-821-R-02-012, United States Environmental Protection Agency, Washington, DC, pp. 1-275.
U.S. EPA (2002b) Short-term methods for estimating the chronic toxicity of effluents and receiving waters to freshwater organisms, EPA-821-R-02-013, U.S. Environmental Protection Agency, Washington, DC, pp. 1-350.
U.S. EPA (United States Environmental Protection Agency) and Environment Canada (1984) Proceedings of the International OECD workshop on biological testing of effluents (and related receiving waters), September 10 through 14, 1984, Duluth, Minnesota, USA, 367 pp.
Vaajasaari, K., Ahtiainen, J., Nakari, T. and Dahlbo, H. (2000) Hazard assessment of industrial waste leachability: chemical characterization and biotesting by routine effluent tests, in G. Persoone, C. Janssen and W. M. De Coen (eds.), *New Microbiotests for Routine Toxicity Screening and Biomonitoring*, Kluwer Academic / Plenum Publishers, New York, pp. 413-423.
van Dam, R.A., Barry, M.J., Ahokas, J.T. and Holdway, D.A. (1998) Effects of water-borne iron and calcium on the toxicity of diethylenetriamine pentaacetic acid (DTPA) to *Daphnia carinata*, *Aquatic Toxicology* **42** (1), 49-66.
van den Heuvel, M.R. and Ellis, R.J. (2002) Timing of exposure to a pulp and paper effluent influences the manifestation of reproductive effects in rainbow trout, *Environmental Toxicology and Chemistry* **21** (11), 2338-2347.
van der Geest, H.G., Greve, G.D., Kroon, A., Kuijl, S., Kraak, M.H.S. and Admiraal, W. (2000) Sensitivity of characteristic riverine insects, the caddisfly *Cyrnus trimaculatus* and the mayfly *Ephoron virgo*, to copper and diazinon, *Environmental Pollution* **109** (2), 177-182.
Van der Wielen, C. and Halleux, I. (2000) Toxicity monitoring of the Scheldt and Meuse rivers in Wallonia (Belgium) by conventional tests and microbiotests, in G. Persoone, C. Janssen and W.M. De Coen (eds.), *New Microbiotests for Routine Toxicity Screening and Biomonitoring*, Kluwer Academic/Plenum Publishers, New York, pp. 295-303.

Van Donk, E., Abdel-Hamid, M.I., Faafeng, B.A. and Källqvist, T. (1992) Effects of Dursban® 4E and its carrier on three algal species during exponential and P-limited growth, *Aquatic Toxicology* **23** (3-4), 181-191.

van Wijngaarden, R.P.A., van den Brink, P.J., Crum, S.J.H., Oude, V.J.H., Brock, T.C.M. and Leeuwangh, P. (1996) Effects of the insecticide Dursban® 4E (active ingredient chlorpyrifos) in outdoor experimental ditches: I. Comparison of short-term toxicity between the laboratory and the field, *Environmental Toxicology and Chemistry* **15** (7), 1133-1142.

VanGenderen, E.J., Ryan, A.C., Tomasso, J.R. and Klaine, S.J. (2003) Influence of dissolved organic matter source on silver toxicity to *Pimephales promelas*, *Environmental Toxicology and Chemistry* **22** (11), 2746-2751.

Vanderbroele, M.C., Heijerick, D.G., Vangheluwe, M.L. and Janssen, C.R. (2000) Comparison of the conventional algal assay and the Algaltoxkit F™ microbiotest for toxicity evaluation of sediment pore waters, in G. Persoone, C. Janssen and W. M. De Coen (eds.), *New Microbiotests for Routine Toxicity Screening and Biomonitoring*, Kluwer Academic / Plenum Publishers, New York, pp. 261-268.

Vasseur, P. and Pandard, P. (1988) Influence of some experimental factors on metal toxicity to *Selenastrum capricornutum*, *Toxicity Assessment* **3**: 331-343.

Vecchi, M., Reynoldson, T.B., Pasteris, A. and Bonomi, G. (1999) Toxicity of copper-spiked sediments to *Tubifex tubifex* (Oligochaeta, Tubificidae): comparison of the 28-day reproduction bioassay with an early-life stage bioassay, *Environmental Toxicology and Chemistry* **18** (6), 1173-1179.

Veith, G.D., Call, D.J. and Brooke, L.T. (1983) Estimating the acute toxicity of narcotic industrial chemicals to fathead minnows, in W.E. Bishop, R.D. Cardwell and B.B. Heidolph (eds.), *Aquatic Toxicology and Hazard Assessment: Sixth Symposium, ASTM STP 802*, American Society for Testing and Materials, Philadelphia, PA, pp. 90-97.

Verge, C., Moreno, A., Bravo, J. and Berna, J.L. (2001) Influence of water hardness on the bioavailability and toxicity of linear alkylbenzene sulphonate (LAS), *Chemosphere* **44** (8), 1749-1757.

Vidal, D.E. and Horne, A.J. (2003) Inheritance of mercury tolerance in the aquatic oligochaete *Tubifex tubifex*, *Environmental Toxicology and Chemistry* **22**(9), 2130-2135.

Viganò, L., Bassi, A. and Garino, A. (1996) Toxicity evaluation of waters from a tributary of the River Po using the 7-Day *Ceriodaphnia dubia* test, *Ecotoxicology and Environmental Safety* **35** (3), 199-208.

Villaescusa, I., Martí, S., Matas, C., Martínez, M. and Ribó, J.M. (1997) Chromium(VI) toxicity to luminescent bacteria, *Environmental Toxicology and Chemistry* **16** (5), 871-874.

Villarroel, M.J., Sancho, E., Ferrando, M.D. and Andreu-Moliner, E. (1999) Effect of an acaricide on the reproduction and survival of *Daphnia magna*, *Bulletin of Environmental Contamination and Toxicology* **63** (2), 167-173.

Villegas-Navarro, A., González, M.C.R., López, E.R., Aguilar, R.D. and Marçal, W.S. (1999) Evaluation of *Daphnia magna* as an indicator of toxicity and treatment efficacy of textile wastewaters, *Environment International* **25** (5), 619-624.

Vujevic, M., Vidakovic-Cifrek, Z., Tkalec, M., Tomic, M. and Regula, I. (2000) Calcium chloride and calcium bromide aqueous solutions of technical and analytical grade in *Lemna* bioassay, *Chemosphere* **41** (10), 1535-1542.

Walker, J.D., Knaebel, D., Mayo, K., Tunkel, J. and Gray, D.A. (2004) Use of QSARs to promote more cost-effective use of chemical monitoring resources. 1. Screening industrial chemicals and pesticides, direct food additives, indirect food additives and pharmaceuticals for biodegradation, bioconcentration and aquatic toxicity potential, *Water Quality Research Journal of Canada* **39**, 35-39.

Walthall, W.K. and Stark, J.D. (1999) The acute and chronic toxicity of two xanthene dyes, fluorescein sodium salt and phloxine B, to *Daphnia pulex*, *Environmental Pollution* **104** (2), 207-215.

Wang, C., Wang, Y., Kiefer, F., Yediler, A., Wang, Z. and Kettrup, A. (2003) Ecotoxicological and chemical characterization of selected treatment process effluents of municipal sewage treatment plant, *Ecotoxicology and Environmental Safety* **56** (2), 211-217.

Wang, Y., Zhang, M. and Wang, X. (2000) Population growth responses of *Tetrahymena shanghaiensis* in exposure to rare earth elements, *Biological Trace Element Research* **75** (1-3), 265-275.

Ward, M.L., Bitton, G., Townsend, T. and Booth, M. (2002a) Determining toxicity of leachates from Florida municipal solid waste landfills using a battery-of-tests approach, *Environmental Toxicology* **17** (3), 258-266.

Ward, T.J., Rausina, G.A., Stonebraker, P.M. and Robinson, W.E. (2002b) Apparent toxicity resulting from the sequestering of nutrient trace metals during standard *Selenastrum capricornutum* toxicity tests, *Aquatic Toxicology* **60** (1-2), 1-16.

Watts, M.M. and Pascoe, D. (1996) Use of the freshwater macroinvertebrate *Chironomus riparius* (diptera: chironomidae) in the assessment of sediment toxicity, *Water Science and Technology* **34** (7-8), 101-107.
Watts, M.M. and Pascoe, D. (1998) Selection of an appropriate life-cycle stage of *Chironomus riparius* Meigen for use in chronic sediment toxicity testing, *Chemosphere* **36** (6), 1405-1413.
Watts, M.M. and Pascoe, D. (2000) Comparison of *Chironomus riparius* Meigen and *Chironomus tentans* Fabricius (Diptera: Chironomidae) for assessing the toxicity of sediments, *Environmental Toxicology and Chemistry* **19** (7), 1885-1892.
Watzin, M.C., McIntosh, A.W., Brown, E.A., Lacey, R., Lester, D.C., Newbrough, K.L. and Williams, A.R. (1997) Assessing sediment quality in heterogeneous environments: a case study of a small urban harbor in Lake Champlain, Vermont, USA, *Environmental Toxicology and Chemistry* **16** (10), 2125-2135.
Wells, P.,K. Lee and C. Blaise (eds.) (1998) *Microscale testing in Aquatic Toxicology Advances, Techniques and Practice*, CRC Lewis Publishers, Boca Raton, Florida, 679 pp.
Wenning, R.J. and Ingersoll, C.G. (eds.) (2002) *Use of sediment quality guidelines and related tools for the assessment of contaminated sediments*, Executive Summary Booklet of a SETAC Pellston Workshop, Society of Environmental Toxiclogy and Chemistry, Pensacola, FL, 48 pp.
Wernersson, A.S. (2004) Aquatic ecotoxicity due to oil pollution in the Ecuadorian Amazon, *Aquatic ecosystem Health and Management* **7**, 127-136.
West, C.W., Mattson, V.R., Leonard, E.N., Phipps, G.L. and Ankley, G.T. (1993) Comparison of the relative sensitivity of three benthic invertebrates to copper-contaminated sediments from the Keweenaw Waterway, *Hydrobiologia* **262**, 57-63.
West, C.W., Phipps, G.L., Hoke, R.A., Goldenstein, T.A., Vandermeiden, F.M., Kosian, P.A. and Ankley, G.T. (1994) Sediment core versus grab samples: evaluation of contamination and toxicity at a DDT-contaminated site, *Ecotoxicology and Environmental Safety* **28** (2), 208-220.
Weyers, A. and Vollmer, G. (2000) Algal growth inhibition: effect of the choice of growth rate or biomass as endpoint on the classification and labelling of new substances notified in the EU, *Chemosphere* **41** (7), 1007-1010.
Wilkes, B.D. and Beatty Spence, J.M. (1995) Assessing the toxicity of surface waters downstream from a gold mine using a battery of bioassays, *Canadian Technical Report of Fisheries and Aquatic Sciences* **2050**, 38-44.
Williams, K.A., Green, D.W.J., Pascoe, D. and Gower, D.E. (1986) The acute toxicity of cadmium to different larval stages of *Chironomus riparius* (Diptera : Chironomidae) and its ecological significance for pollution regulation, *Oecologia (Berlin)* **70** (3), 362-366.
Williams, M.L., Palmer, C.G. and Gordon, A.K. (2003) Riverine macroinvertebrate responses to chlorine and chlorinated sewage effluents - Acute chlorine tolerances of *Baetis harrisoni* (Ephemeroptera) from two rivers in KwaZulu-Natal, South Africa, *Water SA* **29** (4), 483-488.
Williams, T.D., Hutchinson, T.H., Roberts, G.C. and Coleman, C.A. (1993) The assessment of industrial effluent toxicity using aquatic microorganisms, invertebrates and fish, *The Science of The Total Environment* **Supplement**, 1129-1141.
Wong, M.-Y., Sauser, K.R., Chung, K.-T., Wong, T.-Y. and Liu, J.-K. (2001) Response of the ascorbate-peroxidase of *Selenastrum capricornutum* to copper and lead in stormwaters, *Environmental Monitoring and Assessment* **67** (3), 361-378.
Wong, S.L., Wainwright, J.F. and Pimenta, J. (1995) Quantification of total and metal toxicity in wastewater using algal bioassays, *Aquatic Toxicology* **31** (1), 57-75.
Yang, J.-L. and Chen, H.-C. (2003) Effects of gallium on common carp (*Cyprinus carpio*): acute test, serum biochemistry, and erythrocyte morphology, *Chemosphere* **53** (8), 877-882.
Ziehl, T.A. and Schmitt, A. (2000) Sediment quality assessment of flowing waters in South-West Germany using acute and chronic bioassays, *Aquatic Ecosystem Health and Management* **3** (3), 347-357.

Abbreviations

ASTM	American Society for Testing and Materials
AVS	Acid-Volatile Sulphide
CANMET	Canada Center for Mineral and Energy Technology

CISTI	Canada Institute for Scientific and Technical Information
CBR	Critical Body Residue
DDD	dichlorodiphenyldichloroethane
DDE	dichlorodiphenyldichloroethylene
DDT	dichlorodiphenyltrichloroethane
EC	Environment Canada
HAS	Hazard Assessment Schemes
IGETG	Inter-Governmental Ecotoxicological Testing Group
ISO	International Standard Organisation
Kow	octanol-water coefficient
NOEC	no observed effect concentration
NOEL	no observed effect level
PCBs	polychlorinated biphenyls
PGE	Propylene glycol ether
QSAR	Quantitative Structure-Activity Relationships
OECD	Organization for Economic Cooperation and Development
PAH	Polycyclic Aromatic Hydrocarbon
[S,S]-EDDS	trisodium[S,S]-ethylene diamine disuccinate
TBA	Test Battery Approach
TMS	Test Method Standardization
TT	Toxicity Testing
U.S. EPA	U.S. Environmental Protection Agency.

1. MICROTOX® ACUTE TOXICITY TEST

B. THOMAS JOHNSON
Environmental Microbiology
Columbia Environmental Research Center
U. S. Geological Survey
4200 New Haven Road
Columbia, Missouri 65201 USA
btjohnson@usgs.gov

1. Objective, development, and scope

The Microtox Acute Toxicity Test[1], usually identified as Microtox, has played a leading and pivotal role in developing minimalistic microscale toxicity testing. "Speed, simplicity, reproducibility, precision, sensitivity, standardization, cost effectiveness, and convenience" (Isenberg, 1993) were features sought and developed in Microtox. This test uses a specific clonal strain of bioluminescent bacteria prepared in a unique lyophilized vial format. This approach is rapid, simple, cost-effective, and sensitive with large sample throughput capabilities. Microtox is a screening tool and provides an alternate to traditional, complex, and more costly whole animal testing with invertebrates and fish; the manufacturer's suggested applications are listed in Table 1. Microtox uses very few elements[2]: the *Reagent* (a specific bacterial strain of *Vibrio fischeri*), the test sample in compatible carrier solution, the *Diluent* test solutions, a duo-function *Analyzer* that includes an incubator and luminometer, a personal computer, and a data capturing and analyzing *MicrotoxOmni* software package.

"*A simple rapid method for monitoring the toxicity of aquatic samples has been developed*"(Bulich, 1979); thus in 1979, in this short statement, the bacterial toxicity bioassay known as Microtox® ushered in a new far-reaching revolution in bioassays and a paradigm shift in test organisms and, most importantly, introduced a new

---

[1] Use of specific products by USGS and its laboratories does not constitute an endorsement. Columbia Environmental Research Center (CERC) uses Microtox materials and equipment sold by Strategic Diagnostics Inc. (SDI) in Newark, DE, to preserve the Microtox protocol. SDI provides comprehensive instructive guides, manuals and computer software to operate the Microtox test at their Web site (www.azurenv.com). The Microtox protocol described here is a standard USGS SOP.

[2] The USGS as well as others (Environment Canada, 1992) adopted Microtox terminology to reduce confusion. Specific Microtox products are printed in italics with the initial letter in upper case.

microscale biomonitoring tool in environmental toxicology. Over the last twenty-five years bacterial toxicity bioassays have emerged as important screening tools for toxicity assessments, for regulatory compliance, and for use in a battery of tests to rapidly monitor the health hazards and risks of chemicals that enter the nation's aquatic environment (Wells et al., 1998). This chapter describes Microtox, an ecotoxicological screening tool designed to detect aquatic toxicity, to detect changes in toxicity, and to predict expectations of other toxicity tests. The advantages, new and old applications, and limitations of Microtox are explored.

Table 1. Recommended applications for Microtox (SDI Web site, 2003).

- Wastewater treatment plant influent testing for protection of activated sludge.
- Wastewater treatment plant effluent testing for protection of receiving waters.
- Toxicity Reduction Evaluations (TREs) and Toxicity Identification Evaluations (TIEs).
- Surface water monitoring for identification of point source and non-point source pollution.
- Monitoring raw drinking water to detect contamination due to point source or non-point source pollution.
- Bioterrorism.
- Sediment and soil testing.
- Monitoring of remediation processes.
- Biocide monitoring of industrial processed waters.

Water by its very nature is a universal solvent, a natural repository, and a carrier of both biogenic and xenogenic chemicals. The magnitude of this problem is expressed in part in the U. S. Chemical Industry's *Statistical Handbook* (1998) that states the industry annually produces 70,000 chemical products in 12,000 plants. The broad ecological impact of these and other chemicals on the health and well being of aquatic communities presents a very complex problem of hazard and risk assessment for both ecotoxicologists and resource managers.

In the last century analytical chemists have made amazing strides in collecting, separating, and identifying waterborne chemicals at nano- and picogram concentrations (Manahan, 1989). However, ecotoxicologists have only begun to make similar strides in the detection and characterization of environmental toxicants (Wells et al. 1998; Ostrander, 1996; Rand et al., 1995). The unraveling of contaminants (chemicals "out of place") and toxicants (chemicals injurious to ecosystem health) centers on three basic questions: What is the toxicant (qualitative)? How toxic is it (quantitative)? And how does the toxicant move (bioavailability)?

The historical literature is most helpful (Gallo, 1995). Paracelsus[3] told us that all things are toxic and the dosage makes the "toxicant". In this context, a toxicant must be defined both qualitatively (identified) and quantitatively (how much); therefore toxicity is clearly dose-responsive. Following this logic a chemical in the environment may be a contaminant at one concentration and a toxicant at another concentration; dosage makes the difference. The bioassay or bioindicator test, predicated on the dose-response experimental design, has over the last fifty years become a critical element in defining the nature of environmental toxicants (Rand et al, 1995). Today, toxicological bioassays are based upon an experimental design of five elements: the sample, the biota, the duration, the endpoint, and the dose-response. The interaction of these five elements in Microtox (Fig. 1) is the thesis of this chapter.

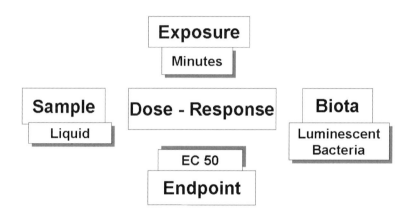

Figure 1. Experimental design: Microtox bioassay template.

2. Summary of test procedure (at a glance)

Microtox determines the acute toxicity of surface waters, ground waters, wastewaters, leachates, organic and aqueous sediment extracts, and passive sorptive device dialysates by measuring the changes of light produced naturally in samples exposed to bioluminescent bacteria under standard conditions. The Microtox

---

[3]Paracelsus (1493-1541) is often considered the father of modern toxicology. He brought empirical evidence into toxicology with his writings *"What is there that is not a poison? All things are poison and nothing without poison. Solely the dose determines that a thing is not a poison".*

Reagent bacteria, a selected strain of *V. fischeri* NRRL B-11177 (Fig. 2), are clonal cultures, which diminish possible genetic differences and ensure quality control of the tester strain and greater assay sensitivity and precision. The test bacteria are stored freeze-dried under vacuum in vials, which eliminates the tedium and cost of continuous culturing of a test organism. Most importantly, Microtox is available on demand because measurable light emission begins immediately after water activation of the lyophilized bacteria strain; bacteria require no preculturing. Aseptic technique is not required because of the short incubation period of the assay. All test media and glassware are pre-packaged, standardized, and disposable; the quantity is minimal, dramatically reducing both the material cost of the test and the disposal expense of toxic waste materials. The test requires minimal laboratory space and limited dedicated equipment: microliter pipetting devices, vortex mixer, incubator, and luminometer with computer assistance, and freezer storage. The test is well defined, computer assisted, and user friendly. Microtox is a unique bacterial bioluminescent inhibition assay.

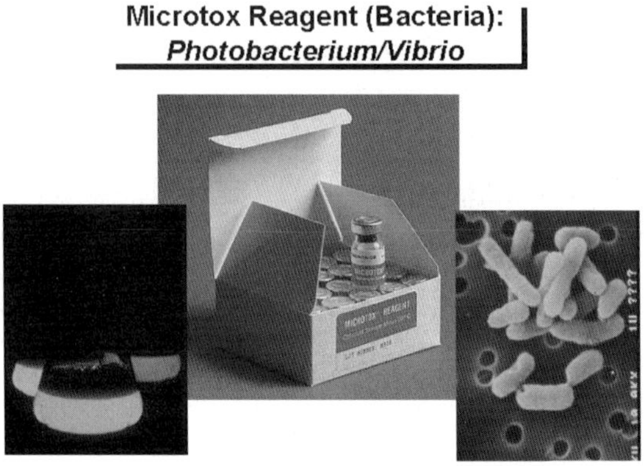

Microtox Reagent (Bacteria):
Photobacterium/Vibrio

Figure 2. Microtox Reagent.

Microtox is microscale; all tests are conducted in microvolumes with microcuvettes. A single reaction cuvette contains *Reagent* bacteria, *Diluent*, and test chemical. Aqueous and organic samples are prepared in the basic dose-response design: 1 control and 4 concentrations in a 1:2 dilution series. Carrier solvents such as DMSO, acetone, and ethanol may be necessary to solubilize certain chemicals; osmotic correction with NaCl may also be necessary with freshwater samples. Freshly prepared glowing luminescent bacteria in stationary growth phase are added to the test sample and placed in a SDI *Model 500 Analyzer* (Fig. 3); readings are taken typically after either five or 15 minutes incubation. The endpoints of all tests are based on light emissions produced by bioluminescent bacteria. The amount of light remaining in the sample is used to determine the sample's relative toxicity, which can then be compared to the standard reference's toxicity. As the toxicant's concentration increases, bacterial light emissions decrease in a dose-dependent manner. Some samples may require an extended range protocol (eight to 10 dilutions with two controls). The luminometer and supporting computer software (*MicrotoxOmni®* software) with a standard log-linear model are used to determine a 50 percent loss of light in the test bacteria, *i.e.*, the effective concentration (EC50) value. All EC50 values are expressed as weight or percent per mL with 95% confidence intervals and reported as the mean of three pseudoreplicates or true replicates; *replicates are a statistical measurement of the test's precision. The lower the EC50 value the greater the toxicity of the sample.* Manufacturer's suggested positive controls are phenol (organic) and zinc sulfate (inorganic). Typically tests are completed and data are available in < 30 min. This rapid response time meets the toxicologist's needs to conduct routine toxicity assays as well as to respond to emergencies such as wastewater effluents, chemical spills, and detection of unstable or transitory toxicants. Microtox protocol and rapid toxicological determination (< 30 min) make throughput capability of large samplings feasible both in the laboratory and in the field.

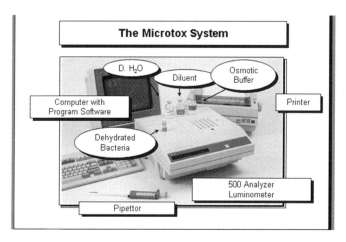

Figure 3. Microtox equipment and supplies.

Box 1. Required materials for testing.

The supplies required to implement the Microtox test are purchased from SDI.
Glassware
Each standard dose-response test (1 Control: 4 concentrations) requires ten disposable borosilicate glass cuvettes: two sizes 12 x 50 mm and 12 x 75 mm. To prevent spills and to make solution mixing easier larger 12 x 75 mm cuvettes can be substituted for the 12 x 50 mm.
Test organism
Microtox *Reagent*, the clonal bacterial isolate *V. fischeri* NRRL B-11177, is lyophilized and packed in 10 mL sealed vials. Each vial will test about 20 samples. Vials are shipped frozen and stored in the freezer compartment of a common refrigerator. Shelf life is 24 months.

Box 1 (continued). Required materials for testing.

Solutions

Microtox *Reconstitution Solution (Recon)* activates the *Reagent* for testing. The *Recon* is tightly sealed in the original container. Shelf life is 12 months.

Microtox Diluent is a 2% NaCl solution used to make dose-response dilutions. The sterile Diluent in sealed bottles is shipped and stored at room temperature. Shelf life is 12 months.

Microtox Osmotic Adjustment Solution is a 22% NaCl solution used to change the salinity of freshwater samples to the 2% required salinity of the assay. The solution is shipped and stored in tightly sealed bottles at room temperature. Shelf life is 12 months.

Box 2. Required equipment for testing.

Toxicity *Analyzer*

The SDI *Model 500 Analyzer* is a dual-purpose instrument serving both as an incubator and luminometer. The incubator is maintained at two temperatures: the thirty cuvette wells for test sample incubation at 15°C and a separate *Reagent Well* for storing one stock culture cuvette of luminous bacteria at 5°C. The luminometer contains a photomultiplier tube that measures the light emissions from bioluminescent bacteria. The *Analyzer* is interfaced with a PC containing the *MicrotoxOmni* software package for collecting, analyzing, and storing test data.

Pipettors

Rapid accurate and precise pipetting is essential for successfully dispensing multiple test solutions. Ergonomic pipettors are desirable because of the highly repetitive action of pipetting necessary for the dose-response experimental design.

Microtox uses several sizes of pipettors: two P-1000 Gilson Pipetman®, variable volume 100 µL – 1000 µL (or comparable); one P-100 Gilson Pipetman®, variable volume 10 µL – 100 µL; one EP-10 EDP-Plus® electronic pipettor, variable volume 1 mL – 10 mL (or comparable); and one EP-100 EDP-Plus® electronic pipettor, variable volume 10 µL – 100 µL.

Freezer

A freezer is essential for storage of the Microtox *Reagent* bacteria at –20°C. Self-defrosting units should be avoided.

Vortex mixer

A standard vortex mixer is used to stir liquids in test cuvettes. The mixer eliminates tedious mixing with pipettes and reduces ergonomic problems with repetitive hand movement. All vortex mixing should be brief measured in a few seconds. Prolonged mixing of test material and bacteria may affect the assay and should be avoided.

Computer

A standard personal computer (PC) with a *MicrotoxOmni* software program interfaced with the *Analyzer* is an essential element in Microtox testing; each step in the test protocol is recognized, controlled, analyzed, and recorded.

Printer

A printer interfaced with the PC makes data more accessible.

Box 3. Laboratory facilities.

> Laboratory facilities for Microtox are consistent with modern Good Laboratory Practices (GLP) protocols. An organized, clean laboratory with limited traffic flow, good lighting and airflow, controlled heating-cooling, electrical outlets, and designated bench space will meet most needs. Because Microtox is a microscale test, laboratory space requirements are comparably small, instrumentation is limited and compact, glassware is microscale, and test solutions are microvolumes.
>
> For health and safety purposes the laboratory must be considered a hazardous zone because the nature of the test substance(s) is usually an unknown and potentially toxic. The user should wear safety glasses, protective outerwear, and disposable gloves. To reduce cross contamination the use of disposable table coverings is recommended. A hooded bench area is useful, but certainly not necessary for all environmental sampling. Closed containers for spent test materials (both liquids and glassware) should be carefully labelled, stored, and monitored for GLP disposal. Although *V. fischeri* are saprophytic bacteria and not known as human pathogens, some laboratories destroy used culture material by heat or a disinfectant (APHA et al., 1998).

3. Overview of development and application of the Microtox toxicity test

An overview of the development and applications of Microtox reveals an intriguing tale of meeting an environmental challenge, of intellectual acuity, of entrepreneurism, and some good luck. In the early 1970s Beckman Instrument Co. (Carlsbad, CA) was asked by the petroleum industry in California to develop an acute toxicity assay, a substitute for the traditional fish and invertebrate tests, to monitor potentially toxic effluents from drilling operations. In formulating the task Isenberg in *The Microtox Toxicity Test: A Developer's Commentary* (1993) states the framework of the Microtox paradigm: "metaphorically …we needed to miniaturize fish, to teach them to talk, to report on their health, and to devise a way for them to be stored in suspended animation" in order to provide on demand availability and convenience. A toxicity bioassay needed "something alive" with "diverse, interdependent enzyme systems controlling a measurable physiological parameter" and an "appropriate measurement system". This toxicity test should be "fast, simple, reproducible, precise, … standardized, cost effective, convenient, and sensitive". The ambitious template for Microtox had been formulated. The question was could it be done?

A bit of serendipity or simply luck occurred when Beckman purchased the North American Rockwell collection of over 200 strains of luminescent bacteria. If luminescent bacteria could function as airborne biosensors of chemical warfare agents, scientists at Beckman (Isenberg, 1993) working on the Microtox Project wondered if these same bacteria could be used in an aquatic matrix. The attraction to luminous bacteria was tantalizing: rapid response time to a toxin and light emission from millions of cells that could be measured and reproduced with high precision. An "enzyme system controlling a measurable physiological parameter" had been found! The task was to find a strain of luminous bacteria with a sensitivity spectrum similar to traditional aquatic test animals.

Isenberg (1993) in reflecting on the Microtox Project years later stated that the work of Johnson et al. (1974) provided "an elaborate and compelling derivation of a general equation for the expression of (acute) toxicity" and formed the mechanistic model for the Microtox acute toxicity bioassay. Johnson et al. (1974) had published seminal work on a reaction rate theory that was based on isolated specific chemical processes and their relationship to complex biological reactions: significantly, the authors had used luminous bacteria to test their theories. Inventively, they expressed this physiological effect as a ratio of the activity lost to the activity remaining and termed this ratio gamma (Γ). Gamma proved to be a precise method when measuring light emissions from luminous bacteria. Gamma calculations permitted Microtox protocol to use simple regression statistics to compute toxicological endpoints: *i.e.*, EC50 values with confidence intervals.

Traditionally, bacteria are stored on agar-slants, frozen in liquid nitrogen or freeze-dried (lyophilized). For the Microtox scientists the obvious method of choice was the lyophilization process because bacteria freeze-dried under vacuum would remain viable and clonal and could be held for long periods of time with minimal care. However, the poor survival rate of bacteria following lyophilization, usually $< 1\%$, was a serious problem. Essentially this meant that luminous bacteria from a freshly opened vial could not emit sufficient light for a bioassay. If bacteria had to be precultured to increase numbers, the "on demand" quality of a microscale bioassay was sacrificed and the clonal integrity of the bacteria would be questionable. This problem was solved when Beckman developed a proprietary technique for the lyophilization of luminous bacteria. This process improved the survival rate of bacteria with cells emitting high luminescence at the moment of reactivation with distilled water. Acceptable concentrations of physiologically active, light-producing bacteria were now available as a biosensor. The Microtox project now had a simple method of storing and shipping a clonal strain of bacteria to scientists around the world. These bacteria would survive, remain clonal, be sensitive, and be available for immediate use (*i.e.*, within minutes of demand). This achievement was pivotal in the development of a successful bioassay.

The next task that faced the Microtox's developers was integrating a device that controlled temperature with an instrument for photochemical measurements into a single laboratory unit. Beckman successfully produced an instrument with an incubator that could hold the test bacteria at optimum temperatures and a photometer to read luminous light emission of bacteria. Beckman's Director of Research, Richard Nesbitt, commented that the Microtox Project was the most complex problem the company had ever undertaken – "not just designing an instrument, but finding the right bugs (bacteria), growing them, preserving them in containers that would not poison them, and arranging to ship them thousands of kilometers, while they retained a product shelf-life of at least one year" (Isenberg, 1993). In 1979, Beckman introduced Microtox in the United States, Canada, and Europe. In 1985 the developers of Microtox formed Microbics Corporation in Carlsbad, CA. In the 1990s the corporation name was changed to AZUR Environmental. In 2000, Strategic Diagnostics Incorporated (SDI) in Newark, DE, purchased AZUR Environmental. SDI now sells all Microtox products. In the last ten years the frequency and volume of publications has nearly doubled, a good indicator of the growing global utilization and acceptance of the Microtox paradigm (Tab. 2).

Table 2. Interest in microscale toxicity testing applications[a] and time-related publications of Microtox[b].

1979-83[a]	1984-89[a]	1990-95[a]	1995-2003[b]	Total publications
29	201	279	891	1400

[a]Redrawn from a table by Wells et al. (1998); data included Mutatox®.
[b]Publications of Microtox Basic only; data derived from multiple Web sites.

4. Advantages of conducting the Microtox toxicity test

Toxicological risk assessments are a growing concern for aquatic resource managers. Increasingly they must address and answer these basic water resource issues: What is toxic? How toxic is it? Where is the toxin? Is it bioavailable? While many good, reliable toxicity bioassays are available to answer these pressing questions, Microtox is a leading choice for a number of reasons. First, and foremost, the protocol is completely standardized and the materials are globally available: 1) the bioluminescent bacteria are cloned, stored in a lyophilized state, and available on demand for immediate testing; no preculturing of test biota is needed; 2) all glassware and test solutions are prepacked and test ready; no premixing is necessary; 3) the *Analyzer* with programmed luminometer and incubator is wired for computer assistance; 4) the computer software package *MicrotoxOmni* directs, computes, stores, and displays data; 5) toxicological results are available in minutes, thus permitting rapid response time to address spills and urban stream monitoring in order to determine hot spots for focusing resources; and 6) technical and material support from the manufacturer is excellent and timely. Furthermore, this test reduces the costs of materials and disposables and minimizes dedicated laboratory space. Short exposure times and microscale supplies provide Microtox with large sampling throughput capabilities not generally possible with animal or other microscale toxicity tests. *Statistical power is predicated on numbers – numbers in terms of sampling sites, numbers in the frequencies of samplings at given site, and numbers of replicates produced for each sample.* Significantly, this sampling protocol and, as a result, the early recognition of areas of concern are attractive features that make Microtox a good environmental monitoring tool.

5. Test species

Marine luminous bacteria are a cosmopolitan group that occurs in planktonic, enteric, saprophytic, parasitic, and symbiotic (in light organs in some marine fish and invertebrates) forms. Using phenotypic and genotypic analyses contemporary bacterial taxonomists Bauman et al. (1983) grouped luminous bacteria into two genera: *Photobacterium* and *Vibrio*. The main components for bacterial bioluminescence have been identified as reduced flavin mononucleotide (FMN), a long chain aldehyde, oxygen, and the enzyme luciferase (McElroy, 1961). These findings suggest that luminous bacteria contain luciferase that catalyzes the oxidation of $FMNH_2$ and aldehyde by oxygen. Significantly, the bacterial luciferase system appears to be coupled to cellular respiration via NADH and FMN. Treatises by Harvey (1952), McElroy (1961), and DeLuca and McElroy (1981) on bacterial bioluminescence offer comprehensive reviews of their findings and the biology of luminous bacteria.

Microtox is a prokaryotic microscale toxicity bioassay with luminescent, gram negative, saprophytic marine bacteria. These bacteria are ubiquitous in marine waters and are easily isolated and cultured from fish and seawater. Early studies (Bulich, 1979) suggested that specific isolates of *Vibrio* (originally taxonomically designated as *Photobacterium phosphoreum*) showed toxicological sensitivity to a broad spectrum of environmental contaminants. Additional investigations using these isolates under carefully standardized conditions revealed that an "on demand" toxicity test could be developed to measure a specific physiological parameter - bioluminescence - in real time. The prokaryotic cells used in Microtox are obtained exclusively from a cloned strain of a marine bacterium, *V. fischeri* NRRL B-11177, isolated, cultured and maintained by the manufacturer (currently SDI). This clonal strain is deposited by SDI at the Northern Regional Research Laboratory, Agricultural Research Service, U.S. Department of Agriculture, Peoria, IL, USA.

6. Culture/maintenance of organism in the laboratory

6.1 PREPARATION OF REAGENTS AND CULTURE MEDIA

The Microtox *Reagent* requires no culturing. No specialized microbiological equipment is necessary. The Microtox *Reagent* bacteria, *V. fischeri* NRRL B-11177, are cultured, freeze-dried under vacuum (lyophilized), sealed in 10 mL vials, shipped in 10 vial lots by SDI, and stored frozen at $-20°C$ to ensure high-level light emissions. Self-defrosting freezers must be avoided. During power outages place the vials in an insulated box containing artificial ice and store in the freezer compartment. Bacterial *Reagent* in this container will remain frozen for several days. For quality assurance and quality control (QA/QC) each vial is dated with the manufacturer's suggested shelf life. The Microtox is an "on demand" acute toxicity bioassay. Biota are available immediately for use whether in the laboratory or in the field. Neither preculturing nor preincubation of cells is necessary.

6.2 WASHING OF GLASSWARE

Protocol for the Microtox assay requires that all cuvettes and pipette tips are disposables and never reused. Beakers for dispensing the *Diluent* are acid-washed and air-dried each day and used only for the *Diluent*. Stock bottles for control chemicals are acid washed and steam sterilized before use; all bottles are stoppered with teflon® liners.

7. Information regarding test samples prior to conducting bioassays

7.1 KNOWN SUBSTANCES

Manufacturer's Material Safety Data Sheets (MMSDS), the Merck Index, and reliable Web sites (Tab. 3) provide valuable information about the compound of interest: its chemical class identification, solvent solubility, hazard identification, stability, primary use(s), disposition, and possible toxicity to vertebrates.

Table 3. Web site generated database sources for Microtox.

Applied Science & Technology Index	General Science
Agricola	Geobase
Aqualine	Georef
Aquatic Sciences & Fisheries Abstracts	Medline
Basic Biosis	OCLC Article First
Biology Sciences	OCLC ECO
Biology Digest	SDI
Chemical Abstract Service	Toxline
Conference Papers Index Abstracts	Water Resources
Environmental Sciences & Pollution Management	

7.2 UNKNOWN SUBSTANCES

All environmental samples are collected in clean containers and held on ice. Prompt testing is most desirable and less likely to introduce experimental errors from microbial activity. If testing is delayed sediment samples for pore-water analyses, organic extractions, and passive membrane dialysates can be stored on ice or refrigerated (3°C). Lipophilic test samples need to be dissolved in a solvent that will solubilize the material in the *Diluent* and also be compatible with the Microtox *Reagent*. Environmental samples are not collected in a complete vacuum of information; the geographical location (urban versus rural), source, season, etc, will provide the user important clues as to probable contaminants in the sample.

7.3 REFERENCE TOXICANT

Reference toxicants are essential elements in a good QA/QC program. The user monitors the relative sensitivity of the Microtox *Reagent* bacteria using reference toxicants under standard conditions in order to note the viability of the activated *Reagent* and to assess pipetting precision. Compound purity, stability, wide availability, aqueous solubility, dose-response profile, and low user hazard are essential components in selecting a good reference toxicant. SDI recommends phenol as an organic reference toxicant and zinc sulfate ($ZnSO_4$) as an inorganic reference toxicant. The 5-min EC50 values for phenol are typically in the 10-30 mg/L range while the 15-min values for $ZnSO_4$ are between 1.5 and 3.0 mg/L (Fig. 4). For years the Microbics sales' force has used Listerine®, a commercial, globally available product, as a reference toxicant to avoid the problem of carrying chemicals aboard airplanes.

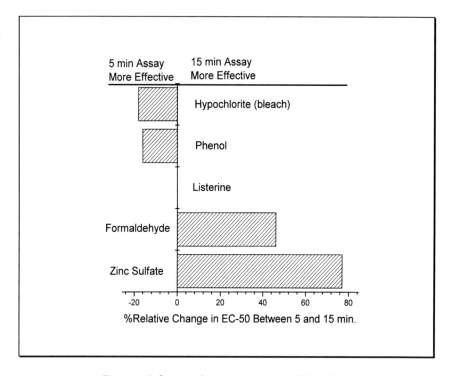

Figure 4. Influence of exposure times on EC50 values.

7.4 PREPARATION OF SAMPLE(S) FOR A TEST RUN

The sample to be tested must be in a liquid form and in an osmotically compatible solution, which may be water or a selected organic solvent. Lipophilic contaminants must be solublized in organic solvents. The compatibility of these solvents with Microtox should be investigated before extensive testing with unknown or pure compounds. Table 4 shows a list of common laboratory solvents and their compatibility with Microtox. Note that acetone, ethanol, and DMSO seemed the most compatible. At CERC we use a high purity grade of DMSO as our universal solvent for lipophilic chemicals; its very low toxicity, solubility range, low vapor pressure, and low freezing point makes it an attractive carrier solvent for Microtox.

Table 4. Influence of carrier solvents on Listerine® toxicity (EC50)[a].

Carrier solvent	EC50[b]	CI[c]
Control (Listerine®)	2.8	2.3 – 3.5
+ Dimethyl sulfoxide (DMSO)	2.5	2.2 – 2.9
+ Dichloromethane (DCM)	0.8	0.2 – 3.0
+ Hexane	0.1	0.1 – 7.1
+ Acetone	1.5	0.9 – 2.3
+ Methanol	3.6	2.6 – 4.7
+ 95% Ethanol	3.1	2.4 – 4.1
+ Isooctane	2.2	1.5 – 3.2

[a]The positive control Listerine® was exposed for 5 min with seven different solvents at concentrations not exceeding 5% of the total volume. Listerine® is a commercially available mouthwash with bactericidal properties. Range finding and definitive test for compound validation consisted of one control and four toxicant concentrations in a 1:2 dilution series.
[b]EC50 = μg/mL; [c] CI = confidence interval.

8. Equipment

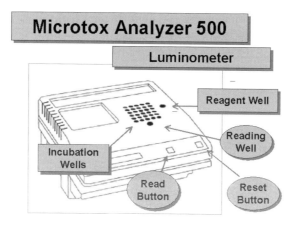

Figure 5. SDI Model 500 Analyzer.

The SDI *Model 500 Analyzer* (Fig. 5) integrates an incubator with a luminometer. On top of the instrument are 30 temperature-controlled incubation wells (15°C) identified as Rows A through F and Columns 1 through 5 and one temperature-controlled *Reagent Well* (5°C). The experimental design (Fig. 6) for the standard configuration of Microtox is 1 control (A1): 4 test concentrations (A2 through A5) with a 1:2 dilution factor. A 1:14 design is the maximum that can be analyzed at one time (one control: A1 and 14 concentrations: A2 through A5, C1 through C5, and E1 through E5). The luminometer measures the light emission remaining after the reagent has been exposed to the test sample. Following PC screen prompting each cuvette is removed from its incubation well and placed in the *Read Well*. The cuvette is depressed in the well, the luminometer reads the light emission from the bacteria, and the *MicrotoxOmni* software computes and records the data.

9. Microtox acute toxicity test: performing the test

First the SDI *Model 500 Analyzer* is turned on for 5 min to allow the incubator to achieve optimal temperatures: 15°C for the incubation wells and 5°C for the *Reagent Well* (Fig. 5).

Second the *Reagent* vial is removed from the freezer and opened. When the seal is broken, the dry culture material will produce a snowflake swirl effect indicating a vacuum was present. Simply adding 1 mL of *Reconstituted Solution* (*Recon*) to the freshly opened vial activates the *Reagent* bacteria; the contents of the *Reagent* vial are immediately transferred to a 12 x 75 mm cuvette and vigorously stirred on a vortex mixer (Fig. 7). The vial is placed in the *Reagent Well* and held for about 5 minutes to stabilize the culture's emission of light prior to testing. The activated *Reagent* normally remains "usable" for about 2 to 4 hrs. The *V. fischeri* are physiologically active and ready for testing. Aliquots of the *Reagent* are removed by micropipettor as needed for each toxicity assay. At CERC the half-life of the freshly activated culture is about two to four hours.

For a standard test with 1 control: 4 concentrations insert five cuvettes (12 x 50 mm) in Row A, five cuvettes (12 x 50 mm) in Row B (Fig. 6), and one 12 x 75 mm cuvette in the *Reagent Well*.

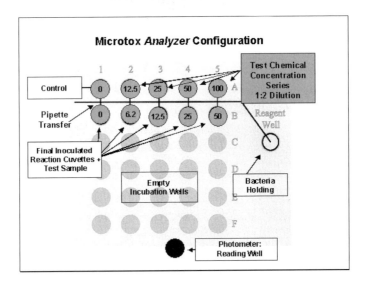

Figure 6. Microtox: dose-response design–1:4 (1 control: 4 test concentrations using 1:2 dilution factor).

Pipette 2.0 mL of *Diluent* into cuvette A5, 1 mL into each of the remaining four cuvettes in row A and 0.5 mL of *Diluent* into each of the 5 cuvettes in Row B.

Pipette the test sample into cuvette A5 and briefly use the vortex mixer to homogenate in the *Diluent*. Using a 1 mL-Pipetman transfer 1 mL from cuvette A5 into cuvette A4 and mix. Similarly transfer 1 mL from A4 to A3 and mix. Next transfer 1 mL from A3 to A2. After mixing discard 1 mL from A2 to bring its final volume to 1 mL. Cuvette A1 remains as a control. This process prepares 4 concentrations of the test sample. Now remove the vial from the *Reagent* well, mix for a few seconds, and load a 100 µL Pipetman with the bacteria from the vial. Dispense 10 µL of this bacterial inoculum into each cuvette in Row B. Place the tip of the pipette inside the cuvette just below the lip of the cuvette. Attempt to direct the inoculum into the *Diluent*, but do not submerge the tip in the *Diluent*. Briefly mix each cuvette to disperse the bacteria.

Figure 7. Vortex mixer.

Boot up the interfaced PC-*Analyzer*, activate the *MicrotoxOmni* program, and select a specific test protocol. Name the sample file and select desired test parameters as prompted: number of controls, number of dilutions, test duplication, initial concentration, units (% or weight per volume), osmotic adjustment, report form, and incubation time. Prompt the PC for desired exposure times - generally 5 or 15 min.

Now use the *Analyzer* to establish a base line reading of light emissions. Following the program's prompting, remove cuvette B5 and place it in the *Reading Well*. Press the *Read Button* and the luminometer will record light emissions. Continuing to follow PC prompting read zero time light levels of all the cuvettes in Row B. Verify that the light levels are reasonable, usually in the 90-100 % range.

Next activate the incubation timer by pressing the PC's space bar and introduce the test sample to the *Reagent* bacteria by transferring 0.5 mL from cuvette A5 to cuvette B5. Similarly, transfer 0.5 mL from cuvette A4 to B4, 0.5 mL from A3 to

B3, 0.5 mL from A2 to B2, and 0.5 mL from A1 to B1. For example, if 10 μg of test material were introduced into cuvette A5, after transferring 1mL to cuvette A4, cuvette A5 would now have only 5 μg of the test material. The transfer of this 0.5 mL from A5 to B5 would yield a final concentration in cuvette B5 of 2.5 μg of test material. Now again press the spare bar to begin a corrected incubation time (note that the software program corrects for the pipetting time).

At the end of the incubation period following PC prompting, place cuvettes from Row B in the *Read Well* and push the *Read Button*. The luminometer will make final light measurements of each cuvette and the *MicrotoxOmni* software will record, compute, and store the data (Fig. 8). The control cuvette is used to correct samples for the time-dependent drift in light output.

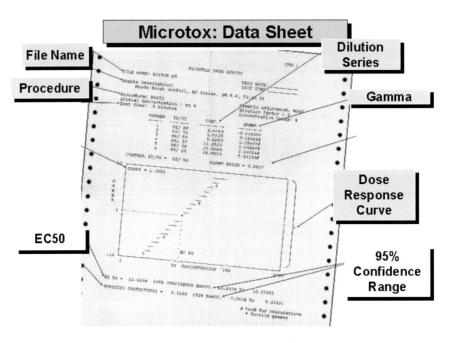

Figure 8. MicrotoxOmni *data sheet.*

The report should contain weekly EC50 data on recognized positive controls that are used as the laboratory's standard with predetermined coefficient of variance (CV) values (usually < 20%). The following questions concerning QA/QC criteria should be addressed: Was the protocol followed? Was the Microtox *Reagent* active with acceptable standard toxicant sensitivity limits for both positive and negative controls?

10. Test sample

10.1 CONCENTRATIONS

To determine the optimal test sample concentration for a definitive test, Microtox protocol suggests using a 1 control: 4 concentrations (1:4) design with a 1:2 dilution factor. The user should seek a concentration series in which the EC50 value is bracketed by at least one concentration on either side. The EC50 values are derived from a graph plotting the dose (the concentration of the test sample) against response (the effect on the test bacteria represented by gamma) on a log-log scale that requires at least three data points to plot a line. These EC50 values should have tight confidence intervals and replicate sampling tests should show coefficient of variance percentages below the manufacturers acceptable 20% CV (Fig. 9).

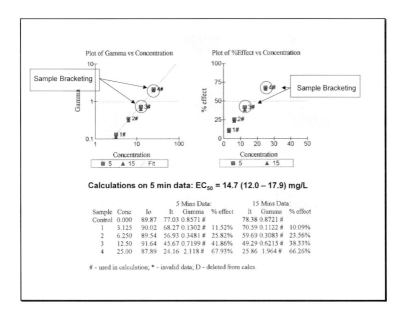

Figure 9. MicrotoxOmni *report on phenol: sample bracketing.*

If the trial range finding assay fails to generate an acceptable EC50 value with the 1:4 1:2 design, the user can probably solve the problem by simply re-testing the sample with the *Extended Range Protocol* (Microbics, 1992): dose concentrations are increased by one or two logs and the ratio of control: concentrations is changed to a 1:8 or even a 1:10 design with either a 1:2 or 1:10 dilution factor. With this expanded protocol a valid estimation of an EC50 value of even a very toxic substance can usually be determined. Figure 10 illustrates the use of the *Extended Range Protocol* to determine the EC50 value of 2,6 dinitrotoluene (2,6 DNT), a munitions by-product of environmental concern. The initial trial assay showed that 2,6 DNT was more toxic than expected. Therefore, the design was expanded to a 1:8 concentration series with a 1:2 dilution factor and, as expected, Microtox produced acceptable, valid results with an EC50 value of 2.5 ± 0.5 mg/L (where n = 5).

10.2 EXPOSURE CONDITIONS

The Microtox software provides three standard exposure options: 5, 15 or 30 min. Figure 4 illustrates that exposure time does significantly influence the EC50 values of phenol, chlorine beach, Listerine, formaldehyde, and zinc sulfate. After a 15 min exposure period the EC50 values for $ZnSO_4$ (a commonly used inorganic positive control) and formaldehyde increased about 80% and 40 % respectively. These data suggest changes in absorption and metabolism of the test material during incubation. However, after a 15 min exposure period, the EC50 values for phenol (a commonly used organic positive control) and Clorox® (a household bleach) decreased about 20%. Interestingly, after a 15 min exposure period, the EC50 value for Listerine did not change (Fig. 4). Obviously, the exposure times must be considered when testing with unknown environmental compounds. For most screening exercises the exposure time is set initially at only 5 min.

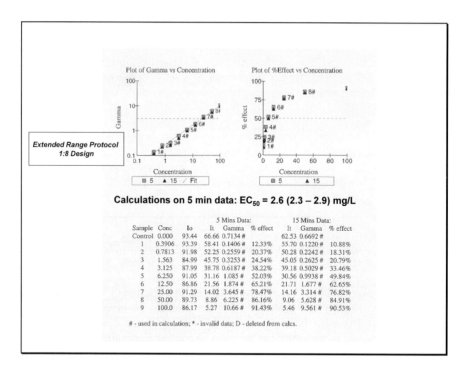

Figure 10. MicrotoxOmni *report on 2,6-dinitrotoluene: extended range protocol.*

11. Post-exposure measurements and endpoint determinations

Light emitted from a bioluminescent culture represents an integrated response of millions of cells. Light lost by bacteria indicates a rate of biological activity as well as an indirect enumeration of organisms affected. Interestingly, light emission by luminous bacteria is a physiological endpoint of respiration (McElroy, 1961), and therefore reflects rapid changes in metabolism due to toxic inhibition; hence, the use of these bacteria makes Microtox a rapid (5 min exposure) response bioassay. The light production of bacteria during actual testing tends to gradually (and slowly) decline over time because the bacteria are stored at 5°C in a buffer and do not grow. The *MicrotoxOmni* software package corrects for these losses. Placing control cuvette B1 in the *Reading Well* and pressing the *Set* and *Read* buttons monitors luminescence in the *Reagent*. If the control cuvette emission reads less than 90%, the *Reagent* has failed and needs to be replaced. The *Reagent* has about a 2 to 4 h window of acceptable physiological activity.

The model for computation of light emissions where toxic effects are expressed as the ratio of activity lost to activity remaining was developed and named gamma (Γ) by Johnson et al. (1974) and adopted by Microtox. Gamma is computed by the formula:

$$\Gamma = I_0/I_t - 1 \tag{1}$$

where: I_0 = light emission of the test bacteria that is lost, and I_t = the final emission produced after exposure time. The concentration of the test chemical that causes Γ to equal 1, that is when the light lost equals the light remaining, is used to compute the EC50 value for the assay. The log transformation in the Γ approach permits simple regression analyses to compute EC50 values and confidence intervals. Although a simple straightforward measurement of light emission lost due to toxicity is feasible in this assay, a precise linear relationship is obtained by plotting the log of Γ against the log of concentration. Microtox software incorporated this feature for test endpoint calculations. With a PC and a *MicrotoxOmni* software package data sets are readily collected, computed, and reported in a clear, succinct format (Fig. 8).

Both negative and positive controls are an integral part of the Microtox protocol and are essential in monitoring the natural changes in light emission by bacteria. Positive and negative controls should be performed at least once for each *Reagent* vial. All EC50 values are recorded and compared as part of QA/QC records. Coefficient of variation deviations of positive controls greater than 20% should be re-evaluated immediately for cause. A laboratory that maintains a CV less than 20% is operating within an acceptable range (Microbics, 1992). Positive controls in Microtox indicate an acceptable performance of (1) the *Reagent* and *Diluent* (2) the *Analyzer* and PC-software, and (3) the test operator skills.

The endpoint of Microtox is the effective concentration value corresponding to the concentration of toxin that produces 50% inhibition of light emission from a specific strain of bioluminescent bacteria. Because Microtox bacteria are essentially a collection of enzymes, the biochemical nature of the toxicological response whether due to a lethal or stasis reaction is unknown; hence, the term effective replaces lethal as the test endpoint designation. The final Microtox report provides an EC50 value and a 95% confidence range that indicates the quality of the data set. This endpoint is designated EC50 in the US and IC50 (Inhibitory concentration) in Europe and Canada.

12. Factors capable of influencing performance of Microtox testing results

As in all environmental toxicological tests, macroscale or microscale, a variety of confounding factors may interfere with an assay's normal functions and compromise its validity. When Microtox malfunctions, the most commonly occurring and expected problems tend to center around sampling, temperature, assay salinity and osmotic regulation, pH, color, turbidity, and organic carrier solvents. A pre-test cleanup of the sample with various chromatographic methods may advantageous. In addition, all organic carrier solvents - negative controls - should be assayed with Microtox before attempting to dissolve and test an environmental sample (see Fig. 9). Use only a high-grade sterile dimethyl sulfoxide (DMSO) that has been stored in tightly stopped dark bottles because this carrier solvent (DMSO) is easily compromised by air, light, etc., resulting in concomitant increases in acute toxicity. Monitoring the *Analyzer's* incubator temperatures can obviate temperature problems. Assay salinity problems are usually corrected with the use of the Microtox *Osmotic Adjustment Solution*. Aqueous samples should be checked to ensure that they are within the acceptable pH range of 6.0-8.5. Color, turbidity, and sampling problems are comprehensively addressed in the Microtox Handbook (Microbics, 1992).

13. Two different applications : toxicant potentiality and toxicant bioavailabilty

The first case study used Microtox as a screening tool to investigate the potential toxicological hazard of sediment contaminants in Pensacola Bay, an estuary that covers about 270 km^2 off the Gulf coast of Florida, USA. Samples for this extensive estuary investigation by USGS and the National Oceanic and Atmospheric Agency (Johnson and Long, 1998) were first concentrated by a standard organic sediment extraction procedure with dichloromethane (APHA et al., 1998), next evaporated, and then transferred to the compatible carrier solvent DMSO. Microtox analyses determined the EC50 values and, as a result, numerous sediment residues were identified as toxic (Tab. 5). While EC50 values determined what is toxic, a toxicity reference index was designed to identify how toxic the area was. Estuary regions were designated acutely toxic when the arbitrary toxicity reference index (TRI) numbers were greater than 1. For example, the Bayou Grande region had a TRI number of 14.1 indicating that the sediment was about 14-fold more toxic than the phenol-spiked reference sediment. (The EC50 value of the phenol-spiked reference sediment divided by the EC50 value of the test sample equals the toxicity reference index number: 5.2/0.37 =14.1). This Index identified areas of toxicological concern in the estuary. Microtox with extracted sediment samples and the TRI was an efficient economical screening tool for this study.

Table 5. Sediment[a] toxicity profile of Pensacola Bay in Florida (adapted from Johnson and Long, 1998).

Location	EC50[b]	TRI[c]
Bayou Grande	0.4	14
Bayou Chico	0.5	11
Bayou Texar	0.7	8
Warrington	7.3	0.7
Bayou Channel	4.7	1
Inner Harbor	2	3
Harbor Channel	10.5	0.5
Lower Bay	10.4	0.5
Central Bay	1.8	3
East Bay	1.1	5
East Bay Extension	2.5	2
Blackwater Bay	3.3	2
Escambia Bay	4.7	1
I-70	1.5	4
River Delta	6.7	0.8
Floridatown	3.4	1
Toxicity Reference	5.2	1

[a]Dichloromethane extracts transferred to DMSO carrier solvent.
[b]Microtox EC50 = mg eq. sediment wet weight per mL.
[c]Toxicity Reference Index (TRI) = EC50 value of a phenol-spiked sediment divided by the EC50 value of the sample.

The second case study used Microtox in the SPMD-TOX paradigm (Box 4) to determine the toxicological hazards of bioavailable contaminants in Lake Tahoe and its tributaries, a large freshwater lake that covers about 500 km^2 in northern California, USA. As part of the USGS's National Water Quality Assessment (NAWQA) program, SPMD-TOX (Johnson et al., 2002), a new tandem microscale monitoring procedure, was employed to determine the effects of diverse and intensive land-use on aquatic communities. The SPMD is a semipermeable membrane device (SPMD) used to collect and concentrate waterborne bioavailable lipophilic chemicals (Huckins et al., 1996) and TOX refers to toxicity tests such as Microtox (Johnson et al., 2000). To assess the lake's potential acute toxicity, SPMD units (Fig. 13) were placed in 15 tributary streams for 30 days. The sequestered samples were recovered and dialyzed with hexane. The dialysates were transferred to DMSO for Microtox analyses. Data strongly suggested that acutely toxic substances were bioavailable in three areas: Incline Creek, North Truckee Drain, and Steamboat Creek (Tab. 6). In these studies EC50 values below 2.5 indicated sample toxicity. This Lake Tahoe study showed that SPMD-TOX was a sensitive, technically simple, and cost-effective assessment tool to monitor urban waterways for bioavailable chemical contaminants.

Box 4. The SPMD-TOX paradigm.

The tendency of organisms to accumulate and concentrate lipophilic chemical contaminants from the aquatic environment is well known (Spacie and Hamelink, 1985). To mimic this bioconcentration process Huckins et al. (1996) designed and patented the semipermeable membrane device (SPMD) as a passive abiotic integrative sampler of waterborne non-polar organic compounds. The SPMD monitors contaminant bioavailability and provides an assessment of organism exposure. The device is a low-density polyethylene lay-flat tube that contains a neutral lipid triolein to passively sample *in situ* bioavailable organic chemical contaminants from water and air (Fig. 11). The SPMD unit is typically mounted in a protective stainless steel container and shipped to and from the sampling site in a sealed metal container (Fig. 12). SPMD as an environmental contaminant-concentrating tool has many advantages: 1. SPMDs are abiotic which means they do not metabolize sequestered products but provide a true reflection of bioavailable contaminants in the environment; 2. They can "survive" in heavily polluted, toxic environments where living organisms may not survive; 3. They are not temperature specific; SPMDs can be used in both cold and warm water environments; 4. They are easily transported to sites of interest for sampling and to laboratories for processing; 5. Their retrieval and subsequent recovery of sequestered contaminants is simple; and, 6. Their use in large monitoring programs is cost-effective. The Microtox Assay with SPMDs as samplers was used in a risk assessment paradigm designated as SPMD-TOX (Fig. 13) by Johnson et al. (2000).

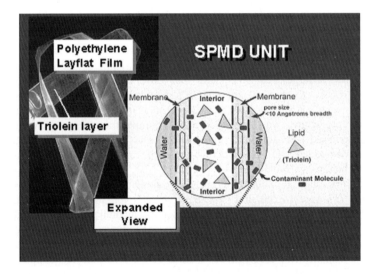

Figure 11. SPMD unit.

Table 6. *Profile of SPMD-TOX dialysates[a] from tributaries of Lake Tahoe, California. EC50 values below 2.5 are designated areas of concern.*

Sites	Locations	EC50[b]	SD
1	Glenbrook Creek at Glenbrook	8.7	2.4
2	Upper Truckee River	9.5	1
3	Taylor Creek	13.8	2.6
4	General Creek	14.1	1.6
5	Blackwood Creek at Hwy 89	15.9	1.3
6	Squaw Creek at Hwy 89	12.5	2.3
7	Incline Creek nr Crystal Bay	1	0.3
8	Truckee River blw Marble Bluff Dam	3.9	1.1
9	Truckee River at Wadsworth	3.9	0.6
10	Truckee River at Clark	4.8	0.6
11	Truckee River at Mogul	2.6	0.8
12	Truckee River nr Sparks	7.6	1
13	Truckee River at Lockwood	6.7	1.4
14	North Truckee Drain at Kleppe Ln	0.5	0.3
15	Steamboat Creek at Cleanwater Way	1.4	0.2
(Cs)	Control SPMD	>24	
(Cd)	Control DMSO	ND	
(Cb)	Control Blank	>20	
(Cp)	Control Phenol	15	2.1

[a]SPMD dialysates recovered in hexane and transferred to DMSO.
[b]Microtox EC50 = mg eq. SPMD/mL; n = 3, mean value ± SD.

These large field studies illustrate the use and the versatility of Microtox. Microtox, in both case studies, presented clear empirical evidence that identified pollutants in the sediment-water column. In the Pensacola Bay study, Microtox demonstrated the acute toxicity potentiality of the contaminant(s) and, in addition, the presence of these contaminants as sediment residue. In the Lake Tahoe study, Microtox again determined acute toxicity at the selected sites and, in addition, the bioavailability of these contaminants in the water column. Thus, this acute toxicity test provided additional information by the simple manipulation of a single element - the sample. Significantly, these simple modifications required minimal use of materials and financial resources. The organic extractions and SPMD dialysates of samples offered sensitive, technically simple, and cost-effective techniques to determine residual and bioavailable chemical contaminants. These two case studies demonstrate how to broaden the scope and breadth of information Microtox produces for environmental monitoring.

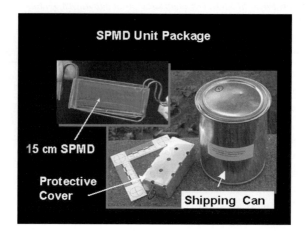

Figure 12. SPMD unit package.

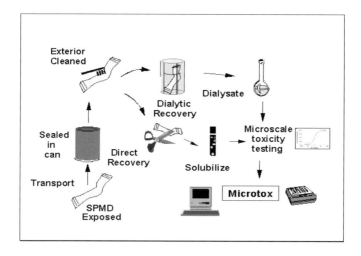

Figure 13. SPMD-TOX sample protocol.

14. Accessory/miscellaneous test information

14.1 LEVEL OF EXPERTISE IN MASTERING TECHNIQUE

Microtox is a user-friendly microscale bioassay to determine acute toxicity of aquatic samples. A modicum of intellectual curiosity, good hand-to-eye coordination, and the ability to read and follow precisely the Microtox protocols are good profiles for success. To guide the Microtox user through every conceivable aquatic test there are well-written manuals, videos, software packages (Microbics, 1992 and *MicrotoxOmni*® software), and a web site (www.azurenv.com). The evolution from novice is, in most instances, for first-time users remarkably rapid. Using live microorganisms, reading bioluminescent emissions, testing unknown toxicants, making precise pipetting measurements and concentration dilutions, and manipulating computer software programs is simple. Academic credentials and/or laboratory experience are naturally helpful but not necessary. To interpret the data in the broad environmental picture however requires additional training and much experience.

14.2 MEDIA FOUND SUITABLE FOR TESTING

Citations covering the twenty-five year life of Microtox usage reflect its direction, depth, evolution, and diversity (Tab. 7). A national and international literature review covering the last ten years reveals nearly 900 peer-reviewed Microtox citations; over 50% of these citations were industrial-domestic wastes and leachate studies. Other studies used Microtox for toxicological assessments of industrial effluents; urban and agricultural storm waters; industrial and agricultural leachates; passive sorptive extracts; domestic and industrial wastewaters; groundwater, river, lake, and marine sediments; drilling mud and fluids; snowmelts, pesticides, oil spills, landfill leachates; soil exudates; and industrial and domestic inorganic and organic chemicals. A comprehensive topic oriented review of Microtox applications is available from the databases listed in Table 4.

14.3 TESTS TO DETERMINE THE SENSITIVITY OF MICROTOX

To assess the sensitivity of bioluminescent bacteria and validate the uses of Microtox for broad acute toxicity monitoring requires testing of many diverse chemical classes under standard protocol conditions in order to collect a valid estimation of a toxicological endpoint(s) – the EC50 value. The relative sensitivity of Microtox, to known and potential environmental contaminants both as pure chemicals and in complex mixtures as determined at CERC over many years, is shown in Tables 8 and 9. Kaiser and Palabrica (1991) provide an extensive compilation of Microtox data for over 1000 compounds. When feasible, specific chemical (or class) sensitivity should be determined before beginning a monitoring study. Intuitively, we know, or at least suspect, that Microtox may work well in some matrices and not so well in others. In many instances a simple pre-concentration step may solve a sensitivity range problem. For test validation, positive controls offer an obvious reference point for what is toxic while negative controls, such as carrier solvents, test for responsiveness and false readings. The bottom line is does the assay's range of sensitivity for potential chemical contaminants adequately meet the interest(s) and needs of the client.

Table 7. Selected literature citations of Microtox applications.

Literature citations	Applications
Bulich, 1979	Genesis: first paper introducing Microtox
Bulich et al., 1981	Toxicity assessment of complete effluents
Curtis et al., 1982	Predicting acute toxicity to fish
Yates & Porter, 1982	Agricultural: detection of mycotoxins
Casseri et al., 1983	Toxicity assessment of industrial waste waters
DeZwart & Sloof, 1983	Toxicity assessment of water pollutants
Plotkin & Ram, 1984	Assessment of land leachates
Bitton & Dutka, 1986	Microtox application review
Ribo & Kaiser, 1987	Test procedures and applications
Blaise et al., 1988	Trends in Canadian Environmental Protection
Kaiser & Ribo, 1988	EC50 data compilations
Mazidji et al., 1990	Waste water studies
Blaise, 1991	Microbiotest review
Munkittrick et al., 1991	Comparative species study
Kaiser & Palabrica, 1991	EC50 data compilations
Microbics Corporation, 1992	Comprehensive test protocols released
Ross, 1993	Progress review of Microtox
Isenberg, 1993	Developers commentary
Richardson, 1993	Ecotoxicology monitoring: comprehensive review
Kaiser, 1993	Comparative species study
Bengtsson & Triet, 1994	Wastewater assessments
Gailli et al., 1994	Soil assessments
Gaggi et al., 1995	Battery of tests assessments
Ghosh et al., 1996	Phenol evaluation
Newman & McCloskey, 1996	Metal toxicity assessments
Qureshi et al., 1998	Current Microtox status from developers
Johnson, 1998	Microscale testing with Microtox test systems
Johnson & Long, 1998	Marine sediment toxicity assessments
Yim & Tam, 1999	Assessment of heavy metals effects on plants
Johnson et al., 2000	SPMD-TOX paradigm
Johnson et al., 2002	NAWQA urban river and stream risk assessments
Johnson et al., 2004	St. Lawrence River oil spill -SPMDs

Table 8. Microtox toxicological evaluation of polychlorinated biphenyls (PCBs), pesticides, and petroleum products (adapted from Johnson and Long, 1998).

Compound	EC50[a]	95% CI[a]
Insecticides: Organochlorine		
Aldrin	0.88	0.75 -1.05
Chlorodane (T)	1.3	1.1 - 1.5
DDE	0.97	0.8 - 1.2
DDT	1.25	1.04 - 1.52
Heptachlor	0.95	0.69 - 1.31
Hexachlorobenzene	0.63	0.54 - 0.73
Kepone	1.41	1.08 - 1.83
Lindane	1.56	1.22 - 1.99
Methoxychlor	0.86	0.78 - 0.94
Mirex	1.2	1.2 - 1.28
Pentachlorophenol	0.83	0.77 - 0.90
Toxaphene	4.9	2.6 - 9.5
Insecticides:Organophosphate		
Dyfonate	2.1	2.0 - 2.1
Malathion	0.85	0.64 - 1.1
Parathion	0.72	0.68 - 1.1
Insecticides: Carbamate		
Carbaryl	0.57	0.52 - 0.62
Carbofuran	0.91	0.74 - 1.11
Insecticides: Pyrethroid		
Permethrin	1.56	1.38 - 1.75
Herbicides: Triazine		
Atrazine	3.8	2.9 - 4.7
Simazine	4.4	3.3 - 5.8
Herbicides: Trifluralin		
Treflan	3.7	2.2 - 6.3
Herbicides: others		
Dacthal	1.3	1.0 - 1.6
Industrial: PCBs		
PCB 1248	0.55	0.51 - 0.59
PCB 1254	1.01	0.87 - 1.2
Industrial: others		
Dihexyl phthalate	82	42.2 - 159.4
Nonylphenol	0.44	0.29 - 0.65
Phenol	15.1	14.2 - 16.3

Table 8 (continued). Microtox toxicological evaluation of polychlorinated biphenyls (PCBs), pesticides, and petroleum products (adapted from Johnson and Long, 1998).

Compound	$EC50^a$	95% CI^a
Petroleum products		
Fuel oil #2	0.06	0.04 - 0.10
Jet fuel JP4	0.12	0.10 - 0.13
Recycled motor oil	1	0.82 - 1.2
Gasoline	0.16	0.12 - 0.21
Crude oil	0.4	0.25 - 0.64

[a] 5 min EC50 = µg/mL; CI = confidence interval; n= 3; DMSO carrier solvent.

Table 9. Microtox toxicological evaluation of complex mixtures containing polychlorinated biphenyls (PCBs), polyaromatic hydrocarbons (PAHs), and pesticides (adapted from Johnson, 1998).

Complex mixtures	$EC50^a$	CI^a
PCBs:1242+1248+1254+1260	0.9	0.85 - 0.95
DDT+DDE+DDD	1.5	1.3 - 1.7
Kepone+Aldrin+Lindane+DDT+PCB1254	1.6	1.4 - 1.7
Phenanthrene+Chrysene+ Anthracene+Benzo(a)pyrene	0.6	0.56 - 0.59
Aminoanthracene+Benzo(a)pyrene+Aminofluorene+ 3-methylcholine	3	2.1 - 4.4
Aminoanthracene+Benzo(a)pyrene+Aldrin+DDT	1.8	1.6 - 2.0
Aldrin+DDT+Heptachlor+Endrin	1.6	1.1 - 2.2
Atrazine+DDT+Aldrin+PCB1254+ Pyrene	1.7	1.4 - 2.1
DDT+Benzo(a)pyrene+PCB1254+1260+Atrazine	2.2	1.6 - 2.9
Carbofuran+Carbaryl+Atrazine+Treflan	1.7	1.4 - 2.1
Carbofuran+Carbaryl+Atrazine+ Permethrin	1.2	0.94 - 1.5
Carbofuran+DDT+Atrazine+ Permethrin	1.6	1.5 - 1.6

[a] 5 min EC50 = µg/mL; CI = 95% confidence interval; complex mixture = weight/weight; DMSO carrier solvent.

14.4 ALTERNATIVE CHOICES OF TEST SPECIES AND TEST METHODS

Toxicity testing of environmental samples may be undertaken with either macro or microscale assays. Whole animal testing with different fish and invertebrate species is usually possible if a sufficient test sample is available to support a traditional invertebrate and fish acute toxicity test. A number of investigators have compared the results obtained using Microtox with those obtained with different fish and invertebrate species. For example, when Munkittrick et al. (1991) reviewed hazard assessments of various chemical groups, sediments, and complex effluents comparing the relative sensitivity of Microtox with tests using daphnid, rainbow trout, and fathead minnow tests, they found sample size, cost, availability, and sensitivity make Microtox the best available choice for rapid toxicological assessment of diverse environmental samples. Qureshi et al. (1998), in a recent comprehensive review of fourteen independent studies, compared the relative sensitivity of Microtox with three commonly used freshwater test species: rainbow trout, fathead minnows, and daphnids; the correlation coefficient values (a value of 1.0 equals perfect correlation) of the data sets for trout, minnows, and daphnids bioassays ranged from 0.74 to 0.89, 0.41 to 1.0, and 0.8 to 0.87 respectively, with an average of 0.85, giving an indication of the degree of similarity in data sets. These studies suggested that the predictive value of Microtox as a prescreening tool was 85% when compared with trout, minnows and daphnids.

A battery of tests could be applied to assay for suspected aquatic contaminants. The premise of this approach (Cairns 1984; Cairns et al., 1997) is that one cannot rely on a single bioassay of a "most sensitive species" to detect all aquatic hazards; different biota have different biological systems, and therefore conceivably different toxicant sensitivities. Ideally the battery would cover several trophic levels and yield no redundant data. For example, Ross (1998) explored this approach with 10 reference compounds (both organic and inorganic) using a battery of four microscale toxicity bioassays: Microtox, a bacterial bioluminescent test; *Selenastrum capricornutum*, an algal photosynthesis test; *Latuca sativa*, a lettuce root elongation test; and *Brachionus calyciflorus*, a freshwater rotifer survival test. Their study found that the four bioassays of this battery were complementary and enhanced sensitivity as well as increased both labor and material costs (argumentatively, do multiple tests really give enough additional information to warrant the increased time and costs?). A recent CERC literature review covering the last twenty years found less than 45 peer-reviewed citations that used a battery of tests for extensive toxicological biomonitoring; this suggests the jury is still out on the wide spread use and acceptance of this approach.

Other microscale acute toxicity tests are available: TOXKITs® (invertebrate on demand assays, Belgium), MetPlate® (a metal-detection test, USA; see Chapter 6, Vol. 2 of this book), ToxAlert® (a bioluminescent bacterial assay, Germany), and ToxScreen®, (a bioluminescent bacterial assay, Israel). In addition, enzyme inhibition tests (Obst et al., 1997) and immunoassays (Dankwardt et al., 1997) can be used to detect aquatic chemical contaminants. When considering an alternative test to monitor environmental toxins, the lack of commercial availability of specific assays, the absence of well-developed standard protocols, the unknown spectrum of sensitivity, the cost-effectiveness, and the absence of supportive literature should forewarn the user of possible problems.

14.5 ARE THERE ALTERNATIVE CHOICES FOR ENDPOINT DETERMINATIONS?

SDI recently introduced *Deltatox®*, a portable luminometer, with greater sensitivity to light emissions from luminescent bacteria than the *Analyzer 500*; however, the

Deltatox data is raw without gamma correction and the system lacks PC software for computation and reporting.

14.6   MICROTOX AUTOMATION POTENTIAL

During the late 1990s in Europe, first with Compagnie Generale des Eaux and later with Siemens Environmental and Yorkshire Water, the *Microtox-OS* On-line System was developed, tested, and implemented. Toxicity samplings of drinking water sources, influents and effluents from water, and sewage treatment plants were made automatically at 15-min intervals. The *Microtox-OS* On-line System had technical problems, little commercial success, and did not remain on the market long. National events and security interests will undoubtedly be a strong catalytic force in developing automated systems to protect the Nation's domestic water resources.

14.7   TEST SAMPLE THROUGHPUT

Microtox can generate large numbers of data points in a day because set up, dilution, exposure, and data reports are completed in < 30 min. One person using the 1:4 Microtox protocol can routinely test about 18 SPMD dialysates (in DMSO carrier) with one *Reagent* vial in about a half day. An individual rarely performs Microtox for a full day due to the tedium of repetition with concomitant error problems. Data analysis requires additional time. The statistical power of toxicity data is based in part on numbers: numbers in terms of sampling sites, numbers in the frequency of samplings at given site, and numbers of replicates of each sample. Resource managers often need large sample numbers to make valid environmental decisions. An attractive feature of Microtox as an environmental biomonitoring tool, yet often overlooked, is the rapid and large test sample throughput.

14.8   RELATIVE COST OF TESTING

Is the Microtox assay "cost-effective", the term frequently used in the literature to describe Microtox? Numbers are necessary for environmental monitoring. Multiple samplings increase data precision, which in turn pinpoint troublesome areas that may need immediate attention. The ecotoxicologist using Microtox can perform more intensive samplings at specific sites and between sites than is possible with other animal or plant toxicity assays. Microtox can be used universally, even in developing countries; its standardized protocol, test sample throughput, its simple technique, its prepackaged supplies, and its reliable equipment provide the numbers needed for data analyses. If a Microtox user tested 18 samples a day for 100 days in a year, 18,000 samples would have been tested in ten years. Numbers make Microtox a cost-effective assay and a simple biomonitoring tool for water resources.

This accounting exercise examines the cost of an environmental sample using a typical 1:4 dose-response series with three replicates; this 1:4 sampling design requires 30 test observations. The Microtox user needs consumables: *i.e., Reagent, Diluen*t*, Recon*, cuvettes, and pipette tips and non-consumables: *i.e.,* a SDI *Analyzer*, pipettors and a vortex mixer. The literature frequently lists the cost of consumables for a Microtox sample generally in the $50 to $100 US range (Ross, 1993). The SDI

Model 500 Analyzer, high quality and durable pipettors, and a vortex mixer are long-term investments. Over the last ten years, my Environmental Microbiology Laboratory has used two *Analyzers,* pipettors, and one vortex mixer and found the quality of the equipment both reliable and durable; the *Analyzers* malfunctioned only twice, one electrical problem and one mechanical, the pipettors needed only minor inexpensive QA/QC care, and the vortex mixer required no attention. The price - $150 - for a Microtox sample analysis includes my costs for Microtox products, equipment, overhead expenses, and labor. The number, collection, volume, transportation, and storage of environmental samples prior to testing will vary with the resource manager's needs, problems, priority and economic resources and directly influence the final cost.

14.9 DEGREE OF ATTAINED TEST STANDARDIZATION

Standardization and validation of toxicological tests are always tedious and time-consuming exercises for both the sponsoring organization and the applicant. Final acceptance and recognition by the national and international scientific communities that Microtox was a valid, reliable assay for environmental risk assessment involved a complex matrix of evaluations: experimental design, sample handling and disposal, sensitivity spectrum determinations, positive-negative control selections, QA/QC incorporations, and interlaboratory round-robin testing. A key element for Microtox occurred in 1984 when the Organization for Economic Cooperation and Development (OECD) accepted Microtox as part of a combined bioassay-chemical paradigm to assess the biohazards of industrial chemical contaminants in aquatic ecosystems (OECD, 1984). L'Association Francaise de Normalisation (AFNOR), American Society for Testing and Materials (ASTM), Deutsches Institut Normung (DIN), International Standards Organization (ISO), and the Organization for Economic Cooperation and Development (OECD) have disseminated Microtox protocols and led, promoted, and contributed to multiple environmental uses of Microtox for toxicological assessments (Tab. 10). Over the last twenty-five years Microtox has drawn both national and international attention as a multifaceted toxicological monitoring tool because of its broad range of sensitivity to known environmental contaminants, its microscale protocol, simplicity, and cost-effectiveness per unit test, its successful use for screening and ranking environmental samples, its support of regulatory compliance, and its ability to predict the outcome of other environmental bioassays.

Table 10. Status of Microtox: regulations and standards (adapted from Qureshi et al., 1998).

Organizations	Applications	Status
Energy Resources Conservation Board, Canada	Drilling waste	Guide
Inter-government Aquatic Toxicity Group, Canada	Effluent	Final
International Standards Organization, France	Effluent	Process
L'Association Francaise de Normalisation, France	Effluent	Standard
Deutsches Institut fur Normung, Germany	Effluent	Standard
National Government Lab. & Research Institute, Italy	Effluent	Process
Netherlands Normalization Institute, The Netherlands	Effluent	Final
Environmental Protection Agency, Mexico	Wastewater	Standard
Environmental Protection Agency, Spain	Leachate	Standard
Environmental Protection Agency, Sweden	Effluent	Issued
Environment Agency, United Kingdom	Effluent	Process
American Society for Testing and Materials, USA	Wastewater	Issued
United States Public Health Service, USA	Wastewater	Issued

Table 11. Microtox® toxicity test system (adapted from Johnson, 1998).

Microtox	Basic	Solid-phase	Chronic	Mutatox
Toxicity test	Acute	Acute	Chronic	Genotoxic
Vibrio fisheri	NRRL B-11177	NRRL B-11177	NRRL B-11177	Dark M169
Sample type	Liquid[a]	Solid[b]	Liquid[a]	Liquid[a]
Test medium	Buffer	Buffer	Nutrients	Nutrients[c]
Growth phase	Stationary	Stationary	Log	Log
Design	Dose-response	Dose-response	Dose-response	Dose-response
Test duration	30 min	30 min	< 24 h	< 24 h
Test endpoint	< Light	< Light	< Light	> Light
Tox designation	EC 50[d]	EC 50[d]	LOEC[e]	Genotoxic[f]
Software	Yes	Yes	Developmental	Yes
Development	In common use	In common use	Introductory	Experimental
Sensitivity	Broad spectrum	Broad spectrum[g]	Experimental	Experimental
Data base	Broad	Expanding	Experimental	Experimental

[a] Wastewater, porewater, dialysates, compatible organic solvent extracts.
[b] Soil or sediment samples.
[c] Rat hepatic S9 fractions added for metabolic activation phase.
[d] EC50 = effective concentration with 50% loss of light.
[e] LOEC = Lowest Observable Effect Concentration.
[f] Genotoxic = two or more positive responses in a dilution series.
[g] Clay and turbidity questions.

14.10   ADDITIONAL USES AND ENDPOINT DETERMINATIONS

Microtox has expanded and the Microtox® Test System (Tab. 11) today includes four toxicity tests: Microtox® Acute Toxicity Test (described here), Microtox® Solid-Phase Toxicity Test (see Chapter 2 of this volume)), Microtox® Chronic Toxicity Test, and Mutatox® Genotoxicity Test. The four bioassays are all based upon the measurement of luminescent bacteria light emissions but differ in the strain selection (wild type versus dark mutant), the sample presentation (liquid versus solid), the growth cycle (stationary versus log), the duration (minutes versus hours), the changes in bioluminescent emissions (decrease versus increase), and the toxicological endpoints (lethality versus genotoxicity) (Johnson, 1998). The Microtox® Test System can be considered a battery of tests, all used as rapid screening assays, to detect the presence of toxic substances in the biosphere - water, soil, sediment, and air.

15. Conclusions

Microtox, a widely used biomonitoring tool for aquatic contaminants, is an acute toxicity test, a screening tool, and a stand-alone bioassay worthy of emulation. Microtox is an on demand test that is simple, rapid, and cost-effective with readily available biota, a sensitivity spectrum clearly defined, a standardized method and comprehensive tutorial protocol software. The Microtox assay is user friendly and easy to run, tabulate, and report data in a timely manner. Monitoring and screening tests such as Microtox are not surrogates; they cannot replace the more expensive bioassays that use native species of interest. However, this biomonitoring test can be viewed as a microscale biosensor expressly designed and used to detect a broad spectrum of environmental chemical contaminants. While the value of environmental relevance and the spectrum of sensitivity favor the macroscale test with native fish and invertebrate species of concern, the microscale test provides large sample capacity, speed, and cost-effectiveness.

   Microtox has been accepted as a toxicological biomonitoring tool with multiple applications as reflected in the nearly 1000 published peer-reviewed reports in the last 10 years. Isenberg (1993) and his founding colleagues did "metaphorically speaking" develop a bioassay in which the biota "could speak" and "could be placed in suspended animation" for "on demand" availability. The success of Microtox ushered in a new far-reaching revolution in microscale bioassays, produced a paradigm shift in test organisms, and, most importantly, introduced a new biomonitoring tool in environmental toxicology. The future uses and directions of Microtox and the microscale concept of toxicity monitoring are essentially a function of the user's creativity.

Acknowledgements

The author wishes to thank the many Microtox users who have contributed over the years to this microscale toxicity paradigm.

References

APHA, AWWA, and WEF (1998) *Standard Methods for the Examination of Water and Wastewater*, Part 5000 Aggregate Organic Constituents, L.D. Clesceri, A.E. Greenberg and A.D. Eaton (eds.), 20thed., Washington, DC.

Bauman, P., Bauman, L., Woolkalins, M.J. and Bank, S.S. (1983) Evolutionary relationships in *Vibrio* and *Photobacterium*: a basis for a natural classification, in *Annual Review of Microbiology*, L.N. Ornstan, A. Balows and P. Balows (eds.), Annual Reviews, Palo Alto, CA, **37**, 369-380.

Bengtsson, B.E. and Triet, T. (1994) Tapioca starch wastewater characterized by Microtox and duckweed tests, *Ambio* **23**, 471-477.

Bitton, G. and Dutka, B.J. (1986) Introduction and review of microbial and biochemical toxicity screening procedures, in *Toxicity Testing Using Microorganisms*, Vol. I., G. Bitton and B.J. Dutka (eds.), CRC Press, Boca Raton, FL, pp. 1-8.

Blaise, C. (1991) Microbiotests in aquatic ecotoxicology: characteristics, utility and prospects, *Environmental Toxicology and Water Quality* **6**, 145-155.

Blaise, C., Sergy, G., Wells, P., Bermingham, N. and Van Coillie, R. (1988) Biological testing – Development, application, and trends in Canadian environmental protection laboratories, *Toxicological Assessment* **3**, 385-406.

Bulich, A.A. (1979) Use of luminescent bacteria for determining toxicity in aquatic environments, in *Aquatic Toxicology: Second Conference*, L.L. Marking and R.A. Kimerle (eds.), ASTM STP 667, American Society for Testing and Materials, Philadelphia, PA, pp. 98-110.

Bulich, A.A., Greene, M.M. and Isenberg, D.L. (1981) Reliability of the bacterial luminescence assay for determination of the toxicity of pure compounds and complex effluents, in *Aquatic Toxicology and Hazard Assessment: Fourth Conference*, D.R. Branson and K.L. Dickson (eds.), ASTM STP 737, American Society for Testing and Materials, Philadelphia, PA, pp. 338-347.

Cairns, J. Jr. (1984) Multispecies toxicity testing, *Environmental Toxicology and Chemistry* **3**, 1-10.

Cairns, J. Jr., Niederlehner, B.R. and Smith, E.P. (1997) Correspondence of a microscale toxicity test to responses to toxicants in natural systems, Microtox Toxicity Test System – new developments and applications, in *Microscale Testing in Aquatic Toxicology*, P.G. Wells, K. Lee and C. Blaise, (eds.), CRC Press, Washington, D.C., pp 539-545.

Casseri, N.A., Ying, W.C. and Sojka S.A. (1983) Use of a rapid bioassay for the assessment of industrial wastewater treatment effectiveness, in *Proceedings of the 38th Purdue Industrial Wastewater Conference*, J.M. Bell (ed.), Butterworth Publishers, Woburn, MA, pp. 867-878.

Curtis, C.A., Lima, A., Lozano S.J. and Veith, G.D. (1982) Evaluation of a bacterial bioluminescent bioassay as a method for predicting acute toxicity of organic chemicals to fish, in *Aquatic Toxicology and Hazard Assessment: Fifth Conference*, J.G. Pearson, R.B. Foster and W.E. Bishop (eds.), ASTM STP 766, American Society for Testing and Materials, Philadelphia, PA, pp. 170-178.

Dankwardt, A., Pullen, S. and Hock, B. (1997) Immunoassays: applications for the aquatic environment in *Microscale Testing in Aquatic Toxicology*, P.G. Wells, K. Lee and C. Blaise, (eds.), CRC Press, Washington, D.C. , pp 13-29.

De Zwart, D. and Sloof, W. (1983) The Microtox as an alternative assay in the acute toxicity assessment of water pollutants, *Aquatic Toxicology* **4**, 129-138.

DeLuca, M.A. and McElroy, W.D. (1981) *Bioluminescence and chemiluminescence: basic chemistry and analytical applications,* M.A. DeLuca, and W.D. McElroy (eds.), Academic Press, New York, 782 pp.

Environment Canada (1992) Biological test method: toxicity test using luminescent bacteria (*Photobacterium phosphoreum*), Environment Canada, Ottawa, ON, *Report Environment Protection Series* 1/RM/24, 61 pp.

Gaggi, C.G, Sbrilli, A.M., Hasab El Naby, Bucci, M., Duccini, M. and Bacci, E. (1995) Toxicity and hazard ranking of triazine herbicides using Microtox®, two green algae species and a marine crustacean, *Environmental Toxicology and Chemistry* **14**, 203-208.

Gailli, R., Munz, C.D. and Scholtz, R. (1994) Evaluation and application of aquatic toxicity tests: use of the Microtox test for the prediction toxicity based upon concentrations of contaminants in soil, *Hydrobiologia* **273**, 179-182.

Gallo, M. A. (1995) *History and scope of toxicology*, Casarett and Doull's Toxicology: The Basic Science of Poisons, C.D. Klassen, M.A. Amdur and J. Doulli (eds.), McGraw-Hill, New York, pp 3-11.

Ghosh, S.K. Doctor, P.B. and Kulkarni, P.K. (1996) Toxicity of zinc in three microbial test systems, *Environmental Toxicology and Water Quality* **9**, 13-19.

Harvey, E.N. (1952) *Bioluminescence*, Academic Press, New York.
Huckins, J.N., Petty, J.D., Lebo, J.A., Orazio, C.E., Prest, H.F., Tillitt, D.E., Ellis, G.S., Johnson, B.T. and Manuweera, G.K. (1996) Semipermeable membrane devices (SPMDs) for the concentration and assessment of bioavailable organic contaminants in aquatic environments, in G.K. Ostrander (eds.), *Techniques in Aquatic Toxicology*, CRC Lewis Publishers, New York, pp. 625-655.
Isenberg, D.L. (1993) The Microtox® Toxicity Test - A Developer's Commentary, in M. Richardson (ed.), *Ecotoxicology Monitoring*, VCH Publishers, New York, pp. 3-15.
Johnson, B.T. (1998) Microtox® Toxicity Test System – new developments and applications, in *Microscale Testing in Aquatic Toxicology*, P.G. Wells, K. Lee and C. Blaise, (eds.), CRC Press, Washington, D.C., pp 201-218.
Johnson, B.T. and Long, E.R. (1998) Rapid toxicity assessment of sediments from large estuarine ecosystems: a new tandem *in vitro* testing approach, *Environmental Toxicology and Chemistry* **17**, 1099-1106.
Johnson, F.H., Eyring, H. and Stover, B.J. (1974) *The Theory of Rate Processes in Biology and Medicine*, John Wiley & Sons, N.Y., 703 pp.
Johnson, B.T., Petty, J.D. and Huckins, J.N. (2000) Collection and detection of lipophilic chemical contaminants in water, sediment, soil and air – SPMD-TOX, *Environmental Toxicology* **15**, 248-252.
Johnson, B.T., Petty, J.D. Huckins, J.N. Burton, C.A., and Giddings, E.M. (2002) Assessment of aquatic habitat quality in urban streams in the Western United States using SPMD-TOX, *23rd Annual Meeting Society for Environmental Toxicology and Chemistry* in Salt Lake City, UT, November 15-20, 2002 (Abstract).
Johnson, B.T., Petty, J.D., Huckins, J.N., Lee, K. and Gauthier, J. (2004) Hazard Assessment of a Simulated Oil Spill on Intertidal Areas of the St. Lawrence River with SPMD-TOX, *Environmental Toxicology* **19**, 329-335.
Kaiser, K.L.E. (1993) Qualitative and quantitative relationships of Microtox data with toxicity data for other aquatic species, in *Ecotoxicology Monitoring*, M. Richardson (ed.), VCH Publishers, New York, pp. 197-211.
Kaiser, K.L.E. and Ribo, J.M. (1988) *Photobacterium phosphoreum* toxicity bioassay, II. Toxicity data compilation, *Toxicological Assessment* **3**, 195-237.
Kaiser, K.L.E. and Palabrica, V.S. (1991) *Photobacterium phosphoreum* toxicity data index, *Water Pollution Research Journal Canada* **26**, 361-431.
Manahan, S.E. (1989) *Toxicological Chemistry*, Lewis Publishers Inc., Chelsea, MI, 317 pp.
Mazidji, C.N., Koopman, B., Bitton, G., Voiland, G. and Logue, C. (1990) Use of Microtox and *Ceriodaphnia* bioassays in wastewater fractionation, *Toxicological Assessment* **5**, 265-277.
McElroy, W.D. (1961) Bioluminescence, in *The Bacteria* Vol. 2, I.C. Gunsalus and R.Y. Stanier, (eds.), Academic Press, New York.
Microbics Corporation (1992) *Microtox Manuel*, Vol. I to V, Microbics Corporation, Carlsbad, CA, 476 pp.
Munkittrick, K.R., Power, E.A. and Sergy, G.A. (1991) The relative sensitivity of Microtox®, daphnids, rainbow trout, and fathead minnows acute lethality tests, *Environment Toxicology and Water Quality* **6**, 35-62.
Newman, M.C. and McCloskey, J.T. (1996) Predicting the relative toxicity and interactions of divalent metal ions: Microtox® bioluminescence assay, *Environment Toxicology and Chemistry* **15**, 275-281.
Obst, U., Wessler A. and Wiegang-Rosinus, M. (1997) Enzyme inhibition for examination of toxic effects in aquatic systems, in *Microscale Testing in Aquatic Toxicology*, P.G. Wells, K. Lee and C. Blaise (eds.), CRC Press, Washington, D.C., pp. 77-94.
OECD (1984) *Organization for Economic Cooperation and Development Guidelines for Testing of Chemicals: Earthworm Acute Toxicity Tests*, OECD Guidelines No. 207, Paris, France.,15 pp.
Ostrander, G.K. (1996) *Techniques in Aquatic Toxicology*, CRC Press, Washington, D.C., 686 pp.
Plotkin, S. and Ram N.M. (1984) Multiple bioassays to assess the toxicity of a sanitary landfill leachate, *Archives of Environmental Contamination and Toxicology* **13**, 197-206.
Qureshi, A.A., Bulich A.A. and Isenberg, D.L. (1998) Microtox Toxicity Test Systems – where they stand today, in *Microscale Testing in Aquatic Toxicology*, P.G. Wells, K. Lee and C. Blaise (eds.), CRC Press, Washington, D.C., pp. 185-195.
Rand, G.M., Wells, P.G. and McCarty, L.S. (1995) Introduction to aquatic toxicology, in G.M. Rand (eds.), *Fundamentals of Aquatic Toxicology*, Taylor & Francis, Washington, D.C., pp. 3-66.
Ribo, J.M. and Kaiser, K.L.E. (1987) *Photobacterium phosphoreum* toxicity bioassay, I test procedures and applications, *Toxicological Assessment* **2**, 305-323.

Richardson, M. (1993) Regulatory status of Microtox, in *Ecotoxicology Monitoring*, M. Richardson (ed.), VCH Publishers, New York, pp. 271-272.
Ross, P. (1993) The uses of bacterial luminescence systems in aquatic toxicity testing, in *Ecotoxicology Monitoring*, M. Richardson (ed.), VCH Publishers, New York, pp. 185-195.
Ross, P. (1998) Role of microbiotests in contaminated sediment assessment batteries, in P.G. Wells, K. Lee and C. Blaise (eds.), *Microscale Testing in Aquatic Toxicology*, CRC Press, Washington, D.C., pp. 549-553.
SDI Web site (2003) www.azurenv.com
Spacie, A. and Hamelink, J. L. (1985) Bioaccumulation, in G.M. Rand and S.R. Petrocelli, (eds.), *Fundamentals of Aquatic Toxicology*, Hemisphere Publishing Corporation, New York, pp. 495-525.
U.S. Chemical Industry (1998) *Statistical Handbook*, Chemical Manufacturers Association, Arlington, VA.
Wells, P.G., Lee, K. and Blaise, C. (1998) Microscale testing in aquatic toxicology: Introduction, Historical Perspective, and Context, in P.G. Wells, K. Lee and C. Blaise (eds.), *Microscale Testing in Aquatic Toxicology*, CRC Press, Washington, D.C., pp. 1-9.
Yates, I.E. and Porter, J.K. (1982) Bacterial bioluminescence as a bioassay for mycotoxins, *Applied Environmental Microbiology* **44**, 1072-1075.
Yim, M. W. and Tam, N.F.Y. (1999) Effects of wastewater-borne heavy metals on mangrove plants and soil microbial activities, *Marine Pollution Bulletin* **1-12**, 179-186.

Abbreviations

AFNOR	Association Française de Normalisation
APHA	American Public Health Association
ASTM	American Society for Testing and Materials
CAS	Chemical Abstracts Service
CERC	Columbia Environmental Research Center
CI	Confidence Interval
CV	Coefficient of Variation
DCM	dichloromethane
DDD	dichlorodiphenyldichloroethane
DDE	dichlorodiphenyldichloroethylene
DDT	dichlorodiphenyltrichloroethane
DIN	Deutsches Institut Normung
DMSO	Dimethyl sulfoxide
2,6,DNT	2,6, dinitrotoluene
FMN	Flavin mononucleotide
GLP	Good Laboratory Practices
ISO	International Standards Organization
MMSDS	Manufacturer's Material Safety Data Sheets
NADP	Nicotinamide adenine dinucleotide phosphate
NAWQA	National Water Quality Assessment
PAHs	Poly Aromatic Hydrocarbons
PC	Personal computer
PCBs	PolyChlorinated Biphenyls
OECD	Organization for Economic Cooperation and Development
QA/QC	Quality Assurance and Quality Control
SDI	Strategic Diagnostics Inc.
SOP	Standard Operating Procedure
SPMD	Semipermeable membrane device

TIEs	Toxicity Identification Evaluations
TREs	Toxicity Reduction Evaluations
USGS	U.S Geological Survey

2. SOLID-PHASE TEST FOR SEDIMENT TOXICITY USING THE LUMINESCENT BACTERIUM, *VIBRIO FISCHERI*

KEN DOE
& PAULA JACKMAN
Environment Canada
Environmental Science Centre
P.O. Box 23005, Moncton
NB, E1A 6S8, Canada
Ken.Doe@ec.gc.ca
Paula.Jackman@ec.gc.ca

RICK SCROGGINS
Environment Canada
Environmental Technology Centre
3439 River Road South, Gloucester
ON, K1A 0H3, Canada
scroggins.richard@etc.ec.gc.ca

DON MCLEAY
McLeay Environmental Ltd.
2999 Spring Bay Road, Victoria
BC, V8N 5S4, Canada
mcleayenvir@islandnet.com

GARY WOHLGESCHAFFEN
Fisheries and Oceans Canada
Bedford Institute of Oceanography
P.O. Box 1006, Dartmouth
NS, B2Y 4A2, Canada
WohlgeschaffenG@mar.dfo-mpo.gc.ca

1. Objective and scope of the method

The solid-phase Microtox™ test for measuring the toxicity of whole sediment using luminescent bacteria (*Vibrio fischeri*) is best run as part of a battery of toxicity tests to estimate the toxic potential of sediment. The endpoint of the test can be used as part of a sediment quality assessment. Because the test is relatively inexpensive, rapid and easy to run, it can be used on its own to screen large numbers of samples, in order to delineate the spatial extent of sediment contamination.

The test system is automated and the bacterial reagent is supplied in a lyophilized (freeze-dried) form, so there is no need for time-consuming culture of the test organisms. As supplied, they are ready for testing at any time that samples might arrive.

The test is most commonly applied to the assessment of freshwater, estuarine, or marine sediment, and to terrestrial soils, but is theoretically applicable to any similar solid material, such as sludges and ore concentrates.

2. Summary of the test procedure

The solid-phase test for measuring the toxicity of whole sediment samples using luminescent bacteria is summarized in Table 1.

Table 1. Rapid summary of the test procedure.

Test organism	*Vibrio fischeri*, strain NRRL B-11177
Type of test	20 minute static test
Test vessels	Disposable glass cuvettes
Volume of test solution	1.5 mL
Inoculum	20 µL reconstituted bacteria
Lighting	Does not apply
Test temperature	$15 \pm 0.5°C$
Experimental configuration	12 test concentrations and three control solutions
Endpoints	Moisture corrected IC50 in mg/L (dry-weight basis), others possible, *e.g.*, IC25
Measurements	Light levels of all test filtrates and controls measured
Reference toxicants	Toxic positive control sediment (*e.g.*, certified reference sediment)

The procedure involves the following steps:
- preparation of the primary dilution of the sediment;
- preparation of sample serial dilutions in diluent using a 50 % dilution series (Fig. 1);

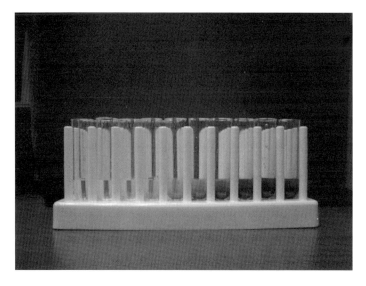

Figure 1. Preparation of whole sediment sample concentrations by serial dilutions in water using a 50 % dilution series.

Figure 2. M500 Microtox photometer, solid-phase tubes, and the computer to run the test, calculate the endpoints, and store the data.

- mixing the dilutions with an inoculum of test organisms (reconstituted *V. fischeri*) and incubation for 20 minutes in test tubes held in a water bath at $15 \pm 0.5°C$;
- filtration of the contents of each test tube;
- stabilization of the filtrate at $15 \pm 0.5°C$ for 10 minutes in a series of cuvettes held within wells of a photometer;
- photometric reading of light produced by the luminescent bacteria remaining in the filtrate (Fig. 2).

The test array consists of 3 controls (comprised of dilution water only) and 12 test concentrations. The maximum test concentration is 197,000 mg/L (19.7%, wt:v), with each successive concentration being 50% of the previous one. A schematic overview of the various stages in the test is shown in Figure 3.

3. Overview of applications reported

A test for measuring the toxicity of aqueous samples to luminescent bacteria was developed in the late 1970s (Anonymous, 1979). Researchers began applying this test to measure the toxicity of contaminated sediments in the 1980s and 1990s by assessing effects in solvent extracts (Schiewe et al., 1985; True and Heyward, 1990) and sediment porewaters (True and Heyward, 1990; Giesy et al., 1988). The different solvent systems, which can be used for the extraction of contaminants from sediments, have different efficiencies, and can even prove toxic to the bacteria (Tay et al., 1992). Furthermore, solvent extracted contaminants will not necessarily represent the bioavailable contaminants. Therefore, a Direct Sediment Toxicity Testing Procedure for measuring sediment toxicity using luminescent bacteria was introduced by Canadian researchers (Brouwer et al., 1990), and an acute Solid-Phase Test for sediment (or soil) toxicity was subsequently adopted and standardized by Microbics Corporation (Carlsbad, CA), as one of several Microtox test methods (Microbics, 1992). Kwan and Dutka (1995) compared these two solid-phase toxicity test methods, and confirmed their suitability as sensitive tests to detect bioavailable toxicants in solid-phase samples. Both tests are practical, reproducible, rapid, and relatively inexpensive compared to solid-phase extraction procedures. There are currently a number of Solid-Phase Test methods using luminescent bacteria, which are in use internationally and have been compared in detail (Environment Canada, 2002).

Since its introduction, the test has been widely utilized. It was employed in correlation studies between a number of solid-phase sediment toxicity tests and *in situ* benthic community structure in freshwater and marine sediments (Day et al., 1995; Porebski et al., 1999; Zajdlik et al., 2000). It can be used to assess the toxicity of sediment being considered for disposal at sea, on land or at any freshwater, estuarine, or marine sites where regulatory appraisals or stringent testing procedures apply. The test has been used to assess the quality of contaminated soils (Qureshi et al., 1998; Environment Canada, unpublished data) and freshwater sediments (Day et

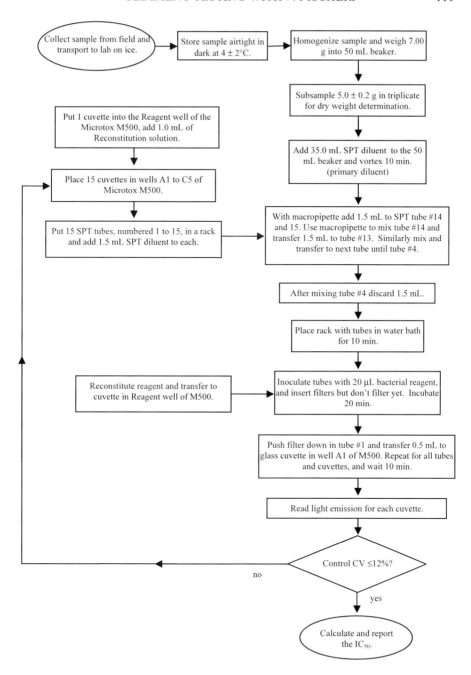

Figure 3. Summary of the Solid-Phase Test using luminescent bacteria.

al.,1995; Carter et al., 1998; Denning et al., 2003). Test design and data interpretation were studied by Ross and Leitman (1995) and Ringwood et al. (1997). Toxicity appraisals of harbour sediments (Halifax, Canada) have been conducted (Tay et al. 1992; Cook and Wells, 1996), and the role of this and other assays in assessing sediment toxicity was examined by Ross (1998) and Bombardier and Bermingham (1999). The interlaboratory precision of a solid-phase test for sediment toxicity using *V. fischeri* was studied by Ross et al. (1999) and McLeay et al. (2001). Mueller et al. (2003) used the assay to determine the effectiveness of sediment bioremediation techniques.

Since its introduction, the test has been widely used by North-Americans and other researchers and regulators for evaluating the toxicity of sediments. Environment Canada (1992) recommended the use of the Microtox^{TM} *solid-phase* test method (Microbics, 1992) for evaluating the toxicity of solid media, while recognizing that the standardization of the test method was in its infancy. In 2002, after marked research efforts and inter-laboratory validation testing, Environment Canada published a Reference Method for determining the toxicity of sediment using luminescent bacteria, *Vibrio fischeri* (Environment Canada, 2002).

This chapter is based on the Environment Canada published Reference Method (Environment Canada, 2002), and describes the procedure for conducting solid-phase tests for measuring sediment toxicity using the luminescent bacterium *Vibrio fischeri*. Further details can be obtained by consulting the Reference Method, which represents one of several regulatory biological test methods recommended as part of sediment assessment under the Canadian Environmental Protection Act Disposal at Sea Regulations (Environment Canada, 1997a; CEPA, 1999; Government of Canada, 2001).

Several organizations have published methodology guideline documents or laboratory standard operating procedures for the solid-phase test for sediment toxicity using luminescent bacteria in which pre-test, test conditions and procedures are summarized: ASTM (1995), Microbics (1995), Environment Canada (1996a), AZUR (1997), AZUR (1998a; 1998b), Environment Canada (1999a), and NICMM (1999). The test has been included as part of several recent reviews concerning sediment toxicity testing (Johnson 1998; Burton et al., 2003).

4. Advantages of conducting the Solid-Phase Test using luminescent bacteria

The solid-phase test for measuring sediment toxicity with the luminescent bacterium *V. fischeri* has several advantages over the more conventional sediment tests used to assess the toxicity of contaminated sediments with invertebrates such as chironomids, amphipods, molluscs, polychaetes, and echinoderms:

- test rapidity: the solid-phase tests using luminescent bacteria can be completed in hours as opposed to days for the other sediment tests;
- small sample volume requirements: this feature makes samples easy to collect and cheaper to ship to testing laboratories. Sample volumes of 25 – 50 mL suffice to

conduct several replicate tests and to measure moisture content of the sample. Other tests might require one L or more of sediment;

- freeze-dried bacterial reagent: no time consuming culturing of test organisms is required since test organisms (*V. fischeri*) are lyophilized and available commercially. Field-collection and continuous culturing of other light-producing micro-organisms would require considerable efforts;
- repeatability: a high quality control source of reagents and supplies ensures standardization and repeatability of results worldwide;
- inter-laboratory validation: the test was validated by inter-laboratory studies conducted by Ross et al. (1999) and McLeay et al. (2001). The latter study involved six testing laboratories where 19 samples were analyzed using the procedure described in this chapter. The inter-laboratory precision was "very favorable and well within the limits considered acceptable in other studies of this nature";
- realistic hazard assessment: test results have shown statistically significant correlation with contaminant concentrations, benthic community structure, and many conventional invertebrate whole sediment bioassays, as discussed by Day et al. (1995) and Zajdlik et al. (2000);
- versatility: the same test procedure can be conducted with (freshwater, estuarine and marine) sediments and soils;
- relevance: bacteria form the basis of many important ecosystem functions such as biodegradation of organic matter, nutrient recycling, etc. They therefore represent an important group of organisms for inclusion in any battery of toxicity tests.

5. Test species

The recommended test organisms come from a standardized culture (strain NNRL B-11177, Northern Regional Research Laboratory, Peoria, IL, USA), and belong to a particular species of luminescent marine bacteria (*i.e.*, *Vibrio fischeri*, formerly classified as *Photobacterium phosphoreum*). This is a bacterium which normally lives in the ocean, and produces blue-green light on a continual basis by a series of enzymatic reactions utilizing metabolic energy obtained from the electron transport system if sufficient oxygen is available (Environment Canada, 1992).

6. Culture/maintenance of organisms in the laboratory

Standard cultures of *V. fischeri* can be purchased from Strategic Diagnostics Inc.[1] Bacteria are marketed as a uniform strain of lyophilized (*i.e.*, freeze-dried under vacuum) bacteria ("*Bacterial Reagent*"), harvested during the exponential phase of

---

[1] This and related products and disposal supplies for performing solid-phase toxicity tests using *V. fischeri* were formerly marketed by AZUR Environmental Ltd. (Carlsbad, CA). Marketing rights for Microtox[TM] products and reagents have now been acquired by Strategic Diagnostics Inc. in Newark, DE. For contact information, see their web site at www.sdix.com, or phone 800 544-8881. The web site lists international distributors for approximately 60 countries.

growth. Production lots are sold in packages containing ≥10 sealed vials. Each vial harbors about 100 million lyophilized organisms. Each lot is suitable for at least two hours (Environment Canada, 1992) and for up to three hours of testing (Gaudet, 1998), after bacteria have been reconstituted to an active state. Because the bacterial reagent (lyophilized *V. fischeri*) is available commercially, there is no time consuming culturing of test organisms. The reagent is stored frozen until required. If desired, the *V. fischeri* can be cultured in the laboratory using methods outlined in ISO (1993).

The number and expiry date of bacterial lots used in each toxicity test should be recorded and this information should be included in the test-specific report together with the species and strain of the test organism. It is recommended that other data specific to the test organisms, including their source, date of receipt, and temperature during storage or holding, should either be included in the test-specific report or held on file for a minimum of five years.

7. Preparation of bacteria for toxicity testing

The bacteria ("*Bacterial Reagent*") are sold in packages containing sealed vials of lyophilized organisms that are stored frozen until use. Once a vial is opened, it is reconstituted by quickly pouring the *Reconstitution Solution* held at 5.5 ± 1°C in a cuvette placed into the reagent well of the *Model 500 Analyzer* or other photometer (see Section 8.7) into the vial. After swirling three times, vial contents are poured back into the same cuvette. The reconstituted bacteria are then held in the reagent well at 5.5 ± 1°C and are ready for use.

8. Testing procedure

8.1 FACILITIES

The test can be conducted in a normal, clean laboratory with standard lighting. The need for any special facilities would be governed by the degree of hazard associated with the samples that are to be tested, and by the risk of sample and apparatus contamination. Facilities must be well ventilated, free of fumes, and isolated from physical disturbances or airborne contaminants that might affect the test organisms.

8.2 APPARATUS AND SUPPLIES

A list of apparatus and supplies required for the test are provided below:

- A Microtox™ *Model 500 Analyzer*[2] or equivalent temperature-controlled photometer (15 ± 0.5°C for ≥15 cuvettes with test solutions; 5.5 ± 1°C for

---

[2] These items are available from Strategic Diagnostics Inc.

single cuvette holding reconstituted bacteria in *Reagent* well) capable of reading light output at a wavelength of 490 ± 100 nm.
- A refrigerated water bath with temperature controlled at 15 ± 0.5°C.
- A test tube rack or incubator block for incubating tubes containing concentrations of test material and *V. fischeri* in the water bath.
- A freezer (not self-defrosting or "frost free" type) for storing lyophilized bacteria (*Bacterial Reagent*).
- Pipettors for delivering volumes of 20, 500, 1000 and 1500 µL, with disposable plastic tips.
- Disposable polystyrene SPT tubes (15.5×56 mm, 7.5 mL capacity, hemispherical bottom).[2]
- Disposable glass cuvettes (borosilicate, 3 mL capacity, 50 mm length × 12 mm diameter, flat bottom).[2]
- Disposable filter columns for SPT test tubes.[2]
- Freeze-dried *Bacterial Reagent*.[2]
- *Reconstitution Solution* (purified non-toxic water).[2]
- Solid-Phase *Diluent* (non-toxic distilled or deionized water plus 3.5% sodium chloride).[2]
- Volumetric borosilicate glassware (acid washed) for processing small aliquots of samples.
- A countdown timer or stopwatch.
- A magnetic plate mixer with Teflon stir bar.
- Balance, accurate to 0.01 g.
- A drying oven (100 ± 5°C).
- Weighing vessels for dry weight determination.
- Metal spoon or spatula for sample homogenization.

The *Bacterial Reagent* should remain in a freezer at -20°C until used. Similarly, Solid-Phase *Diluent* and *Reconstitution Solution* should be stored at room temperature until required.

8.3 MANIPULATIONS, ADJUSTMENTS, AND CORRECTIONS

- Test sediments must not be wet-sieved, and no adjustments of porewater salinity are permitted. Sample pH must not be adjusted. No aeration of samples, test concentrations, or filtrates should be performed.
- Light-emission readings for concentrations of each test material must not be adjusted or corrected.
- The statistical endpoint for the test (*i.e.*, IC50) must be normalized for the moisture content of the sample.

8.4 TEMPERATURE

- The *Bacterial Reagent* is reconstituted to an active state in non-toxic distilled or deionized water and held at 5.5 ± 1.0°C until aliquots are transferred to each

test concentration. Normally this temperature is met by placing the cuvette with the reconstituted bacterial solution in the specified well of the photometer if a Microtox™ *Model 500 Analyzer* is used. Otherwise a temperature-controlled incubator must be used for this purpose.
- All concentrations of test material inoculated with bacteria must be incubated for 20 minutes at $15 \pm 0.5°C$. The temperature is allowed to stabilize for 10 minutes prior to inoculation by *Bacterial Reagent*. A temperature-controlled water bath or room would serve this purpose.
- Following incubation and filtration, all test solutions transferred to cuvettes must be held at $15 \pm 0.5°C$ during the subsequent 10-minute period for stabilization of the filtrates. This temperature control is normally achieved within the wells of the photometer. Alternatively, the cuvettes containing test filtrates can be held within this temperature range in a cuvette holder placed in a temperature-controlled incubator or room.

8.5 TIMING OF EVENTS

- The lyophilized bacteria should be reconstituted immediately before inoculating the test concentrations. This bacterial solution should be used within 2 h, and within a maximum of 3 h after reconstitution. The time of reconstitution should be logged on a bench sheet.
- A primary dilution of sediment is prepared by stirring the sediment and diluent for ten minutes. Then, aliquots are removed to prepare test concentrations. The latter must be allowed to equilibrate to $15 \pm 0.5°C$ for a minimum of 10 minutes before inoculation with bacterial solution. Inoculation should proceed as quickly as possible; all test concentrations should be inoculated within a total time span of ≤ 4 minutes. Record the time of the first inoculation as the start of the test.
- All test concentrations must be incubated for 20 minutes after inoculation of the first tube with bacteria. Once the test concentrations are filtered and transferred to cuvettes, the filtrates must be incubated in cuvettes for 10 minutes in the temperature-controlled wells of the photometer before their light output is measured.
- Total elapsed time for the transfer of filtrates to cuvettes and for reading luminescence of the test filtrates should be similar to that spent inoculating the test concentrations with bacteria (≤ 4 min).

8.6 CONDUCTING THE TEST

The procedures to be followed when performing a solid-phase test involve the simultaneous incubation of three control solutions (comprised of an inoculum of reconstituted *V. fischeri* in Solid-Phase *Diluent*) together with 12 different concentrations of each sample of test material in Solid-Phase *Diluent*. After a prescribed incubation period, the solutions (held in test tubes at a controlled temperature) and test concentrations are filtered, and the resulting filtrates are transferred to cuvettes. After a brief period for stabilization of holding conditions for the filtrates, the light production by test organisms remaining in each filtrate is measured by a photometer. Table 2 provides a checklist of the conditions, apparatus, and procedures recommended for conducting the test. Figure 3 is a flowchart of the entire test procedure.

The procedures applied herein to a photometer assume the use of a Microtox^{TM} *Model 500 Analyzer* or another photometer with similar features. Since the Microtox^{TM} *Model 500 Analyzer* has 30 wells for holding cuvettes containing filtrates of test concentrations, the laboratory analyst using this photometer has the option of performing two tests simultaneously on different test materials. Note that the option to analyze two test materials (*i.e.*, two samples) simultaneously is recommended to save time and *Reagent*.

The following sections describe the procedure by which a sample of test material is processed for assessment of its toxic potential.

8.7 PHOTOMETER, WATER BATH, AND BENCH SHEET

- Switch on the computer, photometer, and balance.
- For the *Model 500 Analyzer,* ensure that the temperature selector switch at the back is set to "Microtox Acute".
- Place 15 cuvettes in the first 3 rows (A-C) of wells. These will be maintained at $15 \pm 0.5°C$. The incubated wells are arrayed in a grid of rows labeled A to C and columns numbered 1 to 5. They are referenced as A1 to C5 (Fig. 4).
- Place one cuvette in the reagent well, and pipette 1.0 mL of *Reconstitution Solution* into it. This will be maintained at $5.5 \pm 1°C$.
- Switch on the water bath incubator. Allow the water temperature to stabilize at $15 \pm 0.5°C$.
- Stir the sample to homogenize it, using a stainless steel spoon.
- Place 15 SPT tubes into a rack.

8.8 SUBSAMPLES FOR MOISTURE

- For each test sediment, label and weigh three empty weighing dishes, and record the weights to the nearest 0.01 g.
- Add 5.0 ± 0.2 g of sediment to the vials and record the weights to the nearest 0.01 g.
- Dry the subsamples by putting the vials into an oven at $100 \pm 5°C$ for 24 h. Record the oven temperature.
- Record the dry weights to the nearest 0.01 g.

Table 2. Checklist of required or recommended test conditions and procedures.

Facilities and equipment	Photometer (*e.g.* Microtox™ *Model 500 Analyzer*, available from Strategic Diagnostics Inc.) reading light output at 490 ± 100 nm; incubator for single cuvette holding reconstituted bacteria at 5.5 ± 1°C; for ≥ 15 cuvettes at the test temperature (15 ± 0.5°C), in an incubator or controlled-temperature room.
Reconstitution Solution	Non-toxic distilled or deionized water or *Reconstitution Solution* supplied by Strategic Diagnostics Inc.
Control/dilution	Solid-Phase *Diluent* (3.5% NaCl solution).
Test temperature	15 ± 0.5°C.
Color correction	No correction.
Aeration	None required.
Subsamples for moisture content	3 replicates of 5.0 ± 0.2 g (precision: ± 0.01 g) dried at 100 ± 5°C for 24 h.
Primary dilution	7.00 ± 0.05 g whole, homogenized test material in 35 mL Solid-Phase *Diluent*, glass or disposable plastic beaker, mixed for 10 min on a magnetic stirrer with Teflon stir bar, at a rate such that the vortex depth is half the height of the liquid level.
Test concentrations	Maximum test concentration normally 197,000 mg/L (19.7% wet wt:vol) with two-fold dilutions, for a total of 12 test concentrations in disposable polystyrene tubes; three control solutions (*Diluent* only); left for 10 min to equilibrate to the test temperature.
Test species	*Vibrio fischeri*, strain NRRL B-11177 (Note: the use of lyophilized *Bacterial Reagent* supplied by Strategic Diagnostics Inc. would increase the standardization of the test procedure), reconstituted by swirling each vial 3-4 times, emptied into a disposable glass cuvette, mixed 10 times with 0.5 mL pipette and held at 5.5 ± 1°C.
Inoculum	20 µL into each test concentration, mixed 3 times with 1.5 mL pipette.
Incubation	20 min at test temperature, filter columns inserted into tops of SPT tubes above surface of test concentration.
Filtrate transfer	500 µL into disposable glass cuvettes at test temperature, left 10 min for equilibration.
Observations	Set light level with first control and then insert cuvettes into photometer read well; light levels of all test filtrates and controls are then measured.
Endpoint	IC50 (mg/L), calculated by software used by each laboratory; normalized for moisture content of sediment (*i.e.*, calculated on dry-weight basis).

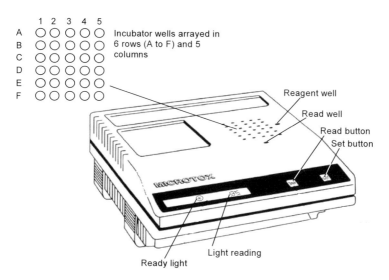

Figure 4. Microtox M500 analyzer showing the operational features and the array of incubator wells into which are placed the glass cuvettes.

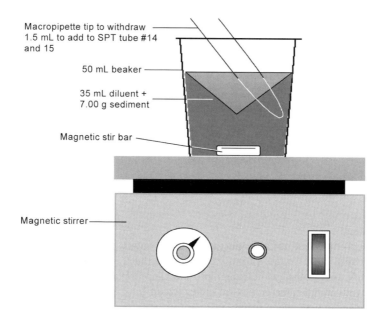

Figure 5. The primary sample dilution is prepared on a magnetic stirrer, agitated for 10 min at a rate such that the vortex is half the height of the liquid level. To aspirate 1.5 mL, the macropipette tip is inserted near the side of the beaker, at about half the depth of the stirring sample.

8.9 PRIMARY DILUTION

- Weigh 7.00 ± 0.05 g of homogenized subsample into a glass or disposable 50 mL plastic beaker.
- Add a 2.5 cm Teflon-coated magnetic stir bar and 35 mL of Solid-Phase *Diluent* to the beaker.
- Stir for 10 min on a magnetic stirrer at a rate to create a vortex that reaches 1/2 the height of the liquid level (Fig. 5).

8.10 TEST CONCENTRATIONS

- Dispense 1.5 mL Solid-Phase *Diluent* into the first 14 tubes in the rack. Tube 15 will be the highest concentration, taken from the primary dilution.
- Following the 10-minute stirring of the primary dilution, use a large-bore macro pipette tip to transfer 1.5 mL of sample suspension from the 50 mL beaker, while it is still stirring, to each of SPT tubes 15 and 14. To do this, insert the pipette tip near the side of the beaker, at about half the depth of the stirring sample. Avoid plugging the tip while aspirating the sample (Fig. 5).
- Beginning at SPT tube 14, which now has 3 mL of solution, make 1:2 serial dilutions as follows. Mix the contents 3 times with the macro pipette, and then quickly draw up a volume of 1.5 mL from about one third depth (to help draw some of the heavier sand grains). Transfer this aliquot to tube 13. Repeat this mixing and transferring from tube 13 to 12, and continue consecutively thereafter from tube 12 to 11, tube 11 to 10, etc. until 1.5 mL of the 1:2 serial dilutions is transferred into tube 4. Finally, discard 1.5 mL from tube 4. Tubes 1-3 contain *Diluent* only, and serve as the controls (Fig. 6).
- Place the rack with SPT tubes containing all test concentrations (including controls) into the water bath at 15 ± 0.5°C. Leave it undisturbed for 10 min for temperature equilibration. The water level of the bath should be just above the liquid level in the SPT tubes.

8.11 RECONSTITUTION OF BACTERIAL REAGENT

- Take a vial of freeze-dried bacteria (Microtox^{TM} *Acute Reagent*) from freezer storage.
- Open the vial and reconstitute its contents by quickly pouring the Reconstitution Solution from the cuvette in the reagent well of the *Model 500 Analyzer* (or other photometer) into the vial, swirling three times and pouring the rehydrated bacteria into the same cuvette.
- Replace the cuvette in the reagent well.
- Using a 500 μL pipette, aspirate any remaining *Reconstituted Bacterial Reagent* from the vial, add it to the cuvette, and mix 10 times using the same pipette and tip.
- Record the reagent lot number, expiry date, and time of reconstitution on the bench sheet.

8.12 INOCULATION AND INCUBATION

- Prepare the following three pipettes: (a) a repeat pipette (such as an Eppendorf or Oxford Nichiryo) with a 0.5 mL syringe fitted with an ultra micro tip; (b) a macro pipette (*e.g.* Oxford 1-5 mL); and (c) a 500 µL pipette.
- Following the 10-min temperature equilibration of the SPT tubes containing the test concentrations set a timer for 20 min but do not start it.
- Make sure the repeat pipette is set to dispense 20 µL per ejection. Place the tip below the surface of the reconstituted bacteria (*Reconstituted Bacterial Reagent*), and draw up sufficient reconstituted bacteria for at least 18 ejections.
- Holding the tip above the *Reagent* and against the cuvette wall, eject 2 times. Wipe the tip with a clean wiper.

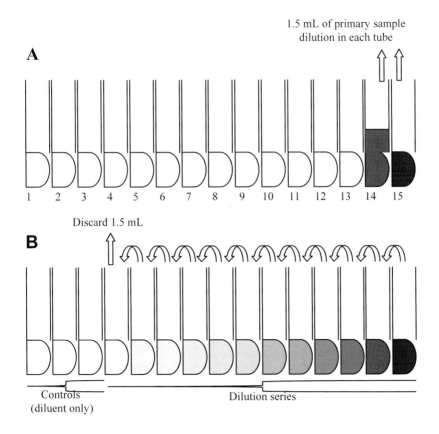

Figure 6. SPT tubes are placed in a rack, then (A) SPT diluent (1.5 mL) is added to each SPT tube (#1 to 15) and 1.5 mL of the primary sample dilution from the beaker on the vortex mixer is added with a macropipette to tubes 14 and 15. The solution in tube 14 is mixed with the macropipette (B) and 1.5 mL is transferred each time to the next tube and mixed. Finally, 1.5 mL is discarded from tube #4.

- Start the 20-min timer, record the time as the "Start" of the test on the bench sheet, and immediately eject 20 µL of *Reagent* into each of the SPT tubes, starting with tube 1 (first control solution) and continuing consecutively to and including tube 15. Resting the collar of the ultra micro tip on the top edge of each tube will assure the tip will be at the surface of the sample and not on the bottom of the tube.
- Remove and discard the ultra micro tip. Eject the remaining *Reagent* from the syringe into the holding cuvette in the reagent well.
- If a second test (*i.e.*, with a second test material) is to be performed concurrently using the remaining 15 wells of the *Analyzer*, refit the syringe with a clean tip, refill it with Bacterial Reagent, and inoculate the next series of test concentrations.
- With the macro pipette set at 1.5 mL, mix each tube twice, beginning with tube 1 (first control) and proceeding consecutively through to and including tube 15. (If performing 2 tests consecutively replace tip between samples).
- Insert a filter column in each tube, with its lower end positioned ~1 cm above the surface of the liquid. Do not wet the filter, since this might adversely affect filtration of the sample.

8.13 PREPARING THE COMPUTER

- Prepare the computer to receive data from the photometer.
- Start the appropriate software program and menus for the solid-phase test.
- Follow the on-screen instructions.
- Refer to the user's manual for additional information.
- Information requested by the software might include the number of controls (3), number of dilutions (12), initial concentration (197,000 mg/L), dilution factor (2), and the test time (10 min).

8.14 FILTRATION AND LIGHT LEVEL READINGS

- When the 20-min timer sounds, respond to the computer software as appropriate.
- Reset the timer for 10 min and start it. If you are using SDI software, this period is automatically initiated. Otherwise, program the computer software to initiate this 10-min period automatically at the touch of a key.
- Gently push the filter in tube 1 (first control) down far enough to obtain slightly greater than 500 µL of filtrate in the filter column. Then, using the 500 µL pipette, transfer 500 µL of filtrate to the cuvette in well A1 of the photometer.
- Repeat this step for all 15 tubes using the same pipette tip, ending with 500 µL of filtrate being transferred from tube 15 to cuvette C5.
- Take note of the time required to complete all the transfers. Respond to the prompt by the software (assuming that this is a component of that used).

- Often with sediments having a high proportion of fines the filter in the highest test concentration (tube 15) will become plugged. Obtain what can be recovered, and transfer it to the designated cuvette (C5). Usually the luminescence from these samples will have decreased to zero before you test the highest concentration. Make a note of such problems on the bench sheet.
- After 10 minutes, set the light level with the first control cuvette and then read light production taking approximately the same amount of time as was taken to filter and transfer the filtrate to the cuvettes. Timing is prompted by the software if SDI software is used.
- The data are then sent from the photometer to the computer, and received by the software which builds a data file.

The procedure to measure light production of the bacteria in the test concentrations will vary depending on the photometer and software used. If a *Model 500 Analyzer* is used, place the first control (cuvette A1) into the read well and press the "set" button. The instrument lowers the cuvette into the well (sometimes 2 or 3 times) to set the zero (dark) and control reading at about 95 and thereby establishes the appropriate sensitivity range for light measurements. A green "ready" light will appear. Then press the "read" button. After reading the cuvette, remove it from the read well and replace it in the incubator block. Proceed to read and record the light emission from all the cuvettes, taking approximately the same average time per cuvette as was required to do the filtering and transferring. Using SDI software, this timing is performed by the computer and prompts occur which indicate when each cuvette should be read.

9. Post-exposure observations/measurements and endpoint determinations

Much of the guidance in this section is based on the document "Solid-Phase Reference Method for Determining the Toxicity of Sediment Using Luminescent Bacteria (*V. fischeri*)" (Environment Canada, 2002). Additional discussion has been provided where necessary to expand on certain concepts.

9.1 DATA ANALYSIS

The mean and standard deviation of the light readings for the control solutions used in the study must be calculated. These values are used to determine the coefficient of variation of the mean for the control solutions, which is used as one of the criteria described herein for judging if the test results are valid (Section 10). The coefficient of variation of the light readings for the control solutions must be $\leq 12\ \%$ for a particular test to be considered valid and for data analysis to proceed. A study performed according to this method should include one or more samples of test sediment together with one or more samples of reference sediment. Additionally, the inclusion of one or more samples of positive control sediment for use as a reference toxicant is required as a measure of sensitivity of the bacterial reagent. Grain size analysis must be performed on each test and reference sediment. Tests involving one or more samples of coarse-grained sediment (*i.e.*, sediment with < 20% fines) must also include one or more samples of negative control sediment (artificial or natural) or clean reference sediment with a percentage of fines content that does not differ by more than 30% from that of the coarse-grained test sediment(s). In each instance, the statistical endpoint to be calculated for each of these test materials is the ICp (inhibiting concentration for a specified percent effect). Unless specified otherwise by regulatory requirements or by design, the endpoint for this method is the concentration causing 50% inhibition of light, *i.e.*, the IC50. The calculations to estimate the IC50 and its 95% confidence limit are included in the most recent (1999 or later) *Omni*^{TM} software packages (Version 1.18 or equivalent) formerly marketed by AZUR Environmental Ltd. and now available from Strategic Diagnostics Inc. (see footnote 1for contact information). In 1999, AZUR released an operating software termed "*MicrotoxOmni*". This software includes a calculation option called "*AutoCalc*". The "*AutoCalc On*" option must be selected if this software is applied; otherwise, erroneous results might occur. *AutoCalc* selects only the data points around the IC50 that have a range of 0.02 < *Gamma* < 200. Using these data points, all the various contiguous series are examined by an iterative convergence regression analysis to determine the tightest 95% confidence limits. If the *MicrotoxOmni* software is not available, guidance for estimating IC50 (together with its 95% confidence limit) is provided in Environment Canada (1992), and other statistical software packages are available which enable this calculation by linear regression (Environment Canada, 1997b; 1997c; 2003).

For each test concentration, *Gamma* (Γ; see definition in the Glossary) is calculated (ASTM, 1995) as:

$$\Gamma = (I_c/I_t) - 1 \qquad (1)$$

where: I_c = the average light reading of filtrates of the control solutions, and I_t = the light reading of a filtrate of a particular test material concentration. Values for each test filtrate that fall within the range of 0.02 < *Gamma* < 200 are plotted. The IC50 is the concentration that corresponds to a *Gamma* of 1. If the *MicrotoxOmni* software is not available, data entered for linear regression should be checked against the observed readings to guard against errors in entry and anomalous estimates of IC50. A manual plot and its estimated IC50 could also be used to check any computer-generated graph and the computer calculation of the IC50 (Environment Canada, 1992). Plots of *Gamma* and % effect versus concentration generated by the *MicrotoxOmni* software are shown in Figure 7 from a solid-phase test on a field-collected sediment sample, and the raw data from the test are provided in Table 3.

A linear regression of logC (concentration, on the ordinate) *versus* log Γ (on the abscissa) is computed according to the following equation (ASTM, 1995):

$$\log \Gamma = b(\log C) + \log(a) \qquad (2)$$

In the above equation, 'b' is the slope and 'log(a)' is the intercept of the regression line with the ordinate (y-axis) at log C = 0. To determine the IC50, solve equation 2 for C, when $\Gamma = 1$.

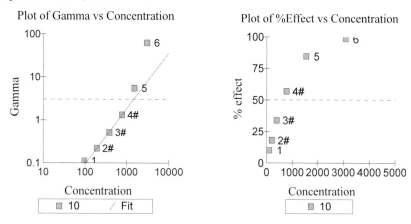

Figure 7. Plots of Gamma and % effect versus concentration, as plotted by the Microtox Omni *software. These plots are from the raw data provided in Table 3 that were also generated using the* Microtox Omni *software from a test performed on a field-collected sediment.*

Table 3. Raw data used in Figure 7 plots. "It" is the light reading of a filtrate of a particular test material concentration. The MicrotoxOmni software automatically selects the data to be used in the calculation.

Test treatment	Concentration (mg/L)	It	Gamma	% Effect
Control	0.000	103.37		
Control	0.000	100.32		
Control	0.000	94.86		
1	96.19	89.50	0.1119 #	10.07
2	192.4	81.70	0.2181 #	17.90
3	384.8	65.78	0.5129 #	33.90
4	769.5	42.99	1.315 #	56.80
5	1539	15.28	5.513 #	84.65
6	3078	1.56	62.79 #	98.43
7	6156	0.68	145.3 *	99.32
8	12310	0.32	310.0 *	99.68
9	24630	0.26	381.8 *	99.74
10	49250	0.09	1105 *	99.91
11	98500	0.08	1243 *	99.92
12	197000	0.07	1421 *	99.93

- used in calculation; * - invalid data.

The IC50 and its associated values for the 95% confidence limits must be converted to (and expressed as) mg/L on a dry-weight basis. This is achieved using the dry-weight data (see Section 8.8) (ASTM, 1995). The IC50 (as well as the upper and lower value of the confidence limits) of the wet sediment is multiplied by the average ratio of the dry-to-wet subsample weights:

$$IC50 = IC50_w \times [(S1_d/S1_w) + (S2_d/S2_w) + (S3_d/S3_w)]/3 \qquad (3)$$

where: $IC50_w$ is the calculated IC50 (or its 95% confidence limits) of the wet sediment sample, $S1_d$ through $S3_d$ are the dry weights of the sediment subsamples (from Section 8.8), and $S1_w$ through $S3_w$ are the corresponding wet weights. These calculations can be expedited by entering the weights and IC50 values into a spreadsheet and using the necessary formulae.

Investigators should consult Environment Canada (2003) for detailed guidance regarding appropriate statistical endpoints and their calculation. The objectives of the data analysis are: to quantify contaminant effects on test organisms exposed to various samples of test sediment; to determine if these effects are statistically different from those occurring in a reference sediment; and to reach a decision as to sample toxicity (Section 10). Initially, ICp (normally, IC50) is calculated for each sample (including those representing the field-collected reference sediment).

Depending on the study design and objectives, an appropriate number (typically, ≥ 5/station, for each depth of interest) of replicates of field and reference sediments should be collected and evaluated. Each series of toxicity tests must include a minimum of three replicate control solutions, and one or more test sediments.

10. Factors capable of influencing performance and interpretation of results

Interpretation of results is not necessarily the sole responsibility of the laboratory personnel undertaking the test. This might be a shared task which includes an environmental consultant or other qualified persons responsible for reviewing and interpreting the findings.

Environment Canada (1999b) provides useful advice for interpreting and applying the results of toxicity tests with environmental samples; and should be referred to for guidance in these respects. Initially, the investigator should examine the results and determine if they are valid. In this regard, the criterion for a valid test must be met (CV ≤ 12 %). Additionally, it is recommended that the dose-response curve for each sample of test sediment be examined to confirm that light loss decreases as test concentration decreases, in an approximately monotonic manner. We recommend that an r^2 value for the regression equation (provided by the *MicrotoxOmni* software, or calculated by the user) be ≥ 0.90. If not (*e.g.*, if one or more data points appear to be "out of place" with respect to the others), consideration should be given to repeating the test for that sample as this type of response is normally caused by pipetting errors. Finally, the results of any reference toxicity test (Environment Canada, 1990; 1995) with a toxic positive control sediment, which was initiated with the same lot of *Bacterial Reagent* as that used in the sediment toxicity test, should be considered during the interpretive phase of the investigation. These results, when compared with historic test results derived by the testing facility using the same reference toxicant and test procedure (*i.e.*, by comparison against the laboratory's warning chart for this reference toxicity test), will provide insight into the sensitivity of the test organisms as well as the laboratory's testing precision and performance for a reference toxicity test with *V. fischeri*. If the results of the reference toxicity test are outside of three standard deviations of the historic mean value of all previous tests with this reference toxicant (the "Control Limit"), all test conditions pertaining to the test should be double-checked thoroughly and consideration should be given to repeating the test (Fig. 8).

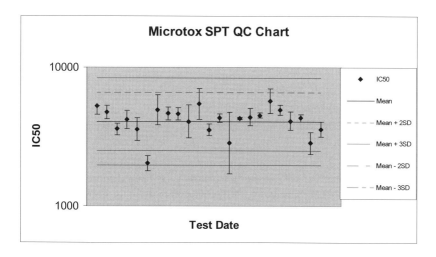

Figure 8. Example of a Quality Control chart for the Microtox Solid-Phase Test.

All data representing the known physico-chemical characteristics of each test material (including that for any samples of reference sediment or negative and positive control sediment included in the study) should be reviewed and considered when interpreting the results. The analytical data determined for whole sediment should be compared with the known influence of these variables on light production by *V. fischeri*, and also compared with sediment quality guidelines for the parameters measured.

Concentrations of porewater ammonia and/or hydrogen sulphide can be elevated in samples of field-collected sediment. This might be due to organic enrichment from natural and/or anthropogenic (man-made) sources. The known influence of ammonia (see, for example: Qureshi et al., 1982; Tay et al., 1998; McLeay et al., 2001) indicates that it is not a major confounding factor in this test. The known influence of hydrogen sulphide (Jacobs et al., 1992; Brouwer and Murphy 1995;

Tay et al., 1998) on the inhibition of light production by *V. fischeri* should be considered together with measured concentrations of these variables in porewater, when considering and interpreting results for field-collected samples and reference sediments.

Observations of turbid or highly colored filtrates analyzed for light emission by *V. fischeri* should be considered when reviewing and interpreting the test results.

A number of variables besides toxicity can interfere with readings of light production by *V. fischeri* surviving in the filtrate of each test concentration, and thus can confound the interpretation of the test results. Investigators performing this method, as well as those interpreting the findings, should be aware of these confounding factors and their implications in terms of judging if test materials are toxic or not. Variables which can interfere with the light production of *V. fischeri* in test filtrates include the following (ASTM, 1995; Ringwood et al., 1997; Tay et al., 1998):

- Sorption of *V. fischeri* to sediment particles (particularly fine-grained ones) retained on the filter; and the resulting loss of transfer of these luminescent bacteria to the filtrate.

- Sorption of *V. fischeri* to the filter, and the resulting loss of transfer of these luminescent bacteria to the filtrate.

- Optical interference of the filtrate, due to color (light absorption) and/or turbidity (light scatter).

The grain size of test sediments can be a significant confounding factor, since an increasing percentage of clay in the test material has been demonstrated to cause a proportionate decrease in resulting IC50s determined for *V. fischeri* recovered in filtrates of uncontaminated sediment. Samples of uncontaminated sediment comprised primarily of sand-sized particles (*e.g.*, 0-5% fines) characteristically yield an IC50 of 28,000 to >100,000 mg/L in a *V. fischeri* solid-phase assay (Cook and Wells, 1996; Ringwood et al., 1997; Tay et al., 1998). IC50s show a "precipitous drop" (Benton et al., 1995; Ringwood et al., 1997; Tay et al., 1998) when the percentage of fines in uncontaminated sediment increases from 5 to ~ 20%, whereupon the IC50 might possibly range from 5000 to 15,000 mg/L depending on the nature of the fines (*e.g.*, % clay and % silt) (Ringwood et al., 1997; Tay et al., 1998; McLeay et al., 2001). Higher percentages of fines in uncontaminated sediment typically show a "leveling off" of further declines in IC50s associated with increasing sediment fines. *V. fischeri* solid-phase tests with 100% kaolin clay have reported IC50s ranging from 1,373 to 2,450 mg/L (Ringwood et al., 1997; Tay et al., 1998). In an interlaboratory study to validate this Environment Canada's Reference Method, IC50s for a sample of 100% kaolin clay ranged from 1,765 to 2,450 mg/L (McLeay et al., 2001). Together, these findings support the Environment Canada interim guidelines for judging samples as toxic or not, according to the *V. fischeri* solid-phase assay (Environment Canada, 2002). These guidelines take into account the percentage of fines in the test sediment and the known sharp inflection of values when their fines content reaches or exceeds 20% (Ringwood et al., 1997), as well as the ability of a test material comprised of 100%

clay to reduce the IC50 to as low as 1,765 mg/L using this reference method (McLeay et al., 2001).

The two following interim guidelines are used for judging the toxicity of test sediment samples. The first one, which has been recommended and applied by Environment Canada in the past (Environment Canada, 1996b; Porebski and Osborne, 1998), is based on the premise that all samples are toxic, according to this biological test method, if their IC50 is < 1,000 mg/L, regardless of grain size characteristics. The second guideline is based on the premise that samples with < 20% fines might be toxic at an IC50 ≥ 1,000 mg/L, since confounding grain size effects are appreciably less in coarse-grained sediment (Environment Canada, 2002).

The first interim guideline should be applied to all samples of test sediment with ≥ 20% fines, as well as to any sample with < 20% fines which has an IC50 < 1,000 mg/L. The second interim guideline should be applied to all samples of test sediment with < 20% fines that have an IC50 ≥ 1,000 mg/L. This second guideline enables toxic coarse-grained sediments to be identified as such when their IC50 is appreciably higher than 1,000 mg/L. It is recommended that the second interim guideline be applied to each sample of test sediment with < 20% fines, except in the instance where the IC50 is < 1,000 mg/L in which case the sample should be judged as toxic and the second guideline does not apply.

Guideline 1: any test sediment from a particular sampling station and depth is judged to have failed this sediment toxicity test if its IC50 is < 1,000 mg/L, regardless of grain size characteristics.

Guideline 2: for any test sediment from a particular sampling station and depth which is comprised of < 20% fines and has an IC50 ≥ 1,000 mg/L. The IC50 of this sediment must be compared with a sample of "clean" reference sediment or negative control sediment (artificial or natural) with a % fines content that does not differ by more than 30% from that of the test sediment.[3] Based on this comparison, the test sediment is judged to have failed the sediment toxicity test if, and only if, each of the following two conditions apply:

---

[3] The following two examples are provided to illustrate how this "must" criterion is to be applied when choosing a negative control sediment or reference sediment with a percentage of fines that does not differ by more than 30% from that of the test sediment. If the test sediment has a fines content of 10%, the percent fines of the reference sediment or negative control sediment must be within the range of 7 to 13%. Similarly, if the test sediment has a fines content of 5%, the percent fines of the reference or negative control sediment must be within the range of 3.5 to 6.5%.

(1) its IC50 is more than 50% lower than that determined for the sample of reference sediment or negative control sediment[4]; and
(2) the IC50s for the test sediment and the reference sediment or negative control sediment differ significantly.

The first condition for Guideline 2 is verified using the following examples for calculations as a guide: if the sample of reference or negative control sediment used to judge the toxicity of the course-grained test sediment has an IC50 of 20,000 mg/L, the IC50 of the test sediment must be < 10,000 mg/L. Similarly, if the sample of reference or negative control sediment used to judge the toxicity of the course-grained test sediment has an IC50 of 5,050 mg/L, the IC50 of the test sediment must be < 2,025 mg/L.

The second condition for Guideline 2 must be verified using a pairwise comparison of values for the two IC50s and their 95% confidence limits, which is described in Sprague and Fogels (1977) as a means of comparing two LC50s.[5]

11. Application of the Solid-phase Test using luminescent bacteria in a case study

In this section we will examine the application of the solid-phase tests for measuring sediment toxicity using the luminescent bacterium *V. fischeri* in a study of toxicity test responses along a known pollution gradient. The test responses will be compared to sediment contamination, other toxicity test responses, and the benthic community living at the test sites.

A study undertaken along a known pollution gradient was conducted in Sydney Harbour, Canada, and the relationships between sediment and porewater chemistry, benthic community structure, and biological toxicity tests were examined (Zajdlik et al., 2001). Major contaminants were PAHs, PCBs, and heavy metals. The toxicity tests employed were: whole sediment 10-day toxicity tests with four species of infaunal amphipods, 14-day survival and growth tests with two species of polychaetes, the solid-phase test for sediment toxicity using the luminescent bacterium *Vibrio fischeri*; and the echinoderm fertilization test on sediment porewater using three species of sea urchins.

---

[4] This condition for judging sample toxicity was derived in light of the findings of two series of interlaboratory studies performed to validate this reference method (McLeay et al., 2001). In one series of tests with four identical subsamples of a contaminated sediment tested in separate assays by each of six participating laboratories, the lowest laboratory-specific IC50 was 14 to 48% lower (mean intralaboratory difference, 31%; n = 6) than its highest IC50. Given this degree of intralaboratory variability in IC50s for the same test sediment, as determined within individual laboratories, it is considered prudent to require that the IC50 for a sample of coarse-grained test sediment, if ≥ 1000 mg/L, must be more than 50% lower than that for the negative control or reference sediment with which it is compared, as one of the conditions for judging the sample as toxic.

[5] This pairwise comparison test delineated in Sprague and Fogels (1977) is thought to be suitable for comparing two IC50s (J.B. Sprague, pers. comm.). A more statistically rigorous pairwise comparison test for IC50s is currently under development by Environment Canada (Environment Canada, 2004).

A brief summary of the study results is presented in Table 4.

The luminescent bacterium solid-phase test performed well, and response (IC50) was negatively correlated with sediment contamination, and positively correlated with biological variables such as 10-day survival of infaunal amphipods in a whole sediment toxicity test, and with benthic community structure (average number of taxa per station). The mean PEL quotient for sediment contaminants (Long et al., 1998) was calculated. Results with mean PEL quotient values < 1 resulted in a 'pass' (i.e., sediment sample was classified as non toxic using interpretation criteria presented in Section 10), while results yielding mean PEL quotient values > 1 resulted in a 'fail' (toxic response). This demonstrates that when established sediment quality guidelines predict an effect, the luminescent bacterium solid-phase test will show an effect, confirming its sensitivity and usefulness as a screening tool. Spearman rank correlation coefficient was used to test if the different biological toxicity tests used in the study ranked the stations in the same way. It was found that the luminescent bacterium solid-phase test correlated highly with the results of the 10-day amphipod sediment toxicity tests.

Table 4. Summary of selected results from Sydney Harbour, NS, Canada, pollution gradient study (Zajdlik et al., 2001).

Parameter	Stations				Reference Station 12	Reference St. Ann's
	1	5	6	9		
Sum of PAHs ($\mu g/g$)	212	86.9	36.9	3.18	0.45	0.37
Total PCBs ($\mu g/g$)	2.1	1.19	0.64	ND[a]	ND[a]	ND[a]
Lead ($\mu g/g$)	286	214	133	32	21	37
Zinc ($\mu g/g$)	516	866	281	91	56	84
Amphipod (Amphiporeia virginiana) % survival	3	52	53	74	79	77
Microtox IC50 (mg/L, moisture corrected)	97	123	145	1,010	13,200	1,730
PEL Quotient[b]	14.9	7.6	3.6	0.44	0.11	0.11
Average # Taxa	2.6	5.0	5.8	17	15.5	3

[a] ND = below detection limit; [b] PEL = Probable Effects Level (Long et al. 1998).

The study demonstrated the sensitivity relevance of the luminescent bacterium solid-phase test, since results correlated with sediment contamination, structure of benthic community present at the sites, and with results of more traditional sediment toxicity tests such as the 10-day whole sediment toxicity tests undertaken with infaunal amphipods.

12. Accessory/miscellaneous test information

The Solid-Phase Test for sediment toxicity using luminescent bacteria was developed for testing samples of whole sediment and similar solid materials. The interlaboratory study described by McLeay et al. (2001) involved six testing laboratories and a total of 19 samples were analyzed using the procedure described in this chapter. The interlaboratory precision was within the limits considered acceptable in other studies of this nature (interlaboratory CV ranged from 12 % to 70 % for the 19 samples), and there was no relationship between laboratory experience with this toxicity test and the precision or validity of the results obtained. This indicates that the test is simple and reproducible. Cross-training new personnel with experienced personal, and close attention to quality control procedures, will ensure valid results.

The test is rapid (results for a sample can be obtained in slightly over an hour). Because two samples can be tested at one time, it is possible to test 6 to 8 samples in a normal working day, and this includes quality control procedures with reference toxicants and occasional duplicate assays. Therefore larger numbers of samples can be screened for toxicity using this test when compared with the traditional invertebrate toxicity test such as marine amphipods, freshwater chironomids, or soil invertebrates such as earthworms. Material costs per test are approximately 60 US$, and there are no labor costs for culturing the organisms as they are purchased in a ready-to-use form.

The small sample size required for the test (refer to Sections 8.8 and 8.9, and Figure 3), and the short test duration, make it easy for safe handling of test substances. Safety measures such as use of protective equipment (lab coat, gloves, goggles), and engineering controls (adequate ventilation and use of fume hoods), will allow the test to be carried out in a safe manner. It is necessary to have proper procedures in place for disposal of highly contaminated samples.

The procedures described in this chapter outline the endpoint for this method as the concentration causing 50% inhibition of light output, compared to unexposed control organisms (*i.e.*, the IC50 and its 95% confidence limit). There is no reason why an investigator could not choose to calculate an alternate more sensitive ICp value, such as an IC25. However, we would advise caution in using lower values of p such as an IC10 or an IC5, because these values may be within the normal variability of the test (Environment Canada, 2004).

13. Conclusions

The solid-phase test for measuring the toxicity of whole sediment using luminescent bacteria has been widely used for the assessment of freshwater, estuarine, or marine sediment, and terrestrial soils, but is theoretically applicable to any similar solid material, such as sludges and ore concentrates. The test is rapid, cost-effective, reproducible and correlated to in-situ toxic effects on benthic communities. Since the test provides information on potential toxic effects to bacteria, it is best used in a battery of toxicity tests (along with invertebrate toxicity tests) to estimate the toxic potential of sediment samples. The endpoint of the test can be used as part of an overall sediment quality assessment. Because the test is relatively cheap and easy to run, it can be used on its own to screen large numbers of samples, in order to delineate the spatial or temporal extent of sediment contamination.

The test is subject to a number of confounding factors, in particular the grain size of test sediments, and sulfide content of the sediments. Ammonia is not a strong confounding factor for this test, which could make the test useful for sediment toxicity identification evaluations. Interpretation criteria outlined in this chapter will provide guidance to the reader on the significance of the test results. More research is encouraged to validate or improve the interpretative guidance provided.

Acknowledgements

The following participated in the Environment Canada interlaboratory study to refine and validate the Environment Canada Reference for a Solid-phase test using luminescent bacteria Method: Ms. Paula Jackman, Environment Canada, Moncton, NB; Ms. Janet Pickard and Ms. Karen Kinnee, BC Research Inc., Vancouver, BC; Mr. Garth Elliott and Ms. Wendy Antoniolli, Environment Canada, Edmonton, AB; Mr. Sylvain Trottier, Environment Canada, Montreal, PQ; Mr. Graham van Aggelen and Mr. Craig Buday, Environment Canada, North Vancouver, BC; Dr. Kenneth Lee and Mr. Gary Wohlgeschaffen, Fisheries and Oceans Canada, Dartmouth, NS.

AZUR Environmental Inc. graciously supplied Bacterial Reagent and supplies enabling an inter-laboratory study that resulted in the refinement and validation of the Environment Canada Reference method for a Solid-phase test using luminescent bacteria.

References

Anonymous (1979) Flash ! Instrument Replaces Fish, *Environmental Science and Technology* **13**, 646.
ASTM (American Society for Testing and Materials) (1995) Standard Guide for Conducting Sediment Toxicity Tests with Luminescent Bacteria, Draft No. 8, ASTM, Philadelphia, PA, USA.
AZUR (1997) Basic Solid-Phase Test, Draft 6-11-96, AZUR Environmental, Carlsbad, CA, USA, 4 pp.
AZUR (1998a) Basic Solid-Phase Test (Basic SPT), AZUR Environmental, Carlsbad, CA, USA, 16 pp.
AZUR (1998b) Solid-Phase Test (SPT), AZUR Environmental, Carlsbad, CA, USA, 19 pp.
Benton, M.J., Malott, M.L., Knight, S.S., Cooper, C.M. and Benson, W.H. (1995) Influence of sediment composition on apparent toxicity in solid-phase test using bioluminescent bacteria, *Environmental Toxicology and Chemistry* **14** (3), 411-414.

Bombardier, M. and Bermingham, N. (1999) The sed-tox index: toxicity-directed management tool to assess and rank sediments based on their hazard - concept and application, *Environmental Toxicology and Chemistry* **18** (4), 685-698.

Brouwer, H. and Murphy, T. (1995) Volatile Sulfides and their Toxicity in Freshwater Sediments, *Environmental Toxicology and Chemistry* **14** (2), 203-208.

Brouwer, H., Murphy, T. and McArdle, L. (1990) A Sediment-contact Bioassay with *Photobacterium phosphoreum*, *Environmental Toxicology and Chemistry* **9**, 1353-1358.

Burton, G.A., Jr., Denton, D. L., Ho, K. and Ireland, D.S. (2003) Sediment Toxicity Testing: Issues and Methods, Chapter 5, in D.J. Hoffman, B.A. Rattner, G.A. Burton, Jr., and J. Cairns, Jr. (eds.), *Handbook of Ecotoxicology, 2nd Edition*, CRC Press LLC, pp. 111-150.

Carter, J. A., Mroz, R. E., Tay, K.L. and Doe, K.G. (1998) An evaluation of the use of soil and sediment bioassays in the assessment of three contaminated sites in Atlantic Canada, *Water Quality Research Journal of Canada* **33** (2), 295-317.

CEPA (Canadian Environmental Protection Act) (1999) Disposal at Sea, Part 7, Division 3, Sections 122-127 and Schedules 5 and 6, Statutes of Canada, Chapter 33.

Cook, N.H. and Wells, P.G. (1996) Toxicity of Halifax Harbour Sediments: an Evaluation of the Microtox Solid-phase Test, *Water Quality Research Journal of Canada* **31**(4), 673-708.

Day, K.E., Dutka, B.J., Kwan, K.K., Batista, N., Reynoldson, T.B. and Metcalfe-Smith J.L. (1995) Correlations Between Solid-phase Microbial Screening Assays, Whole-sediment Toxicity Tests with Macroinvertebrates and *In Situ* Benthic Community Structure, *Journal of Great Lakes Research* **2** (2), 192-206.

Denning, A., Doe, K., Ernst, B. and Julien, G. (2003) Screening Level Ecological Risk Assessment. Stephenville Upper Air Station, Stephenville, Newfoundland, Surveillance Report EPS-5-AR-03-02, Environment Canada, Dartmouth, NS, 47 pp. plus appendices.

Environment Canada (1990) Guidance Document on the Control of Toxicity Test Precision Using Reference Toxicants, Conservation and Protection, Ottawa, ON, Report EPS 1/RM/12, 85 pp.

Environment Canada (1992) Biological Test Method: Toxicity Test Using Luminescent Bacteria (*Photobacterium phosphoreum*), Conservation and Protection, Ottawa, ON, Report EPS 1/RM/24, 61 pp.

Environment Canada (1995) Guidance Document on Measurement of Toxicity Test Precision Using Control Sediments Spiked with a Reference Toxicant, Environmental Protection Service, Ottawa, ON, Report EPS 1/RM/30, 56 pp.

Environment Canada (1996a) Procedure for Conducting a Microtox Solid Phase Test, Standard Operating Procedure #43, prepared January 1996 by G. Wohlgeschaffen, Atlantic Environmental Science Centre, Moncton, NB, 10 pp.

Environment Canada (1996b) 1996 National Compendium Monitoring at Ocean Disposal Sites, Unpublished Report, Disposal at Sea Program, Marine Environment Division, Ottawa, ON.

Environment Canada (1997a) 1996-97 Discussion Paper on Ocean Disposal and Cost Recovery, Unpublished Report, Disposal at Sea Program, Marine Environment Division, Ottawa, ON.

Environment Canada (1997b) Biological Test Method: Test for Survival and Growth in Sediment Using the Larvae of Freshwater Midges (*Chironomus tentans* or *Chironomus riparius*), Environmental Protection Service, Ottawa, ON, Report EPS 1/RM/32, 131 pp.

Environment Canada (1997c) Biological Test Method: Test for Survival and Growth in Sediment Using the Freshwater Amphipod *Hyalella azteca*, Environmental Protection Service, Ottawa, ON, Report EPS 1/RM/33, 123 pp.

Environment Canada (1999a) Standard Operating Procedure for the Solid-Phase Toxicity Test Using Luminescent Bacteria (*Vibrio fischeri*), SOP No. IC50MS10.SOP, first draft, prepared October 1999 by C. Buday, Pacific Environmental Science Centre, North Vancouver, BC, 18 pp.

Environment Canada (1999b) Guidance Document for the Application and Interpretation of Single-species Data from Environmental Toxicology Testing, Environmental Protection Service, Ottawa, ON, Report EPS 1/RM/34.

Environment Canada (2002) Biological Test Method: Solid-Phase Reference Method for Determining the Toxicity of Sediment Using Luminescent Bacteria (*Vibrio fischeri*), Environment Canada, Environmental Protection Series, EPS 1/RM/42.

Environment Canada (2004) Guidance Document on Statistical Methods for Environmental Toxicity Tests, Fifth Draft, March 2003, Environmental Protection Service, Ottawa, ON, Report EPS 1/RM/xx, in preparation.

Gaudet, I.D. (1998) Effect of Microtox Reagent Reconstitution Age on the Variability of Analytical Results from the Microtox Assay, Final Report to Western Canada Microtox Users Committee by Alberta Research Council, Edmonton, AB.

Giesy, J.P., Graney, R.L., Newsted, J.L., Rosiu, C.J., Benda, A., Kreis, R.G. Jr. and Horvath, F.J. (1988) Comparison Of Three Sediment Bioassay Methods Using Detroit River Sediments, *Environmental Toxicology and Chemistry* **7**, 483 – 498.

Government of Canada (2001) Disposal at Sea Regulations, SOR/2001-275 under the Canadian Environmental Protection Act, *Canada Gazette Part II,* **135** (17), 1655-1657, August 1, 2001, Ottawa, ON.

ISO (1993) Water quality – Determination of the Inhibitory Effect of Water samples on the Light Emission of *Vibrio fischeri* (Luminescent bacteria test), International Organization for Standardization, ISO/TC 47/SC 5N107 (Draft), 18 pp.

Jacobs, M.W., Delfino, J.J. and Bitton, G. (1992) The Toxicity Of Sulfur To Microtox From Acetonitrile Extracts Of Contaminated Sediments, *Environmental Toxicology and Chemistry* **11**, 1137 – 1143.

Johnson, B.T. (1998) Microtox Toxicity Test System – New Developments and Applications, Chapter 14, in P.G. Wells, K. Lee, and C. Blaise (eds.), *Microscale Testing in Aquatic Toxicology - Advances, Techniques, and Practice*, CRC Press, New York, NY, USA, pp. 201-218.

Kwan, K.K. and Dutka, B.J. (1995) Comparative Assessment Of Two Solid-Phase Toxicity Bioassays: The Direct Sediment Toxicity Testing Procedure (DSTTP) and The Microtox Solid-Phase Test (SPT). *Bulletin of Environmental Contamination and Toxicology* **55**, 338 – 346.

Long, E.R., Field, L. J. and MacDonald, D. D. (1998) Predicting toxicity in marine sediments with numerical sediment quality guidelines, *Environmental Toxicology and Chemistry* **17**, 714–727.

McLeay, D., Doe, K., Jackman, P., Ross, P., Buday, C., van Aggelen, G., Elliott, G., Antoniolli, W., Trottier, S., Pickard, J., Kinnee, K., Lee, K. and Wohlgeschaffen, G. (2001) Interlaboratory Studies to Validate Environment Canada's New Reference Method for Determining the Toxicity of Sediment Using Luminescent Bacteria (*Vibrio fischeri*) in a Solid-Phase Test, Technical Report Prepared by McLeay Environmental Ltd. (Victoria, BC) and Environment Canada's Atlantic Environmental Science Centre (Moncton, NB) for the Method Development and Applications Section, Environment Canada, Ottawa, ON.

Microbics (1992) Detailed Solid-phase Test Protocol, in *Microtox Manual - A Toxicity Testing Handbook*, Microbics Corporation, Carlsbad, CA, USA, pp. 153-178.

Microbics (1995) Microtox Acute Toxicity Solid-phase Test, Microbics Corporation, Carlsbad, CA, 18 pp.

Mueller, D.C., Bonner, J.S., McDonald, S.J., Autenrieth, R.L., Donnelly, K.C., Lee, K., Doe, K. and Anderson, J. (2003) The use of toxicity bioassays to monitor the recovery of oiled wetland sediments, *Environmental Toxicology and Chemistry* **22** (9), 1945-1955.

NICMM (National Institute for Coastal and Marine Management/RIKZ) (1999) Standard Operating Procedures, Marine: Microtox Solid-phase *(Vibrio fisheri)* Sediment Toxicity Test, in C.A. Schipper, R.M. Burgess, M.E. Schot, B.J. Kater, and J. Stronkhorst, *Project SPECIE-BIO*, RIKZ/AB-99.107x, Ver.1.1, June 1999, NICMM, Middelburg, The Netherlands, 23 pp.

Porebski, L.M. and Osborne, J.M. (1998) The application of a Tiered Testing Approach to the Management of Dredged Sediments for Disposal at Sea in Canada, *Chemistry and Ecology* **14**, 197-214.

Porebski, L.M., Doe, K.G., Zajdlik, B.A., Lee, D., Pocklington, P. and Osborne, J. (1999) Evaluating the techniques for a tiered testing approach to dredged sediment assessment - a study over a metal concentration gradient, *Environmental Toxicology and Chemistry* **18** (11), 2600-2610.

Qureshi, A.A., Flood, K.W., Thompson, S.R., Janhurst, S.M., Innis, C.S. and Rokosh, D.A. (1982) Comparison of a Luminescent Bacterial Test with Other Bioassays for Determining Toxicity of Pure Compounds and Complex Effluents, in J.G. Pearson, R.B. Foster, and W.E. Bishop, (eds.), *Aquatic Toxicology and Hazard Assessment: Fifth Conference, ASTM STP 766*, American Society for Testing and Materials, Philadelphia, PA., USA, pp. 179-195.

Qureshi, A.A., Bulich, A.A. and Isenberg, D.I. (1998) Microtox Toxicity Test Systems – Where They Stand Today, Chapter 13, in P.G. Wells, K. Lee, and C. Blaise (eds.), *Microscale Testing in Aquatic Toxicology -- Advances, Techniques, and Practice*, CRC Press, New York, NY, USA, pp. 185-199.

Ringwood, A.H., DeLorenzo, M.E. Ross, P.E. and Holland, A.F. (1997) Interpretation of Microtox Solid-phase Toxicity Tests: the Effects of Sediment Composition, *Environmental Toxicology and Chemistry* **16** (6), 1135-1140.

Ross, P. and Leitman, P.A. (1995) Solid Phase Testing of Aquatic Sediments Using *Vibrio fischeri*: Test Design and Data Interpretation, Chapter 6, in M. Richardson (ed.), *Environmental Toxicology Assessment*, Taylor and Francis, London, UK, pp. 65-76.

Ross, P. (1998) Role of Microbiotests in Contaminated Sediment Assessment Batteries, Chapter 38, in P.G. Wells, K. Lee, and C. Blaise (eds.), *Microscale Testing in Aquatic Toxicology -- Advances, Techniques, and Practice*, CRC Press, New York, NY, USA, pp. 549-556.

Ross, P., Burton, G.A. Jr., Greene, M., Ho, K., Meier, P., Sweet, L., Auwarter, A., Bispo, A., Doe, K., Erstfeld, K., Goudey, S., Goyvaerts, M., Henderson, D., Jourdain, M., Lenon, M., Pandard, P., Qureshi, A., Rowland, C., Schipper, C., Schreurs, W., Trottier, S. and van Aggelen, G. (1999) Interlaboratory Precision Study of a Whole Sediment Toxicity Test with the Bioluminescent Bacterium *Vibrio fischeri, Environmental Toxicology* **14**, 339-345.

Schiewe M.H., Hawk, E.G., Actor, D.I. and Krahn, M.M. (1985) The Use Of Bacterial Bioluminescence Assay To Assess Toxicity of Contaminated Marine Sediments, *Canadian Journal of Fisheries and Aquatic Sciences* **42**, 1244 – 1248.

Sprague, J.B. and Fogels, A. (1977) Watch the Y in Bioassay, in *Proceedings of the 3rd Aquatic Toxicity Workshop*, Halifax, NS, November 2-3, 1976, Environment Canada, Environmental Protection Service Technical Report No. EPS-5-AR-77-1, pp. 107-118.

Tay, K.-L., Doe, K.G., Wade, S.J., Vaughan, J.D.A., Berrigan, R.E. and Moore, M.J. (1992) Sediment bioassessment in Halifax Harbour, *Environmental Toxicology and Chemistry* **11**, 1567-1581.

Tay, K.L., Doe, K.G., MacDonald, A.J. and Lee, K. (1998) The Influence of Particle Size, Ammonia, and Sulfide on Toxicity of Dredged Materials for Ocean Disposal, Chapter 39, in P.G. Wells, K. Lee, and C. Blaise (eds.), *Microscale Testing in Aquatic Toxicology - Advances, Techniques, and Practice*, CRC Press, New York, NY, USA, pp. 559-574.

True, C.J. and Heyward, A.A. (1990) Relationships between Microtox test results, extraction methods, and physical and chemical compositions of Marine sediment samples, *Toxicology Assessment* **5**, 29 - 45.

Zajdlik, B.A., Doe, K.G. and Porebski, L.M. (2000) Report on Biological Toxicity Tests Using Pollution Gradient Studies - Sydney Harbour, Environment Canada, Marine Environment Division, Ottawa. Report EPS 3/AT/2, July, 2000, 104 pp.

Abbreviations

CV	coefficient of variation
ICp	inhibiting concentration for a (specified) percent effect
IC50	50% inhibiting concentration
PAHs	Polycyclic Aromatic Hydrocarbons
PCBs	polychlorinated-biphenyls
PEL	Probable Effects Level
SD	standard deviation
SOP	standard operating procedure
SPT	solid-phase test
wt:v	weight-to-volume.

3. ALGAL MICROPLATE TOXICITY TEST

CHRISTIAN BLAISE
St Lawrence Centre, Environment Canada
105 McGill Street, Montreal
Quebec H2Y 2E7, Canada
christian.blaise@ec.gc.ca

PAULE VASSEUR
Université Paul Verlaine, ESE, CNRS FRE 2635
Campus Bridoux, rue du Général Delestraint
57070 METZ, France
vasseur@sciences.univ-metz.fr

1. Objective and scope of test method

The test method described below can be employed on its own, or as part of a *battery of tests approach*, to estimate the phytotoxicity potential of diverse types of liquid samples. Because it is conducted in a 96-well microplate format, the technique is simple and offers other advantages which include cost-effectiveness and space-saving features.

It is theoretically applicable to any liquid. Testing can be undertaken, for example, with samples representing the following media:

(1) domestic and industrial wastewaters, treated or untreated;
(2) surface, groundwater or leachates;
(3) sediment interstitial waters;
(4) any chemical that is soluble in water;
(5) any water-insoluble chemical that can be rendered soluble by means of an organic solvent or other techniques (*e.g.*, sonication or emulsification).

2. Summary of test procedure (at a glance)

The test system makes use of exponentionally growing cells of *S. capricornutum* (the biological reagent or test organism) that are exposed for 72 h in a 96-well microplate to varying concentrations of a liquid solution (a chemical or water/wastewater sample), under controlled experimental conditions of temperature and light (Tab. 1). In such a compact microplate format, each well can be viewed as one of 96 independent *miniaturized flasks* which are available for undertaking algal toxicity testing. Employing a standard microplate experimental configuration (Fig. 1, details given later on in Section 8.2), a specific test solution concentration (200 µL) is introduced into a pre-defined well to which are also added nutrient spike (10 µL) and algal inoculum (10 µL).

Table 1. Rapid summary of test procedure.

Test organism	—	Physiologically active cells of *Selenastrum capricornutum* taken from a culture in exponential phase of growth (4-8 d old cells) are harvested as inoculum for testing.
Type of test	—	Chronic toxicity test (72-h exposure); static.
Test format	—	96-well polystyrene microplate with round-bottomed wells.
Well volume contents	—	200 µL test solution; 10 µL nutrient spike; 10 µL algal inoculum.
Algal inoculum	—	10,000 ± 1,000 cells/mL.
Lighting	—	24-h (uninterrupted) vertical fluorescent illumination ("cool-white") having 4 klx at the surface of the microplate lid and a quantal flux of 60-80 $\mu E.m^{-2}/s^{-1}$.
Temperature	—	24 ± 2°C.
Experimental configuration	—	10 control wells; 10 serial dilutions of test solution, each with five replicates.
Measurements	—	Cell enumeration after 72 h with an electronic particle counter or hemacytometer.
Endpoints determined	—	IC50, IC25, NOEC/LOEC, based on cell yield in relation to controls.
Reference toxicants	—	Zn^{2+} ($ZnSO_4$), Cu^{2+} ($CuCl_2$) [or Cu SO_4], $K_2Cr_2O_7$ or NaCl.

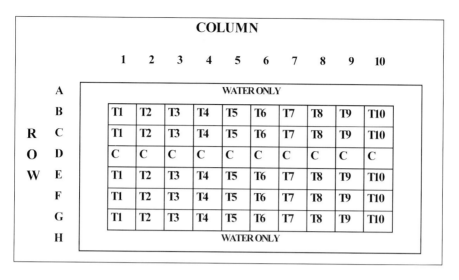

Figure 1. Suggested experimental configuration of a 96-well microplate for phytotoxicity testing (see Section 8.2 for details).

After filling of microplate wells is completed, the nutrient spike addition will have enabled (unexposed) control algal cells to reproduce and to attain an acceptable biomass over the incubation period. Growth inhibition resulting in wells containing specific test solution concentrations can then be attributed to a toxic effect stemming from the presence of bio-available chemical(s) in the test solution and not to a nutrient deficiency. Comparing cell yield of control algae to that of test solution-exposed algae at the end of the exposure period will allow the calculation of growth inhibition percentages from which endpoints of interest (*e.g.*, IC50) can be determined.

3. Overview of applications with microplate-based phytotoxicity tests

Since the appearance of the first report describing a microplate-based phytotoxicity test investigation conducted on oil dispersants over two decades ago (Heldal et al., 1978), more than 50 articles have been published describing methods, applications and data comparisons with such procedures (Blaise et al., 1998). Because microplate-based phytotoxicity tests are alternatives to earlier flask-based tests, some research groups felt it was important to demonstrate concordance of toxicity responses between the two procedures. Several studies were, in fact, conducted in this perspective and did confirm interprocedural comparability. Others have developed several acute and chronic exposure tests employing various micro-algal species, with a diversity of assessment and measurement endpoints, and have generated data to study the toxicity of a wide array of chemicals and complex samples. Microplate testing protocols, however, reflect notable differences insofar as experimental variables are concerned and toxicity responses have been shown to vary accordingly (Blaise et al., 1998). While comparing similarities and differences (and relative value) of microplate assays is beyond the scope of this chapter, the reader will appreciate that research initiatives aimed at developing and applying miniaturized algal tests have been, and continue to be, an effervescent part of the field of small-scale aquatic toxicology (Tab. 2).

Table 2. *Summary of types of tests, micro-algae, endpoints and testing media, linked to freshwater single species phytotoxicity testing conducted with microplates (adapted from Blaise et al., 1998).*

Test types and assessment/measurement endpoints

Acute tests (1-5 h) :

- Assessment endpoints: esterase inhibition; ATP energy loss; cell growth recovery; motility inhibition.
- Measurement endpoints: IC20, IC50; NOEC/LOEC.
- Testing media: metals, organics.

Chronic tests (2-30 d) :

- Assessment endpoints: growth inhibition (cell counts; fluorescence/absorbance/ATP as biomass indicator; visual observation of impaired growth); cell death (esterase inhibition and chlorophyll impairment by flow cytometry measurement).
- Measurement endpoints: *toxic threshold effect concentration*; IC20, IC50, IC100; LC50.
- Testing media: metals, organics, herbicides, oil dispersants, effluents, solid waste leachates, groundwater, organic extracts.

Test species

Blue-green algae: *Anabaena cylindrica*

Green algae: *Chlamydomonas variabilis*; *Chlamydomonas reinhardti*; *Chlorella vulgaris*; *Chlorella sp.*; *Chlorella pyrenoidosa*; *Chlorella kessleri*; *Monorhaphidium pusillum*; *Scenedesmus quadricauda*; *Scenedesmus subspicatus*; *Selenastrum capricornutum.*

4. Advantages of conducting microplate-based phytotoxicity tests

Undertaking phytotoxicity assays in a 96-well microplate format has several advantages which have been outlined in previous reports (Heldal et al., 1978; Blaise et al., 1986; Thellen et al., 1989). The more important ones are integrated and highlighted in Table 3. Clearly, some of these advantages will also hold true for microplate-based toxicity tests conducted with different cells, micro-organisms or small organisms, as other chapters in this book indicate.

Table 3. Main advantages of using microplates for undertaking toxicity tests with micro-algae.

Advantages	*Remarks*
Small sample volume requirement	10 mL for a microplate test suffice. Relevance: complex environmental samples (*e.g.* an industrial effluent) often have considerable loads of suspended particles and must be filtered (0.45 µ) prior to testing. Filtering a markedly smaller effluent volume is a real time-saver. For testing chemicals, some of which may be hazardous or costly, reduced quantities are used for preparing test solutions.
Incubator space economy	A flask-based test using 3 replicates for each of 10 test concentrations + 3 control flasks (similar to the experimental configuration described in the microplate technique) would require placing 33 × 125 mL Erlenmeyer flasks in the incubator: this corresponds to placing one 96-well microplate in the incubator.
Disposable microplates and pipette tips	One-time use of (recyclable) polystyrene microplates eliminates post-experimental washing of glassware (*i.e.*, flasks). This saves time and effort, as well as preventing potential contamination and/or toxicity problems resulting from re-use of glassware. With time, wear and tear can affect optical properties of flasks (*e.g.*, through aging and scratching) and contribute to decreased test precision (*i.e.*, by augmenting cell count variability of test and/or control replicates).
Increased bioanalytical output	Owing to the availability of 1) efficient micro-pipetting devices (*e.g.*, multi-channel and repetitive micro-pipettes) and 2) robotic instrumentation (*e.g.*, automated dilutors), microplate tests can be initiated more rapidly than flask tests. This can contribute to increased laboratory productivity, particularly if large numbers of samples have to be processed for phytotoxicity studies.

5. Test species

The recommended test species for the microplate procedure described is the chlorophyte (green alga) *Selenastrum capricornutum*, Printz, 1914 (average width: 2-4 µ; average length: 10-15 µ; average volume: 40-60 µm^3). In studying growth curves of this alga, cells have been shown to be smaller (15-20 µm^3) during exponential growth and larger (60-70 µm^3) as the stationary phase is reached (Skulberg, 1967). Mean cell volumes can increase even more (*e.g.*, circa 200 µm^3), particularly when this alga is grown in test waters likely to contain chemicals (*e.g.*, metals and pesticides) which are known to inhibit cell division (Miller et al., 1978). Minimum and maximum temperatures enabling growth of *S. capricornutum* have been reported to be 10 and 35°C, respectively (Haaland and Knutson, 1973). Optimal temperature for growth, with close to doubling of the population every 12 hours, has been shown to be 24°C (Reynolds et al., 1975).

S. capricornutum is a chlorophyll-containing eukaryotic cell whose taxonomical position is as follows:

- Phylum: *Chlorophyta*
- Sub-phylum: Chlorophycophyta
- Class: Chlorophycea
- Sub-class: *Chlorophycidea*
- Order: Chlorococcales

It is crescent-shaped, unicellular, non motile, non polymorphic (*i.e.*, retaining the same shape throughout its cell cycle) and is representative of both eutrophic and oligotrophic freshwater environments (Fig. 2). Aggregation of cells is uncommon in *S. capricornutum* which, on top of its other characteristics, makes it a very suitable alga for enumeration via electronic particle counters. Finally, its demonstrated sensitivity to a variety of hazardous substances favors its use as a reliable indicator of phytotoxicity (Blanck et al., 1984; Blaise et al., 1986; Blaise et al. 1987; Hickey et al., 1991; St-Laurent et al., 1992; Wängberg et al., 1995).

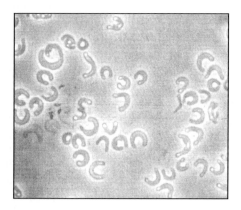

Figure 2. The green alga, Selenastrum capricornutum, *under phase contrast microscopy at 400 x.*

Isolated in 1959 from the Nitelva River (Akershus county, Norway) by the Norwegian scientist Olav M. Skulberg, *S. capricornutum* was preserved at the Norwegian Institute of Water Research (NIVA) under code name NIVA-CHL 1 and employed as an indicator to estimate the fertility potential of freshwater bodies (Skulberg, 1964). Used again in North America since the 1960's to undertake similar eutrophication studies (U.S. EPA, 1971; Miller et al., 1978), it quickly became popular as a test species for phytotoxicity assessments as soon as such procedures were developed (Chiaudani and Vighi, 1978; Heldal et al, 1978; Joubert, 1983; Blanck et al., 1984).

The alga, *S. capricornutum* (Printz), has undergone two recent taxonomical changes. It was first renamed *Raphidocelis subcapitata* (Nygaard et al., 1986) and later became *Pseudokirchneriella subcapitata* (Hindak, 1990). Following these taxonomical tribulations, therefore, the correct appellation for NIVA-CHL 1 algal strain is purported to be *"Pseudokirchneriella subcapitata* (Korshikov) Hindak*"*. Whereas the latter name is more prevalent in Europe, the name *S. capricornutum* still remains popular in North America. Although the procedure detailed below is performed with *S. capricornutum*, let the reader be reminded that it could be undertaken with other algal species as well (Tab. 2).

This alga can be purchased from different sources, some of which are listed in Box 11 (Section 14). When received, axenic strains obtained from these sources may be on an agar slant, in liquid culture or frozen in an ampoule as a dried pellet.

The test species can also be obtained as algal beads preserved in an alginate matrix from the source given in Box 12 (Section 14). As they have been cultured and trapped in this matrix while in their exponential phase, they are immediately usable for testing once they have been de-immobilized from their matrix (see details of procedure in Section 7.3 *Algal preservation techniques* below).

6. Culture / maintenance of organism in the laboratory

6.1 LAB FACILITIES REQUIRED

Laboratory facilities (with standard materials and equipment) where temperature and lighting can be controlled and monitored continuously are essential to ensure adequate conduct of algal bioassays with the microplate procedure. Chambers/incubators employed for culturing algae must be separate from those designated for phytotoxicity testing. All culturing, maintenance and toxicity testing areas should be well-ventilated, free of potential toxicant input (*e.g.*, toxic dust and/or vapours from chemical compounds which may not be properly contained or may be in too close proximity) and protected from any disturbing external factor (*e.g.*, sound vibrations or contamination from other incompatible microbiological activities).

6.2 MATERIALS

Materials required for conducting the microplate phytotoxicity test are listed below. While relatively complete, this list is meant as a guide to readers and not all materials may be absolutely necessary to carry out the assay. Use of various items will depend on options open to specific laboratories. For example, if an electronic particle counter is not available and algal cell counts are estimated with a hemacytometer, accuvettes do not have to be purchased, but hemacytometer cover glasses do.

Box 1. Required material for testing.

— *S. capricornutum* micro-algae.	— Graduated cylinders: 25, 50, 100, 500 and 1000 mL capacity.
— Microplates: sterile disposable, rigid, polystyrene, 96-well (capacity ≈ 0.25 mL), U-shaped or round-bottomed with cover).	— Beaker: 1 L.
	— Filtration membranes (0.45 μm porosity).
— Graduated glass (or sterile disposable serological) pipettes: 1 and 10 mL.	— Weighing spatula.
	— Weighing dishes.
— Adjustable digital multichannel pipette (10-50 μL capacity).	— Glass Pasteur pipettes.
	— Hemacytometer cover glasses.
— Adjustable digital multichannel pipette (50-200 μL capacity).	— 10 mL disposable plastic cups (*e.g.*, Coulter accuvettes or Sarstedt # 73.1056 cups) for enumerating algae with an electronic particle counter.
— Adjustable digital micropipette (10-100 μL capacity).	
— Adjustable digital micropipette (100-1000 μL capacity).	— Magnetic stirring bars.
	— Volumetric flasks: 100, 500, 1000 mL capacity.
— Disposable pipette tips to accommodate pipetting devices (as above).	— Tube rack for 20-mm tubes.
	— Tube rack for 40-mm tubes.
— Disposable plastic reservoirs (for multichannel pipetting needs).	— Inoculating loop and holder.
	— Wash bottle (for buffered water solution: see Section 6.5.5).
— Disposable glass test tubes (16 × 150 mm).	— Erlenmeyer flasks: 125 mL, 500 mL, 1 L, 2 L, 4 L capacity (size dependent on importance of algal biomass required for inoculating purposes).
— Disposable centrifuge tubes with screw caps (15 and 50 mL capacity).	
— Disposable Petri dishes (15 × 100 mm).	
— Sealable transparent (polyester) plastic bags (≈ 16 × 20 cm).	— Aluminium foil.

6.3 EQUIPMENT

Just as for the materials listed above, not all pieces of equipment listed below may be essential to carry out the assay, and those used may depend on laboratory facilities and available apparatus. All equipment employed for measurements should, of course, be adequately maintained and calibrated regularly based on good laboratory practices. Moreover, any equipment in contact with algae, reagent water, nutrient solutions, growth/enrichment media, test solutions, etc., must be made of chemically inert material (*e.g.*, glass, stainless steel, plastic), clean and free of substances which could interfere with testing.

Box 2. Required equipment for testing.

— Autoclave (used to sterilize everything other than growth medium).	— Centrifuge (benchtop) with 2000 g speed capacity.
— Electronic particle counter for enumerating algal cells (*e.g.*, Beckman Coulter ZM model).	— Calculator.
	— Gas burner and gas source.
— Hemacytometer (a Neubauer counting chamber for enumerating algal cells if particle counter is not available).	— Metric balance for weighing chemicals.
	— pH meter.
— EC (environmental chamber) or I (incubator) with built-in light and temperature controls, as well as 100 rpm orbital shaking capacity.	— Filtering device and accessories: vacuum pump and tubing; 1-L filtering flask; 47-mm stainless steel filter holder.
	— Magnetic stirrer.
— Millipore Super - Q^{TM} water purification system (or equivalent such as glass distilled): ensures water quality that is free of ions, organics, particles and micro-organisms greater than 0.45 µm.	— Heat sealer.
	— Vortex.
	— Glass or electronic thermometer.
	— Illumination meter (0-10 klx range): to verify light intensity at the surface where algae are grown or tested.
— Refrigerator.	
— Microscope (400 × magnification capacity) ideally with phase contrast capability.	

6.4 WASHING OF GLASSWARE

All reusable glassware (flasks, graduated cylinders, beakers, etc. as listed in *Materials* section above) must be cleaned to remove all substances (trace metals, organics, nutrients) capable of interfering with algal growth according to the method outlined below (Environment Canada, 1992). It is also recommended that new glassware be conditioned in the same manner to avoid potential toxicity and/or enrichment problems from chemicals leaching into solution from inner walls of such. Reusable material made of products other than glass must also be washed similarly if it can withstand the specified acid treatments.

Box 3. Method for washing of glassware.

- Wash material with a non phosphate detergent solution.
- If required, loosen any visible (particulate) matter attached to the inner wall of glassware with a brush (stiff-bristle).
- Rinse three times with tap water.
- Rinse with a cleaning solution (*e.g.*, 10% HCl or equivalent); for large glassware, fill partially but make sure entire inner wall has been soaked (by swirling) with cleaning solution.
- Rinse three times with deionised water.
- Oven dry at 105°C.
- Cover opening of each glassware piece with aluminium foil and store until use.

6.5 PREPARATION OF REAGENTS AND ALGAL CULTURE MEDIA

- Micronutrient (solution 1)
- Macronutrient (solutions 2-5)
- Algal culture medium (1×)
- Algal test 13.75× enrichment medium
- Buffered water
- Algal cell counting diluent

All chemical compounds (reagents) used must be of analytical grade quality (*e.g.*, American Chemical Society specifications) or at least of 99% purity. Glass distilled or Millipore Super Q water is recommended as reagent water.

The Algal culture medium is composed of both micro- and macro-nutrients essential to ensure proper growth of algae for culture and maintenance in the laboratory (Miller et al., 1978). Five stock solutions are prepared with the reagents listed below. To start off, label five 500 mL volumetric flasks (solutions 1-5) and add 350 mL of water to each.

6.5.1 Micronutrient stock solution (Solution 1)
Weigh each chemical and add individually to solution 1 flask in the order shown below. Ensure that each chemical is dissolved prior to adding the next chemical by proper mixing of the flask.

Box 4. Solution 1 ingredients.

Magnesium chloride	$MgCl_2 \cdot 6H_2O$	6.08 g
Calcium chloride	$CaCl_2 \cdot 2H_2O$	2.20 g
Boric acid	$(H_3BO_3)_3$	92.8 mg
Manganese chloride	$MnCl_2 \cdot 4H_2O$	208.0 mg
Zinc chloride	$ZnCl_2$ [a]	1.64 mg
Iron chloride	$FeCl_3 \cdot 6H_2O$	79.9 mg
Cobalt chloride	$CoCl_2 \cdot 6H_2O$ [b]	0.714 mg
Sodium molybdate	$NaMoO_4 \cdot 2H_2O$ [c]	3.63 mg
Copper chloride	$CuCl_2 \cdot 2H_2O$ [d]	0.006 mg
Tetraacetic acid	$Na_2EDTA \cdot 2H_2O$	150.0 mg

- [a] Using a zinc-labelled 100 mL volumetric flask, add 70 mL of water to it. Next, weigh out 164 mg of zinc chloride ($ZnCl_2$) and transfer into the flask. Complete the volume to 100 mL with water and mix well. From this zinc solution, withdraw 1 mL with a pipet and transfer to solution 1 flask.

- [b] Using a cobalt-labelled 100 mL volumetric flask, add 70 mL of water to it. Next, weigh out 71.4 mg of cobalt chloride ($CoCl_2 \cdot 6H_2O$) and transfer into the flask. Complete the volume to 100 mL with water and mix well. From this cobalt solution, withdraw 1 mL with a pipet and transfer to solution 1 flask.

- [c] Using a molybdate-labelled 100 mL volumetric flask, add 70 mL of water to it. Next, weigh out 363 mg of sodium molybdate ($NaMoO_4 \cdot 2H_2O$) and transfer into the flask. Complete the volume to 100 mL with water and mix well. From this molybdate solution, withdraw 1 mL with a pipet and transfer to solution 1 flask.

- [d] Using a copper-labelled 100 mL volumetric flask, add 70 mL of water to it. Next, weigh out 60.0 mg of cupper chloride ($CuCl_2 \cdot 2H_2O$) and transfer into the flask. Complete the volume to 100 mL with water and mix well. Transfer 1 mL of this solution into another volumetric flask containing 70 mL of water. Complete this volume to 100 mL with water and mix well. From this second copper solution, withdraw 1 mL with a pipet and transfer to solution 1 flask.

After all compounds have been added, adjust the volume of solution 1 to 500 mL with water.

6.5.2 Macronutrient stock solutions (Solutions 2-5)
Weigh each chemical and add to designated flasks for solutions 2-5. Ensure that each chemical is well dissolved by proper mixing of each flask. After all compounds have been added, adjust the volume of solutions 2-5 to 500 mL with water.

Box 5. Ingredients of solutions 2 to 5.

Solution #	Chemical		Quantity (500 mL^{-1})
2	Sodium nitrate	(NaNO$_3$)	12.75 g
3	Magnesium sulfate	(MgSO$_4$·7H$_2$O)	7.35 g
4	Potassium phosphate	(K$_2$HPO$_4$)	0.552 g
5	Sodium bicarbonate	(NaHCO$_3$)	7.5 g

6.5.3 Algal culture medium 1× (ACM-1×)

This normal strength (1×) algal medium serves to culture and maintain healthy algae in the laboratory as a source of inocula for toxicity tests (details are given in Section 7). It is prepared by adding 1 mL of each of the five 1000× stock nutrient solutions described above (*i.e.*, solutions 1-5) to a 1-L volumetric flask filled with ≈ 900 mL of reagent water. Mix well between each addition and complete to 1000 mL afterwards. By then adding a magnetic bar into the volumetric flask and moderate swirling over a magnetic stirrer, the final pH is adjusted to 7.5 ± 0.1 with 1N HCl or 1N NaOH. The ACM-1× is next filter-sterilized at a vacuum not exceeding 50.7 kPa (380 mm Hg) by pouring into a (sterile) stainless steel filtering funnel (linked to a 47-mm stainless steel filter holder) equipped with a (reagent water) pre-washed 0.45μ membrane. Sterilization of ACM-1× by autoclave is not advised as it may reduce algal growth potential, possibly by altering bioavailability of nutrients to algae and/or rendering cells less able to assimilate them (Environment Canada, 1992). The filter-sterilized ACM-1× solution is poured into one (or several) sterile Erlenmeyer flask(s), capped with sterile tops, such as pieces of aluminium foil or aluminium weighing boats, and can be refrigerated at 4°C in the dark for up to six months.

As it stands to reason, the amount of ACM-1× growth medium prepared will be dictated by the number of toxicity tests that a laboratory will undertake on a daily or weekly basis. Dependent on such logistics, laboratory personnel will prepare enough medium to ensure sufficient quantity of algal cells for testing purposes (an example of ACM-1× volume calculation is given in Section 8.4.2). On this basis, ACM-1× will be prepared accordingly and may be poured into 125, 250, 500, 1000-mL (or larger) Erlenmeyer flasks. To avoid suboptimal algal growth owing to potential carbon dioxide limitation, a volume-to-flask ratio of 20% is essential (Miller et al., 1978). To respect this ratio, one would place 25 mL of ACM-1× medium in a 125 mL flask, 50 mL of ACM-1× medium in a 250 mL flask, and so on.

6.5.4 Algal test 13.75× enrichment medium (ATEM)

In phytotoxicity testing, addition of a specific quantity of nutrients to experimental containers (*i.e.*, flasks, vials or microplate wells as in this technique) is necessary to confirm that algal growth inhibition, which may result from exposure to a particular sample, is genuinely due to a toxic effect and not to a lack of nutrients. This is the reason for preparing a concentrated ATEM solution as described below, a small volume of which will be dispensed as a *nutrient spike* to each microplate well at the onset of a test. The rationale behind preparing this solution as a 13.75× concentrate of the ACM-1× is explained in Section 8.4.2.

Since all five stock solutions (*i.e.*, solutions 1-5) are 1000× concentrates of ACM-1×, 13.75× test enrichment medium is prepared by adding 13.75 mL of each stock solution (in the order 1, 2, 3, 4 and 5) to 800 mL of reagent water and completing to 1 L. Mix well, adjust final pH to 7.5 ± 0.1 and filter-sterilize as before (Section 6.5.3). The ATEM solution is then poured into an (inert material) clean container (*e.g.*, 1 L polypropylene bottle with screw cap top) and can afterwards be kept refrigerated at 4°C in the dark for up to six months.

6.5.5 *Buffered water*

A water-bicarbonate solution is essentially used as a reagent during the centrifugation of algal cells in order to concentrate their numbers and to wash them so as to prepare an algal inoculum for testing. From stock solution 5 (15 g/L $NaHCO_3$), prepare a 1/1000 dilution by pipetting 1 mL of solution 5 into a 1 L volumetric flask containing 900 mL of water. Mix well and adjust the volume to 1 L with water. This solution can be poured into a 1 L plastic (squeezable) wash bottle equipped with a spout for easy dispensing into centrifuge tubes used for centrifuging algae.

6.5.6 *Algal cell counting diluent (electrolyte solution)*

If algae are to be enumerated with an electronic particle counter (*e.g.*, for monitoring cell growth in culture, preparing an algal inoculum for testing or measuring cell yield after testing), an algal cell counting diluent can be purchased (*e.g.*, Isoton: Fisher Scientific CS606-20) or can be prepared as follows:

NaCl	0.15 mole. L^{-1}
KCl	3.0 m mole. L^{-1}
Phosphate buffer (pH = 7.5)	15.0 m mole. L^{-1}

7. Preparation of micro-algae for testing

7.1 ENSURING AXENIC CULTURES

Axenic cultures of *S. capricornutum* are, by definition, free of bacteria and other micro-organisms. Algal cells used in toxicity tests should be axenic when obtained initially and remain as such when cultured and maintained in the laboratory.

Although micro-algal cultures purchased from a culture collection (see Section 5) should be certified as axenic, it is recommended that this be verified prior to commencing toxicity testing. This is accomplished by inoculating algae into a 20-26°C ("room temperature")-conditioned flask containing ACM-1×, as explained below.

Box 6. Method for ensuring axenic cultures.

- If the algal culture is in colony form on an agar slant or frozen in an ampoule as a dried pellet (purchased as such from a culture collection), a small portion of cell growth can be aseptically removed with a bacteriological loop and inoculated into a 125-mL flask containing 25 mL of ACM-1×. Alternatively, if the algal culture is in liquid form (purchased as such from a culture collection or obtained from another laboratory or stems from a laboratory undertaking a recommended monthly check for culture purity), a small volume ($\approx$ 1 mL) of liquid culture can be aseptically removed with a sterile pipette and inoculated into a 125-mL flask containing 25 mL of ACM-1×.

- The inoculated flask is then incubated under continuous lighting (60-80 $\mu E.m^{-2}.s^{-1}$) at 24 ± 2°C and with agitation (manual swirling of a culture flask, for a few seconds at a time, three times daily or at 100 rpm if an incubator-rotator chamber is available). Depending on their physiological state, algae may take 7-14 days before a visible greenness (indicative of the exponential growth phase) becomes apparent. At this stage, an initial appraisal of culture purity can be performed by transferring a loop of algal culture onto a microscope slide and checking for the presence of contaminating micro-organisms (magnification between 400-1000×). Concurrently, 1 mL of the algal culture is withdrawn with a sterile pipette and dispensed into a Petri dish containing solid ACM-1× medium in 1% agar* where algal cells are then uniformly distributed with a sterile glass *hockey stick* (small piece of *"∠- shaped"* bent glass used to spread liquid inocula on the surface of solid bacteriological media). After incubating the Petri plate upside down (with lighting and temperature conditions as above), visible colonies will appear after approximately two weeks.

- An isolated (axenic) colony of *S. capricornutum* can then be picked (aseptically with a bacteriological loop) and inoculated into a fresh flask containing ACM-1× medium. After incubating as before, a visible greenness (indicative of the exponential growth phase) should once again become apparent after 4-7 days. Several inoculations of this algal culture should then be made by streaking with a bacteriological loop into either solid 1% agar ACM-1× medium Petri dishes or into prepared agar slants containing the same medium. When visible colonies appear in Petri dishes (or slants) after several days of incubation, they can be stored in the dark at 4°C for up to three months. These colonies thus constitute a laboratory's reserve of healthy axenic cells for future use.

* To prepare solid 1% agar ACM-1× medium, add 1% agar to ACM-1× and heat to dissolve. Sterilize by autoclaving at 98 kPa (1.1 kg/cm^2) and 121°C for 30 min. Pour into Petri dishes, cover partially at first until medium has cooled somewhat and then cover lid totally. After medium has solidified, store upside down at 4°C in dark. Solid ACM-1× medium can be used up to three months afterwards.

7.2 EXPERIMENTAL REQUIREMENTS FOR PREPARING LOG PHASE CELLS

Under controlled conditions of light, temperature and nutrients, *S. capricornutum* cells (as with those of other algal species) follow a set course characterized by several phases. After inoculation into growth medium with a fixed number of physiologically-active cells, population growth patterns normally include a lag phase, an exponential (or log) phase, one of declining growth rate, a stationary phase and, finally, a death phase (Fig. 3). For the successful conduct of phytoxicity tests, it is imperative to use as inoculum cells from algal stocks which are in exponential phase. This ensures a shortened lag phase and optimal growth rates leading to cell densities that will allow adequate comparisons between control and sample-exposed growth at the end of the exposure period (Walsh, 1988).

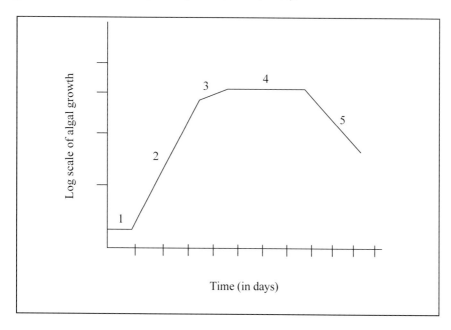

Figure 3. Growth phases of algal cultures: Lag (1), Exponential (2), Declining growth rate (3), Stationary (4), Death (5), according to Fogg (1965).

Once axenic cultures of algae have been obtained (Section 7.1 above), they can (if phytotoxicity testing is conducted on a regular basis) be routinely sub-cultured once a week in ACM-1× medium by transferring 1-2 mL of liquid growth from a 4-8 d old culture flask to a new culture flask. This ensures a steady supply of logarithmic phase cells for bioanalytical inoculations. Algal cultures, for example, can be grown for this purpose by placing 500 mL of ACM-1× into a 2 L flask under optimal temperature and light conditions (Tab. 1) with orbital shaking at 100 rpm.

Periodically (4 times per year), it is important to determine an algal growth curve from a newly-inoculated growth medium flask as a check for health and performance of cells which are used for toxicity testing. At T = 0 and on subsequent days, approximately 5 mL of algal culture are aseptically withdrawn from the 4-8 d old culture flask and placed in a test tube. Then, 3 × 1 mL aliquots (or portions thereof) are respectively enumerated (with a particle counter or hemacytometer). Plotting daily values graphically indicates the approximate time at which the stationary phase is reached (usually between 8-10 days) and when the growth curve experiment can end, as shown in Figure 4. Also observable is the time period of the exponential phase (*circa* 4-8 days post-inoculation), occurring when controlled temperature, lighting and nutrient regimes are adhered to, and during which cells should be harvested as inocula for toxicity testing.

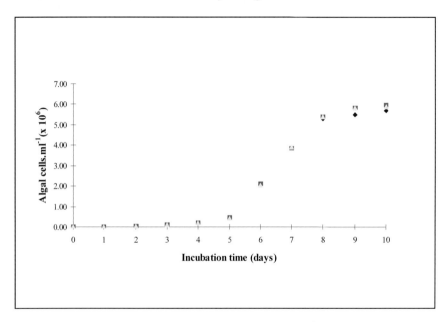

Figure 4. Typical growth curve of Selenastrum capricornutum *inoculated (with 100 μL of a 4-8 d old culture flask growth) into a 2 L flask containing 500 mL of ACM-1x.*

7.3 ALGAL PRESERVATION TECHNIQUES

Readily available biological reagents can confer advantages for toxicity testing by eliminating the need for constant culturing, thereby contributing to reduction of labour and increased cost-effectiveness. Examples include lyophilisation of luminescent bacteria (Bulich et al., 1981) and the use of cryptobiotic (dormant) stages of micro-invertebrates (Persoone, 1991). Immobilization of exponentially-growing micro-algae in an alginate bead matrix is an effective means of preserving cells for prolonged periods (Mohammad, 1991). Laboratories may find it more convenient to periodically prepare batches of algae preserved in this manner instead of having to culture them on a continuous basis. *S. capricornutum* cells coated in this way have been shown to retain full viability for several months and can be employed as reliably for toxicity testing as laboratory-cultured algae (Persoone, 1998). The major steps involved in the immobilization (preservation) and de-immobilization (preparation for testing) procedure of *S. capricornutum* are described below. Prepared algal beads preserved in an alginate matrix, as well as a matrix dissolving medium, can also be purchased commercially (see Section 5).

Box 7. Method for (de)immobilizing algae.

Immobilization of algae

− Working under sterile conditions, prepare a 5% sodium alginate solution (Alginic acid sodium salt, A-7128 or Protonal, PROTAN Ltd., 3000 Drammen, Norway) in a ACM-1× algal culture medium (quantity to prepare will depend on frequency of testing, but will generally vary from 100-1000 mL). Homogenize in an Oster^{TM} glass blender and pour 50 mL portions in as many 250 mL Erlenmeyer flasks as required. Autoclave flasks (20 min; 120°C; 15 psi). Afterwards, let cool until contents reach room temperature.

− Prepare a 2% $CaCl_2·2H_2O$ solution and autoclave (20 min; 120°C; 15 psi). Refrigerate between 4-8°C.

− Pour 50 mL of a 4-7 day old algal culture (algal density will generally be between 1-3 million cells per mL) into a sodium alginate solution flask and stir (vigorously and constantly) for 5 min to mix well. After introducing volumetric portions of this (alginate-algal) mixture in a burette, a syringe or a micropipette, allow drops to fall into a beaker (or flask) containing the $CaCl_2·2H_2O$ solution. The diameter of the falling drops can be changed depending on the diameter of the outlet used and depending on flow speed of the falling drops. Leave the algae (which are now immobilized as spherical beads) immersed for 50-60 min in the $CaCl_2·2H_2O$ solution. Then, rinse the beads three times with ACM-1× culture medium.

− Keep the algal beads in the culture medium at 4°C in darkness until it is time to use them for testing.

****

De-immobilization of algae

− If algae have been preserved in an alginate matrix, take 5-6 algal beads (making sure to pour off any ACM-1× medium liquid with which they may be associated and taking care not to lose any of the algal beads in the process) and transfer into a 50 mL conical tube.

− Add 5 mL of 0.1M NaCitrate (an alginate matrix dissolving solution) to this tube. Cap the tube and handshake vigorously for 30 s every 2 min until the matrix immobilizing the algae is totally dissolved (20-30 min). Using a vortex shaker (moderate intensity) will speed up the process (5 min).

− Centrifuge the tube at 1000 g for 10 min in a benchtop centrifuge. Pour off the supernatant and replace it with 10 mL of buffered water. Cap the tube and shake it vigorously by hand (or vortex) to resuspend the algae. Centrifuge the tube at 1000 g for 10 min, pour off the supernatant and resuspend the algae in 5 mL of buffered water.

− Alginate-immobilized algae are thus rapidly de-immobilized in ≈ 30 min, after which they are immediately available as inoculum for initiating toxicity testing.

8. Testing procedure

8.1 INFORMATION / GUIDANCE REGARDING TEST SAMPLES PRIOR TO CONDUCTING BIOASSAYS

8.1.1 Chemicals

For the health and safety of laboratory personnel, physical and chemical properties of the substance undergoing investigation should be obtained and safety data sheets should be consulted when these are available. Important information having relevance for test procedure and data interpretation include the following: water solubility, vapour pressure, structural formula, dissociation constants, K_{OW} (n-octanol: water partition coefficient); degree of purity; nature and amounts of impurities and/or additives; stability and persistence in freshwater.

Volumetric flasks should be employed for preparation of stock and test solutions. Stock solutions of chemicals not readily soluble in water can be prepared by one of various techniques which include ultrasonic dispersion or use of organic solvents, emulsifiers or surfactants. Laboratory operators engaged in biological testing of such substances should consult competent colleagues and select the most appropriate method available or recommended for this purpose. If a carrier substance is used to dissolve a chemical in preparation for testing, an additional *carrier control solution* must be incorporated into the microplate experimental design at the highest (non phytotoxic[1]) concentration at which the solubilizing agent is employed in the test.

8.1.2 Complex environmental samples

It is assumed that aqueous samples such as effluents, receiving waters and leachates, as well as sediments from which elutriates or interstitial waters might be extracted must first be collected according to standard methods to ensure the representativity of their subsequent phytotoxicity responses. Samples are usually placed in clean labelled containers of inert material (*e.g.*, glass or polypropylene/polyethylene vessels), filled to the brim (minimal headspace) and transported in the dark to the laboratory on ice ($\approx$ 4-7°C). While less than 10 mL are sufficient for conducting the microplate assay, 1 L of sample, particularly in the case of effluents, is usually collected as further toxicity testing (*e.g.*, range-finding test followed by a definitive test or having to redo testing owing to laboratory error/problem) or other observations (*e.g.*, color, odor) and measurements (*e.g.*, temperature, turbidity, suspended solids, pH, various chemical analyses) may warrant additional volume. Effluents and leachates should preferably be tested as soon as possible but no longer

---

[1] If the highest non toxic concentration of the solubilizing agent is unknown, the microplate test as described below must first be conducted with this carrier by exposing algae to different concentrations in order to determine its NOEC (No observed effect concentration). Usually, carriers such as acetone and methanol (St-Laurent et al., 1992) are used at concentrations which are less than their NOEC in order to avoid any possible interactions (*e.g.* additivity or synergistic exacerbation of the phytotoxic response) which might occur between an agent and a test chemical (Stratton and Smith, 1988).

than three days after collection. Extraction of elutriates or interstitial (pore) waters from collected sediments should be performed within 14 days of collection and liquids tested within three days.

Because complex environmental (waste)waters often contain varying amounts of suspended solids and micro-organisms such as bacteria and micro-algae, the procedure described herein must be carried out on 0.45 µm membrane-filtered samples. This is necessary to offset major interferences that would be imposed on instruments such as electronic particle counters, photometers or fluorometers used to measure cell yield at the end of the exposure period. Since toxic effects can also be linked to suspended solids (Pardos and Blaise, 1999; White et al., 1996a), the filtering process may not enable this technique to appraise the full toxic potential inherent in a complex sample. This is clearly a limitation of most phytotoxicity tests and of other bioassay procedures that need to be performed on filtered liquid media to eliminate sample interferences which would otherwise confound test results and make them uninterpretable. Hence, the algal growth inhibition microplate procedure presented here can only report on the toxicity of complex samples associated with bio-available contaminant(s) that are present in their soluble fraction but not for those which may be adsorbed on their particulate fraction.

Algal growth inhibition occurs when the pH of ACM-1× culture medium is outside of a 6-9 pH range (Miller et al., 1978). Hence, test samples with pH values beyond these two limits will inhibit growth of test algae owing to pH-based toxicity. If test objective is to assess phytotoxicity considering that pH is an integral part of a sample's total hazard potential, then testing can proceed without sample pH adjustment. If, on the other hand, there is an interest in knowing the toxic potential of the sample without the influence of pH, then sample pH can be adjusted to 6.5 (for samples whose initial pH was ≤ 6) and to 8.5 (for samples whose initial pH was ≥ 9). In such a case, the sample is therefore tested with and without pH adjustment. Because of the important role that it may play, as a toxic agent on its own or in modulating the toxicity of test sample contaminant(s), the pH of all samples must be measured prior to testing.

8.2 EXPERIMENTAL CONFIGURATION OF MICROPLATE

With 96 wells available as individual test containers in a microplate, it is obvious that several experimental configurations can be chosen based on testing objectives (*e.g.*, acute or chronic toxicity assessment of one or more substances with one or more species). Selection of a specific configuration must also aim to minimize any foreseeable interferences that can be associated with toxicity testing conducted in microplates (*e.g.*, edge effects or test substance volatility discussed further on). Figure 1 displays an experimental disposition, initially used to evaluate the phytotoxicity of herbicides (St-Laurent et al., 1992), and which is now recommended by Environment Canada to perform standardized toxicity testing with *S. capricornutum* (Environment Canada, 1992). Since a phenomenon commonly associated with 96-well microtitration plates is the so-called "edge effect", whereby the evaporation rate of circumferential wells tends to be greater than that of centrally-located wells, the experimental disposition excludes peripheral wells from testing because they increase variability among replicates. Hence, for chronic tests of one day or more, experimental designs are usually built around the 60 internal wells, as seen in Figure 1. While peripheral wells are not directly employed for testing, they are nevertheless always filled with water to increase humidity inside the microplate. In turn, this also acts to minimize evaporative losses from the inner wells used for testing. In this sense, peripheral wells contribute positively to the overall experimental protocol.

Typically, the assay is run with five replicates for each of 10 test concentrations (Fig. 1) with 10 control wells (row D) *sandwiched* in between two (rows B, C) and three (rows E, F, G) test concentration rows, respectively. The central insertion of control wells in this manner (*i.e.*, parallel to a gradient of decreasing test concentrations from T1 to T10) is meant to check for the presence of volatile toxicants which may emanate from adjacent test wells (owing to the fact that microplate lids are loose-fitting and do not hermetically seal off wells from one another). If this occurs, heterogeneity in post-exposure cell yields may be observed among control wells with those located next to the higher test concentration wells (*i.e.*, T1-T5) displaying lesser algal densities than their counterparts located next to the lower test concentration wells (*i.e.*, T6-T10). Indeed, skewness in control well counts was observed when the phytotoxicity of phenol, a volatile compound, was assessed with the microplate technique (Thellen et al., 1989). Hence, noting control heterogeneity likely signals the presence of a volatile toxicity effect and, in a certain sense, actually confers an advantage to this microplate configuration in being able to give this type of relevant information. The downside may mean that too much variability observed in control replicates will require the microplate assay to be repeated (see Section 9.3 on test validity), this time with lower starting concentrations of the test sample or substance, so as to reduce or eliminate volatile effects. Another option would be to separate individual test well concentration columns (*e.g.*, the five T1 test wells) by filling adjacent columns with reagent water (*e.g.*, the five T2 test wells would be filled with H_2O) and running the assay with two microplates so as to allow testing the full ten test concentrations as before.

If, as discussed above in Section 8.1.1, it is necessary to use a carrier substance to dissolve a hydrophobic chemical in preparation for testing, an additional *carrier control solution* (at the highest concentration at which the solubilizing agent is employed in the test) would then be incorporated into the microplate experimental design and placed in row E thereby excluding the normal series of T1-T10 test concentrations for that row. In this event, the microplate assay is carried out with four replicates of each test concentration instead of five.

8.3 TEST SAMPLE CONCENTRATIONS

Under ideal conditions, a phytotoxicity test should include 1) a concentration that will have no effect on algal growth (0% growth inhibition), 2) a concentration that will induce an intense or total algistatic effect (90-100% growth inhibition), 3) two concentrations below a 50% growth effect and 4) two concentrations above a 50% growth effect. In this situation, four growth-inhibiting toxicity values are available for plotting a *percent growth inhibition* (y-axis) versus *test concentration* (x-axis) graph from which measurement endpoints (*e.g.*, IC50) can be estimated. When the toxicity of a substance or (waste)water is known, as in the case of a reference toxicant routinely used to verify the adequacy of the algal reagent or an industrial effluent periodically monitored to check its conformity to environmental regulations, use of a standard range of test concentrations will enable proper estimation of an IC50 with one single test. This is not always possible when the toxicity of an environmental sample or substance is unknown. In this event, more than one test (see below) may be necessary to properly circumscribe the IC50 within the right range of test concentrations. Because the suggested microplate experimental configuration can accommodate 10 concentrations for toxicity testing (Fig.1), double the number of test concentrations that quite a few bioassays commonly propose in their procedures, it is often possible to estimate an IC50 from the conduct of a single test. A definitive test conducted with a proper range of test concentrations, however, must sometimes be preceded by one or more *range-finding tests* as indicated hereafter.

8.3.1 Selecting test concentrations for samples whose toxicity is known
If the toxicity of a test sample to *S. capricornutum* is known, test concentrations are prepared that encompass a range of responses from 0% to 90-100% growth inhibition. Since these samples likely belong to the category of reference toxicants, frequently monitored (waste)waters or substances whose phytotoxicity has been reported elsewhere, the test performed, more often than not, is usually sufficient to enable the estimation of the desired endpoint.

8.3.2 Selecting test concentrations for samples whose toxicity is unknown
For chemicals whose toxicity to *S. capricornutum* is unknown, preparation of a stock solution at a concentration close to its solubility limit in water is first called for. A similar stock solution can be prepared for hydrophobic chemicals with the help of an appropriate carrier as discussed in Section 8.1.1.

Starting out with as high a test concentration as possible from the stock solution, a range-finding test can be performed to broadly estimate the IC50 with widely separated concentrations that should allow the establishment of substance concentrations for the definitive test. Most bioassays could recommend, for example, using a minimum of five concentrations at 100, 10, 1, 0.1 and 0.01% of the highest possible test concentration. With 10 test concentrations available with the microplate format, a more than adequate range-finding test could call for testing 100, 50, 10, 5, 1, 0.5, 0.1, 0.05, 0.01 and 0.005% of the test chemical. Additional to identifying the response range of concentrations entrapping the IC50, the range-finding test can also serve to determine whether the test substance, if toxic, may have volatile properties as discussed in Section 8.2.

Clearly, range-finding tests can make use of several types of arithmetic (*e.g.*, 100, 50, 25, 12.5, 6.25 ... 0.195 mg/L for a chemical or 100, 50, 25, 12.5, 6.25 ... 0.195% v/v for an industrial effluent sample) or geometric series (*e.g.*, 100, 60, 36, 22, 13 ... 1 mg/L or % v/v based on a dilution factor of 0.6) of test concentrations to get a first estimate of the IC50 and some may or may not give expected results the first time around. A second range-finding test, with a different range of test concentrations may sometimes be required before a definitive test can be undertaken to report a final and precise IC50.

If a chemical substance is toxic to algae in the range-finding test (and the IC50 has not been able to be determined with sufficient reliability), a definitive test can be undertaken by making sure that the estimated IC50 will be encompassed by at least five concentrations of toxicant. For example, if the range-finding test has shown a test substance to have had no growth inhibition and total growth inhibition effects at 10 mg/L and 100 mg/L, respectively, test concentrations of 100, 60, 36, 22 and 13 mg/L (based on a dilution factor of 0.6) should provide acceptable growth inhibition percentages to calculate the IC50. In some cases, this sequence may prove inappropriate for a chemical having a very narrow phytotoxic effect concentration range. In the example just given, it may be that the IC50 is actually circumscribed between 10 mg/L (no effect) and 25 mg/L (100% effect) and another *definitive test* would have to be conducted with a series of concentrations lying within this 10-25 mg/L zone.

From the above, the reader will appreciate that there is no fool-proof method for determining test concentrations that will ensure determination of a statistically sound IC50 in one single definitive test. Algal responses to toxicant effects, obviously, are key in guiding the ecotoxicologist to what will eventually turn up to be the appropriate series of concentrations for this purpose.

Range-finding tests for (waste)waters may end up being less complicated than for chemicals. In running the microplate assay, Canadian laboratories usually prepare 10 serial dilutions of test concentrations (dilution factor of 0.5) such that algae will initially respond to effects lying between 100% v/v (highest concentration) and 0.195% v/v (lowest concentration) of sample concentration. In most cases, this range of test concentrations will encompass the IC50 and a definitive test may not be necessary. In countries where environmental statutes have not yet enacted regulations to control the quality of industrial effluent discharges based on bioassays, the above series of test concentrations may be too high to determine an IC50 and a more diluted series of effluent concentrations may be called for.

8.4 DISPENSING SAMPLE, ALGAE AND NUTRIENTS TO A TEST MICROPLATE

All media required for testing should be at room temperature in preparation for testing.

8.4.1 Preparation and dispensing of sample test concentrations
Once a series of test concentrations has been selected based on sample type and rationale presented above in Section 8.3, the sample dilution process can commence. It is explained in Box 8 and schematized in Figure 5.

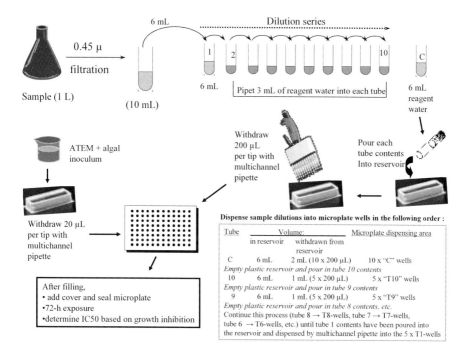

Figure 5. Sample dilution procedure for undertaking a microplate toxicity test.

Box 8. How to fill a test microplate.

- If a (waste)water sample is being tested, filter ≈ 10 mL through a 0.45 μm membrane and place the filtered sample (or highest chemical concentration to be tested in case a prepared chemical solution is being assayed for toxicity) in a 20 mm test tube.

- Prepare the selected series of dilutions by combining appropriate volumes of sample and reagent water in designated 20 mm test tubes. The easiest way to prepare serial dilutions is with a dilution factor of 0.5 (as pictured in Fig. 5). In this scenario, tubes 2 to 10 are first filled with 3 mL[2] of reagent water. Then, 3 of the 6 mL of sample contents of tube 1 (highest test concentration) are withdrawn (*e.g.*, with a 5-mL pipette) and dispensed into tube 2 containing 3 mL of reagent water. After this transfer, mix tube contents by manual shaking or by vortex. In tube 2, the sample is now diluted two-fold and is at 50% of its original concentration.

---

[2] Volume present in each sample concentration test tube must be <u>no less than 3 mL</u> when later poured into the plastic reservoir as explained here. Subsequent multichannel pipetting could not otherwise adequately withdraw multiple microvolumes for dispensing into microplate wells if smaller volumes were employed.

Box 8 (continued). How to fill a test microplate.

- Continue this process (3 mL withdrawn from tube 2 dispensed into tube 3, mixing; 3 mL withdrawn from tube 3 dispensed into tube 4, mixing; etc.) until tube 10. Mix tube 10 (lowest test sample concentration now at 0.195% of tube 1 sample concentration) which now contains a total volume of 6 mL. For each volumetric transfer from tube to tube, use separate (clean) pipettes, since the dilution gradient is in the direction of *higher to lower test sample concentrations*.

- Fill a separate 20 mm test tube with 5-6 mL of reagent water. This is the control tube (C-tube) from which reagent water will be added to designated microplate wells.

- Starting with the control tube (C-tube), pour its contents into a sterile disposable plastic reservoir. It is specially designed for subsequent multichannel[3] pipette uptake of microliter volumes. Place 10 tips on the multichannel pipette and withdraw 10×200 µL volumes from the reservoir. Dispense into the $10 \times$ C-wells of the microplate.

- Void reservoir of remaining reagent water and pour tube 10 (lowest prepared sample concentration) contents into it. Leave 5 of the previous 10 tips on the multichannel pipette and withdraw 5×200 µL volumes from the reservoir. Dispense into the $5 \times$ T10-wells of the microplate. Make sure pipette tips on the multichannel pipette are properly aligned (two adjacent tips - no tip -three adjacent tips) since T10-wells (and other test wells) are separated by a row of control (C-wells) in the experimental microplate configuration.

- Continue this process (tube 10 → T10-wells, tube 9 → T9-wells, tube 8 → T8-wells, etc.) until tube 1 contents have been poured into the reservoir and dispensed by multichannel pipette into the $5 \times$ T1-wells. Since sample dilutions are microplated in the direction of *lower to higher test sample concentrations* (*i.e.*, T10 to T1 wells), the same reservoir can be used for the entire process. The reservoir, of course, must be emptied of its previous content of sample concentration volume prior to pouring in the next highest sample concentration.

- Before or after filling microplate wells with sample dilutions, fill each of the 36 peripheral wells with 220 µL of reagent water.

---

[3] If a multichannel pipette is not available, a repetitive pipette can be used or a single channel pipette. In this event, the sterile disposable plastic reservoir is not employed and microvolumes are directly withdrawn from tubes 10 to 1 and dispensed into microplate wells.

8.4.2 Preparation/dispensing of algae and nutrients and sealing of microplate
Preparation of the algal inoculum solution. Whether taken from a 4-8 day old growth culture or from a de-immobilized alginate matrix (see Section 7.3), the algal inoculum is prepared no more than 2-3 h prior to testing. To estimate the concentration of algae required for testing (based on number of tests to be conducted, volume of algal inoculum that will have to be dispensed into microplate wells and minimum volume considerations for multichannel pipetting), one needs to measure the concentration of cells present in the stock culture available. Stock cultures in 4-8 day old flasks usually vary from 0.5 - 4 × 10^6 cells/mL. The following table describes a method for preparing an algal inoculum from a stock culture of physiologically-active cells found to be at a concentration of 1.2 × 10^6 cells/mL, based on counts obtained with a particle counter.

Box 9. How to prepare an algal inoculum for testing.

- Withdraw 10 mL of the stock culture algae with a 10 mL pipette and dispense into a 15 mL plastic centrifuge tube. At a concentration determined to be 1.2 × 10^6 cells/mL, this volume is amply sufficient to initiate several microplate tests if required. Harvested cells are then centrifuged at 2000 *g* for 15 min, after which the supernatant is poured off and replaced with a few mL (*e.g.*, 5 mL) of buffered water. Cap the tube and shake it vigorously by hand (or vortex) to resuspend the algae, and determine the concentration of cells with a particle counter or hemacytometer.

- With the concentration now found to be 2 × 10^6 cells/mL, it is evident that some (normal) loss of cells has occurred owing to the pipetting and centrifugation steps just undertaken. While we started off with 1.2 × 10^7 cells (*i.e.*, 1.2 × 10^6 cells/mL in 10 mL = 1.2 × 10^7 cells), we now end up with 1 × 10^7 cells/mL (*i.e.*, 2 × 10^6 cells/mL in 5 mL = 1 × 10^7 cells/mL). This corresponds to a loss close to 17% and indicates the importance of re-counting algal cells after the steps performed so that the right algal inoculum can be calculated.

- Keeping in mind that the required inoculum for testing is set at 10,000 cells/mL (Table 1) and that we intend to dispense 2200 cells/10 µL with a multichannel pipette to each microplate well, we now need to prepare a cell concentration of 220,000 cells/mL for this purpose.

- With our stock solution now at 2 × 10^6 cells/mL, we can calculate that 0.11 mL of this solution contains 220,000 cells: *i.e.*, 2 × 10^6 cells/mL = 220,000 cells/x mL and x mL = 0.11 mL.

Box 9 (continued). How to prepare an algal inoculum for testing.

- To prepare the 220,000 cells/mL inoculum solution, withdraw 0.11 mL from the stock solution and add it to 0.89 mL of buffered water in a 15 mL centrifuge tube: *i.e.*, [0.11 mL of stock solution ÷ [(0.89 mL of buffered water + 0.11 mL of stock solution)] × [2 × 10^6 cells/mL] = 220,000 cells/mL.

- Bearing in mind that dispensing 10 µL of this solution in each of 60 wells per microplate will use up 600 µL (0.6 mL) and that a multichannel pipette will be used to withdraw this amount from a plastic reservoir, one mL of our 220,000 cells/mL solution is clearly insufficient to ensure adequate volume withdrawal by the tips of the multichannel pipette.

- Depending on our test needs, we can simply remove more volume from the 2 × 10^6 cells/mL stock solution to prepare our inoculum solution. For example, removing 0.44 mL (0.11 mL × 4) from the stock solution and adding this volume to 3.56 mL (0.89 mL × 4) of buffered water will now give us 4 mL (0.44 mL + 3.56 mL) of inoculum solution.

Preparation of the nutrient spike solution. The nutrient spike, which should already have been prepared ahead of time, corresponds to the ATEM solution (algal test 13.75× enrichment medium) described in Section 6.5.4. Just as a volume of 10 µL of the algal inoculum solution described above is dispensed into each of the 60 microplate wells used for testing via a multichannel pipette, so does a similar volume of the nutrient spike solution (10 µL) have to be introduced into the same wells.

Dispensing of algae and nutrients. While the algal inoculum solution and the nutrient spike solution can individually be introduced into the test microplate wells (*i.e.*, 10 µL of nutrient spike micropipetted with a 10-tipped multichannel pipette followed by 10 µL of algal inoculum solution micropipetted similarly), they can also be introduced simultaneously. Assuming, as above, that 4 mL of algal inoculum solution have been placed into a 15 mL centrifuge tube, then 4 mL of nutrient spike (ATEM solution) is simply added to this tube. After proper mixing, tube contents are carefully poured into a plastic reservoir with a 10-tipped multichannel pipette now set to withdraw 20 µL of this (algal inoculum + nutrient spike) mixture for dispensing into test microplate wells. Combining the two solutions in this way is not only a time-saver, but also insures better micropipetting precision because a larger micro-volume (20 µL) is being dispensed.

Purpose of ATEM (algal test 13.75× enrichment medium) solution. Readers may question the rationale behind the preparation of a 13.75× enrichment medium. The 1× algal culture medium described herein (see Section 6.5.3) contains 0.186 mg.L^{-1} P (from 1.04 mg.L^{-1} of K_2HPO_4). Since P is the first limiting factor for growth of *S. capricornutum* in this medium, this amount of P will allow a cell yield corresponding to a biomass of 80 mg.L^{-1} of algae (dry weight) if maximum standing crop were to be measured after 7-8 days of growth under experimental conditions described for the microplate technique (Keighan, 1977; Miller et al., 1978). When equal volumes of ATEM (13.75× enrichment medium) and algal inoculum solution are combined as explained above, its resulting concentration then becomes 6.875×. Afterwards, 20 µL of the (algal inoculum and 6.875× enrichment medium) mixture are dispensed into each well containing a 200 µL sample volume. In the wells, ATEM then has a concentration of 0.625× (since 20 µL/ [20 µL + 200 µL] × 6.875 × ATEM = 0.625 ATEM). Ultimately, this specific concentration of enrichment medium is able to sustain a maximum algal growth yield of 50 mg.L^{-1} of biomass in dry weight (*i.e.*, 0.625 ATEM × 80 mg.L^{-1} = 50 mg.L^{-1}). In phytotoxicity testing, an enrichment medium spike such as this one is essential to confirm that algal growth inhibition, which may result from exposure to a specific sample, is due to a toxic effect and not to a lack of nutrients.

Sealing microplate. Once filled, the microplate is covered with its lid and placed in a transparent plastic bag (*e.g.*, Hot Sealable Pouches, Kapak/Scotchpak, 16.5 × 20 cm, Fisher Scientific). This bag is then heat-sealed (Kapak Sealer, Fisher Scientific) so as to minimize evaporation of well contents during the subsequent 72-h exposure period.

8.5 EXPOSURE CONDITIONS

Sealed microplates are placed (unshaken) in an incubator (*e.g.*, New Brunswick^{MC}) or designated room with temperature and light control. Vertical (uninterrupted) fluorescent ("cool-white") illumination having 4 ± 10% klx (= 4 kilolux = 4000 lumens = 400 foot-candles) at the surface of the microplate lid and a quantal flux of 60-80 µE.m^{-2}.s^{-1} are recommended to insure optimal growth of *S. capricornutum* at a temperature of 24 ± 2°C. Total exposure time is 72 h.

9. Post-exposure observations / measurements and endpoint determinations

9.1 POST-EXPOSURE CELL ENUMERATION WITH A PARTICLE COUNTER

Post-exposure microplates are unsealed and the lid is removed. By placing the microplate over a white background (*e.g.* over a white sheet of paper), the resulting algal biomass can be readily observed. This will help to determine which of the test wells should be used for cell enumeration in order to determine the IC50, as algae may not necessarily have to be counted in all experimental wells. For example, T1 wells (refer to suggested experimental configuration in Fig. 1) may show total absence of growth resulting from the 72-h exposure whereas T8, T9 and T10 wells may visually appear to show as much growth as the C (control) wells. In this case, the IC50 is likely circumscribed between experimental wells T2 to T8 meaning that it is not necessary to count cells in T1, T9 and T10 wells.

After proper calibration of your electronic particle counter according to standard operating procedures, enumeration of *S. capricornutum* cells can begin with an aperture tube having an operative aperture of 70 µ. Counting principle is based on the fact that each particle (algal cell) passing through the aperture causes a voltage drop proportional to its volume. Each drop in voltage is then registered by the instrument to subsequently yield a total cell count specific for each sample (= algal cells present in each microplate well). Algal cells, which may have settled at the bottom of wells because the microplate is unagitated during exposure, must first be resuspended by rapidly drawing and releasing contents back into the wells with either a micropipette or multichannel pipette with tips set to withdraw 170 µL. After drawing and releasing 10 consecutive times to ensure homogeneous resuspension of algal cells in wells, a volume of 170 µL is withdrawn from each well (starting with controls C and then with T10, T9, T8 ... T1 wells) and placed in corresponding 10 mL counting beakers (plastic cups). Afterwards, these are filled with 7 mL of electrolyte solution, the composition of which is described in Section 6.5.6.

With the microplate method described herein, the particle counter is usually set to count particles (algal cells) from a 100 µL (0.1 mL) volume of the counting beaker solution. The resulting cell density per well is calculated according to the formula below.

$$\text{Cell density/mL} = [(V_{ES} + V_W) \div V_W] \times mL/V_{EPC} \times ACC \quad (1)$$

Where:

V_{ES}	=	volume of electrolyte solution;
V_W	=	volume withdrawn from each well;
V_{EPC}	=	designated sample volume of "($V_{ES} + V_W$)" passing through aperture of the electronic particle counter;
ACC	=	algal cell count registered by electronic particle counter.

Box 10. *Calculating algal cell density from cell numbers obtained with a particle counter.*

Volume of electrolyte solution (V_{ES})	Volume withdrawn from each well (V_W)	Designated sample volume of "($V_{ES} + V_W$)" passing through aperture of the electronic particle counter (V_{EPC})	Algal cell count registered by electronic particle counter (ACC)	Calculated algal cell density/mL
7 mL	0.170 mL	0.10 mL	2300	970,060*
7 mL	0.170 mL	0.10 mL	990	417,550
7 mL	0.170 mL	0.10 mL	150	63,260

* i.e., [(7 mL + 0.170 mL) ÷ 0.170 mL] × mL/0.1 mL × 2300 = 421.765 × 2300 = 970,060.

9.2 MEASUREMENT ENDPOINT DETERMINATION

Determining the estimated sample concentration at which a specified percent reduction in growth occurs compared to control algae is a recommended endpoint. This ICp (inhibiting concentration for a specified percent) is typically sought for a 50% effect (IC50), although other ICps can also be measured (*e.g.*, IC20, IC25). Table 4 gives an example of laboratory data generated with Cu^{2+}. Algal cells in microplate wells are first enumerated electronically (see Section 9.1). Mean cell yield of control and individual cell yield of sample-exposed algae from each well are then calculated from which growth inhibition percentages are derived based on the formula below:

$$I = [(R_c - R) \div R_c] \times 100 \qquad (2)$$

where:
I = the percent inhibition in algal growth for each sample concentration replicate;
R_c = the mean cell density (number/mL) for control algae;
R = the cell density (number/mL) for each sample concentration replicate.

Table 4. Phytotoxicity data generated for Cu^{2+} (test concentrations were prepared with a $CuSO_4$ solution) with the Algal Microplate Method.

Cu^{2+} test concentration (µg/L)	Algal concentration (cells $\times 10^5$)	Mean algal concentration (cells $\times 10^5$)	Growth inhibition in relation to control algae (%)
0 (control)	8.94, 10.68, 11.01, 8.11, 12.53, 12.73, 10.54, 9.67 (n = 8)[a]	10.52	---
8	10.92, 10.31, 9.95 (n = 3)	10.39	1.2
16	10.09, 8.79, 9.14 (n = 3)	9.34	11.1
24	5.43, 6.14, 6.07 (n = 3)	5.88	43.7
32	3.13, 3.25, 2.68 (n = 3)	3.02	70.7
40	1.20, 1.09, 0.82 (n = 3)	1.04	89.3
48	0.44, 0.51, 0.41 (n = 3)	0.45	94.8
56	0.30, 0.38, 0.33 (n = 3)	0.33	96
64	0.21, 0.26, 0.19 (n = 3)	0.22	97
72	0.20, 0.16, 0.31 (n = 3)	0.22	97
80	0.21, 0.19, 0.12 (n = 3)	0.17	97.5
Calculated 72h-IC50 = 26.30 (21.52 – 31.14)[b] µg.L^{-1}			

[a.] While there are ten control wells (*i.e.*, wells D1 to D10, as per Fig. 1, eight (D1 to D4 and D7 to D10) are normally counted routinely.
[b.] 95% confidence intervals.

These data should then be plotted (percentage inhibition values, y-axis, *versus* sample concentration values, x-axis) and a line of best fit drawn around the data points (Fig. 6). This provides a visual evaluation of the data which is helpful in detecting trends or patterns in the responses, as well as an aid in result interpretation. The line drawn perpendicularly from the 50% inhibition point on the y-axis to the regression line and again perpendicularly downwards until it meets the y-axis gives identifies the 72h-IC50 value (Fig. 6).

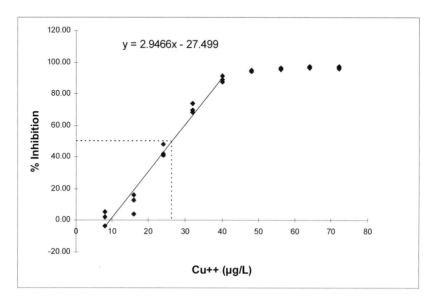

Figure 6. Concentration-response curve indicating percentage growth inhibition of S. capricornutum after a 72-h exposure to a range of Cu^{2+} concentrations. The 50% effect concentration, or IC50, is close to 27 µg/L of Cu^{2+}.

A linear interpolation method is then used to determine a precise point estimate of the test sample (toxicant solution, effluent, etc.) that produces a specific percent reduction (*e.g.*, 20, 25 or 50%) in algal growth. A particular ICp can be obtained with the computer program *ICPIN* (Norberg-King, 1993; U.S. EPA, 2002), which determines several quantitative parameters associated with the measured endpoint (*e.g.*, control and test sample means and coefficients of variation, ICp confidence limits). A copy of this program, and supporting documentation, can be obtained by written request from the following address: EMSL-Cincinnati, 3411 Church Street, Cincinnati, OH 45244, USA.

Toxicity results can also make use of hypothesis testing to report a NOEC (*no observed effect concentration*) and a LOEC (*lowest observed effect concentration*). These may be useful endpoints to determine when a concentration-response trend, for example, is not apparent (or absent) in the data. Calculations are based on the same data for algal cell density in each control and test sample well replicate. Whenever possible, reporting both ICp and NOEC/LOEC endpoints is recommended. NOEC and LOEC values from data presented in Table 4 above are 16 and 24 µg.L^{-1}, respectively. A disadvantage of reporting NOEC/LOEC values is linked to the fact that variance cannot be calculated which does not allow the determination of confidence limits. Statistical procedures for estimating NOEC/LOEC endpoints are explained in several papers (Newman, 1995; Environment Canada, 2001; U.S. EPA, 2002).

The commercial software package called *TOXSCALC* includes the ICPIN program and also allows determination of NOEC/LOEC endpoints. It can be purchased from Tidepool Scientific Software (see Box 13).

9.3 CONDITIONS FOR TEST VALIDITY AND BUILT-IN QUALITY CONTROL

9.3.1 Conditions for test validity

Cell yield in control wells must indicate an absence of trend (*e.g.*, a type of skewed gradient whereby control cell yield in wells adjacent to higher sample concentrations might be less than in control wells adjacent to lower sample concentrations: this might suggest the presence of volatile toxicants, as discussed in Section 8.2). A trend analysis test (*Mann-Kendall test;* $\alpha = 0.05$) performed on cell yields of the ten control wells will indicate presence or absence of trend (Gilbert, 1987).

The coefficient of variation (c.v.) for cell yield in the control wells must be $\leq 20\%$ (U.S. EPA, 2001). The c.v. for eight recorded controls in Table 5 data is 15.3%.

The algal cell density in control wells must have increased by a factor of more than 16 after 72 h (Environment Canada, 1992). The factor of increase for Table 5 data is 105.2 (*i.e.*, 1,052,000 cells/mL $\div$ 10,000 cells/mL = 105.2).

9.3.2 Built in quality control

Deviation from normalcy (in the case of a test result with a reference toxicant) may indicate change(s) in laboratory performance (health of test organism, contamination, faulty test media, procedural error) and merit investigation. To ensure adequacy of results, a toxicity test with a reference toxicant (*e.g.*, Zn^{2+}: $ZnSO_4$, NaCl or Cu^{2+}: $CuSO_4$) should be conducted once monthly and preferably performed within 14 days before or after the conduct of toxicity testing. Individual IC50 values should then be used to construct a control chart (Fig. 7) to determine whether the new IC50s are within ± 2 standard deviations (= warning limits) or ± 3 standard deviations (= control limits) of values obtained in previous tests using the same reference toxicant and test procedure. Preparation and update of this chart is an essential part of quality control to ensure algal toxicity test performance (Environment Canada, 1990; Environment Canada, 2001). Reference toxicant data generated with Zn^{2+} indicate overall good adequacy of the algal microplate toxicity test (Fig. 7), as only one IC50 test value (test number 6) was borderline with the UWL (Upper Warning Limit) during 15 consecutive tests. Had other IC50s shown a similar trend (*e.g.*, test numbers 7, 8, 9 onwards displaying IC50s close to or above the UWL), sensitivity of the algal culture, as well as test performance and precision, would have been questioned and triggered a thorough check of all culturing and test conditions.

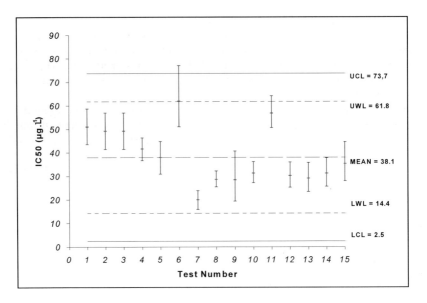

Figure 7. Example of a warning/control chart constructed with Zn^{2+}, a reference toxicant, as a check for algal toxicity test performance. UWL (Upper Warning Limit = + 2α), LWL (Lower Warning Limit = - 2α), UCL (Upper Control Limit = + 3α), LCL (Lower Control Limit = - 3α).

10. Factors capable of influencing algal growth and testing results

Toxicity tests, regardless of their degree of validation or standardization, can be subject to various factors linked to experimental procedure that will limit their applications if they are not properly addressed. Some such factors can also affect phytotoxicity tests, whether they are microplate-based or undertaken differently (*e.g.*, tests carried out in flasks or vials). Awareness of any obstacles to proper conduct of testing is clearly paramount and must be dealt with accordingly. Factors having the potential to adversely influence testing results for microplate phytototoxicity tests have been reviewed recently (Blaise et al., 1998). The more important ones are briefly recalled and commented below.

10.1 WELL EVAPORATION

Because microplate lids are not air-tight, excessive evaporation can occur in microplate wells under chronic exposure (>1 d). When test microplates are properly sealed inside their transparent polyester lining (Section 8.4.2), however, evaporation is adequately controlled. Filling peripheral wells (Fig. 1) with water will also offset evaporation, as will ensuring high humidity levels in test incubators.

10.2 ADHERENCE OF CHEMICALS TO WELLS

Because plastic (polystyrene) microplates are used for testing, as well as owing to the greater area to volume ratio of microplate wells, greater toxicant adhesion can theoretically occur on wells as opposed to that which might occur in glass flasks. Based on adequate agreement of results reported thus far between flask and microplate comparative studies, there is no real evidence suggesting increased uptake of test chemicals by microplate wells, although this could occur with specific chemicals whose affinity for plastic is elevated. Since affinity for one or the other medium (*i.e.*, plastic or glass) will likely be chemical-dependent, neither will always be the most convenient.

10.3 VOLATILE SUBSTANCES

Again, because microplate lids are loose-fitting and do not hermetically seal off wells from one another, algal growth may be affected by the presence of volatile contaminant(s) in aqueous samples being tested. The issue of well cross-contamination owing to volatile toxicants has been addressed in Section 8.2 and ways of minimizing and/or eliminating their impact are indicated.

10.4 COLORED SAMPLES

Colored samples can theoretically filter or alter light rays beneficial for algal cells during chronic exposure phytotoxicity tests based on growth inhibition. In this situation, it may not always be possible to discriminate between intrinsic toxicity owing to bioavailable contaminant(s) and color when a sample is tested in a concentration range where color prevails. One school of thought would certainly hold that color is a genuine part of toxicity (*e.g.*, in the case where the tested sample is a complex industrial wastewater) and that there is no need to try to differentiate between chemical toxicity and color interference. If, however, it is deemed important to distinguish between toxicity and color (*e.g.*, for toxicity reduction assessment of an effluent plant), then acute exposure (five hours or less) phytotoxicity tests using endpoints other than growth (*e.g.*, enzyme or motility inhibition) may provide useful alternatives (Kusui and Blaise, 1995; Snell et al., 1996).

10.5 GAS EXCHANGE

Unlike flasks or vials used for phytotoxicity tests, for example, microplates are seldom agitated during exposure. This is the case for the present protocol (see Section 8.5). Since rotatory movement of experimental vessels insures adequate gas exchanges for algal cells, CO_2 limitation has been an issue of concern for microplate-based tests. One study, essentially utilizing the same micro-technique described herein, compared algal growth and ensuing toxicity results for Cd^{2+} and phenol under regimes of passive (undisturbed microplate) and active (repetitive multichannel pipetting of well contents on days 1, 2 and 3 of a 96-h exposure period) gas exchange (Thellen et al., 1989). While toxicity results were unaffected (*i.e.*, comparative 96-h IC50s were similar), algal growth was significantly higher under an active gas regime suggesting that CO_2 limitation may have occurred during the latter stages of the 96-h incubation period, as cell biomass and pH increase. Since the chronic exposure time of the present procedure is shorter (3-d *versus* 4-d for Thellen et al., 1989) and calls for a more modest starting inoculum (10,000 cells/mL *versus* 20,000 cells/mL for Thellen et al., 1989), CO_2 limitation is no longer a preoccupying factor.

10.6 pH

As algal growth increases in flask-, vial- or microplate-based tests, the pH will tend to increase because of greater CO_2 consumption (Miller et al., 1978). As pH changes can alter toxicity responses (*e.g.*, speciation effects on metal ions), standard phytotoxicity protocols require that initial pH of test solutions should be fixed between 7.0 - 7.5 and that test samples, when outside a pH range of 6 - 9, should be adjusted to 6.5 or 8.5, respectively. Furthermore, it is also stipulated that pH should not have varied by more than 1.5 units in controls during the test exposure period to comply with test validity criteria. Since contemporary standard tests are now run with lower starting inocula (10,000 cells/mL) and shorter exposure times (3-d) than before, the resulting algal biomass is reduced and pH shifts owing to CO_2 depletion are minimal. In short, 3-d chronic exposure tests coupled with an initial inoculum of 10,000 cells/mL have essentially solved gas exchange (Section 10.5) and pH interference problems.

10.7 MISCELLANEOUS FACTORS

All types and brands of 96-well polystyrene microplates may not prove to be adequate for phytotoxicity testing. Microplates designed for tissue culture work, for example, may have undergone some type of pre-treatment and present toxicity to algae, thereby rendering them unusable. It is highly recommended that algal growth be monitored in first-time use of new microplates and compared to flask growth over a 3-4 d period to ensure their adequacy for testing.

Finally, specific types of complex samples (*e.g.*, industrial/municipal effluents, surface waters, sediment elutriates) may harbour both toxic and auxinic compounds and lead to numerous types of interactive effects. In some instances, resulting algal growth in wells of some test concentrations may be more abundant than in controls wells. Stimulatory effects of this nature may confound toxicity results and render impossible the reporting of a 50% toxic effect concentration. In such cases, it is nevertheless important to report percent growth stimulation in relation to controls at each test concentration where this occurs, as calculated below. Micro-algal stimulation may thus signal an adverse enrichment effect capable of contributing to increased eutrophication of receiving waters (U.S. EPA, 2002).

$$S\ (\%) = (T-C)/C \times 100 \tag{3}$$

Where:

- S = percent stimulation of algal growth of a particular test concentration over control growth;
- T = mean algal growth of a particular test concentration (in cells.mL^{-1});
- C = mean algal growth of controls (in cells.mL^{-1}).

11. Application examples (case studies) with the algal microplate toxicity test

Micro-algae are useful in contributing to the knowledge base of contaminant-related toxic effects. As an example, the toxicity responses of *S. capricornutum* were appraised following phytotoxicity appraisal of 46 effluent samples of the Pulp and Paper sector - a major industrial force in the province of Quebec, Canada (Blaise et al., 1987). At the time of this study, microplate tests were conducted with exposure times of eight days. Hence, chronic 8-day IC50 values were first determined for each of the effluents, whose wastewater treatment was either based on 1) primary, 2) primary and secondary and 3) Best Practicable Technology (BPT) installations. The BPT mill was the only one, at the time, exemplifying optimal treatment capabilities for this industrial sector. Individual 8 day-IC50s were then transformed into toxic units (TU), summed up for each of the effluent treatment categories and an average TU value calculated for each (see Fig. 8). Results showed that the algal bioassay was able to readily discriminate effluent toxicity on the basis of treatment application. The microplate phytotoxicity test results also demonstrated the efficiency of combined primary and secondary treatment coupled to BPT capacity in markedly reducing the toxic potential of these liquid wastes, which were at the time unquestionably toxic toward primary producers.

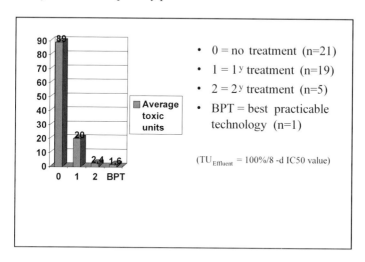

Figure 8. Toxicity responses of S. capricornutum *exposed to pulp and paper effluents having different waste treatment facilities (from Blaise, 2002).*

As a second example, the microplate technique was employed to appraise the phytotoxicity of nine herbicides reflecting different chemical classes, modes of action and uses (St-Laurent et al., 1992). Because of the expanding use of herbicides in Canada and elsewhere, concerns over non target plant phytotoxicity are certainly justified and hazards posed on primary producers must be addressed. One objective of this study was therefore to compare the relative toxicity of these herbicides on *Selenastrum capricornutum*, a representative green alga commonly used in North America for diverse environmental applications. As shown in Table 5, more than three orders of magnitude separated the most toxic (16.9 µg/L) and the least toxic (91.1 mg/L) herbicides, thereby yielding information important for individual herbicide registration. Overall, the algal microplate assay proved to be advantageous as a cost-effective screen for herbicide toxicity assessment.

Table 5. Toxicity of herbicides reported with the S. capricornutum *microplate assay (adapted from St-Laurent et al., 1992).*

Herbicide (chemical class affiliation)		96-h IC50 (95% confidence intervals)
Less toxic herbicides (IC50s in mg/L)		
Imazametabenz	(Imidazolinone)	91.1 (78.1 - 104.1)
2,4-D	(Phenoxy compound)	24.2 (23.7 - 24.7)
Picloram	(Picolinic acid)	22.7 (18.5 - 27.0)
Glyphosate	(Aliphatic acid)	7.8 (3.0 - 12.7)
Bromoxynil	(Benzonitrile)	3.4 (2.8 - 4.0)
More toxic herbicides (IC50s in µg/L)		
Metolachlor	(Acetamide)	50.9 (44.8 - 56.8)
Diquat dibromide	(Pyridine)	4.9 (2.4 - 7.3)
Hexazinone	(Triazine)	27.7 (22.7 - 32.5)
Cyanazine	(Triazine)	16.9 (12.6 - 21.2)

12. Accessory/miscellaneous test information

Laboratory personnel should be knowledgeable in the field of environmental microbiology and be specifically trained in the conduct of bioanalytical techniques. Rigorous application of the experimental protocol and careful attention to detail and quality control will ensure the successful undertaking of phytotoxicity testing.

Considerations for safety must be in place prior to carrying out algal testing with toxic chemicals and any solvent carriers. Consulting manufacturer data sheets on the potential effects that test chemicals could exert on human health is a necessary precaution, as is that of using care when dealing with specific types of complex liquid wastes. In all cases, proper safety measures must be applied before testing. Among others, this will include protective wear (laboratory coat, gloves, eye-glasses) and other suitable means of defence (*e.g.,* respirator and fume hoods to guard against volatile toxicants). It stands to reason that a laboratory should also possess an approved safety plan describing ways of handling hazardous chemicals, as well as one for their disposal and that of other contaminated material (*e.g.,* pipettes, gloves, spent solvent).

In terms of applicability, algal microplate procedures can theoretically be carried out on any liquid medium (section 1). Their use has been reported for effluents, surface waters, solid waste leachates, sediment pore waters, groundwaters, organic extracts, various types of organics (*e.g.,* herbicides, oil dispersants) and metals. Furthermore, they generally elicit sensitive responses to toxic (in)organic contaminants with IC50s lying in the $\mu g.L^{-1}$ to $mg.L^{-1}$ range (Blaise et al., 1998). As pointed out in Table 2 (Section 3), several types of micro-algae (other than *S. capricornutum* employed in our method) can be used in the conduct of microplate tests.

There are, of course, alternatives to microplates as experimental substrates for undertaking algal bioassays. Flask tests developed ever since the late 1970's (*e.g.,* Chiaudani and Vighi, 1978) are still in use (*e.g.,* U.S. EPA, 2002 and Chapter 4 of this volume), as are more recent tests conducted in small vials (Arensberg et al., 1995; Blaise et al., 2000), in spectrophotometric polystyrene cells with 10-cm pathlengths (Persoone, 1998) and in tubes (Léonard et al., 2003). Again, algal growth inhibition endpoints can be measured in different ways besides that of enumerating cells with a particle counter, as described in this chapter. Alternative choices of cell yield determinations can include the use of a Neubauer counting chamber (*e.g.,* Blaise et al., 2000), spectrophotometry (*e.g.,* Persoone, 1998) and fluorometry (*e.g.,* Chapter 6 of this volume).

The initiation of the algal microplate test can be automated via existing robotic instruments. While we have not yet attempted to do so, the technological capacity clearly exists. One example of a successful 96-well microplate technique automation has been reported for the SOS Chromotest, a screening test for genotoxic agents present in environmental samples (White et al., 1996b). Automated initiation of microplate techniques can be advantageous, particularly when laboratories are confronted with large sample throughputs.

When undertaken manually with the (micro)pipettes and multichannel pipettes described herein, an experienced operator can easily initiate 30 microplate tests in one work day (*i.e.,* process 30 samples). This is offset, however, by the more laborious post-exposure cell enumeration procedure requiring a particle counter (Section 9.1) where, conservatively, 10 microplates would normally be counted in one work day. Hence, a batch of 10 samples might be considered an optimal number to process, where sample dilution preparations and filling of the microplates could be done in ½ of a work day. Seventy-two hour post-exposure counting would demand another work day to which must be added another ½ day for data treatment, interpretation and report writing. Material costs to process 10 samples are estimated at 15US (1US for each microplate × 10 microplates + 5US for pipette tips, disposable plastic reservoirs, chemicals/reagents, electricity, etc.). In short, undertaking 10 phytotoxicity tests requires 2-d of labor and 15US in operation and maintenance costs.

The algal microplate toxicity test is a standardized biological test method of Environment Canada (Environment Canada, 1992) having undergone some amendments in 1996. To our knowledge, it is presently employed by some 20 international laboratories for phytotoxicity testing. The test was recently involved in an inter-calibration exercise seeking to compare the responses of three phytotoxicity procedures (algal microplate test, OECD flask test and the micro-algal automated fluorometric assay employed to test 12 chemical substances representing varied classes of chemicals. Results indicated good agreement among all three procedures (Léonard et al., 2003).

13. Conclusions/prospects

Ever since its development (Blaise et al., 1986) and later optimization (Thellen et al., 1989) and standardization (Environment Canada, 1992), the algal microplate assay has proved valuable to assess the potential toxicity of chemicals and varied environmental media. Along with other types of available phytotoxicity tests, it comprises one choice for ecotoxicology laboratories desirous of screening, ranking and prioritizing chemicals, singularly or interactively (e.g., via complex wastewater samples) for vested interests linked to the safeguard of the aquatic environment. As an adjunct to other small-scale tests conducted at different trophic levels in test battery approaches, the algal microplate assay can contribute to enhanced cost-efficiency and diagnostic capacity enabling improved decision-taking to protect specific receiving waters. One example lies in the PEEP strategy to evaluate the toxic potential of industrial effluents (Chapter 1, volume 2 of this book).

Exploring new endpoints allowing measurable toxicity responses after exposures of only several hours may also be a way forward for microplate phytotoxicity techniques. Possibilities can come from assessment endpoints including physiological, biochemical, enzymatic and immunological responses, coupled with new advances in instrumentation. Earlier work undertaken in the area of rapid effects measurements following acute exposures of micro-algae to contaminants may well stand to be re-examined in this context (e.g., Vasseur et al., 1981; Hickey et al., 1991; Snell et al., 1996). Assuming that such acute tests can be shown to provide sufficient sensitivity and to be as relevant as a chronic endpoint based on growth inhibition, they would certainly prove to be attractive bioanalytical tools. For one, they would be exempt of the potential problems or interferences that chronic exposure tests can suffer from. For another, users would enjoy being able to initiate and complete testing in a single work day.

Until these new breakthroughs occur, the algal microplate assay described in this chapter will continue to play a useful role in appraising the toxic effects that contaminants can exert on phototrophic micro-organisms.

ADDRESSES AND WEBSITES OF INTEREST

Box 11. Addresses to purchase algal cultures.

1.	**University of Toronto Culture Collection (UTCC)** Dept. of Botany, University of Toronto Toronto, Ontario Canada, M5S 3B2 Telephone: (416) 978-3641 Fax: (416) 978-5878 jacreman@botany.utoronto.ca http:/www.botany.utoronto.ca/utcc/index.html *Selenastrum capricornutum*: UTCC 37
2.	**American Type Culture Collection (ATCC)** 12301 Parklawn Drive Rockville, Maryland, U.S.A. 20852 Telephone: (301) 881-2600 Fax: (301) 816-4361 sales@atcc.org http://www.atcc.org/ *Selenastrum capricornutum*: ATCC 22662
3.	**Culture Collection of Algae (UTEX)** Botany Dept. University of Texas, Austin, Texas U.S.A. 78712 Telephone: (512) 471-4019 Fax: (512) 471-3878 jeff_n_judy@mail.utexas.edu http://www.bio.utexas.edu/research/utex/ *Selenastrum capricornutum*: UTEX 1648
4.	**Culture Centre for Algae and Protozoa (CCAP)** CCAP Administration, CEH Windermere The Ferry House, Far Sawrey AMBLESIDE, Cumbria LA22 0LP United Kingdom Telephone: +44 (0)15394 42468 Fax: +44 (0)15394 46914 ccap@ceh.ac.uk http://www.ife.ac.uk/ccap/ *Selenastrum capricornutum*: CCAP 278/4

Box 12. Address to purchase preserved algal beds.

Microbiotests Inc.
Industriezone «De Prijkels»
Venecoweg 19
9810 Nazareth, Belgium
Telephone: 32 (0) 9 380 8545
Fax: 32 (0) 9 380 8546
microbiotests@skynet.be
http://www. microbiotests.be

Box 13. Address to purchase TOXCALC^{TM} software.

Tidepool Scientific Software
P.O. Box 2203
McKinleyville, CA
U.S.A. 95519
tidesoft@aol.com
Phone: 707-839-5174
Fax: 707-839-5174
http://www.members.aol.com/tidesoft/toxcalc

Acknowledgements

The assistance of Manon Harwood, Sylvain Trottier and Geneviève Farley, Saint-Lawrence Centre, is appreciated for providing phytotoxicity data and assistance in graphics preparation. The authors also wish to thank copyright holders for permission to reproduce the following:

Figure 1: Wiley-Liss, a division of John Wiley & Sons Inc., New York.

St-Laurent, D., Blaise, C., MacQuarrie, P., Scroggins, R. and Trottier B. (1992) Comparative assessment of herbicide phytotoxicity to *Selenastrum capricornutum* with microplate and flask bioassay procedures, *Environ. Toxicol. Water Qual.* **7**, 35-48 (Figure 1, slightly modified).

Figure 8: John Wiley & Sons Inc., New York.
Blaise, C., 2002. Use of microscopic algae in toxicity testing, *in* G. Bitton (ed.), *Encyclopedia of Environmental Microbiology*, Vol. 6, Wiley Publishers, New York, NY, USA, , pages 3219-3230 (Figure 7, unaltered).

References

Arensberg, P., Hemmingsen, V.H., and Nyholm, N. (1995) A miniscale algal toxicity test, *Chemosphere* **30**, 2103-2115.
Blaise, C. (2002) Use of microscopic algae in toxicity testing, in G. Bitton (ed.), *Encyclopedia of Environmental Microbiology*, Vol. 6, Wiley Publishers, New York, NY, USA, pp. 3219-3230.
Blaise, C., Legault, R., Bermingham, N., van Coillie, R. and Vasseur, P. (1986) A simple microplate algal assay technique for aquatic toxicity assessment, *Toxicity Assessment* **1**, 261-281.
Blaise, C., van Coillie, R., Bermingham, N. and Coulombe, G. (1987) Comparaison des réponses toxiques de trois indicateurs biologiques (bactéries, algues, poissons) exposés à des effluents de fabriques de pâtes et papiers, *Revue Internationale des Sciences de l'Eau* **3**, 9-17.
Blaise, C., Férard, J.F. and Vasseur, P. (1998) Microplate toxicity tests with microalgae: a review, in P.G. Wells, K. Lee and C. Blaise (eds.), *Microscale Testing in Aquatic Toxicology: Advances, Techniques and Practice*, CRC Press LLC, Boca Raton, Florida, USA, pp. 269-288.
Blaise, C., Forget, G. and Trottier, S. (2000) Toxicity screening of aqueous samples using a cost-effective 72-h exposure *Selenastrum capricornutum* assay, *Environmental Toxicology* **15**, 352-359.
Blanck, H., Wallin, G. and Wängberg, S.A. (1984) Species dependent variation in algal sensitivity to chemical compounds, *Ecotoxicology and Environmental Safety* **8**, 339-351.
Bulich, A.A., Greene M.W. and Isenberg, D.L. (1981) Reliability of the bacterial luminescence assay for determination of the toxicity of pure compounds and complex effluents, in D.R Branson and K.L. Dickson (eds.), *Aquatic Toxicity and Hazard Assessment*, Fourth Conference of the American Society for Testing and Materials, STP 737, Philadelphia, PA., pp. 338-347.
Chiaudani, G. and Vighi, M. (1978) The use of *Selenastrum capricornutum* in batch cultures in toxicity studies, *Verhandlungen Internationale Vereinigung Limnologie* **21**, 316-329.
Environment Canada (1990) Guidance document for control of toxicity test precision using reference toxicants, Report EPS 1/RM/12, Environmental Protection Series, Environment Canada, Ottawa, Ontario, 85 pp.
Environment Canada (1992) Biological test method: growth inhibition test using the freshwater alga *Selenastrum capricornutum*, Report EPS 1/RM/25, Environmental Protection Series, Environment Canada, Ottawa, Ontario, 41 pp.
Environment Canada (2001) Guidance Document on Statistical Methods to Determine Endpoints of Toxicity Tests, Fourth draft, March 2001, Report EPS 1/RM/, Environmental Protection Service, Environment Canada, Ottawa, Ontario, 153 pp.
Fogg, G.E. (1965) *Algal Cultures and Phytoplankton Ecology*, University of Wisconsin Press, Madison.
Gilbert, R.O. (1987) *Statistical Methods for Environmental Pollution Monitoring*, Van Nostrand Reinhold Co., New York, NY, 320 pp.
Heldal, M., Norland, S., Lien, T. and Knutsen, G. (1978) Acute toxicity of several oil dispersants toward the green algae *Chlamydomonas* and *Dunaliella*, *Chemosphere* **3**, 247-255.
Hickey, C., Blaise, C. and Costan, G. (1991) Microtesting appraisal of ATP and cell recovery end-points after acute exposure of *Selenastrum capricornutum* to selected chemicals, *Toxicity Assessment* **6**, 383-403.
Hindak, F. (1990) *Biologicke Prace* **5**, 209.
Haaland, P.T. and Knutson, G. (1973) Growth experiments with *Selenastrum capricornutum* Printz, in *Algal Assays in Water Pollution Research*, Proceedings of Nordic Symposium, Secretariat of Environmental Sciences, Oslo, Norway, Nordforsk, pp. 69-72.
Joubert, G. (1983) Detailed method for quantitative toxicity measurements using the green algae *Selenastrum capricornutum*, in J.O. Nriagu (ed.), *Aquatic Toxicology - advance in environmental science and technology*, John Wiley & Sons, Vol. 13, pp. 467-485.
Keighan, E. (1977) Caractérisation du Niveau d'Enrichissement et de la Toxicité des Eaux du Bassin du Fleuve St-Laurent, Comité d'étude sur le Fleuve St-Laurent, Environnement Canada, Service de la Protection de l'Environnement, Région du Québec, Rapport technique No. 6: 153 pp.
Kusui, T. and Blaise, C. (1995) Acute exposure phytotoxicity assay based on motility inhibition of *Chlamydomonas variabilis*, in M. Richardson, (ed.), *Environmental Toxicology Assessment*, Taylor & Francis, London, pp. 125-135.
Léonard, M., Fel, J.P., Pandard, P., Poulsen, V., Blaise, C. and Harwood, M. (2003) Comparison of MAAFA (Micro-algal automated fluorometric assay) with two standardized micro-algal toxicity assays, Poster presentation, 11th International Symposium on Toxicity Assessment, Vilnius, Lithuania, 2003.

Miller, W.E., Greene, J.C. and Shiroyama, T. (1978) The *Selenastrum capricornutum* Printz Algal Assay Bottle Test: Experimental Design, Application and Data Interpretation protocol, U.S. EPA (United States Environmental Protection Agency), Report Number EPA-600/9-78-018, Corvallis, Oregon, 126 pp.

Mohammad, A.H. (1991) A Practical Method for Immobilization and De-immobilization of Freshwater Algae for Toxicity Screening and Water Quality Studies, NIVA (Norwegian Institute for Water research), Report number E-88427, 13 pp.

Newman, M.C. (1995) *Quantitative Methods in Aquatic Toxicology*, Lewis Publishers, Boca Raton, FL.

Norberg-King, T.J. (1993) A linear interpolation method for sublethal toxicity: the inhibition concentration (ICp) approach (Version 2.0), Technical Report 03-93, U.S. Environmental Protection Agency, Duluth, Minnesota, 25 pp.

Nygaard, G., Komarek, J. Kristiansen, J. And Skilberg, O. (1986) Taxonomic designations of the bioassay alga, NIVA-CHL I (*Selenastrum capricornutum*) and some related strains, *Opera Botanica* **90**, 1-46.

Pardos, M. and Blaise, C. (1999) Aspects of toxicity and genotoxicity assessment of hydrophobic organic compounds in wastewater, *Environmental Toxicology* **14**, 241-247.

Persoone, G. (1998) Development and first validation of a "stock-culture free" algal microbiotest: the AlgalToxkit, in P.G. Wells, K. Lee and C. Blaise (eds.), *Microscale Testing in Aquatic Toxicology: Advances, Techniques and Practice*, CRC Press LLC, Boca Raton, FL, USA, pp. 311-320.

Persoone, G. (1991) Cyst-based toxicity tests I. A promising new tool for rapid and cost-effective toxicity screening of chemicals and effluents, Zeitschrift für Angewandte Zoologie **78**, 235-241.

Reynolds, J.H., Middlesbrooks, E.J., Porcella, D.B. and Grenney, W.J. (1975) Effects of temperature on growth constants of *Selenastrum capricornutum*, *Journal of the Water Pollution Control Federation* **47**, 2420-2436.

Skulberg, O.M. (1964) Algal problems related to the eutrophication of European water supplies and a bioassay method to assess fertilizing influences of pollution on inland waters, in D.F. Jackson (ed.), *Algae and Man*, Plenum Press, pp. 262-299.

Skulberg, O.M. (1967) Algal cultures as a means to assess the fertilizing influence of pollution, in O. Jaag and H. Liebman (eds.), *Advances in Water Pollution Research*, Water Pollution Control Federation, Washington, D.C., Vol. 1, pp. 113-138.

Snell, T., Mitchell, J.L. and Burbank, S.E. (1996) Rapid toxicity assessment with microalgae using *in vivo* esterase inhibition, in G.K. Ostrander (ed.), *Techniques in Aquatic Toxicology*, CRC Press/Lewis Publishers, Boca Raton, FL, U.S.A., pp. 13-22.

St-Laurent, D. Blaise, C., MacQuarrie, P., Scroggins, R. and Trottier, B. (1992) Comparative assessment of herbicide phytotoxicity to *Selenastrum capricornutum* using microplate and flask bioassay procedures, *Environmental Toxicology* **7**, 35-48.

Stratton, G.W. and Smith, T.M. (1988) Interaction of organic solvents with the green alga, *Chlorella pyrenoidosa*, *Bulletin of Environmental Contamination and Toxicology* **40**, 736-742.

Thellen, C., Blaise, C., Roy, Y. and Hickey, C. (1989) Round robin with the *Selenastrum capricornutum* microplate toxicity assay, *Hydrobiologia* **188/189**, 259-268.

U.S. EPA (United States Environmental Protection Agency) (1971) Algal Assay Procedure Bottle Test, National Eutrophication Research Program, Corvallis, Oregon, 82 pp.

U.S. EPA (United States Environmental Protection Agency) (2002) Short-Term Methods for Estimating the Chronic Toxicity of Effluents and Receiving Waters to Freshwater Organisms, Fourth edition, October 2002, Report Number EPA-821-R-02-013, Office of Water, Washington, DC 20460, 335 pp.

Vasseur, P., Jouany, J.M., Férard, J.F. and Toussaint, B. (1981) Intérêt du dosage de l'ATP en tant que critère d'écotoxicité aiguë chez les algues, *Institut National de la Santé et de la Recherche Médicale*, **106**, 207-226.

Walsh, G.E. (1988) Methods for toxicity tests of single substances and liquid complex wastes with marine unicellular algae, Report Number EPA/600/8-87/043, Environmental Research Laboratory, U.S. Environmental Protection Agency, Gulf Breeze, Florida, U.S.A., 64 pp.

Wängberg, S.A., Bergström, B., Blanck, H. and Svanberg, O. (1995) The relative sensitivity patterns of short-term toxicity tests applied to industrial wastewaters, *Environmental Toxicology and Water Quality* **10**, 81-90.

White, P., Rasmussen, J. and Blaise, C. (1996a) Sorption of organic genotoxins to particulate matter in industrial effluents, *Environmental and Molecular Mutagenesis* **27**, 140-151.

White, P., Rasmussen, J. and Blaise, C. (1996b) A semi-automated, microplate version of the SOS Chromotest for the analysis of complex environmental extracts, *Mutation Research* **360**, 51-74.

Abbreviations

ACC	Algal cell count
ACM-1×	Algal culture medium at normal strength (= 1×)
ATCC	American Type Culture Collection
ATEM	Algal test enrichment medium (at 13.75 times the strength of ACM-1×)
ATP	Adenosine tri-phosphate
BPT	Best Practicable Technology
CCAP	Culture Centre for Algae and Protozoa
c.v.	coefficient of variation
2,4-D	(2,4-dichlorophenoxy) acetic acid
EC	Environmental chamber
EDTA	ethylenediamine tetraacetate ($C_{10}H_{14}O_8N_2$)
IC25	25% effect inhibitory concentration
IC50	50% effect inhibitory concentration
ICp	inhibitory concentration for a particular percent effect
klx	kilolux
kPa	KiloPascal
LCL	Lower Control Limit
LOEC	lowest observed effect concentration
LWL	Lower Warning Limit
NIVA	Norwegian Institute of Water Research
NOEC	no observed effect concentration
OECD	Organization for Economic Cooperation and Development
PEEP	Potential Ecotoxic Effect Probe
TU	Toxic Units
UCL	Upper Control Limit
UTCC	University of Toronto Culture Collection
UWL	Upper Warning Limit
$\mu E.m^{-2}.s^{-1}$	micro-Einstein per square meter per second.

4. ALGAL TOXICITY TEST

JANE P. STAVELEY
ARCADIS G&M, Inc.
4915 Prospectus Drive, Suite F, Durham
North Carolina, 27713, USA
jstaveley@arcadis-us.com

JERRY C. SMRCHEK
U.S. Environmental Protection Agency
Office of Prevention, Pesticides and Toxic Substances
1200 Pennsylvania Ave., N.W.
Washington, D.C., 20460, USA
smrchek.jerry@epamail.epa.gov

1. Objective and scope of test method

Algae are included in many hazard assessment schemes as representatives of the aquatic plant community. Algae are ubiquitous in aquatic ecosystems, where they incorporate solar energy into biomass, produce oxygen, function in nutrient cycling, and serve as food for animals. Because of their ecological importance and sensitivity to many substances, especially herbicides and metals, algae are often used in toxicity testing.

The test method described below has been widely used for many years to determine the toxicity of test materials to various species of microalgae. It is derived from a method originally developed in the late 1960's and early 1970's for an "algal assay bottle test" to examine the eutrophication potential of surface waters (U.S. EPA, 1971; Miller et al., 1978). The "bottle test" was subsequently adapted for the purpose of determining toxicity to algae. In this method, which appeared in the mid-1980's (U.S. EPA, 1982; U.S. EPA, 1985; U.S. EPA, 1986), the test material is added to nutrient medium, an inoculum of a single species of algae is added, and the test vessels are incubated under appropriate conditions to examine differences in population growth between treated cultures and controls. This method has been used extensively to determine the toxicity of a variety of test materials, including pesticides (U.S. EPA, 1982; U.S. EPA, 1986; Boutin et al., 1993), industrial chemicals (U.S. EPA, 1985), and effluents (U.S. EPA, 2002). It is sometimes referred to as the "flask method" to distinguish it from scaled-down algal tests conducted in vials or microplates. The advantages of this test include the relatively short duration; high replicability and repeatability; minimal requirements for instrumentation and facilities; and the availability of sufficient aqueous sample for analytical confirmation of test concentrations.

2. Summary of test procedure (at a glance)

Organisms of a particular species of microalgae are maintained under static conditions in test vessels containing nutrient medium alone (controls) and nutrient medium to which the test material has been added. In preparation for the test, appropriate volumes of nutrient medium and/or test solution are placed in the test vessels (Erlenmeyer flasks), with replicates for each treatment. Algae are then introduced into the flasks, which are subsequently placed in a growth chamber, which provides standardized light and temperature conditions. Each test vessel is inoculated at an initial population density to provide for growth sufficient to allow accurate quantification without resulting in nutrient or carbon dioxide limitation under the test conditions. Data on population growth during the test are obtained on a daily basis for 96 hours. The results of the test are expressed as the 96-h $IC50$, based upon final population density and the average specific growth rate. The NOEC (no observed effect concentration) should also be determined. Test results are usually based upon measured test concentrations. Unlike scaled-down test methods, the flask method employs enough test solution for most chemical analytical procedures. The test method is summarized in Table 1.

3. Overview of applications of the algal toxicity test

The flask-based method is the basis of toxicity test methods published by numerous organizations, including the U.S. Environmental Protection Agency (U.S. EPA, 1971; U.S. EPA, 1974; Miller et al., 1978; U.S. EPA, 1978; U.S. EPA, 1982; U.S. EPA, 1985; U.S. EPA, 1986; U.S. EPA, 1996; U.S. EPA, 2002), the Organization for Economic Co-operation and Development (OECD, 1984), and the American Society for Testing and Materials (ASTM, 2003a). These organizations periodically revise their standardized methods, and some changes are anticipated to the cited documents. However, the procedures discussed below reflect the basic test principles that have been in use for over 25 years for a wide variety of toxicity assessment and regulatory purposes. The specific procedures described in this chapter most closely reflect current U.S. EPA approaches to conducting algal toxicity tests with pesticides (under the U.S. Federal Insecticide, Fungicide and Rodenticide Act) and industrial chemicals (under the U.S. Toxic Substances Control Act). For a description of similar Agency methods using algae to determine the toxicity of effluents and receiving waters, refer to U.S. EPA, 2002.

Table 1. Rapid summary of test procedure.

Test type	Static
Test duration	96 hours
Test matrix	Synthetic growth medium appropriate for the test species
Temperature	24°C for *P. subcapitata* and *N. pelliculosa*; 20°C for *S. costatum*
Light quality	Cool-white fluorescent
Light intensity	60 μmol.m^{-2}.s^{-1}
Photoperiod	Continuous light for *P. subcapitata* and *N. pelliculosa*. 14 h light:10 h dark for *S. costatum*
Shaking	Continuous at 100 oscillations/minute for *P. subcapitata* and *N. pelliculosa*. Manual, once or twice daily, for *S. costatum*
Salinity (for saltwater species)	30 ± 5 ppt (for *S. costatum*)
Test vessel size	125 - 500 mL Erlenmeyer flasks
Test solution volume	≤ 50% of the test vessel volume
Age of inoculum	From logarithmically-growing stock cultures (typically 3 - 7 days old)
Inoculum concentration	10 000 cells/mL for *P. subcapitata* and *S. costatum*. At least 10 000 cells/mL for other species. Inoculum volume < 2 mL
Number of replicates	Four test vessels per concentration (recommended minimum)
Test concentrations	Unless performing a limit test (Section 8.3), a minimum of 5 test concentrations plus appropriate controls
Test concentration preparation	Aqueous solutions prepared by adding test material to synthetic nutrient medium, directly or via carrier
Measurement endpoints	IC50 based upon final population density (yield) and average specific growth rate; NOEC should be observed

4. Advantages of the algal toxicity test

The flask-based method has been in widespread use for many years and has stood "the test of time". It requires only simple equipment that is common in most laboratories, and technicians need minimal training in its use. It employs ecologically relevant organisms that are at the base of the food chain. The test duration is short, although it is inappropriate to term the test an "acute" test, since most test species will undergo several population doublings during the 96-hour exposure period. The algal test thus has an advantage over tests with organisms such as fish and invertebrates, because it measures a population-level response. The basic flask method has been adapted for use with a variety of sample types (including effluents) and test organisms (from cyanobacteria to diatoms). One distinct advantage of the flask method is that it provides a sufficient amount of test solution to allow analytical confirmation of test concentrations, which is often not possible with scaled-down test methods.

5. Test species

Species of algae recommended as test organisms are the freshwater green alga *Pseudokirchneriella subcapitata* (formerly known as *Selenastrum capricornutum* and also as *Raphidocelis subcapitata*); the marine diatom, *Skeletonema costatum*; and the freshwater diatom, *Navicula pelliculosa*. Additional species that have been used include the freshwater green alga *Scenedesmus subspicatus* (recently renamed *Desmodesmus subspicatus*), the marine diatom *Thalassiosira pseudonana,* the marine golden-brown alga *Phaeodactylum tricornutum,* and the marine dinoflagellate *Dunaliella tertiolecta.* Other species, formerly classified as blue-green algae but currently considered cyanobacteria (*Anabaena flos-aquae* and *Microcystis aeruginosa*) can also be tested using these procedures but with a reduced light intensity (see ASTM, 2003a). Additional potential test species are listed by Boutin et al. (1993). The recommended species have been used successfully and have been demonstrated to be sensitive to a variety of test substances. The responses of algal species vary and there is no single "most sensitive" species. Therefore, testing of several species may be needed. For pesticide registration, the U.S. EPA requires (depending upon the use pattern of the pesticide) testing with four species: *P. subcapitata*, *A. flos-aquae*, *S. costatum* and a freshwater diatom such as *N. pelliculosa*.

6. Culture/maintenance of organism in the laboratory

6.1 SOURCE, AGE AND CONDITION

Algae to be used in toxicity tests may be initially obtained from commercial sources and subsequently cultured using sterile technique. Commercial sources include the American Type Culture Collection, 12301 Parklawn Drive, Rockville, MD, 20852 and the University of Texas Algal Collection, Botany Department, Austin, TX,

78712. Upon receipt of an algal culture not previously maintained in a facility, a period of six weeks culturing is recommended to establish the ability to successfully maintain a healthy, reproducibly-growing culture. Information on culturing algae can be found in the references listed in the ASTM Guide E-1218 (ASTM, 2003a). Aseptic stock transfer should be performed on a regular schedule (*e.g.*, once or twice weekly) to maintain a supply of cells in or near the logarithmic growth phase. Long-term maintenance of cultures on a solid medium containing 1% agar in sterile Petri plates or test tubes may be desirable. However, the algal inoculum used to initiate toxicity testing must be from a liquid culture shown to be actively growing (*i.e.*, capable of logarithmic growth within the test period) in at least two subcultures lasting 7 days each prior to the start of the definitive test.

6.2 APPARATUS AND FACILITIES

Normal laboratory equipment and especially the following are necessary:

- Equipment for determination of test conditions (*e.g.*, pH meter and light meter).
- Containers for culturing and testing algae. Erlenmeyer flasks should be used as test vessels. The flasks may be of any volume between 125 and 500 mL as long as the same size is used throughout testing and the test solution volume does not exceed 50 percent of the flask volume. To permit gas exchange but prevent contamination, the flasks should be covered with foam plugs, stainless steel caps, glass caps or screw caps. (The acceptability of foam plugs should be investigated prior to use because some brands have been found to be toxic). All test vessels and covers in a test must be identical.
- A growth chamber or a controlled environment room that can hold the test vessels and will maintain the air temperature, lighting intensity, and photoperiod specified in this test guideline. If necessary for the species, a mechanism for continuously shaking the test vessels.
- Apparatus for preparing sterile nutrient media.
- Apparatus for sterilizing glassware and maintaining aseptic technique during culturing and testing.
- Microscope capable of 100 to 400 X magnification.
- Apparatus for enumerating algae, *e.g.*, hemacytometer, plankton counting chamber, or electronic particle counter. An alternative method to performing cell counts is to determine the chlorophyll *a* concentration through spectrophotometric or fluorometric methods.
- Facilities should be well ventilated and free of fumes that may affect the test organisms. Construction materials and equipment that may contact the stock solution, test solution, or nutrient medium should not contain substances that can be leached or dissolved into aqueous solutions in quantities that can affect the test results. Construction materials and equipment that contact stock or test solutions should be chosen to minimize sorption of test materials.

6.3 CLEANING AND STERILIZATION OF GLASSWARE

New test vessels may contain substances which inhibit growth of algae. They are therefore to be cleaned thoroughly and used several times to culture algae before being used in toxicity testing. All reusable glassware employed in algal culturing or testing is to be cleaned and sterilized prior to use. Wash glassware using a non-phosphate detergent and a stiff bristle brush to remove residues. This is followed by thorough rinsing with water, a rinse with a water-miscible solvent (such as acetone), additional rinsing with water, a rinse with acid (such as 10% hydrochloric acid), and at least two final rinses with reagent grade water. These procedures are generally suitable to remove test material residues from previous toxicity testing, but additional procedures may be required depending upon the nature of the test material.

Glassware may be dried in an oven at 50 to 100°C, capped with flask closures or covered loosely with foil, and sterilized by autoclaving for 20 minutes at 121°C and 1.1 kg/cm^2.

6.4 PREPARATION OF NUTRIENT MEDIA

Water used for preparation of nutrient medium should be of reagent quality (*e.g.*, ASTM Type I water). Freshwater algal nutrient medium (AAP or "Algal Assay Procedure" medium, as described by Miller et al., 1978) is prepared by adding specified amounts of reagent-grade chemicals to reagent water. Marine algal nutrient medium is prepared by adding reagent grade chemicals to synthetic salt water (see Walsh and Alexander, 1980) or to filtered natural salt water, or by preparing a complete saltwater medium. Salinity for saltwater medium should be 30 ± 5 ppt.

Formulation and sterilization of nutrient medium used for algal culture and preparation of test solutions should conform to those currently recommended by ASTM for freshwater and marine algal toxicity tests (see Tables 2 and 3). Chelating agents (*e.g.* EDTA) are included in the nutrient medium for optimum cell growth. Nutrient medium should be freshly prepared for algal testing or may be stored under refrigeration for several weeks prior to use. Nutrient medium should be sterilized by autoclaving or filtering (0.22 µm filter). At the start of the test, the pH of the nutrient medium should be 7.5 ± 0.1 for freshwater algal medium and 8.0 ± 0.1 for marine algal medium. The pH may be adjusted prior to addition of the test material with 0.1N or 1N sodium hydroxide or hydrochloric acid.

Table 2. Preparation of medium for freshwater algae.

This medium (often referred to as AAP medium) is prepared by adding 1 mL of each macronutrient stock solution and 1 mL of the micronutrient stock solution listed below to approximately 900 mL reagent grade water and then diluting to 1 L.

Each of the <u>six macronutrient stock solutions</u> is prepared by dissolving each of the following chemicals into 500 mL of reagent grade water:

1) $NaNO_3$ — 12.750 g
2) $MgCL_2 \cdot 6H_2O$ — 6.082 g
3) $CaCl_2 \cdot 2H_2O$ — 2.205 g
4) $MgSO_4 \cdot 7H_2O$ — 7.350 g
5) K_2HPO_4 — 0.522 g
6) $NaHCO_3$ — 7.500 g

For diatom species only, add $Na_2SiO_3 \cdot 9H_2O$ as another macronutrient. May be added directly (202.4 mg) or by way of a stock solution to give a final concentration of 20 mg/L Si in medium.

<u>The micronutrient stock solution</u> is prepared by dissolving the following chemicals into 500 mL of reagent water:

H_3BO_3 — 92.760 mg
$MnCl_2 \cdot 4H_2O$ — 207.690 mg
$ZnCl_2$ — 1.635 mg
$FeCl_3 \cdot 6H_2O$ — 79.880 mg
$CoCl_2 \cdot 6H_2O$ — 0.714 mg
$Na_2MoO_4 \cdot 2H_2O$ — 3.630 mg
$CuCl_2 \cdot 2H_2O$ — 0.006 mg. (Typically must be prepared by serial dilution).
$Na_2EDTA \cdot 2H_2O$ — 150 mg. [Disodium (Ethylenedinitrilo) tetraacetate].
($Na_2SeO_4 \cdot 5H_2O$ — 0.005 mg. Used only in medium for stock cultures of diatom species).

Adjust pH to 7.5 ± 0.1 with 0.1 N or 1.0 N NaOH or HCl.

Filter all media into a sterile container through a 0.22 µm membrane filter if a particle counter is to be later used for enumerating algal cells otherwise through a 0.45-µm filter. Store medium in the dark at approximately 4°C until use.

Table 3. Preparation of medium for saltwater algae.

The Micronutrient Mix is prepared by adding the specified amount of chemicals in the order listed below to 900 mL reagent water and diluting to 1 L.

Micronutrient Mix:
$FeCl_3 \cdot H_2O$ — 0.048 g
$MnCl_2 \cdot 4H_2O$ — 0.144 g
$ZnSO_4 \cdot 7H_2O$ — 0.045 g
$CuSO_4 \cdot 5H_2O$ — 0.157 mg
$CoCl_2 \cdot 6H_2O$ — 0.404 mg
H_3BO_3 — 1.140 g
$Na_2EDTA.2H_2O$ — 1.0 g

The Minor Salt Mix is prepared by adding the specified amounts of the chemicals listed below to 900 mL reagent water and diluting to 1 L.

Minor Salt Mix:
K_3PO_4 — 0.3 g
$NaNO_3$ — 5.0 g
$NaSiO_3 \cdot 9H_2O$ — 2.0 g

The Vitamin Mix is prepared by adding the specified amount of chemicals in the order listed below to 900 mL reagent water and diluting to 1 L.

Vitamin Mix:
Thiamine Hydrochloride — 500 mg
Biotin — 1 mg
B_{12} — 1.0 mg

The stock solutions are added to a sterile recipient containing either natural salt water that has been filtered through a 0.22 μm membrane filter or reconstituted salt water. Add the amounts given below to prepare medium used for toxicity testing. Add twice the amounts given to prepare medium for use in maintenance of stock cultures.

Add 15 mL of Micronutrient Mix/L of medium
Add 10 mL of Minor Salt Mix/L of medium
Add 0.5 mL of Vitamin Mix/L of medium. (Add 1 mL of vitamin mix if *Thalassiosira* is used).

Adjust pH to 8.0 ± 0.1 with 0.1 N or 1.0 N NaOH or HCl. Store medium in the dark at approximately 4°C until use.

7. Preparation of test species for toxicity testing

The cultures used as the source of inoculum should be maintained under the same conditions as used for testing. The algal inoculum to begin the toxicity test should be from logarithmically-growing stock cultures (typically 3 to 7 days old). All algae used for a particular test should be from the same source and the same stock culture. Also, the clone of all species should be specified. Test algae must not have been used in a previous test, either in a treatment or a control. A culture should not be used for starting a test if it is not in logarithmic growth phase, if microscopic examination at 400 X shows contamination by fungi or other algae, or if the health of the culture is doubtful in any respect.

Each test vessel should be inoculated at an initial population density to allow sufficient growth under the test conditions without resulting in nutrient or carbon dioxide limitation. The primary criterion for the initial cell concentration is that accurate estimates of population density can be obtained with the chosen method of measurement during the test. For *P. subcapitata* and *S. costatum*, the initial cell concentration should be 10 000 cells/mL. Higher concentrations may be necessary for other species, but the upper limit should be no more than 100 000 cells/mL. It is not usually necessary to concentrate the algal cells as part of inoculum preparation. The volume of inoculum to be added to each test vessel is calculated based upon the cell concentration in the stock culture, the volume in the test vessel, and the desired initial cell concentration. It is important to maintain aseptic technique in all culturing and testing procedures.

8. Testing procedure

8.1 RANGE-FINDING TEST

A range-finding test is usually conducted to establish the appropriate test solution concentrations for the definitive test. In the range-finding test, the test organisms are exposed to a series of widely-spaced concentrations of the test material, *e.g.*, 0.1, 1.0, 10, 100 mg/L, etc. (Note that for effluents, range-finding tests may not be practical due to limitations on holding times of samples). In a range-finding test, no replicates are required and nominal concentrations of the test material are acceptable.

8.2 DEFINITIVE TEST

The goal of the definitive test is to determine concentration-response curves and IC50 values (with 95 percent confidence intervals and standard error) for algal population growth for each species tested. In addition, the slopes of the concentration-response curves, the associated standard errors and the 95% confidence intervals of the slopes should be determined. For this determination, a minimum of five concentrations of the test material, plus appropriate controls, are required. The range of concentrations tested should bracket the expected IC50 value. Analytical confirmation of test concentrations should be performed using an acceptable validated analytical method. At the end of the exposure period, algistatic and algicidal effects can be determined as described in Section 9.4.

8.3 LIMIT TEST

In some situations, it is only necessary to ascertain that the IC50 is above a certain limit. A limit test has also been referred to as a Tier I test or Maximum Challenge Concentration test. In a limit test, at least three replicate test vessels are exposed to a single "limit concentration," with the same number of test vessels containing the appropriate control solution(s). If the IC50 is greater than the limit concentration, multiple-concentration definitive testing may be waived. Acceptable limit tests must meet all the requirements for acceptable multi-concentration definitive tests, with the exception of the number of test concentrations and endpoint determinations. Acceptable limit tests require analytical confirmation of the limit concentration.

8.4 PREPARATION OF TEST MATERIAL

8.4.1 Basic information
Basic information about the test material should be known prior to testing. This includes the following: chemical name; CAS number; molecular structure; source; lot or batch number; purity and/or percent active ingredient (a.i.); identities and concentrations of major ingredients and major impurities; date of most recent assay and expiration date for sample. In addition, it is important to know the appropriate storage and handling conditions for the test material to protect the integrity of the test material and the solubility and stability of the test material under test conditions. Physico-chemical properties of the test material can affect the design and interpretation of the test, and should be considered carefully. These include: solubility in water and various solvents; vapor pressure; hydrolysis at various pH, etc.

8.4.2 Preparation of stock solution
In some cases, test solutions are prepared by adding the test material directly to the growth medium on a weight/volume or volume/volume basis. More often, a stock solution of the test material is prepared and aliquots of the stock solution or secondary stock solutions are added to the growth medium. The preferred practice is to make a bulk preparation of each test solution and distribute portions to each replicate test vessel. Samples are taken from the bulk preparations for analytical confirmation of initial test concentrations.

The preferred choice for preparation of the stock solution is to use reagent water (deionized, distilled or reverse osmosis water), providing the test material can be dissolved in water and does not readily hydrolyze, and providing that the amount of stock solution added to the growth medium will be less than 10% of the total volume (in order to avoid changes in the growth medium). To avoid alterations in the growth medium (*e.g.*, unacceptable change in salinity or in concentration of nutrients), the stock solution may also be prepared in growth medium.

If the test material cannot be dissolved in reagent water or growth medium, carriers are often used. If a carrier, *i.e.*, a solvent and/or a dispersant, is absolutely necessary to dissolve the test material, the amount used should not exceed the minimum volume necessary to dissolve or suspend the test material in the growth medium. If the test material is a mixture, formulation or commercial product, none of the ingredients is considered a carrier unless an extra amount is used to prepare the stock solution. The preferred solvent for algal toxicity tests is N,N-dimethylformamide, as solvents such as acetone can cause stimulation of bacterial growth (Hughes and Vilkas, 1983). The concentration of solvent should preferably be the same in all test treatments and should not exceed 0.1 mL/L.

Solvent use should be avoided if possible. If a carrier is employed, a carrier control must be included in the test, in addition to the growth medium control. The selected carrier should not affect the test organisms at the concentration used. The carrier (solvent) control must be prepared from the same batch of solvent as that used to prepare the test treatment solutions.

The pH may be adjusted in stock solutions to match that of the medium if pH change does not affect the stability of the test material in the stock solution or test solution. Hydrochloric acid and sodium hydroxide may be used for this adjustment if warranted. The pH should generally not be adjusted after the addition of the test material or stock solution into the test medium. If the test material is highly acidic and reduces the pH of the test solution below 5.0 at the first measurement, or is highly basic and increases the pH of the test solution similarly, appropriate adjustments should be considered, and the test solution measured for pH on each day of the test. If the pH of the test solutions is altered, a concurrent test without pH adjustment of the test solutions is recommended.

8.4.3 Test concentrations

A toxicity test designed to allow calculation of a regression-based estimate such as an IC50 usually consists of one or more control treatments and at least five test solution concentrations. The test solution concentrations are usually selected in a geometric series in which the ratio is between 1.5 and 3.2. The selection of test concentrations depends upon the expected slope of the dose-response curve, which can be determined based upon the results of the range-finding test. Some methods for calculating the IC50 require that the test concentrations be equally spaced, while some methods do not.

8.5 ENVIRONMENTAL CONDITIONS

The test temperature is 24°C for *P. subcapitata* and *N. pelliculosa*, and 20°C for *S. costatum*. Excursions from the test temperature should be no greater than ± 2°C. Test vessels containing *P. subcapitata* and *N. pelliculosa* should be illuminated continuously; those containing *S. costatum* are to be provided a 14 h light:10 h dark photoperiod. Cool-white fluorescent lights providing 60 $\mu mol.m^{-2}s^{-1}$ should be used (for cool-white fluorescent lighting, this is approximately equivalent to 4300 lux). A PAR (photosynthetically active radiation) sensor should be used to measure light quality and measurements should be made at each test vessel position at the approximate level of the test solution. The light intensity should not vary more than ± 15% from the selected light intensity at any test vessel position in the incubator or growth chamber. Additional information on the use of lighting in plant toxicity tests can be found in ASTM E-1733 (ASTM, 2003b).

Stock algal cultures of *P. subcapitata* and *N. pelliculosa* should be shaken on a rotary shaking apparatus. Test vessels containing these species should also be placed on a rotary shaking apparatus and oscillated at approximately 100 cycles/min during testing. The rate of oscillation should be determined at the beginning of the test or at least once daily during testing if the shaking rate is changed or changes. Culture and test vessels containing *S. costatum* should be shaken by hand once or twice daily. If clumping of cells is not experienced, *S. costatum* may be continuously shaken at approximately 60 cycles/min.

9. Observations/measurements and endpoint determinations

9.1 MEASUREMENT OF TEST MATERIAL

Analytical confirmation of test concentrations should be performed at test initiation and at test termination. The analytical method used to measure the amount of test material in a sample should be validated before beginning the test. Samples for analysis of initial test concentrations should be collected from the bulk preparations used to begin the test. At the end of the test (and after aliquots have been removed for algal growth-response determinations, microscopic examination, mortal staining, or subculturing), the replicate test containers for each chemical concentration may be pooled into one sample. An aliquot of the pooled sample may then be taken and the concentration of test chemical is determined after all algal cells have been removed, either by centrifugation or filtration. The effect of centrifugation or filtration upon recovery of the test material should be determined during method validation. As an additional procedure, the concentration of test material associated with the algae alone may be determined, if desired. To do this, separate and concentrate the algal cells from the test solution by centrifuging or filtering the remaining pooled sample and measure the test material concentration in the cell concentrate.

Observations on test material solubility should be recorded. The appearance of surface slicks, precipitates, or material adhering to the sides of the test vessels should also be recorded.

9.2 MEASUREMENT OF ENVIRONMENTAL CONDITIONS

It is impractical to measure the temperature of the solutions in the test vessels while maintaining axenic conditions. Therefore, one or two extra test vessels may be prepared for the purpose of measuring the solution temperature during the test. Alternatively, hourly measurements of the air temperature (or daily measurements of the maximum and minimum) are acceptable. Because vessels are placed in an environmental chamber or incubator, the air temperature is more likely to fluctuate than the water temperature.

The pH in control and test solutions should be measured at the beginning and end of the test. It can be measured in the bulk test solutions at test initiation and in samples of pooled replicates of each test treatment at test termination (provided none of the replicates appear to be "outliers" with respect to growth, in which case individual pH measurements should be made).

As testing begins, light intensity (light fluence rate) should be monitored at the approximate level of the test solution at each test chamber position in the growth chamber. Random repositioning of the test vessels on a daily basis during the test is recommended to minimize spatial differences in temperature and lighting.

9.3 BIOLOGICAL OBSERVATIONS

The test is based upon the increase in algal biomass observed in exposed cultures compared to that in the control. Because biomass (*e.g.*, the dry weight of living matter present in a given volume) is difficult to measure accurately, surrogate measures of biomass are typically used in this test. The most common measure is to determine algal population density by counting the number of cells in a given volume. Cell counts in each test vessel should be determined at 24, 48, 72 and 96 hours. Performing cell counts using direct microscopic observation or using an electronic particle counter are both acceptable methods for determining population density. Chlorophyll *a* (measured spectrophotometrically or fluorometrically) or other measurements may also be used. Dry weight, although a direct measure of biomass, is a destructive measure that can only be used at test termination and must be accomplished carefully to obtain accurate results.

Microscopic counting of cells can be performed using a hemacytometer or an inverted microscope with settling chambers. Precision is proportional to the square root of the number of cells counted. For microscopic counting, two samples should be taken from each test vessel and two counts made of each sample. Whenever feasible, at least 400 cells per test vessel should be counted in order to obtain ± 10% accuracy at the 95% confidence level.

An alternative method to enumerate large numbers of cells very rapidly is to use an electronic particle counter. It is recommended that the laboratory develop data demonstrating the correlation between electronic particle counts and microscopic counts for each algal species. Automated particle counting, although the most rapid and sensitive method, has limitations, some related to particle interferences. If the test solution does not have a low background in the particle size range of the test species, masking errors will result. An additional test vessel at each concentration containing test material and growth medium without algae can allow measurement of potential particle interference.

Microscopic observations at test termination should be performed to determine whether the altered growth response between controls and test algae (at the concentrations of test material demonstrating an effect) was due to a change in relative cell numbers, cell sizes, or both. Noting any unusual cell shapes, color differences, differences in chloroplast morphology, flocculations, adherence of algae to test vessels, or aggregation of algal cells is also recommended. While these observations are qualitative and descriptive, they are independent of endpoint calculations. They can be useful, however, in demonstrating additional effects of test materials.

Other measurements that may be useful include determination of mean cell volume, organic carbon content of the cells, and dry weight. These measurements are not routinely required but may provide important information if the test material has an effect upon algal biomass that is not reflected in cell counts.

9.4 DETERMINATION OF ALGISTATIC AND ALGICIDAL EFFECTS

At the end of the 96-hour exposure period, determination of algistatic and algicidal effects may be performed, if desired (Payne and Hall, 1979). If the test material is algicidal, the algae have been killed and the population is unable to recover. If the test material is algistatic, population growth is inhibited in the presence of the test material but resumes once it is removed. In test concentrations where growth is maximally inhibited, algistatic effects may be differentiated from algicidal effects by either of the following two methods.

(1) Add 0.5 mL of a 0.1 percent solution (weight/volume) of Evans blue stain to a 1-mL aliquot of algal suspension from a control vessel and to a 1-mL aliquot of algae from the test vessel having the lowest concentration of test material which completely inhibited algal growth. Complete inhibition of algal growth is demonstrated if the algal population density at 96 hours is approximately the same as the initial population density. If algal growth was not completely inhibited, select an aliquot of algae for staining from the test vessel having the highest concentration of test material where at least some algal growth inhibition has occurred. Wait 10 to 30 min, examine microscopically, and determine the percent of the cells which stain blue (indicating cell mortality). A staining control is to be performed concurrently using heat-killed or formaldehyde-preserved algal cells; 100 percent of these cells should stain blue. This method will work for *S. costatum* (as it was initially developed with this species) and possibly *Navicula* spp., but it may not work with *P. subcapitata*.

(2) Remove 0.5 mL aliquots of test solution containing growth-inhibited algae from each replicate test vessel having the lowest concentration of test material which completely inhibited algal growth. If algal growth was not completely inhibited, select aliquots from the highest concentration of test material indicating algal growth inhibition. Combine these aliquots into a new test vessel and add a sufficient volume of fresh nutrient medium to dilute the test material to a concentration which does not affect growth (using the original test vessel size and solution volume is generally appropriate). Aliquots from the control test vessels are also transferred to clean medium. Incubate these subcultures under the environmental conditions used during the exposure period for up to 9 days, and observe periodically (*e.g.*, every other day) for algal growth to determine if the algistatic effect noted after the 96-h exposure is reversible. This subculture test may be discontinued as soon as growth occurs.

9.5 TREATMENT OF RESULTS

Algal population density is the biomass measurement normally used to evaluate the inhibitory and stimulatory effects of the test material. Two response variables are calculated: final population density, also referred to as yield, and average specific growth rate. The IC50 value is determined (with 95 percent confidence interval and standard error, as well as slope of the concentration-response curve, standard error, and 95 percent confidence interval) for each of these response variables. The NOEC and LOEC should also be determined. The calculation of measurement endpoints at 72 h, in addition to 96 h, is desirable, provided growth is sufficient for analysis at these earlier time periods.

9.5.1 Use of measured concentrations

Results are expressed based upon measured concentrations of the test material, if available. If analytical verification of test concentrations has not been performed, the nominal values are used.

One of the advantages of this test design over scaled-down tests is that sufficient sample volume is usually available to measure the test concentrations at the beginning and end of the test. Thus, the flask-based test is the method of choice where analytical confirmation is needed. It is not uncommon, however, for test concentrations to decline during the exposure period, usually due to inherent properties of the test material, although uptake and adsorption by algal cells can also occur. Analysis of the concentration in a "blank" test vessel (prepared and incubated as the other replicates for a particular concentration, but not inoculated with algae) can shed light on these phenomena. If the test material concentrations decline during the exposure period, it may be possible to determine the rate of decline and use this to calculate the actual exposure concentrations. Otherwise, the mean of the initial and final measured concentrations is used as an approximation. Alternatively, if concentrations decline by less than an amount set by the precision of the analytical method (typically about 20%), the initial concentrations may be used. Because this test is a static toxicity test, there is little that can be done to maintain test concentrations during the exposure period. Conducting flow-through and renewal exposure procedures with microalgae are currently impractical, which can be a disadvantage to this and other phytotoxicity tests.

9.5.2 Final population density

Final population density at test termination (96 h) for each test vessel, or more correctly, yield, is used to calculate the IC50. To correctly represent yield, the initial population density values should be subtracted from the final population density values for each test vessel. Since the initial values are extremely small relative to the final values, this correction has a small impact upon the test results but is nonetheless recommended. Population densities at the end of 24, 48 and 72 hours can also be used to calculate IC50s for those time periods, if desired, and if growth is sufficient.

9.5.3 Average specific growth rate

Average specific growth rate is also used to calculate the IC50. It represents the growth rate calculated over the entire test period. In addition, the specific growth rate during the course of the test (days 0-1, 1-2, 2-3, etc.), also called the section-by-section growth rate, should be calculated to assess effects of the test material, such as an increased lag phase, occurring during the exposure period. Substantial differences between the section-by-section growth rates and the average growth rates indicate deviation from theoretical exponential growth and that close examination of these data are warranted. In this instance, the recommended approach is to compare specific growth rates from exposed cultures during the time period of maximum inhibition to those for controls during the same period. The same time interval should be used for each test vessel in all treatments. The growth rate for each test vessel over the selected time interval is calculated as follows:

$$\mu = \frac{\ln N_2 - \ln N_1}{t_2 - t_1} \quad (1)$$

where:
μ = growth rate, in day^{-1}
N_1 = population density at the beginning of the selected time interval
N_2 = population density at the end of the selected time interval
t_2 = time at end of interval (in days)
t_1 = time at start of interval (in days).

9.5.4 Calculation of IC50

The IC50 and 95% confidence interval are determined using an appropriate statistical model to establish the concentration-response curve for the response variables. The values for each test vessel, not the mean for each concentration, should be used as the response variable in fitting the model.

Statistical procedures for modeling continuous toxicity data are available and should be used (Bruce and Versteeg, 1992; Nyholm et al., 1992; VanEwijk and Hoekstra, 1993). Regarding terminology, the term ICx is often used for non-quantal endpoints, rather than ECx.

Percent inhibition (%I) at each test concentration is calculated as follows:

$$\%I = \frac{C - X}{C} \times 100 \quad (2)$$

where: C = the average value of the response variable in the control test vessels and X = the average value of the response variable in the test treatment. Stimulation is reported as negative percent inhibition.

9.5.5 Calculation of NOEC and LOEC

Hypothesis testing procedures can be used to determine the NOEC and LOEC for each of the measured response variables. Assumptions of statistical procedures should be examined and verified as met prior to their use. Results of hypothesis tests should be reported along with some measure of the sensitivity of the test (either the minimum significant difference or the percent change from the control that this minimum difference represents).

9.6 TEST ACCEPTABILITY

Validity criteria for the test include acceptable growth in the controls and acceptable variation between control replicates. During 96 hours, cell counts in the controls should increase by a factor of at least 100 times for *P. subcapitata* and a factor of at least 30 times for *S. costatum*. The appropriate increase within 96 hours for *N. pelliculosa* has not been determined at this time. For any algal species, the coefficient of variation for yield in the control should be calculated and should generally be less than 20%. For growth rate, which is a logarithmically-transformed variable, the coefficient of variation should be substantially less than 20% (*e.g.*, < 12%).

9.7 REPORTING

The reported results of the test should include the following:

- Test facility, dates and personnel.

- Identification of test material and purity.

- Description of the preparation of the synthetic growth media used, the concentrations of all media constituents, and the initial pH.

- Methods of stock solution and test solution preparation and the concentrations of test material and solvent, if applicable, used in definitive testing.

- Detailed information about the test organisms, including the scientific name, method of verification, strain, and source. Information about the culture practices and conditions. Description of preparation of inoculum used to begin test.

- A description of the growth chamber and test vessels, the volumes of solution in the test vessels, the way the test was begun (*e.g.*, conditioning, test material additions, etc.), the number of replicates, the temperature, the lighting, and method of incubation, oscillation rates, and type of apparatus. Specific modifications in test procedures due to using species other than those recommended must be noted.

- The concentration of the test material in the control(s) and in each treatment at the beginning and end of the test and the pH of the solutions at the beginning and end of the test.

- The number of algal cells per milliliter in each test vessel (or other biomass surrogate data) and the method used to derive these values at the beginning, at 24, 48, and 72 h, and at the end of the test; calculated mean values with standard deviation; the percentage of inhibition or stimulation of growth relative to controls (based upon means); and other adverse effects in the control and in each treatment.

- The 96-h IC50 values, and when sufficient data have been generated, the 24-, 48-, and 72-h IC50s and 95 percent confidence limits. The IC50 should be determined based upon final population density (yield) and average specific growth rate. The slopes of the concentration-response curves, associated standard errors and the 95% confidence intervals of the slope should be reported as well. NOEC/LOEC values should also be reported.

- Methods of statistical analysis, including software used, should be described.

- Methods used in the analysis of concentrations of test material should be described. The accuracy of the method, method detection limit, and limit of quantification should be given.

- Microscopic appearance of algae, size or color changes, and any other observed effect.

- If determined, report the algistatic and algicidal concentrations.

- For a limit test, report the percent effect upon the measured response variables at the tested concentration.

- Any protocol deviations or occurrences which may have influenced the final results of the test.

10. Factors capable of influencing algal growth and test results

Test solutions that are highly colored or opaque can reduce or prevent light transmission, affecting algal photosynthesis due to a physical effect rather than a toxicological effect. Test materials that are highly volatile can escape from the test system, since the flask stoppers permit gas exchange (and thus allow photosynthesis). It is possible to modify the test design to accommodate highly volatile materials by adding supplemental carbon and eliminating the head space (*e.g.*, using a BOD bottle), but such procedures are not part of the typical method. Some test materials (*e.g.*, some anionic polymers) cause chelation of the trace nutrients needed for algal growth. Since the nutrient medium for freshwater algae has a low hardness, growth inhibition can be observed in these circumstances and interpreted as toxicity. However, when sufficient calcium (as divalent cation) is added to satisfy the ionic charge of the polymer, toxicity to algae is mitigated (Nabholz et al., 1993).

11. Application of the algal toxicity test in a case study

Several algal species were used to evaluate the toxicity of the herbicide atrazine in a study that reported the IC50, NOEC, and algistatic and algicidal endpoints (Hughes et al., 1988). This study used a 5-day exposure period, and there were some minor differences in the methods relative to those described in this chapter. However, the approach taken and the comparison of the test endpoints are illustrative of the principles of the flask-based algal toxicity test. The test species included the freshwater diatom *Navicula pelliculosa*, the marine flagellate *Dunaliella tertiolecta*, and the cyanobacteria *Anabaena flos-aquae*. The results are presented in Table 4.

Table 4. Effects of atrazine on three species in the algal toxicity test.

Species	NOEC mg/L	IC50[1] mg/L	Algistatic concentration[1] mg/L	Algicidal concentration mg/L
Anabaena flos-aquae	< 0.1	0.23 (0.12 – 0.38)	4.97 (2.39 – 14.2)	> 3.2
Dunaliella tertiolecta	< 0.1	0.17 (0.11 – 0.26)	1.45 (0.44 – 6.72)	3.2
Navicula pelliculosa	< 0.1	0.06 (0.002 – 0.21)	1.71 (0.40 – 13.2)	> 3.2

[1] The 95% confidence limits are given in parentheses.

Each species was significantly affected by the lowest test concentration of atrazine, thus the NOEC was below 0.1 mg/L. The IC50 values were calculated based upon final population density, and ranged from 0.06 mg/L for *N. pelliculosa* to 0.23 mg/L for *A. flos-aquae*. (IC50 values based upon average specific growth rate were not determined). The algistatic concentration was determined as the concentration of test material at which the population density on day 5 was the same as the initial population density. This value ranged from 1.45 mg/L for *D. tertiolecta* to 4.97 mg/L for *A. flos-aquae*. *D. tertiolecta* was unable to recover from exposure to a concentration of 3.2 mg/L atrazine, while the other species did recover, indicating that atrazine was not algicidal to *N. pelliculosa* and *A. flos-aquae*. Atrazine prevented photosynthesis, but since all of the algal cells were not killed, the population of these two species was able to recover in the absence of atrazine. This can be useful information for a variety of risk assessment applications, especially if the test substance is expected to have a short duration of use or limited stability in the environment. Although a continuous exposure to an algistatic concentration of a test substance would cause complete inhibition of growth, in the absence of continuous input, recovery of the algal population would be expected as the test substance degrades. Due to the additional time and effort required to determine algistatic and algicidal effects, however, the use of the IC50 is an acceptable and conservative way to express toxicity to algae. It should be emphasized that an IC50 for algae represents a population effect and that it is not analogous to similar endpoints (*e.g.*, EC50 or LC50) for aquatic animals.

12. Miscellaneous test information

Algal species vary in their sensitivity and no single species is the most sensitive to all toxicants (Blanck et al., 1984; Peterson et al., 1993). For this reason, it is recommended to test several species to appropriately define potential hazard (Lewis, 1990; Swanson et al., 1991). The flask-based algal toxicity test is adaptable to use with various species, although changes in the initial inoculum concentration, test duration, incubation conditions and medium may be necessary.

13. Conclusions

The algal toxicity test described in this chapter has been widely used for at least 25 years to determine the toxicity of a variety of test substances to microalgae. The method has undergone standardization by groups including ASTM and OECD. It is a practical means to evaluate toxicity to organisms that are considered to be the basis of the food web in most aquatic systems. However, it is unlikely that algae can reliably serve as surrogates for higher aquatic plants, and additional developmental work is needed on test methods with submersed and emergent aquatic macrophytes to examine the relative sensitivity of all of these organisms. Assessing the response of aquatic plants is critical to the risk assessment process, particularly for chemicals such as herbicides. For example, U.S. EPA's new draft ambient water quality criterion for atrazine was derived based upon changes in aquatic plant community structure, as this was the most sensitive response observed (U.S. EPA, 2003).

The algal toxicity test examines the response of only one species at a time and thus does not consider interactions within the algal community, which can be an important influence upon the overall productivity of the aquatic ecosystem. New multi-species test procedures (Franklin et al., 2004) show promise in elucidating these types of interactions; however, they require the use of sophisticated equipment not currently in routine use in most laboratories.

Although the procedures for conducting the algal toxicity test are straightforward, there is room for improvement in understanding, interpreting and using the results in risk assessment. Probably due to its long history as one of the basic tests in a "tiered" risk assessment, an algal IC50 is too often equated with a measure of mortality in an acute exposure for an aquatic fish or invertebrate. Research to establish the linkage of laboratory tests with microalgae to responses in field situations would advance the utility of algal toxicity test data. As pointed out by Lewis (1990), the significance of reductions in algal growth observed in a laboratory test must be interpreted in light of ecological factors such as adaptation and compensation to improve the utility of laboratory results in risk assessment.

Acknowledgements

The constructive comments of three reviewers of this chapter are greatly appreciated, as is the assistance of the Technical Publications staff in the Durham, NC office of ARCADIS, Inc.

Disclaimer

The contents of this chapter are based on results from U.S. EPA contract no. GS-10F-0226F. These results have been reviewed and approved by the Agency. This approval neither signifies that the contents necessarily reflect the views and policies of the Agency nor does mention of tradenames or commercial products constitute endorsement or recommendation for use.

References

ASTM (American Society for Testing and Materials) (2003a) Standard guide for conducting 96-h toxicity tests with microalgae, in *Annual Book of ASTM Standards*, Vol. 11.05, ASTM E 1218-97a. West Conshohocken, PA., Current edition approved October 10, 1997.

ASTM (American Society for Testing and Materials) (2003b) Standard guide for use of lighting in laboratory testing, in *Annual Book of ASTM Standards*, Vol. 11.05, ASTM E 1733 -95. West Conshohocken, PA., Current edition approved September 10, 1995.

Boutin, C., Freemark, K.E. and Keddy, C.J. (1993) Proposed Guideline For Registration Of Chemical Pesticides: Nontarget plant testing and evaluation, Tech. Rpt. Series No. 145, Canadian Wildlife Service, Environment Canada.

Blanck, H., Wallin G. and Wangberg, S. (1984) Species-dependent variation in algal sensitivity to chemical compounds, *Ecotoxicology and Environmental Safety* **8**, 339-351.

Bruce, R.D. and Versteeg, D.J. (1992) A statistical procedure for modeling continuous toxicity data, *Environmental Toxicology and Chemistry* **11**, 1485-1494.

Franklin, N.M, Stauber, J.L. and Lim, R.P. (2004) Development of multispecies algal bioassays using flow cytometry, *Environmental Toxicology and Chemistry* **23**, 1452 – 1462.

Hughes, J. S., Alexander, M.M. and Balu, K. (1988) An evaluation of appropriate expressions of toxicity in aquatic plant bioassays as demonstrated by the effects of atrazine on algae and duckweed, in W.J. Adams, G.A. Chapman, and W.G. Landis, (eds.), *Aquatic Toxicology and Hazard Assessment: 10th Volume*, ASTM STP 97, American Society for Testing and Materials, Philadelphia, pp. 531 - 547.

Hughes, J.S. and Vilkas, A.G. (1983) The Toxicity of N,N-dimethylformamide used as a solvent in toxicity tests with the green alga, *Selenastrum capricornutum*, *Bulletin of Environmental Contamination and Toxicology* **31**, 98-104.

Lewis, M.A. (1990) Are laboratory-derived toxicity data for freshwater algae worth the effort? *Environmental Toxicology and Chemistry* **9**, 1279-1284.

Miller, W.E., Greene, J.C. and Shiroyama, T. (1978) The *Selenastrum capricornutum* Printz Algal Assay Bottle Test: Experimental Design, Application, and Data Interpretation Protocol, EPA-600/9-78-018, Corvallis, OR.

Nabholz, V.N, Miller P. and Zeeman, M. (1993) Environmental risk assessment of new chemicals under the Toxic Substances Control Act (TSCA) Section Five, in W.G. Landis, J.S. Hughes, and M.A. Lewis (eds.), *Environmental Toxicology and Risk Assessment.*, ASTM STP 1179, American Society for Testing and Materials, Philadelphia, PA., pp. 40-55.

Nyholm, N., Sorenson, P.S., Kusk, K.O. and Christensen, E.R. (1992) Statistical treatment of data from microbial toxicity tests, *Environmental Toxicology and Chemistry* **11**, 157-167.

OECD (Organization for Economic Co-operation and Development) (1984) OECD Guidelines for Testing of Chemicals, TG 201, Alga, Growth Inhibition Test.

Payne, A.G. and Hall, R.H. (1979) A method for measuring algal toxicity and its application to the safety assessment of new chemicals, in L.L. Marking and R.A. Kimerle (eds.), *Aquatic Toxicology*, ASTM STM 667, American Society for Testing and Materials, Philadelphia, PA., pp. 171-180.

Peterson, H.G., Boutin, C., Freemark, K., Martin, P.A., Ruecker N.J. and Moody, M.J. (1993) Aquatic phytotoxicity of 23 pesticides applied at expected environmental concentrations, *Aquatic Toxicology* **28**, 275-292.

Swanson, S.M., Rickard, C.R., Freemark K.E. and MacQuarrie, P. (1991) Testing for pesticide toxicity to aquatic plants: Recommendations for test species, in J.W. Gorsuch, W.R. Lower, M.A. Lewis and W. Wang, (eds.), *Plants for Toxicity Assessment*, Vol. 2, STP 1115, American Society for Testing and Materials, Philadelphia, PA., pp. 77-97.

U.S. EPA (U.S. Environmental Protection Agency) (1971) Algal Assay Procedure Bottle Test. National Eutrophication Research Program, Office of Research and Development, Environmental Research Division, Western Ecology Division, Corvallis, OR., 82 pp.

U.S. EPA (U.S. Environmental Protection Agency) (1974) Marine Algal Assay Procedure: Bottle Test, Eutrophication and Lake Restoration Branch, Pacific Northwest Environmental Research Center, Corvallis, OR.

U.S. EPA (U.S. Environmental Protection Agency) (1978) Bioassay Procedures for the Ocean Disposal Permit Program, EPA-600/9-78-010, Environmental Research Laboratory, Gulf Breeze, FL.

U.S. EPA (U.S. Environmental Protection Agency) (1982) Pesticide Assessment Guidelines, Subdivision J Hazard Evaluation: Non-target plants, EPA 540/9-82-020, Washington, DC.

U.S. EPA (U.S. Environmental Protection Agency) (1985) Toxic substances control act test guidelines; Final rules. Federal Register 50: 39252 – 39516 (Part 796 – Chemical fate testing guidelines, Part 797 – Environmental effects testing guidelines).

U.S. EPA (U.S. Environmental Protection Agency) (1986) Hazard Evaluation Division Standard Evaluation Procedure, Non-target Plants: Growth and Reproduction of Aquatic Plants, Tiers 1 and 2, EPA 540/9-85-134.

U.S. EPA (Environmental Protection Agency) (1996) OPPTS Series 850 Ecological Effects Test Guidelines, Public Draft. At: http://www.epa.gov/opptsfrs/OPPTS_Harmonized/850_Ecological_Effects_Test_Guidelines/index.html

U.S. EPA (U.S. Environmental Protection Agency) (2002) Short-term Methods for Estimating the Chronic Toxicity of Effluents and Receiving Waters to Freshwater Organisms, Fourth Edition, Office of Water, Washington, DC, EPA-821-R-02-013, 335 pp.

U.S. EPA (U.S. Environmental Protection Agency) (2003) Ambient Aquatic Life Water Quality Criteria for Atrazine – Revised Draft, October, 2003, EPA-R-03-023.

VanEwijk, P.H. and Hoekstra, J.A. (1993) Calculation of the IC50 and its confidence interval when subtoxic stimulus is present, *Ecotoxicology and Environmental Safety* **25**:25-32.

Walsh, G.E. and Alexander, S.V. (1980) A marine algal bioassay method: results with pesticides and industrial wastes, *Water, Air and Soil Pollution* **13**:45-55.

Abbreviations

AAP	Algal Assay Procedure
a.i.	active ingredients
ASTM	American Society for Testing and Materials, also known as ASTM International.
CAS number	Chemical Abstracts Service (Registry) number
EDTA	ethylenediamine tetraacetate
LOEC	lowest observed effect concentration
OECD	Organization for Economic Co-operation and Development
PAR	photosynthetically active radiation
NOEC	no observed effect concentration
ppt	parts per thousand
U.S. EPA	United States Environmental Protection Agency.

5. MICROALGAL TOXICITY TESTS USING FLOW CYTOMETRY

JENNIFER STAUBER
Centre for Environmental Contaminants Research
CSIRO Energy Technology
Private Mail Bag 7,Bangor, NSW 2234, Australia
jenny.stauber@csiro.au

NATASHA FRANKLIN
Department of Biology, McMaster University
1280 Main Street West, Hamilton
Ontario L8S 4K1, Canada
nfrank@univmail.cis.mcmaster.ca

MERRIN ADAMS
Centre for Environmental Contaminants Research
CSIRO Energy Technology
Private Mail Bag 7,Bangor, NSW 2234, Australia
merrin.adams@csiro.au

1. Objective and scope of tests

These tests determine the inhibitory effects of liquid samples on the growth rate or enzyme activity in freshwater microalgae using flow cytometry as the detector. Flow cytometry is a rapid method for counting and measuring fluorescence and light scattering properties of algal cells at low cell densities, typical of that found in freshwater environments. The test is based on standard phytotoxicity tests (see Chapter 3 of this volume; OECD, 1984; U.S. EPA, 2002) and is applicable to all liquid samples including:
- wastewaters and effluents, filtered or unfiltered;
- sediment pore waters;
- surface waters, groundwaters or leachates;
- chemicals.

The test may be used for screening or definitive tests, alone, or as part of a battery of tests approach. Two tests with the green alga *S. capricornutum* are described in this chapter:
- a chronic toxicity test measuring growth rate inhibition using flow cytometry to count algal cells at low cell densities in the presence of particulates;

- an acute toxicity test measuring enzyme (esterase) inhibition using flow cytometry.

2. Summary of test procedure

Two toxicity tests with exponentially growing cells of the chlorophyte (green microalga) *S. capricornutum* are described in this chapter.

The first test is a modification of the standard growth rate inhibition bioassay over 72 h, a chronic test in minivials using flow cytometry to count the cells each day. Algae are exposed in minivials containing 6 mL of culture medium, to a range of toxicant concentrations under controlled laboratory conditions. One unique feature of this bioassay is that the initial inoculum can be lowered to 100 or 1000 cells/mL, more representative of cell densities in aquatic systems. This prevents changes in toxicant speciation and subsequent bioavailability, which can lead to an underestimation of toxicity in standard bioassays. Cells are counted each day using the technique of flow cytometry, which is sufficiently sensitive to count cells at these low cell densities. Light scatter and fluorescence measurements collected simultaneously allow several endpoints, together with growth inhibition, to be assessed. A comparison of growth rates in the controls and the test solution-exposed algae enables the calculation of IC50 (Inhibition Concentration 50), LOEC (Lowest Observed Effect Concentration) and NOEC (No Observed Effect Concentration) values.

The second test is an acute enzyme toxicity test that uses flow cytometry to measure inhibition of esterase activity after a 3 or 24-h exposure to toxicant. Cells are exposed to toxicant for 3 or 24 h in minivials, after which a substrate fluorescein diacetate (FDA) is added. Cell suspensions are incubated for 5 min, and then analysed by flow cytometry. Healthy control cells take up FDA, which is cleaved by esterases, releasing a fluorescent product, the fluorescein that is retained in the cells. This is measured as an increase in algal cellular fluorescence in the green region of the spectrum. Toxicants decrease FDA cleavage by esterases and subsequent green fluorescence, and this is measured as a shift in fluorescence intensity from the healthy control region towards the unhealthy (dead) cell region. This shift, detectable by flow cytometry, is quantified and the percentage shift out of the control region is calculated. This enables typical toxicity test endpoints such as IC50, NOEC and LOEC to be determined.

Both tests are sensitive to a range of toxicants and are highly reproducible.

3. Overview of development and applications of flow cytometry-based toxicity tests

Flow cytometry is a rapid method for the quantitative measurement of individual cells in a moving fluid. Thousands of cells are passed through a light source (usually a laser) and measurements of cell density, light scatter and fluorescence are collected simultaneously. Although this technique has been widely applied to biomedical and oceanographic studies, flow cytometry has only been applied to ecotoxicology for slightly more than a decade.

Microalgae are ideally suited to flow cytometric analysis as they contain photosynthetic pigments such as chlorophyll a, which autofluoresce when exited by blue light. In addition, specific fluorescent dyes can be used and detected by flow cytometry to provide information about the physiological status of algal cells in response to toxicants (Jochem, 2000; Molecular Probes, 2003).

A preliminary study by Premazzi et al. (1989) showed that algal cell numbers and cell size measured by flow cytometry were similar to those obtained by conventional counting techniques. Other groups demonstrated the usefulness of flow cytometry in algal physiological studies by investigating the effect of copper on marine diatoms (Cid et al., 1995; 1996). These studies, however, used very high metal concentrations to obtain a detectable algal response. More recently, flow cytometry has been used to developespace more environmentally relevant toxicity tests capable of detecting multiple effects on algal cells at low toxicant concentrations, more typical of that found in natural waters (Franklin et al., 2001a,b; 2002; Stauber et al., 2002). These tests can determine growth rate inhibition, changes in algal cell size, effects on chlorophyll a fluorescence, cell viability and inhibition of enzyme activity. In addition, tests using low cell densities (100-1000 cells/mL) close to that found in surface waters are now possible, and these avoid changes in chemical speciation and bioavailablity often encountered when high cell densities are used in standard toxicity tests (Franklin et al., 2002).

Tests with both marine and freshwater microalgae have been developed using flow cytometry as the detector and applied to testing wastewaters, chemicals and sediment porewaters (Stauber et al., 2002; Hall and Cumming, 2003). Toxicity tests with whole sediments have also been developed for both freshwater (Blaise and Ménard, 1998) and marine applications (Adams and Stauber, 2004).

4. Advantages of conducting flow cytometry-based phytotoxicity tests

Algal bioassays have been widely used to evaluate the potential impact of contaminants in marine and freshwater systems. Standard tests use population growth and measure inhibition of growth rate or cell yield over 48-96 h. Because cells divide daily over this period, substantial losses of toxicants to the cells and test containers (up to 20%) may occur, particularly when high cell densities (> 10^4 cells/mL) are used as the initial inoculum (Stauber and Davies, 2000; Franklin et al., 2002). Algal metabolism over this time can also cause an increase in pH (>1 pH unit) and subsequent chemical alteration of the test medium (Nyholm and Kallqvist, 1989). For this reason, tests of short duration using low cell densities (~10^3 cells/mL) are preferred. The problem in the past has been that traditional methods for counting cells require high cell densities to obtain a measurable response. The application of flow cytometry to algal toxicity testing is in its infancy, but this sensitive technique is able to overcome many of the problems associated with standard growth tests.

Undertaking phytotoxicity tests using flow cytometry to count and analyse algal cells has many advantages over conventional analysis techniques. The important ones are highlighted in Table 1.

Table 1. Advantages of using flow cytometry in toxicity tests with microalgae.

Feature	Remarks
Can distinguish between live and dead algal cells	Conventional automated counting techniques such as Coulter and laser counters cannot distinguish between live and dead cells. Flow cytometry can detect live cells based on their chlorophyll *a* fluorescence, and can exclude counts of dead cells with low fluorescence.
Can count cells at low cell densities (>100 cells/mL)	Standard algal growth inhibition tests use high initial cell densities (10^4-10^5 cells/mL) that can alter toxicant speciation, bioavailability and toxicity. Toxicity tests using flow cytometry can use initial cell densities as low as 100 cells/mL, more representative of cell densities in natural waters.
High sample throughput	Automatic loaders mean that cell counts/analysis on many different samples can be done automatically.
Cell sorting capabilities	Cells can be physically sorted from other species and sediment particles for further analysis.
Can measure multiple effect parameters simultaneously	Cell density (for growth inhibition tests) can be determined at the same time as cell size (light scatter), chlorophyll *a* fluorescence, metabolic state, DNA content, membrane permeability, enzyme activity.
Can count algae in the presence of particulate material	In standard algal toxicity tests, effluents and other samples have to be filtered to remove particulate material, which may contribute to the sample's toxicity. Flow cytometry can be used to count cells in unfiltered effluents and in sediments.
Multi-species bioassays possible	Standard toxicity tests are single species tests. Because flow cytometry can distinguish between different algae on the basis of size and pigment fluorescence, multi-species assays with two or more species together can be carried out (Franklin et al., 2004).
Small sample volumes	Volumes as small as 300 µL only are required for cell counts. Recent availability of a microplate reader on the front end of the flow cytometer means that counts can now be made directly from microplates.
Initial counts possible without dilution	Freshwater samples can be counted on Day 0 and 1 of the toxicity test, as dilution with high conductivity solutions (such as Isoton® for impedance counters) is not necessary.

One disadvantage is that currently flow cytometers are quite expensive and require skilled operators to carry out the toxicity tests. Until flow cytometers are more widely used in ecotoxicological testing, interlaboratory studies will be limited.

Short duration tests that detect acute sub-lethal endpoints, such as inhibition of enzyme activity, also show promise for overcoming some of the limitations of standard algal growth tests. Inhibition of esterase activity in algae has been shown to relate well to metabolic activity and cell viability (Gala and Giesy, 1990). Esterases are a group of enzymes involved in phospholipid turnover in cell membranes and can be measured *in vivo* using fluorogenic stains such as fluorescein diacetate (FDA). FDA is a lipophilic non-fluorescent dye that diffuses freely across the plasma membrane. Esterases hydrolyse FDA in the cytoplasm, producing fluorescein, which is retained by viable cells. Fluorescein fluorescence can be detected by flow cytometry and reflects both esterase activity and cell membrane integrity, both of which indicate cell viability (Dorsey et al., 1989; Franklin et al., 2001b).

5. Test Species

Any non-chain forming marine or freshwater microalga can be used as the test species. For the purposes of this chapter, we will only describe two tests with the freshwater chlorophyte (green alga) *S. capricornutum* (now called *Pseudokirchneriella subcapitata* Hindak, 1990) as this species has been most widely used (Figure 1). The former name remains popular in North America and is therefore adopted throughout the text. Further details on this species are given in Chapter 3 of this volume.

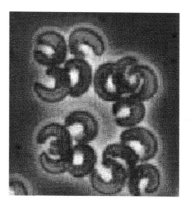

Figure 1. S. capricornutum *(approximate magnification 1000x).*

6. Culture maintenance of organism in the laboratory

6.1 LABORATORY FACILITIES REQUIRED

Culture cabinets with controlled temperature and lighting or temperature-controlled rooms with lighting are required for maintenance of the algal stock culture and for carrying out the toxicity tests.

6.2 MATERIALS

Materials required for conducting the toxicity tests and analysing the cells by flow cytometry are given below in Table 2. This is not an exhaustive list but rather summarizes the main materials and consumables required.

Table 2. Materials required for testing.

S. capricornutum microalgae	Graduated glass (or sterile disposable serological pipettes) 2 mL
Glass centrifuge tubes-30 mL and centrifuge rack	Glass Pasteur pipettes
Fluorescent calibration beads for flow cytometer (CaliBRITE beads, BD Biosciences, USA)	GF/F filters (alternative filters for filtering effluent if required)
Flow cytometer sample tubes (5-mL Falcon or equivalent)	Parafilm or equivalent laboratory sealing film
TruCount™ counting tubes (BD Biosciences, USA)	Polyethylene wash bottles and storage containers (1-10 L)
Glass graduated measuring cylinders and beakers	Adjustable automatic pipettes (5 µL to 5 mL)
Glass volumetric flasks	Disposable pipette tips
Membrane filter (pore size 0.45 µm diameter)	Chemicals and reagents including:
Glass Erlenmeyer flasks (200 or 250 mL) with loose-fitting glass caps	- salts for U.S. EPA medium (AR grade)
Glass scintillation minivials (20-30 mL capacity) with screw cap plastic lids	- acids and bases for pH adjustment (HCl, NaOH),
	- non-phosphate detergent and HNO_3 for glassware washing.
Weighing trays and spatula	- Fluorescein diacetate (FDA) (Sigma)
Magnetic stirrers	- Acetone (AR grade)

6.3 EQUIPMENT

Equipment required for the algal toxicity test with flow cytometric analysis is listed in Table 3 below. All equipment must be regularly maintained and calibrated according to the manufacturer's instructions. Any equipment that makes contact with the test organisms, media, control water or test solutions must be made of chemically inert materials (*e.g.*, glass, polyethylene) and be thoroughly cleaned before use.

Table 3. Equipment required for testing.

Flow cytometer (*e.g.*, Becton Dickinson FACSCalibur™, Beckman Coulter Epics XL or equivalent) with computer and colour printer	Environmental chamber/incubator with built in light and temperature controls, and shaking platform
Milli-Q water or equivalent water purification system	Centrifuge (benchtop) – 4 x 30 mL capacity with swing-out buckets (700 *g*)
Refrigerator for storing stock solutions (4°C)	Vortex mixer to mix algal suspension prior to inoculation
Biohazard cupboard or laminar flow cabinet for aseptic algal culturing	Mechanical shaking platform to fit inside environmental chamber
Autoclave	Analytical balance for weighing chemicals
Filter apparatus – 47 mm filter holder, vacuum pump and tubing, 1L flask	Magnetic stirrer
pH meter and buffers	Thermometer
Conductivity meter	Timers
Light meter	

6.4 WASHING AND SILANISING OF GLASSWARE

All the reusable glassware must be washed with a non-phosphate detergent solution (*e.g.*, Extran MA03) and rinsed at least three times with de-ionized water. The glassware should then be soaked overnight in 10% nitric acid (AR grade) and rinsed five times with de-ionized water and five times with Milli-Q water or equivalent. New glassware should be washed in the same way before use. Glassware used for culturing and stock solutions should be kept separate to glassware used in the toxicity test.

If toxicity tests with metals are to be conducted, glassware used for the toxicity test must be thoroughly dried and coated with a silanising solution (such as Coatasil, BDH) or equivalent to reduce adsorption of the toxicant to the glass test vessel. This procedure must be carried out in a fume cupboard. The glassware (either Erlenmeyer flasks or minivials) should be dried overnight in the fume cupboard then acid-washed in 10% nitric acid and rinsed thoroughly with Milli-Q water or equivalent before use. Glassware should be re-silanised every month if in constant use.

6.5 PREPARATION OF REAGENTS AND ALGAL CULTURE MEDIUM

All chemicals used in the preparation of the algal culture medium must be of analytical grade quality. Solutions should be prepared in Milli-Q water or equivalent.

6.5.1 Preparation of algal culture medium

S. capricornutum stock cultures are maintained in clean 250 mL glass Erlenmeyer flasks in 100 mL U.S. EPA Stock medium with EDTA ($Na_2EDTA.2H_2O$). To make this medium, five concentrated stock solutions are prepared and stored at 4°C until use. Details of the stock solution preparation are given in Table 4. Each stock is prepared by weighing the appropriate amount of chemical salt into a weighing tray and rinsing with Milli-Q water into a clean 250 mL volumetric flask. For stock solution 1, each salt should be dissolved prior to adding the next chemical according to the specific instructions in Table 4.

Table 4. Liquid growth medium for the stock algal culture.

Stock solution	Compound	Amount dissolved in 250 mL Milli-Q water
1	$MgCl_2.6H_2O$	3.04 g
	$CaCl_2.2H_2O$	1.10 g
	H_3BO_3	46.4 mg
	$MnCl_2.4H_2O$	104.0 mg
	$ZnCl_2$	0.82 mg[a]
	$FeCl_3.6H_2O$	40 mg
	$CoCl_2.6H_2O$	0.36 mg[b]
	$Na_2MoO_4.2H_2O$	1.82 mg[c]
	$CuCl_2.2H_2O$	0.003 mg[d]
	$Na_2EDTA.2H_2O$	150 mg
2	$NaNO_3$	6.375 g
3	$MgSO_4.7H_2O$	3.675 g
4	K_2HPO_4	0.261 g
5	$NaHCO_3$	3.75 g

[a] $ZnCl_2$ - 164 mg is weighed out and diluted to 100 mL and 0.5 mL is added to Stock #1.
[b] $CoCl_2.6H_2O$ - 71.4 mg is weighed out and diluted to 100 mL and 0.5 mL is added to Stock #1.
[c] $Na_2MoO_4.2H_2O$ - 36.4 mg is weighed out and diluted to 10 mL and 0.5 mL is added to Stock #1.
[d] $CuCl_2.2H_2O$ - 60.0 mg is weighed out and diluted to 1000 mL. Then dilute 1 mL of this solution into a 10 mL volumetric flask and 0.5 mL is added to Stock #1.

One millilitre of each stock solution is then added to approximately 900 mL of Milli-Q water. After this is mixed well, the solution is diluted to 1 L and adjusted to a pH of 7.5 ± 0.1, by the dropwise addition of 0.1 M HCl or NaOH. The final concentration of nutrients in the culture medium is given in Table 5.

Table 5. Final concentration of nutrients in stock algal culture (U.S. EPA medium).

Macronutrient	Concentration (mg/L)
$NaNO_3$	25.5
$MgCl_2 \cdot 6H_2O$	12.2
$CaCl_2 \cdot 2H_2O$	4.41
$MgSO_4 \cdot 7H_2O$	14.7
KH_2PO_4	1.04
$NaHCO_3$	15.0
Micronutrient	**Concentration (µg/L)**
H_3BO_3	185
$MnCl_2 \cdot 4H_2O$	416
$ZnCl_2$	3.27
$CoCl_2 \cdot 6H_2O$	1.43
$CuCl_2 \cdot 2H_2O$	0.012
$Na_2MoO_4 \cdot 2H_2O$	7.26
$FeCl_3 \cdot 6H_2O$	160
$Na_2EDTA \cdot 2H_2O$	300
pH	7.5 ± 0.1

The algal culture medium is then filter-sterilized by pouring the medium into a sterilized glass filter funnel with a 47 mm sterile membrane filter (0.45 or 0.22 µm) and sterile filtrate receiving flask. Aliquots of 50 mL of the filter-sterilized medium are then dispensed into each of three sterile 250-mL Erlenmeyer flasks, capped with loose fitting glass lids (or equivalent sterile tops) and placed in the algal culture cabinet ready for algal inoculation. The culture media can be stored at 4°C for up to 6 months.

6.6 MAINTENANCE OF ALGAL STOCK CULTURES

All handling and transfer of algae during culturing procedures is carried out using aseptic techniques, in a Class II biohazard cabinet (*i.e.*, a sterile environment).

Stock cultures of *S. capricornutum* are prepared weekly. Each of three flasks is aseptically inoculated with 1 mL of the previous week's culture of *S. capricornutum*, using sterile glass 2 mL pipettes. The flasks are capped and stored in an incubation cabinet at 24 ± 1°C under continuous "cool white" fluorescent light with an intensity of 65 ± 5 $\mu mol.m^{-2}.s^{-1}$. Incubation conditions for the culture of *S. capricornutum* are summarized in Table 6.

Table 6. Summary of culture conditions for the freshwater alga S. capricornutum.

Temperature:	24 ± 1°C
Light quality:	"Cool White" fluorescent lighting
Light intensity:	65 ± 5 $\mu mol.\ m^{-2}.s^{-1}$
Illumination:	Continuous
pH:	7.5 ± 0.1

7. Preparation of microalgae for toxicity testing

Although the toxicity test procedure described in this chapter is not carried out under sterile conditions, stock cultures of *S. capricornutum* should be maintained axenically. Details of how to check that the algal stock culture is bacteria-free are given in Chapter 3 of this volume, together with algal preservation techniques and growth characteristics.

Immediately prior to the test, the algal inoculum should be prepared and used within 2 hours. Decant an exponentially-growing stock culture (usually 4-5 days old) of *S. capricornutum* into two glass centrifuge tubes (about 25 mL in each) and centrifuge at low speed (700 g). Pour off the supernatant in each tube and gently resuspend the algae in about 25 mL of Milli-Q water, mix with a vortex mixer for several seconds and then centrifuge again. The centrifuging and rinsing process should be repeated two more times, resulting in a concentrated algal suspension. Finally resuspend the algae in about 15 mL of Milli-Q water, ready for counting and inoculating into the toxicity test containers.

8. Flow cytometry – general description and instrument settings

A variety of flow cytometers are currently on the market and suitable for analysing microalgae. In this chapter we describe the use of the BD-FACSCalibur™ (Becton Dickinson BioSciences, San Jose, CA, USA) flow cytometer. It should be noted that between different flow cytometers the instrument settings and methods of analysis may vary.

The FACSCalibur™ flow cytometer is a four-colour, dual-laser benchtop instrument capable of both cell analysis and cell sorting. It is equipped with an air-cooled Argon-ion laser providing 15 mW at an excitation wavelength of 488 nm (blue light) and with standard filter setup. Dual excitation is possible as it also has a diode capable of excitation in the red region of the spectrum (635 nm).

Cells are presented to the flow cytometer and are hydrodynamically focused in a sheath fluid as they pass through blue light. The resulting fluorescence and light scatter characteristics of the cells are collected in photomultiplier tube detectors (Fig. 2). Sheath fluid is high purity Milli-Q water (Millipore Corp) or equivalent.

The instrument has two light-scatter detectors, which serve to identify the morphology of the cell. The forward angle light scatter (FSC <15°) detector provides information on cell size, while the side angle light scatter (SSC, 90°) detector provides information on internal cell complexity/granularity. Fluorescence is collected at a range of wavelengths by photomultiplier tubes (PMTs) with different fluorescence emission filters. Chlorophyll a or autofluorescence (present in all algae) is detected as red fluorescence in the 650-nm long-pass filter band (FL3). Green fluorescence from cells stained with FDA is collected in the FL1 channel (530 ± 15 nm) and orange fluorescence from cells stained with propidium iodide (PI) is collected in the FL2 channel (564-606 nm).

A detailed description of the initial set-up, calibration, acquisition and analysis of cells using flow cytometry can be found in the FACSCalibur™ instrument manual and CellQuest Pro™ and FACSComp™ software manuals (BD Bioscience).

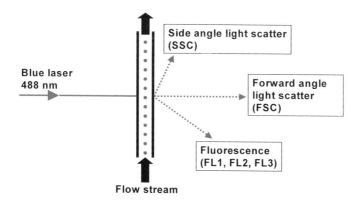

Figure 2. Schematic diagram of a flow cytometer showing the light scatter and fluorescence detectors.

8.1 INSTRUMENT SETTINGS

Because calibration procedures are generally designed for mammalian cell analysis, it is necessary to re-optimize (adjust) the instrument settings for analysing microalgae. Typical instrument settings for the analysis of *S. capricornutum* cell profiles are listed in Table 7. Once the instrument settings have been established, they are saved and can be reloaded for future analysis.

Table 7. Flow cytometry instrument settings for the analysis of S. capricornutum.

Parameter	Setting			
Threshold				
Primary parameter	SSC	Value	200[a]	
Secondary parameter	None	Value	-	
Compensation				
FL1	0.0% of FL2			
FL2	0.0% of FL1			
FL2	0.0% of FL3			
FL3	0.0% of FL2			
Detectors/Amps		*Voltage*	*AmpGain*	*Mode*
P1	FSC	E-1	3.93	Log
P2	SSC	320[b]	1.00	Log
P3	FL1	470[b]	1.00	Log
P4	FL2	470[b]	1.00	Log
P5	FL3	370[b]	1.00	Log
P6	FL1-A	-	1.00	Lin
P7	FL1-W	-	1.00	Lin
Four-Colour	OFF			

[a] The threshold must remain on the left of the distribution of algal cells to ensure that all algal cells are captured for analysis.

[b] These values are a guide only. The operator can adjust these values to alter the position of the algal distribution along the FSC, SSC, FL1, FL2 and FL3 axis. Analysing healthy cells harvested from a stock culture is the best way to gauge this movement. Generally, algal populations are positioned in the centre of the FSC, SSC and FL3 axis so that shifts (both increase and decrease in intensity) can be observed. For the FL1 and FL2 axis it is necessary to position the algal distribution (unstained cells) in the first decade of the logarithmic scale so that high fluorescence intensities can be observed when cells are stained with FDA and PI respectively.

Typical parameter plots used to determine the cell density of *S. capricornutum* include:
- a cytogram (*i.e.*, a two parameter plot) of FL3 versus SSC, to identify the algal population with a region (R2) capturing the algal population and excluding low chlorophyll fluorescing cells and particles;
- a cytogram of FL1 versus FL2, to identify the population of standard beads used to determine cell density.

For the determination of esterase activity, additional plots are prepared including:
- a cytogram, of SSC versus FL3, gated on the algal population (R2) identified above (*i.e.*, displaying only those cells that are captured by the R2 region);
- a histogram of FL1, also gated on R2 to visually observe shifts in FL1 (FDA) fluorescence intensity.

8.2 COUNT TIME/EVENTS

Data should be analysed using a standard flow cytometric software package (*e.g.*, CellQuest™). Although flow cytometry is able to count cells at very low cell densities, a minimum of 1000 events (cells) per sample should typically be analysed to achieve a coefficient of variation (CV) of < 10% (Li, 1990). For bioassays using cell inocula of 1×10^4 cells/mL, a preset acquisition time of 120 s will enable sufficient cell numbers to be obtained. However, at recommended low initial cell densities of 10^2-10^3 cells/mL, longer acquisition times (*e.g.*, 300 s) may be required. To avoid unnecessarily long counting times when cell numbers have increased over the course of the bioassay, a feature of this flow cytometer allows data acquisition to be stopped when the number of cells (events) in a specified region (*e.g.*, *S. capricornutum* cells) reaches >1000. We recommend that all parameters be collected as logarithmic signals and analysis performed at a high flow rate (60 µL/min).

9. Test samples

9.1 CHEMICALS

Individual chemicals for toxicity testing should be supplied with material safety data sheets to ensure appropriate handling by laboratory personnel. Stock solutions of the test chemical should be prepared as closely as possible to the time of testing, particularly unstable compounds that degrade in solution. Stock solutions of metals may be acidified to enable storage and to reduce metal losses to the container walls, however, care must be taken that addition of the acidified stock does not alter the pH in the toxicity test treatments.

If carrier solvents (*e.g.*, acetone, ethanol) are necessary to dissolve chemicals that are poorly soluble in water, additional solvent controls must be included in the toxicity test design. If possible, the same amount of solvent should be added to each test vial at a concentration below that known to cause a toxic effect. If different amounts of solvent are used at each chemical test concentration, then solvent controls at each solvent concentration must be prepared. Even if a carrier solvent is not toxic at this concentration, it has the potential to increase cell membrane permeability and hence allow toxicants to enter more freely into the cell causing a greater toxic effect on cell growth rate or enzyme activity. The determination of cell membrane permeability is discussed in more detail in Section 12.2.3.

9.2 COMPLEX ENVIRONMENTAL SAMPLES

Complex effluents, receiving waters and leachates should be sampled into pre-rinsed cleaned containers, preferably glass or polyethylene, filled with no headspace and transported at 4°C (not frozen). About 500 mL of sample is sufficient for the toxicity test. Effluents are stored at 4°C until testing and should be tested within 3 days of sample collection.

Physico-chemical measurements on the sample as received should include pH, conductivity and dissolved oxygen. If the sample pH is outside the optimal pH range for the test (6-9 for the growth inhibition test and 7.8-8.5 for the esterase inhibition test), it can be pH adjusted or left unadjusted, depending on the purposes of the test. If the conductivity of the sample is > 2000 µS, then conductivity controls should be included in the test design. The sample may be filtered through either a GF/F filter (approximate pore size 0.7 µm) or a 0.45 µm cellulose acetate filter. However, the advantage of the flow cytometry-based toxicity test is that the samples do not have to be filtered prior to testing. Testing of unfiltered samples allows detection of toxicants associated with particulate material. However, while bacteria and other algae in the sample may be distinguished from *S. capricornutum*, they may alter the growth of the test algae and affect the toxic response. Careful interpretation of results from unfiltered samples is required.

10. Chronic growth rate inhibition test

The growth inhibition test measures the decrease in growth rate of *S. capricornutum* over 72 h. Unique features of the test include the use of lower initial cell densities if required and counting by flow cytometry. Growth rates in test solutions are compared statistically to that of controls, enabling calculation of NOEC, LOEC and IC50 values.

The test is based on the OECD Test Guideline 201 (OECD, 1984) and the US EPA protocol (U.S. EPA, 2002) and is summarized in Table 8.

10.1 SUMMARY OF TEST PROCEDURE

This summarized step-by-step procedure is followed when conducting a *S. capricornutum* growth inhibition bioassay and contains references to later sections of the protocol for more details and/or instruction.

Tests may be conducted in either glass Erlenmeyer flasks (250 mL) with loose fitting glass lids or in glass 20 mL minivials (scintillation vials) with plastic screw-on lids. Minivials have the advantage that only small sample volumes are required, making it an ideal method for testing complex environmental samples (*e.g.*, industrial effluents) that often need a time-consuming filtering step prior to testing. Due to the reduced size of the test vessel, the minivial bioassay also requires much less incubator space, allowing multiple tests to be conducted at the same time. Only the minivial procedure will be described here.

Table 8. Summary of toxicity test conditions for the freshwater algal S. capricornutum growth inhibition test.

Test type:	Static
Temperature:	$24 \pm 1°C$
Light quality:	"Cool white" fluorescent lighting
Light intensity:	$4000 \pm 10\%$ lux ($65 \ \mu mol.m^{-2}.s^{-1}$)
Illumination:	Continuous
Test chamber size:	20 mL minivial
Test solution volume:	6 mL
Renewal of test solutions:	None
Age of test organisms:	4-5 days
Growth phase:	Exponential
Initial cell density:	10 000 cells/mL (standard test) or 1000 cells/mL in low cell density test
No. replicate vessels / concentration:	6 vials (3 vials for daily cell counts, 2 vials for pH measurement, 1 vial for chemical analysis)
Shaking rate:	100 rpm
Dilution water:	Algal culture medium (with/without EDTA)
pH range:	6-9
Test duration:	72 h
Effect measured:	Cell growth inhibition, measured as inhibition of exponential growth rate
Test acceptability:	Cell density in the control to increase by a factor of 16 after 72 h, corresponding to a specific growth rate of 0.9/day. Variability in the growth rate of controls not to exceed 10%.

S. capricornutum test medium (used for diluent water and controls) consists of the standard US EPA media (see Section 6.5). This medium has an alkalinity of 9 mg $CaCO_3$/L and a water hardness of 15 mg $CaCO_3$/L. For metal toxicity tests or samples expected to contain metal toxicants, EDTA may be omitted from the test medium. However without EDTA, metals such as copper may co-precipitate with iron hydroxides, leading to a decrease in dissolved metal in the test medium. *S. capricornutum* growth can also be more variable in medium without EDTA. The test procedure can be summarized as follows:

(1) Decide how many vessels are required for the test (according to test design and type). Wash and then label the vessels.

(2) Start to prepare the inoculum (Section 7). While the algae are being centrifuged, continue setting up the bioassay. The repeated washing and centrifuging required for the inoculum preparation can be done in between test set-up steps.

(3) Prepare a fresh batch of U.S. EPA media with/without EDTA (Section 6.5). A fresh batch of U.S. EPA media without sterilization (filtered to 0.22 μm) can be prepared and used on the initial day of the test. Alternatively, sterilized media can be stored at 4°C for up to 6 months.

(4) Prepare the test solutions and fill each test vial with the appropriate volume of test solution (6 mL) (Section 10.2).

(5) Finalize the inoculum preparation, by determining and recording the volume of algal suspension to be inoculated into each vial.

(6) Inoculate each test vial, ensuring that the algal suspension is stirred on the vortex mixer between every 3-4 vials.

(7) Confirm and record the Day 0 cell density, by determining the final cell density in the first control replicate.

(8) pH measurements: these are recorded on Day 0 and at test completion on Day 3. For a minivial test, one of the two extra minivials for pH is used for measurement on Day 0, and the other is used for Day 3. Neither minivial is used for cell counts.

(9) Incubation: place and secure the minivials in racks in a random order on the electronic shakers in an environmental cabinet set at the conditions defined in Table 6 and leave shaking for 72 h.

(10) Subsample 0.5 mL from each vial on each day (1, 2 and 3 days) for cell counts by flow cytometry.

(11) At test conclusion after 72 h, measure pH in the appropriate vials.

(12) Conduct calculations and analyses for relevant statistical data required.

10.2 PREPARATION OF TEST SOLUTIONS

10.2.1 Test solutions

Immediately prior to use, the culture medium and test solutions should be allowed to equilibrate to room temperature.

For the **minivial bioassay,** six replicate controls are prepared by dispensing 6 mL of culture medium into each vial. Three vials are for daily counts while two vials are for physicochemical measurements at the beginning and end of the test. The remaining vial is for chemical analysis and not inoculated with algae. Several additional vials can be prepared in the same way and used as "counting vials" on Day 0.

For chemical testing, at least five concentrations (also with six replicates) are prepared by dispensing 6 mL of culture medium into each of 30 minivials (5 × 6). The chemical is then spiked directly into each test vial or a serial dilution prepared depending on the pH of the stock solution of the chemical. If the chemical is then spiked directly into each test vial, ensure that the spike volume is < 60 µL. Test concentrations are chosen with the aim of encompassing a range of responses from 0% to 90-100% growth inhibition and should be in a geometric series with a dilution factor of 2 (*i.e.*, 2.5, 5, 10, 20, 40 µg/L). For chemicals whose toxicity to *S. capricornutum* is unknown, a range-finding test should initially be performed using widely separated concentrations (*e.g.*, dilution factor of 10) to broadly estimate the IC50 for later definitive tests (see Chapter 3 of this volume).

If the test chemical is a metal (from an acidified stock solution), the highest metal concentration should be prepared directly in culture medium without EDTA (100 mL) in a beaker. This should be pH adjusted if necessary and then serially diluted in culture medium using a dilution factor of two. To do the serial dilution, prepare four beakers each containing 50 mL of culture medium without EDTA. Sub-sample 50 mL from the top test concentration and add to beaker 1. Mix well, then sub-sample 50 mL and add to beaker 2. Repeat the procedure to obtain five test concentrations (100%, 50%, 25%, 12.5% and 6.3%), where 100% is the highest metal concentration.

If the sample is an effluent or leachate, dispense 100 mL into a beaker. Spike this solution with nutrients (0.1 mL of each of the five culture medium nutrient stocks described in Table 4). Mix well. For metal-containing effluents, EDTA should be omitted from the test medium. Adjust the pH if necessary and then prepare 1:2 serial dilutions using culture medium with EDTA.

Dispense 6 mL of each test solution concentration into each of six minivials – three vials for daily counting and two vials for physicochemical measurements at the beginning and end of the test. The sixth vial is for chemical analysis of the test solution at each concentration if appropriate. For example, in a test with copper, this vial would be used to measure the actual copper concentration by either inductively coupled plasma atomic emission spectrometry or graphite furnace atomic absorption spectrometry. The vial should be acidified with 12 µL of concentrated nitric acid (*e.g.*, Normatom or Suprapur grade) prior to analysis. For analysis of other toxicants, more than 6 mL of sample may be required. Additional vials should be set up at each test concentration for this purpose.

10.2.2 Quality assurance – reference toxicant
Each test should also include a reference toxicant *e.g.*, copper sulfate to ensure that the algae are responding in a reproducible way to a known toxicant. A dilution series of copper (40, 20, 10, 5 and 2.5 μg Cu^{2+}/L) prepared from a copper sulfate stock solution should be set up in the same way as a test sample above (in culture medium without EDTA). A reference toxicant test should be run with each batch of test samples (see Chapter 3 of this volume).

10.3 TEST PROCEDURE

Day 0 (*i.e.*, starting day of test t = 0 h)

(1) The concentrated algal suspension is prepared as described in Section 7.

(2) Using the flow cytometer (see Section 10.4), the density of the concentrated algal suspension is determined. The volume (x μL) needed to add to each minivial to obtain 1 or 10×10^3 cells/mL is then calculated and checked by inoculating a "counting vial" with x μL of the algal suspension. The "counting vial" is a vial containing exactly the same volume and solution as the controls. The density of algae in the counting vial is then determined using the flow cytometer. From this, the volume of algal inoculum required to give a recommended starting cell density of 1 or 10×10^3 cells/mL in the test vessel can be determined.

(3) Finally, each test vessel is inoculated with x μL of algal suspension. To ensure the suspension remains homogenous, it is covered and stirred on a vortex mixer between every 3–4 inoculations. The volume of test inoculum added to each vessel must not exceed 0.5% of the total volume in the vessel, so for a minivial bioassay, the volume used to inoculate should be no more than 30 μL.

The vials are placed randomly on a shaker platform (100 rpm) in an environmental cabinet at the specified test conditions for *S. capricornutum* (Tab. 4; identical to culture conditions).

Days 1–3

(1) Each vial is gently agitated and a sub-sample (0.5 mL) taken for counting by flow cytometry at 24, 48 and 72 h after beginning the test (when t = 0 h). Cell counts recorded at the end of the 48-h period are designated as Day 2 and at the end of the 72-h period as Day 3 observations.

(2) The pH, temperature and conductivity of one replicate vial is measured and recorded at the end of the test (Day 3).

10.4 DETERMINATION OF CELL COUNTS BY FLOW CYTOMETRY

For most commercially available flow cytometers, absolute cell counts are obtained by adding a known amount of reference beads into the sample (ratiometric counting). By comparing the algal cell count with the bead count, the cell concentration can be calculated. This approach has been shown to be accurate, particularly when using primary reference bead solutions (*i.e.*, Becton Dickinson TruCount™ tubes, described below) and also provides an internal standard that can be used to assess the performance of the instrument (*e.g.*, standardization of light scatter and fluorescence).

10.4.1 Preparation of cells for counting

Before counting, the minivial should be well mixed. An appropriate aliquot of cells (usually 0.5 mL) is immediately taken and placed directly into a TruCount™ tube*. The tube is mixed well and checked for the presence of air bubbles under the metal retainer, which can interfere with the analysis (air bubbles are removed by gentle tapping on the tube). To ensure accurate counts are obtained, it is essential that all pipettes used for dispensing solutions be well calibrated prior to use. Pipette calibration can be performed using distilled water (1 µL distilled water = 1 mg at 25°C) and a precision weighing scale.

---

* Becton Dickinson TruCount™ Absolute Count Tubes (# 340334) contain a lyophilized pellet of 4.2 µm fluorescent-dyed beads. The pellet is restrained in the bottom of the tube by a stainless-steel retainer. The number of beads in each pellet (beads per test) varies among lots and is printed on the foil pouch. Tubes are packaged in two foil pouches, each pouch containing 25 tubes. Store tubes in the foil pouch at room temperature and use within 1 hour after removal from the foil pouch. Reseal foil pouches immediately after each use. Once the pouch has been opened, the tubes are stable for 30 days.

10.4.2 Acquisition and analysis of data

Acquire the sample on the flow cytometer using the appropriate instrument settings for *S. capricornutum* (see Table 7).

To analyze the data, draw a gate around the TruCount™ bead population (see Fig. 3A) from a dot plot of FL1 vs FL2. Proceed to remove (*i.e.*, gate out) this population from a new plot of SSC vs FL3 to obtain the *S. capricornutum* population alone (Fig. 3B). View the region statistics (Fig. 3) to determine the number of bead and algal events within each region.

Calculate the absolute number of *S. capricornutum* cells in the sample using the following equation:

$$\frac{\text{\# of events in region containing cell population}}{\text{\# of events in bead region}} \times \frac{\text{\# of beads per test}^*}{\text{test volume}} = \text{Concentration of algal population} \quad (1)$$

e.g.,

$$\frac{4187 \text{ (\# of events in R2)}}{1310 \text{ (\# of events in R1)}} \times \frac{5243^*}{0.5 \text{ mL of sample}} = 3.35 \times 10^4 \text{/mL}$$

* This value is found on the foil pouch label.

Box 1. Modification of the counting method using a diluted TruCount™ tube bead stock.

> Most fluorescent beads can be used as a bead reference stock after they have been standardized by determining the absolute bead number. Most laboratories standardize their own bead stocks for use in determining cell counts. To help reduce costs associated with running the bioassay, we have included a modification to the method of counting cells described above. This is based on diluting a TruCount™ tube and using this tube as a bead reference stock. An aliquot is then added to the algal sample to be counted rather than using a new TruCount™ tube for every sample (Franklin et al., 2004). TruCount™ tubes have the advantage of being pre-standardized by the manufacturer.
>
> The fluorescent bead stock solution is prepared by pipetting 1 mL of Milli-Q water directly into a TruCount™ tube containing a known quantity of fluorescent beads. The tube is then mixed thoroughly and 200 µL of bead stock added to a sample tube containing an appropriate aliquot of algal sample (*e.g.*, 0.5 mL). It is recommended that the pipette tip be rinsed with the sample solution by taking up and dispensing the solution several times. It should be noted that the high-precision and accuracy of this counting method is limited only by the pipetting steps. Due to the viscous nature of the bead solution, a positive-displacement pipette is recommended.
>
> Absolute cell counts are determined using the above equation, with the exception that the number of beads per test is now the number of beads in 200 µL of the bead stock solution (*e.g.*, 52 445 beads in 1mL = 10489 beads in 200 µL), and the test volume is increased to include the total volume of the sample (*e.g.*, 0.5 mL algae + 0.2 mL bead spike = 0.7 mL). A dilution factor (*e.g.*, 0.7/0.5 = 1.4) is also required to account for the bead volume. As with all populations measured by flow cytometry, at least 1000 bead events must be acquired to ensure the accuracy of this technique.

10.5 ENDPOINT DETERMINATION

Step 1. The $\log_{10}$ cell density for each replicate in each treatment should be plotted versus time in days. Lines of best fit (linear regression) are then calculated for each test treatment, and the slope of the line is equivalent to the growth rate. The specific growth rate is calculated by multiplying the slope by 2.303 (= ln 10) for each treatment and control. Alternatively the specific growth rate equation can be used:

$$N_t = N_o exp[\mu_t(t-t_0)] \qquad (2)$$

where:
N_t is the cell density at time t (days)
N_0 is the cell density at time t_0
μ is the growth rate at time t.

This equation can be rewritten as:

$$\mu = (lnN_t - lnN_0)/(t-t_0) \quad (day^{-1}) \qquad (3)$$

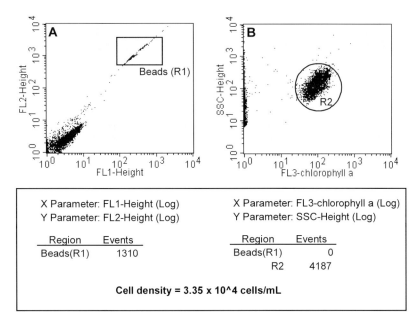

Figure 3. Determination of absolute cell counts of S. capricornutum *using TruCount™ tubes A) dot plot of FL1 vs FL2 fluorescence identifying bead population, R1, B) dot plot of FL3 (chlorophyll a) vs SSC (side-angle light scatter/cell complexity) (excluding bead population) identifying healthy algal population, R2.*

Step 2. The next step is to calculate growth rates in each treatment as a percentage of the mean control growth rate, according to the following formula:

$$Growth\ rate\ (\%\ of\ control) = \mu_T/\mu_C \times 100 \qquad (4)$$

where:
μ_C is the mean value for average specific growth rate in the control,
μ_T is average specific growth rate for the treatment replicate.

Step 3. Growth rate as a percentage of control in each individual replicate should be plotted against the logarithm of the test substance concentration. This is the concentration-response curve.

Step 4. A number of statistical procedures can be used to calculate the ICp *i.e.*, the inhibitory concentration to cause a p% effect (usually 50% effect or IC50). For the growth rate inhibition test, the IC50 is calculated using linear interpolation and is the toxicant concentration that causes a 50% reduction in growth rate compared to the control. This can be calculated using commercially available software such as

ToxCalc (Tidepool Software) or U.S. EPA software described in Chapter 3 of this volume. The 95% confidence limits should be reported with each IC50 value.

Step 5. Toxicity results can also make use of hypothesis testing to report a NOEC (no observed effect concentration) and a LOEC (lowest observed effect concentration). The data is first tested for normality (Shapiro-Wilk's Test) and homogeneous variance (Bartlett's Test) and if these assumptions are met, Dunnett's multiple comparison test can be used to determine NOEC and LOEC values (U.S. EPA, 2002). NOEC and LOEC values are dependent on the concentration range selected for the test and the test precision.

10.6 QUALITY ASSURANCE

For the test to be valid, a number of performance criteria should be met.

The cell density in the control should have increased by a factor of 16 within 72 h. This corresponds to a specific growth rate of 0.9 day^{-1}. The coefficient of variation of the average specific growth rate in replicate control cultures should not exceed 10%.

Water quality parameters throughout the test must be within acceptable limits, with a pH change of not more than 1.5 pH units. For metal toxicants, pH changes of < 0.5 pH units are desirable.

Reference toxicant IC50s should be within two standard deviations of the mean IC50 calculated from the running quality control chart.

11. Acute esterase inhibition test

The esterase inhibition test measures the decrease in cellular esterase activity of *S. capricornutum* after a 3- and 24-h exposure to a toxicant. Algal esterase activity is measured by flow cytometry as FDA fluorescence (FL1 fluorescence) after incubation with the fluorescent substrate FDA. The test protocol is summarized in Table 9.

11.1 EXPERIMENTAL CONFIGURATION

Tests are carried out in 20-mL glass vials treated with a silanising solution to reduce the adherence of toxicants, especially metals, to the vessel walls. Alternatively, tests can be carried out in 30-mL polycarbonate vials without the silanising treatment, however plastic vessels are not recommended for samples where organic toxicants are of concern. Toxicant losses to test vessels and cell biomass should not be as great as the growth inhibition test due to the short duration of the esterase inhibition test.

A typical esterase bioassay consists of a positive and negative control, a reference toxicant and the test sample at various dilutions.

Table 9. Summary of toxicity test conditions for the freshwater algal S. capricornutum esterase inhibition test.

Test type:	Static
Temperature:	$24 \pm 1°C$
Light quality:	"Cool white" fluorescent lighting
Light intensity:	$4000 \pm 10\%$ lux (65 $\mu mol.m^{-2}.s^{-1}$)
Illumination:	Continuous
Test chamber size:	20 mL minivial
Test solution volume:	10 mL
Renewal of test solutions:	None
Age of test organisms:	4-5 days
Growth phase:	Exponential
Initial cell density:	10 000 cells/mL
No. replicate vessels / concentration / exposure time:	4 vials (3 vials, 1 vial for pH measurement)
Shaking rate:	0 rpm (static)
Dilution water:	Algal culture medium (without EDTA)
pH range:	7.8-8.5
Test duration:	3- and/or 24-h
Effect measured:	Cell esterase activity, measured as inhibition of FDA fluorescence
Test acceptability:	> 90% of control algal cells in the healthy FDA fluorescence region (S2), < 10% overlap in FDA fluorescence intensity of control and negative control cells, reference toxicant EC50 within quality control chart limits.

Positive and negative controls are incorporated into the test to ensure that the esterase activities of both healthy and unhealthy algal cells are easily distinguishable from each other based on their FL1 (FDA) fluorescence intensity (*i.e.*, level of esterase activity). The negative control vials are treated just like controls until 1 h prior to post-exposure analysis when the cellular esterase enzymes are inactivated by heat treatment.

A reference toxicant, *e.g.*, copper added as copper sulfate, is included to ensure that the algae are responding to a known toxicant in a reproducible way. Because microalgae are highly sensitive to copper, it is commonly used as a reference toxicant (Stauber and Davies, 2000). At least three concentrations of the reference toxicant, each in duplicate, are included to allow for the estimation of an EC50 value. Prior to carrying out tests on samples, a quality control chart should be established using the EC50 values calculated after testing the reference toxicant at five concentrations (in triplicate). Tests on samples should only be attempted once the laboratory is confident of obtaining reproducible results. A running cumulative summation of the EC50 values along with the standard deviation defines the acceptable EC50 range (see Chapter 3 of this volume).

Two test vials are prepared per control and sample concentration, one for 3-h analysis and one for 24-h analysis. If the sample volume is limited, one replicate per concentration can be prepared with sub-samples taken at 3- and 24-h for analysis.

Due to the short exposure period of this acute test, it is important to keep the total number of vials in each test to a maximum of thirty per exposure time point. The greater the number of vials to analyze (each one taking 3 min) and hence the longer analysis time, the greater the possibility of obtaining results not representative of that particular time point.

For reference toxicants or when one chemical is being tested, it is advisable to measure each concentration in the test vial to confirm the nominal concentrations. Additional test vials can be prepared for this purpose.

11.2 TEST SAMPLE CONCENTRATIONS

To select appropriate test concentrations, follow the method described in Chapter 3 of this volume. Ideally the majority of test concentrations should fall within 16-85% of control, with at least one test concentration that will have no effect.

For samples whose toxicity is unknown, a range finding test using a dilution factor of 10 is recommended, prior to carrying out a definitive test using a dilution factor of 2 or 3.

For samples whose toxicity is known, a different concentration range may be used for the 3-h and 24-h exposure. This is generally applied when the sensitivity of the algae increases with increasing exposure time. For example, for the reference toxicant copper (from a copper (II) sulfate stock solution), concentrations of 25, 50, 100, 150 and 200 µg Cu/L are used for the 3-h exposure, and 12.5, 25, 50, 100 and 150 µg Cu/L are recommended for the 24-h exposure.

11.3 DISPENSING SAMPLE, ALGAE AND NUTRIENTS TO TEST VIALS

Once the chemical or complex effluent has been prepared (*i.e.*, filtered, pH adjusted to 7.8-8.5) and the concentrations to be tested have been determined, sample dilutions are prepared and dispensed into test vials.

11.3.1 Control and diluent water
To eliminate the possibility of metal complexation, EDTA is not used in the control and diluent water. Prepare an additional micronutrient stock solution (#1) without EDTA. Half fill a 1-L volumetric flask with Milli-Q water and pipette 1 mL of each

of the five nutrient stock solutions into the volumetric flask, shaking after each addition. Measure the pH of the medium and adjust to pH 7.8 by the dropwise addition of NaOH or HCl (0.1 or 1 M).

11.3.2 Adding nutrients, diluting and dispensing sample

Three replicates are prepared per exposure time for each test concentration plus one additional vial per test concentration for pH measurements at the beginning and end of the test.

Complex effluents. For complex effluents, wastewaters or other types of samples that require serial dilution, add 0.2 mL of each of the five nutrient stock solutions, in order, to 200 mL of sample and mix well. A stock solution #1 prepared without the addition of EDTA is used to reduce metal complexation in the bioassay.

To prepare the required dilutions for testing, dilute the sample with dilution water into glass beakers. A minimum of 100 mL of each concentration is required. The easiest and simplest way to prepare serial dilutions is to dilute the sample using a dilution factor of 0.5 (*i.e.,* 1:2 dilutions). For example, to obtain sample concentrations of 50, 25, 12.5 and 6.25%, first transfer 100 mL of media into 4 glass beakers using a measuring cylinder. Transfer 100 mL of 100% sample into one beaker and mix with a stirring rod. This is the 50% dilution. To the next beaker, transfer 100 mL of 50% sample and stir. This is the 25% sample. Continue until the lowest concentration has been prepared (6.25%). In this concentration there will be 200 mL sample. Pipette 10 mL of each of the sample concentrations into the test vials.

Single toxicants (with stock solution). For single toxicants, pipette a small aliquot (10-100 µL) of the stock solution directly into a test vial containing 10 mL of dilution water to obtain the desired concentration. For example, to prepare a test vial of 50 µg Cu/L, pipette 10 mL of dilution water into the test vial followed by 100 µL of the 5 mg/L copper stock solution. Additional stock solutions may need to be prepared by diluting the stock solution with Milli-Q water.

11.3.3 Control, negative control and reference toxicant

Three controls are prepared by pipetting 10 mL of dilution water into each test vial. Two additional control vials (one for 3-h analysis and one for 24-h analysis) are prepared and labeled appropriately for use as negative controls. The reference toxicant, *e.g.*, copper (Stauber and Davies, 2000) is prepared in the same manner as a single toxicant. At least three concentrations, in duplicate, are prepared by spiking the appropriate volume of the stock solution into 10 mL of dilution water.

11.3.4 Algal inoculum

A washed algal inoclum is prepared as outlined in Section 7. Algal cells are added to each test vial to give a final cell density of 10 000 cells/mL ± 10%. Stir on the vortex the washed algal inoculum and pipette 50 µL into a test vial with 10 mL dilution water. Shake the vial and measure the cell density using flow cytometry (see Section 10.4). If the cell density is 10 000 cells/mL, 50 µL is the inoculum volume that is added to each test vial. If the cell density is less than or greater than 10 000 cells/mL, calculate the volume of algal inoculum required to give the desired cell density* and prepare and count a new vial with this inoculum volume.

> *To limit pipetting errors and the volume of inoculum added to each vial, ensure that the inoculum volume is in the range of 10-100 µL. If the volume calculated is less than 10 µL, dilute the algal inoculum with Milli-Q water or if the volume is greater than 100 µL, concentrate the algal inoculum by centrifugation and start the process again. Continue until the volume required to give an initial cell density of 10 000 cells/mL ± 10% is 10-100 µL.

11.3.5 Dispensing test organism

To dispense the algae, stir on the vortex the algal inoculum and pipette the required volume (determined above) into each test vial. To ensure that a reproducible aliquot is dispensed each time (i.e., that each vial has the same initial cell density), the algal inoculum must be stirred on the vortex immediately before inoculating vials. A maximum of 3-4 vials can be inoculated before the inoculum should be mixed again.

11.3.6 Preparation of test vials for incubation

The test vials are incubated immediately after the algae have been added (time 0 h). Loosely cap (sufficient to allow for gas exchange) and gently shake each vial. Record the pH of each test solution using the additional vials set up for pH measurements.

11.4 EXPOSURE CONDITIONS

The test vials are randomly placed in an environmental chamber or temperature-controlled room and incubated for 3 or 24 h at 24 ± 1°C under continuous "cool white" fluorescent illumination of 65 µmol. $m^{-2}.s^{-1}$ (the same as those used for maintaining cultures, Table 6). Due to the short duration of the test, the exposure period is static (i.e., no shaking) however if facilities are available to enable continuous shaking they can be used. The temperature should be monitored throughout the test using an automated temperature logger and the pH should be measured in one replicate of each sample at the beginning and end of the test.

11.5 MEASURING CELL ESTERASE ACTIVITY USING FLOW CYTOMETRY

11.5.1 Preparation of the FDA working stock solution

The 100 mM FDA stock solution is prepared daily by weighing 0.0104 g of FDA into a small glass weigh-boat and rinsing the FDA into a 25 mL volumetric flask with acetone (AR grade). After the FDA has completely dissolved, the volumetric flask can be made up to volume with acetone, labeled and stored in a freezer (-4°C).

11.5.2 Negative controls

Two types of negative controls are analysed for FDA fluorescence;

(1) control (healthy cells/untreated cells) that have not been stained with the FDA dye and,

(2) cells that have had their esterase enzymes inactivated by heat treatment (unhealthy cells) and have been stained with FDA.

To analyse the first negative control, tightly cap and shake a control vial and pour 1-2 mL of the sample into a flow cytometer tube and analyse for 2 minutes.

Inactivate the algal esterase enzymes in the negative controls approximately 1 hour prior to carrying out the 3- and 24-h measurements of esterase activity by flow cytometry. To inactivate algal esterase enzymes, take one of the control vials that was set aside for use as a negative control and place the vial in boiling water for 10 min. Afterwards, remove the sample from the boiling water and set aside to cool to room temperature.

Transfer 4.88 mL of the inactivated algal solution into a clean glass vial. The vials used must be glass as the addition of acetone in the dye reacts with some plastics. Add 125 μL of FDA stock solution to the vial and immediately start a timer for 5 min. Afterwards, analyse the sample by flow cytometry for 2 minutes.

11.5.3 Control, reference toxicant and sample
Analyse all of the controls followed by the reference toxicant and the test sample. Each sample and replicate are analysed as described below.

Tightly cap and shake each vial to resuspend the algae prior to transferring 4.88 mL into a clean glass vial. Pipette 125 μL of the FDA stock solution into the vial and start the timer for 5 min. Then, shake the vial and pour approximately 1 mL of the FDA-stained sample into a flow cytometer sample tube and analyse for 2 minutes.

To increase the sample throughput and to shorten the analysis time, samples are incubated in a staggered arrangement by adding FDA to each vial in 3-minute intervals. If there is poor separation (>10% overlap in FL1 fluorescence intensity) between the control and negative control, the test should be terminated.

11.6 FLOW CYTOMETRIC ANALYSIS

Because the data has been saved, they can be analysed immediately after the samples have been run or at a later date. Typical plots and regions used to determine the esterase activity (FDA fluorescence) of the cells are shown in Figure 4.

11.6.1 Negative controls (unstained or stained heat-treated cells)
For the two negative controls analysed (healthy unstained cells and FDA-stained heat-treated cells), check that the R1 region incorporates the whole algal population on the SSC versus FL3 plot. The FL1 (FDA) fluorescence intensity of these two samples should be similar and of low intensity (ideally in the first decade of the log scale) (Fig. 4).

11.6.2 Positive controls
For the positive control samples (healthy cells stained with FDA), position the marker S2 manually so that it captures at least 90% of the algal cells, for each control replicate, along the FL1 fluorescence intensity axis (Fig. 5). Focus the assignment of the marker around the distinct normal distribution of the majority of the cells and

ignore the small percentage of cells that may appear on the left of the histogram, as it is normal to observe some cells with low fluorescence intensity in the control treatment.

The marker, S1, is positioned to capture the area less than S2 and marker S3 to capture the area greater than marker S2. Algal populations that shift into regions S1 and S3 represent an inhibition and stimulation of esterase activity compared to the control respectively.

For each control, reference toxicant and test sample, record the % of gated cells in each region.

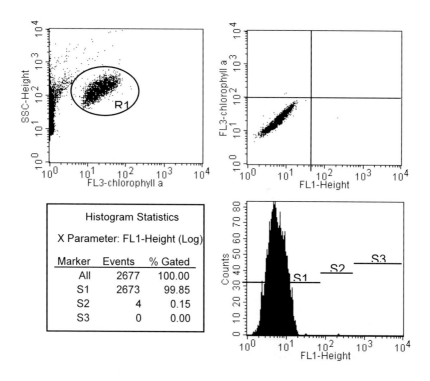

Figure 4. Flow cytometric analysis of negative control data. Algal cells in R1 are stained heat treated cells (weak chlorophyll a fluorescence). Plots of chlorophyll a fluorescence (FL3) versus FDA fluorescence (FL1), show that the cells appear in the bottom left hand quadrant. When plotted as a histogram, cells appear in region S1 indicating weak FDA fluorescence (unhealthy/dead cells) as expected in a negative control.

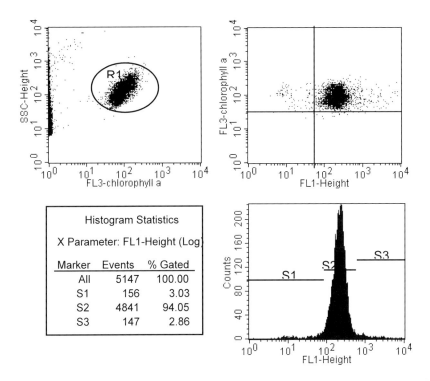

Figure 5. Flow cytometric analysis of healthy algal cells. Healthy FDA-stained cells (R1) appear in the upper right hand quadrant, indicating strong chlorophyll a and FDA fluorescence. When plotted as an FL1 (FDA) fluorescence histogram cells appear in the healthy region S2.

Healthy cells occur in region S2 (Fig. 6). Unstained or heat-treated controls fall to the left into region S1. Cells with inhibited FDA fluorescence (inhibited esterase activity) shift from S2 towards region S1. This shift is quantified and used to determine inhibition of esterase activity. Typical shifts in FL1 fluorescence observed for the reference toxicant copper, are shown in Figure 7. Occasionally stimulation of esterase activity is observed and cells shift in fluorescence to the right from region S2 into region S3.

A batch analysis can be initiated to automatically export all of the statistics and for importing into a spreadsheet program (*e.g.,* Excel). However, ensure that the same R1 region captures the algal population in all of the data files as some toxicants may alter FSC, SSC and FL3 characteristics.

11.7 ENDPOINT DETERMINATIONS

The test measures a decrease in FDA fluorescence in cells, seen as a shift of cells from region S2 left into region S1. For each exposure period, results are expressed as a percentage decrease in S2/S3 compared to the control according to the following equation:

$$\% \text{ of control} = (100 - \%S1_t) \div (100 - \%S1_c) \times 100 \quad (5)$$

where:
%$S1_t$ is the percentage of treated cells in S1;
%$S1_c$ is the percentage of control (untreated) cells in S1.

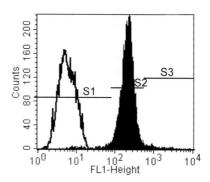

Figure 6. Flow cytometric analysis of FDA-stained control (healthy) cells and negative control cells. FDA-stained healthy cells occur in region S2. Negative control cells (unstained healthy cells and FDA-stained heat-treated cells) fall to the left into region S1. Cells with inhibited FDA fluorescence (inhibited esterase activity) shift from S2 towards region S1.

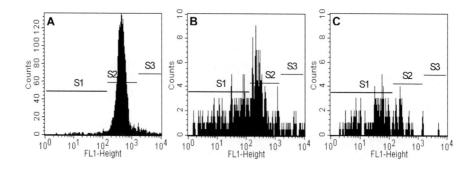

Figure 7. FL1 (FDA) fluorescence intensity of S. capricornutum after a 24-h exposure to A) 0 µg Cu/L (control) with 91% of cells in S2, B) 115 µg Cu/L with 38% of cells in S2 and C) 200 µg Cu/L with 20% of cells in S2.

A spreadsheet software package such as Microsoft Excel is useful to create a worksheet to carry out the calculations automatically. This is particularly useful when the flow cytometric statistics data has been imported into the same spreadsheet software.

Plot the % of control versus the logarithm of the toxicant concentration to give the concentration-response curve. The proportional data is arc sine transformed and tested for normality of distribution (Shapiro-Wilk's test) and homogeneity of variance (Bartlett's test) prior to hypothesis testing using Dunnett's multiple comparisons test if the assumptions of normality and homogeneity of variance are met. Dunnett's multiple comparison test determines which sample concentrations are significantly different to the controls in order to estimate LOEC and NOEC values.

The 3- and 24-h EC50 value (the effective concentration to cause a 50% shift in algal cells out of the control fluorescence region) is calculated using the Trimmed Spearman-Karber method. There are many statistics packages available that are capable of calculating these endpoints, for example, ToxCalc (Tidepool Software), which has the advantage of being compatible with Microsoft Excel software.

11.8 QUALITY ASSURANCE

The use of positive controls (untreated algal cells with healthy esterase activity) and negative controls (cells with inactivated esterase activity) defines the expected FL1 fluorescence intensity for the no effect test concentrations and 100% effect (inhibition) in esterase activity respectively. A separation with < 10% overlap in FL1 fluorescence intensity is required to ensure that shifts in FL1 fluorescence can be quantified.

Control FDA-stained cells should display a normal distribution for the algal population on the FL1 histogram. The control region, S2, can then be confidently defined around > 90% of the cells.

The reference toxicant copper is included in each test to ensure that the algae are responding to a known toxicant in a reproducible way. At least three concentrations of copper, each in duplicate, are included to allow for the estimation of an EC50 value.

A quality control chart should be established with the EC50 values calculated after testing the reference toxicant at five concentrations (in triplicate). A running cumulative summation of the EC50 values along with the two standard deviation value defines the acceptable EC50 range. Test samples should only be attempted once the laboratory is confident in obtaining reproducible results.

12. Factors affecting algal toxicity tests

The toxicity response of any laboratory bioassay is dependent on the procedure used (light, temperature, pH, nutrient medium, exposure time, inoculum size and pre-exposure), the assessment endpoint (effect parameter) chosen, the species used, the laboratory (operator performance) and the nature of the test sample. Even with

12.1 FACTORS AFFECTING THE GROWTH RATE INHIBITION TEST

12.1.1 Colored samples

Colored samples can alter light quality and quantity available to the algal cells in the growth rate inhibition test, potentially causing decreased algal growth, unrelated to sample toxicity. To overcome this problem, additional color correction controls at various dilutions of sample can be prepared. Minivials containing control water (culture medium only) are placed in a beaker and the beaker filled with the colored sample so that the colored sample just covers the control water in the minivials. Cells are counted daily as usual and any effect of light reduction on algal control growth can be compared to normal controls.

For some algal species, light intensity in the bioassay can be increased to help overcome any light reduction due to the color of the sample. This is not recommended for the standard protocol with *S. capricornutum*.

12.1.2 Adsorption losses

Adsorption losses of the toxicant to the test container and algal biomass in the standard growth bioassay can be substantial, particularly for metal toxicants such as copper and uranium (Stauber and Davies, 2000; Charles et al., 2002). Static renewal, in which test media is renewed daily, is difficult in algal tests because centrifuging cells each day may lead to reduced growth rates and subsequent failure of the test to meet acceptability criteria.

Adsorptive losses to test containers may be reduced by pre-silanising glass test containers with solutions such as Coatasil (BDH) or pre-conditioning glass surfaces to the test solution. However, metal losses may still exceed 20% and pre-conditioning was also reported to be of limited use (Stauber and Davies, 2000).

Metal losses to polycarbonate test containers and microplates are lower, however some algal species are unable to grow in these containers due to release of inhibitory plasticizers (Arensberg et al., 1995). Fortunately *S. capricornutum* grows well in polystyrene microplates (Chapter 3 of this volume). A semi-static test with this species, in which periodic renewal of media occurs in microplates with membrane-bottomed wells, has also been developed (Radetski et al., 1995).

Reducing toxicant loss to increasing algal biomass over the test is also possible if lower cell densities are used as the initial test inoculum. This is discussed further in Section 12.1.4.

12.1.3 pH

Increases in pH in the test medium, particularly in controls, occur as the algae grow and utilize carbon dioxide, bicarbonate and nitrogen. A change in pH of 1 unit may change the speciation and subsequently the toxicity of metals by a factor of 10 (Peterson et al., 1984; Franklin et al., 2000). This limitation of algal growth bioassays has been discussed in detail by Nyholm and Kallqvist (1989).

To overcome this problem, short test durations (48-72 h) are now more commonly used than 96 h. Low initial cell densities (10^2-10^3 cells/mL) now possible with the flow cytometry bioassay, also prevent nutrient limitation and pH increases over the test duration.

12.1.4 Initial cell density
Standard test protocols use high initial cell densities ($\geq$ 10 000 cells/mL) compared to those found in aquatic systems in order to obtain a measurable algal response. These high cell densities can result in changes in chemical speciation, bioavailability and toxicity through toxicant losses to the algal biomass, changes in pH and production of algal exudates, which may bind toxicants. Laboratory tests may therefore underestimate contaminant toxicity compared to natural waters containing lower cell numbers. Franklin et al. (2002) showed that copper concentrations required to inhibit the growth rate of *S. capricornutum* by 50% increased from 6.6 µg/L to 17 µg/L as the initial cell density increased from 10^2 to 10^5 cells/mL (Fig. 8). Even though in Figure 8 there appears to be little difference in copper toxicity at 10^2-10^4 cells/mL initial inoculum, there was a significant decrease ($p < 0.05$) in the IC50 value at initial cell densities of 10^3 and 10^4 cells/mL (6.2 and 7.2 µg Cu/L respectively).

The advantage of flow cytometry is that toxicity tests with low cell densities can now be carried out to avoid the problems of toxicant losses to cell biomass and preventing a decline in toxicant concentration in solution over the course of the test. These low cell densities are much more typical of algal concentrations in natural waters, further improving the environmental relevance of the test.

12.1.5 Test medium
For samples in which metal toxicants are of concern, EDTA added to the algal culture medium may complex metals and reduce their toxicity. For this reason it has been recommended that EDTA is omitted from the test medium for metal toxicity bioassays. Omission of EDTA however, can lead to precipitation of iron and other trace metals in the test medium, making control growth rates lower and more variable than tests with EDTA.

12.2 ESTERASE INHIBITION TEST

12.2.1 pH
The most important criteria for testing samples using the esterase bioassay is that the pH of the media and each sample concentration must be 7.8-8.5. The pH is a critical factor controlling FDA conversion to fluorescein and test solutions within the pH range of 7.8-8.5 show a reasonable separation (< 10% overlap) between FDA-stained control cells and negative control populations (Fig. 6). Solutions with pH values of 5.8, 6.8 and 7.3, cause a considerable overlap (> 50%) in the fluorescence intensity of control and negative control populations making the allocation of esterase activity states (S1, S2 and S3) very difficult (Franklin et al., 2001b). Compared to other algal bioassay endpoints, such as growth, this is a relatively small pH range and limits the application of this test to industrial wastewaters.

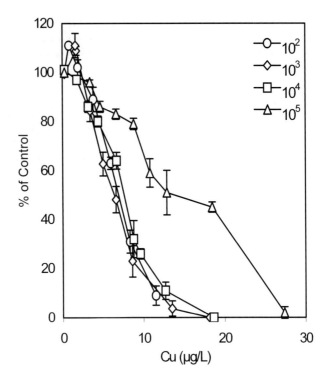

Figure 8. Effect of initial cell density on copper toxicity to S. capricornutum *(initial cells/mL) (from Franklin et al., 2002).*

12.2.2 Stimulation effects

Stimulation in *S. capricornutum* esterase activity compared to controls can sometimes be observed for certain toxicants and/or for some toxicants at low concentrations (*e.g.*, copper at concentrations of around 3-13 µg/L after a 24 h exposure). Stimulation in esterase activity should be noted when reporting results.

12.2.3 Cell membrane permeability

A decrease in FDA fluorescence in *S. capricornutum* can be due to either inhibition of esterase activity or due to a change in cell membrane permeability, both of which have been regarded as indicators of cell viability (Dorsey et al., 1989). Berglund and Eversman (1988) found that the amount of fluorescein that can accumulate in cells is dependent on the amount that leaks out of the cells due to the permeability of the cell membrane. If membranes are made more permeable by toxicants, then it is possible that a decrease in FDA fluorescence could be due to reduced uptake of FDA or leakage of fluorescein out of the cells, rather than an effect on esterase activity.

To ensure that a decrease in FDA fluorescence is due to an effect on esterases within the cell, membrane integrity can be measured by flow cytometry using a nucleotide-binding stain propidium iodide (PI) (Franklin et al., 2001b). A fully intact membrane is impermeable to PI, and DNA will only be stained in cells that are dead or that have compromised membranes.

For the esterase test described here, a decrease in FDA fluorescence in *S. capricornutum* after exposure to copper was due to inhibition of intracellular esterases and not a consequence of changes in membrane permeability and FDA uptake (Franklin et al., 2001b).

Staining may be undertaken on a separate sample of toxicant-exposed cells, or together with the FDA staining of cells. However, fluorescence compensation is required for dual staining (FDA and PI in the same sample at the same time) and should only be attempted by experienced flow cytometry operators. Brief details of this method are given in the box below, but the reader is referred to Franklin et al. (2001b) for more information.

Box 2. Measuring cell membrane permeability using propidium iodide and flow cytometry.

Prepare a 100 μM stock solution of PI (Sigma) in Milli-Q water. Add an aliquot to 5 mL of toxicant-exposed *S. capricornutum*, to give a final concentration of 7.5 μM. After 5 min, analyse the sample by flow cytometry. The orange fluorescence emission is detected in FL2. Cells with compromised membranes should have a higher orange fluorescence than healthy intact cells, and this is seen as a shift to the right on a histogram plot of count versus FL2. If no cells have higher FL2 fluorescence, then any decrease in FDA fluorescence in FL1 is due to true inhibition of esterase activity. Cells killed by heat treatment (100°C for 10 min) or formaldehyde treatment (4% for 24 h) should be included in each experiment as negative controls.

---

13. Case studies with the flow cytometer toxicity tests

Flow cytometry-based toxicity tests with several freshwater species (*S. capricornutum*, *Chlorella* sp.) and several marine species (*Nitzschia closterium*, *Phaeodactylum tricornutum*, *Entomoneis* cf *punctulata*) have been widely applied to testing the toxicity of industrial effluents, mining-impacted waters, chemicals and sediments. One example of the use of the flow cytometry-based growth inhibition test with *S. capricornutum* involved a field trial of a lanthanum-modified clay (Phoslock) in the Canning River, Western Australia. Nuisance cyanobacterial blooms are a common occurrence in the Canning River during low-flow summer conditions. When Phoslock is applied to water bodies it reduces phosphate concentrations and subsequent cyanobacterial blooms. As part of a field trial to determine if Phoslock itself was toxic to other aquatic organisms including green algae, cladocerans and fish, samples of surface and bottom waters were collected from the Canning River before and after Phoslock application. Over 100 samples were collected from four sites in two separate field trials. Unfiltered and filtered river water samples were tested for toxicity to *S. capricornutum* using the 72-h growth inhibition bioassay with counting by flow cytometry. Other algae were potentially present in the unfiltered samples at low concentrations, so flow cytometry was essential to separate any native algal signal from the added *S. capricornutum*, and to accurately count the algae in the presence of particulate material.

River water samples from control sites and from sites prior to Phoslock application were not toxic to *S. capricornutum* growth over 72 h. Significant stimulation of algal growth was observed in most samples. One day after Phoslock application, one unfiltered surface water and one unfiltered bottom water from the same site were toxic, reducing algal growth by 61% and 52% respectively. Toxicity was related to the high total lanthanum concentrations in these samples, but toxicity did not persist, with no effect on algal growth detectable one week after Phoslock application. None of the filtered samples were toxic on any occasion.

The standard growth inhibition test with *S. capricornutum* (which requires filtered samples) would not have detected the lanthanum-related toxicity, thereby underestimating Phoslock toxicity. On the basis of the flow cytometry bioassay, the Phoslock formula was modified to reduce the amount of lanthanum released from the clay. Further field trials of the modified Phoslock showed that no riverwater samples were toxic. Phoslock has since passed chemical registration and is now marketed as a water treatment to control cyanobacterial blooms.

A second example was an investigation of the effect of copper on *S. capricornutum*, using both the flow cytometry esterase test and the growth rate inhibition test. Initial cell densities for both tests were 2×10^4 cells/mL, with no EDTA in the test medium. As shown in Table 10, the growth inhibition test was more sensitive to copper than the esterase test.

Table 10. *Inhibitory effect of copper on esterase activity (FDA fluorescence) and growth rate of* S. capricornutum *(from Franklin et al., 2001b).*

Metal	FDA fluorescence inhibition (µg/L)		Growth rate inhibition (µg/L)	
	3-h EC50	24-h EC50	48-h IC50	72-h IC50
Cu	112 (88-143)[a]	51 (38-70)	4.9 (4.1-5.8)	7.5 (6.8-8.2)

[a] 95% confidence limits.

14. Miscellaneous test information

Flow cytometry techniques have only recently been applied to algal toxicity testing, so the test protocols described in this chapter have not yet been standardized and interlaboratory studies are limited. However, the basic test set up and experimental design is similar to that outlined in OECD and Environment Canada protocols. The main differences are the potential to use lower initial cell densities, unfiltered samples, multiple effect parameters and multiple species in the test.

Although this chapter describes two tests with *S. capricornutum*, flow cytometry test protocols have also been developed in our laboratory for other freshwater species (*e.g.* a tropical *Chlorella* sp.), several marine species (*e.g., Nitzschia closterium)* and benthic (sediment dwelling) species (*e.g., Entomoneis* cf *punctulata*) (Franklin et al., 2001a,b; Adams and Stauber, 2004). The benthic algal test (an esterase inhibition test) has been widely used in the Australasian region to determine the toxicity of marine whole sediments in contaminated harbours, as part of a battery of sediment tests with benthic invertebrates. These aquatic and sediment tests have been shown to be sensitive to metals, ammonia and PAHs.

Multispecies tests have also been developed using mixtures of three freshwater (*Microcystis aeruginosa, S. capricornutum* and *Trachelomonas* sp.) or three marine species (*Micromonas pusilla, Phaeodactylum tricornutum* and *Heterocapsa niei*) (Franklin et al., 2004). Flow cytometry enabled the algal species to be separated and counted on the basis of their different size and pigment content. The effect of copper on these species in single-species versus multi-species tests (with equivalent surface areas) was investigated. Single species freshwater bioassays were shown to underestimate the toxicity of copper, whereas the marine single species tests overestimated copper toxicity. Flow cytometry has great potential to further develop more environmentally relevant bioassays.

The flow cytometry toxicity tests described above all require considerable investment in equipment and skilled operators. However, once the individual test protocols are established, it is relatively easy to train operators in their routine use. Different flow cytometers have different advantages and disadvantages and instrument settings, and methods of analysis will vary with the different instruments used. In this chapter we have described the use of the Becton Dickinson FACSCalibur™ instrument, but adapting the methods to other flow cytometers should be relatively straightforward. In our laboratory we have also used a BIO-RAD Bryte HS flow cytometer, which is particularly simple to operate. This instrument takes a known sample volume so direct algal counts are possible without the need for internal calibration beads. This is a much cheaper alternative as the cost of consumables for routine counting can be quite high. Unfortunately this instrument has been discontinued, so protocols with this instrument were not described in this chapter.

Flow cytometry-based toxicity tests are very amenable to automated testing. With automatic sample loaders, sophisticated software for multiple samples, and the introduction of a microplate front-end loader, hundreds of samples can automatically be counted each day. Analysis and calculation can be done manually later on stored data.

15. Conclusions and prospects

The use of flow cytometry as a tool in ecotoxicology has only begun to be explored. This technique has limitless potential in the development of more environmentally relevant aquatic and sediment toxicity tests with unicellular algae. Multi-parameter, multi-species tests at low cell densities are now available to better assess the bioavailability of contaminants in aquatic systems. Not only can growth inhibition tests be carried out, but simultaneous measurements of algal size, cell complexity, physiology and metabolic activity can also provide additional information on the mode of action of toxicants on algal cells.

Future applications of flow cytometry in toxicity testing may include the development of natural phytoplankton and periphyton bioassays. Neither does the technique have to be limited to microalgae, as it may also prove useful for developing bioassays with bacteria and microbial microcosms for assessing sediment and water quality.

Acknowledgements

The authors would like to thank Monique Binet for writing the CSIRO laboratory standard operating procedure for the *S. capricornutum* growth inhibition bioassay, part of which is used in this chapter.

The authors wish to thank the copyright holder for permission to reproduce Figure 8, Franklin, et.al., (2002) *Environmental Toxicology and Chemistry* 21, 742-751, Copyright SETAC, Pensacola, Florida, USA.

References

Adams, M.S. and Stauber, J.L. (2004) Development of a whole-sediment toxicity test using a benthic marine microalga, *Environmental Toxicology and Chemistry* (in press).

Arensberg, P., Hemmingsen, V.H. and Nyholm, N. (1995) A miniscale algal toxicity test, *Chemosphere* **30**, 2103-2115.

Berglund, D.L. and Eversman, S. (1988) Flow cytometric measurements of pollutant stresses on algal cells, *Cytometry* **9**, 150-155.

Blaise, C. and Ménard, L. (1998) A micro-algal solid phase test to assess the toxic potential of freshwater sediments, *Water Quality Research Journal of Canada* **33**, 133-151.

Charles, A.L., Markich, S.J., Stauber, J.L. and De Filippis, L.F. (2002) The effect of water hardness on the toxicity of uranium to a tropical freshwater alga (*Chlorella* sp), *Aquatic Toxicology* **60**, 61-73.

Cid, A., Herrero, C., Torres, E. and Abalde, J. (1995) Copper toxicity on the marine microlaga *Phaeodactylum tricornutum*: Effects on photosynthesis and related parameters, *Aquatic Toxicology* **31**, 165-174.

Cid, A., Fidalogo, P., Herrero, C. and Abalde, J. (1996) Toxic action of copper on the membrane system of a marine diatom measured by flow cytometry, *Cytometry* **25**, 32-36.

Dorsey, J., Yentsch, C., Mayo, S., and McKenna, C. (1989) Rapid analytical technique for the assessment of cell metabolic activity in marine microalgae, *Cytometry* **10**, 622-628.

Franklin, N.M., Stauber, J.L., Markich, S.J. and Lim, R.P. (2000) pH-dependant toxicity of copper and uranium to a tropical freshwater alga (*Chlorella* sp.), *Aquatic Toxicology* **48**, 275-289.

Franklin, N.M., Stauber, J.L. and Lim, R.P. (2001a) Development of flow cytometry-based algal bioassays for assessing toxicity of copper in natural waters, *Environmental Toxicology and Chemistry* **20**, 160-170.

Franklin, N.M., Adams, M.S., Stauber, J.L. and Lim, R.P. (2001b) Development of an improved rapid enzyme inhibition bioassay with marine and freshwater microalgae using flow cytometry, *Archives of Environmental Contamination and Toxicology* **40**, 469-480.

Franklin, N.M., Stauber, J.L., Apte, S.C. and Lim, R.P. (2002) The effect of initial cell density on the bioavailability and toxicity of copper in microalgal bioassays, *Environmental Toxicology and Chemistry* **21**, 742-751.

Franklin, N.M., Stauber, J.L., and Lim, R.P. (2004) Development of multispecies algal bioassays using flow cytometry, *Environmental Toxicology and Chemistry* **23**, 1452-1462.

Gala, W.R. and Giesy, J.P. (1990) Flow cytometric techniques to assess toxicity to algae, in W.G. Landis and W.H. Van der Schalie (eds.), *Aquatic toxicology and risk assessment*, Vol. 13, American Society for Testing and Materials, Philadelphia, USA, pp. 237-246.

Hall, J. and Cumming, A. (2003) Flow cytometry in aquatic science, *Water and Atmosphere* **11**, 24-25.

Hindak, F. (1990) Pseudokirchneriella subcapitata Korshikov, F.*Biologicke Prace* **5**, 209.
Jochem, F.J. (2000) Probing the physiological state of phytoplankton at the single-cell level, *Scientia Marina* **64**, 183-195.
Li, W.K.W. (1990). Particles in "particle-free" seawater: growth of ultraphytoplankton and implications for dilution experiments, *Canadian Journal of Fisheries and Aquatic Sciences* **47**, 1258–1268.
Molecular Probes Incorporated (2003) *Handbook of Fluorescent Probes and Research Chemicals*, Richard P. Haugland (ed.) 7th edition, Molecular Probes Incorporated, Eugene, OR, USA.
Nyholm, N. and Kallqvist, T. (1989) Methods for growth inhibition toxicity tests with freshwater algae, *Environmental Toxicology and Chemistry* **8**, 689-703.
OECD (Organization for Economic Cooperation and Development). (1984) Alga, growth inhibition test, in *OECD Guidelines for Testing of Chemicals*, Guideline 201, Vol 1, Paris, France, pp 1-14.
Peterson, H.G., Healey, F.P. and Wagemann, R. (1984) Metal toxicity to algae: a highly pH dependent phenomenon, *Canadian Journal of Fisheries and Aquatic Science* **41**, 974-979.
Premazzi, G., Buonaccosi, G. and Zilio, P. (1989) Flow cytometry for algal studies, *Water Research* **23**, 431-442.
Radetski, C.M., Férard, J.F. and Blaise, C. (1995) A semistatic microplate-based phytotoxicity test, *Environmental Toxicology and Chemistry* **14**, 299-302.
Stauber, J.L. and Davies, C.M. (2000) Use and limitations of microbial bioassays for assessing copper bioavailability in the aquatic environment, *Environmental Reviews* **8**, 255-301.
Stauber, J.L., Franklin, N.M. and Adams, M.S. (2002) Applications of flow cytometry to ecotoxicity testing using microalgae, *Trends in Biotechnology* **20**, 141-143.
U.S. EPA (2002) Short term methods for estimating the chronic toxicity of effluents and receiving waters to freshwater organisms, in United States Environmental Protection Agency Report EPA-821-R-02-013, Washington, DC, USA, 4th Ed., 350 pp.

Abbreviations

CV	Coefficient of variation
EC50	Effective concentration to cause a 50% effect
EDTA	ethylenediamine tetraacetic acid
FDA	fluorescein diacetate
FL1	fluorescence collected by detector 1 (530 ± 15 nm)
FL2	fluorescence collected by detector 2 (564-606 nm)
FL3	fluorescence collected by detector 3 (>650 nm)
FSC	forward angle light scatter
g	gravitational constant at the surface of the Earth. It is equal to 9.8 m/sec^2.
h	hour(s)
IC50	Inhibitory concentration to cause a 50% effect
LOEC	lowest observable effect concentration
NOEC	no observable effect concentration
PI	propidium iodide
PMTs	photomultipier tubes
SSC	side angle light scatter
μmol s^{-1} m^{-2}	micro-mole of photons per second per square meter.

6. ALGAL MICROPLATE TOXICITY TEST SUITABLE FOR HEAVY METALS

HANS G. PETERSON
WateResearch Corp.
11 Innovation Boulevard, Saskatoon
Saskatchewan S7N 3H5, Canada
hanspeterson@sasktel.net

NIELS NYHOLM
Environment and Resources
Technical University of Denmark
Bld 113, DK-2800
Lyngby, Denmark
nin@er.dtu.dk

NORMA RUECKER
WateResearch Corp.
11 Innovation Boulevard, Saskatoon
Saskatchewan S7N 3H5, Canada
N.Ruecker@provlab.ab.ca

1. Objective and scope of test method

This test method was specifically designed to assess phytotoxicity of samples containing heavy metals (synthetic solutions, effluents, elutriates, leachates). Metal toxicity can be greatly affected by media composition and by changes caused by algal growth, such as the increase in pH and the release of dissolved organics. These problems can be severe in many test systems, but have been minimized in the procedures described here. The testing specifically addresses the following:
- Metal complexing compounds, which can affect toxicity, have been reduced or eliminated from the test media.
- Short duration testing results in small increases in algal biomass causing no significant changes in prevailing environmental conditions (pH, production of organics, *etc.*).
- A battery of test organisms, varying in morphology, nutrient requirements and ecological relevance is used.
- A microplate format makes the test cost, space and time effective.

- Stringent validity criteria ensure quality results and reproducible testing.

2. Summary of test procedure

The test system allows for better characterization of toxicity of a test material by examining three different media scenarios: 1) minimal medium (no chelators or trace elements), which provides maximum sensitivity towards toxic metals; 2) a standard medium modified for heavy metals by reducing added chelator and iron (aimed for comparative studies); 3) a naturally derived medium to assess site-specific toxicity of heavy metals in the sample (Table 1).

The 96-well microplate format allows for ease of handling, ability to increase replicate numbers for improved statistical results, efficient utilization of space and automated endpoint reading. The toxic response of the metal is assessed for five different species of algae during log-phase growth. The test organisms used represent diatoms, green algae and cyanobacteria (blue-green algae) with varying morphology (unicellular and filamentous), nutritional requirements (nitrogen and trace elements) and ecological relevance.

The short duration of the test (24 hours or less for minimal medium and 48 hours for other test media), results in a final algal biomass, which has minimal impact on the testing environment. Strictly controlled environmental conditions (temperature, pH, light, humidity) provide a reproducible testing environment. Fluorescence readings are used to calculate cell numbers to represent biomass. The toxic response is calculated as a percent (%) reduction of growth rate. The toxicity results for a test treatment are curve-fit using parametric statistical analysis and the 25% inhibitory concentration (IC25) is predicted. Minimal criteria for growth rates and toxic response to a reference toxicant must be met for each algal species for the test to be considered valid.

3. Culture/maintenance of organisms in the laboratory

3.1 TEST ORGANISMS

A five species test battery is used for the characterization of the tested samples. Morphology, nutrition, ease of handling, sensitivity to a range of compounds, as well as past performance, were all considered in the selection of these test species.

The five organisms are from three taxonomic groupings:

> *Chlorophyta, Chlorophyceae* (green algae)
> *Selenastrum capricornutum*, UTEX1648
> *Nannochloris* sp. UTEX2291
> *Cyanophyta, Cyanophyceae* (blue-green algae)
> *Microcystis aeruginosa* FWI22
> *Anabaena flos-aquae*, UTCC64
> *Bacillariophyta, Bacillariophyceae* (diatoms)
> *Nitzschia* sp. FWI110

Table 1. Rapid summary of test procedure.

Test medium	Select from: minimal (TM 1), comparative (TM 2), natural (TM 3)
Test vessel	96-well round bottom, sterile, non-tissue culture treated microplate
Number of test replicates	8 replicates per control and concentration tested
Light	70-90 µE/m^2/s, Vitalite ultra high output, full spectrum
	20-40 µE/m^2/s for nitrogen fixing species (*Anabaena flos-aquae*)
Time	24 or 48 hours depending upon medim selection
Test volume	240 µL
Test organisms	Five species test battery (*Selenastrum capricornutum*[1], *Nannochloris* sp., *Microcystis aeruginosa*, *Anabaena flos-aquae*, *Nitzschia* sp.)
Growth phase of algal inoculum	Log–phase (40 to 48 hour incubation with conditions identical to test conditions)
Initial cell number	10^4 or 10^5 cells/mL depending upon algal species
Temperature	23 – 27°C
Humidity	40-60%
Shaker speed	400 rpm
Initial pH	Synthetic media 8.00; natural water 7.5 to 8.5
Final pH	Initial pH ± 0.5
Measured variable	Fluorescence
Response variable	Growth rate
Response	Reduction in growth rate calculated as a percentage of control
Reported key result	IC25 (concentration of sample which inhibits growth by 25% calculated by using parametric statistical analysis)
Validity criteria	Control growth rate of test organisms, response to reference toxicant (potassium dichromate) and less than 0.5 units pH drift

[1] *Selenastrum capricornutum*'s taxonomic name has been changed several times and it is currently described as *Pseudokirchneriella subcapitata*. *Selenastrum capricornutum* remains, however, commonly used is indicated as such by the authors.

3.1.1 Sources of algae

UTEX		Culture Collection of Algae
		Botany Department, University of Texas
		Austin, Texas, USA 78712
		www.bio.utexas.edu/research/utex/
UTCC		University of Toronto Culture Collection
		Attn: Judy Acreman
		Department of Botany, University of Toronto
		Toronto, Ontario, Canada, M5S 3B2
		www.botany.utoronto.ca/utcc/index.html
FWI		Freshwater Institute
		Attn: Len Hendzel
		Culture Collection of Freshwater Algae
		501 University Crescent
		Winnipeg, Manitoba, Canada R3T 2N6
		Email: HendzelL@dfo-mpo.gc.ca

3.2 LABORATORY FACILITIES REQUIRED

Controlled environment settings are required to adequately conduct phytotoxicity testing. Controlled environment (CE) chambers or incubators used must allow for the control of temperature, light and humidity. The testing environment should be free of any factors, which may affect testing (*i.e.,* presence of volatile substances in the atmosphere).

3.3 EQUIPMENT

All laboratory equipment and consumable supplies used to conduct testing are listed in Tables 2 and 3. All instruments for routine measurements of the basic chemical, physical, and biological variables must be maintained properly and calibrated regularly. Calibrations and quality control measures must be documented for each instrument.

Suitable plate closures were fabricated by using a drill press to make five holes (0.8 mm diameter) in a standard microplate lid at positions indicated in Figure 1. The layout allows any given well to be a maximum of three wells from a hole or the edge of the plate. The lid/plate interface is sealed with transparent tape. Parafilm has been used for testing, but any transparent tape would be suitable. If there is a concern that the tape may interfere with the test, Teflon tape could be used and possibly secured by a more adhesive tape. This lid design allows adequate CO_2 exchange (important for algal growth and the maintenance of pH) and an even rate of evaporation over all wells for the test duration.

Table 2. Non-consumable equipment and supplies used in growth inhibition test.

- Modified microplate lids (see Figure 1 and explanation below)
- Controlled environment chamber or incubator
- Vitalite full spectrum light bulbs
- Microplate shaker
- Rotational mixer
- Millipore Milli-Q water purification system (or equivalent)
- Refrigerator
- Microscope with phase contrast providing 100 to 1000 magnification
- Haemocytometer
- Centrifuge: max 4000 rpm 4 x 30 mL capacity
- Laminar flow hood
- Burner and gas source
- Adjustable 5 to 50 µL multi-channel pipettor
- Adjustable 50 to 300 µL multi-channel pipettor
- Adjustable 2 to 20 µL pipettor
- Adjustable 20 to 200 µL pipettor
- Adjustable 200 to 1000 µL pipettor
- Adjustable 2 to 5 mL pipettor
- Eppendorf repeater pipettor (up to 50 mL capacity)
- Test tube racks
- Analytical balance
- Teflon coated weighing spatula
- Wash bottle
- Volumetric flasks: 100, 500, 1000, 2000 mL capacity
- pH meter
- Filtration apparatus: 47 mm glass filter holder and flask
- Vacuum source and tubing
- Automated dispenser with autoclavable tubing and attachments
- Glass beakers: 250 and 500 mL
- Microplate fluorometer
- Various sizes of Pyrex brand bottles segregated for solutions and media usage
- Vortex mixer
- 50 mL (Falcon brand) centrifuge tubes
- 25 x 100 mm glass Corex brand high speed centrifuge tubes

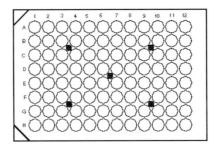

Figure 1. Configuration of holes in microplate lid.

Table 3. Consumable supplies used in the algal growth inhibition test.

- Microplates: disposable rigid polystyrene, 96-well round bottom microplates (must be non-tissue culture treated)
- 1 litre containers or equivalent
- Disposable 2 to 5 mL pipette tips
- Disposable 200 to 1000 µL pipette tips
- Disposable 2 to 200 µL pipette tips
- Various sized tips for repeater pipette
- Disposable plastic reservoirs
- Acid-washed disposable borosilicate glass test tubes (16 x 125 mm)
- Sterile (autoclaved) and non-sterile disposable borosilicate glass test tubes (16 x 125 mm)
- Disposable borosilicate glass test tubes (25 x 150 mm)
- Nylon filtration membranes (0.2 µm pore size)
- Weighing dishes
- Cover slips
- Transparent tape
- Sterile (autoclaved) glass disposable Pasteur pipettes
- 500 mL polyethylenephtalate (PET) bottles

3.4 WASHING OF GLASSWARE

All new or re-used glassware, including disposable test tubes, are washed using a stringent washing procedure. They are first soaked overnight in 4% detergent (Contrad 70) and rinsed five times in reverse osmosis (RO) purified water, then soaked overnight in 10% HCl, rinsed five times in Milli-Q water, and oven-dried at 58°C. The microplate lids and any re-usable containers used for algae are washed as previously described. Bottles used for media and stock solutions and volumetric flasks may be reused without washing providing they are segregated for a particular use. When not in use, they are stored filled with Milli-Q water.

3.5 PREPARATION OF ALGAL CULTURE AND TEST MEDIA

Careful consideration of the test medium is an integral part of this procedure. The media used is a modified version of a fresh water inhibition test developed by the International Standards Organization (ISO, 1989). The objectives for the modifications were:
- Testing for maximum sensitivity using a medium devoid of chelator, iron and trace metals (TM 1).
- Testing with synthetic reference media with low chelating capacity (TM 2).
- Testing with natural receiving water as the dilution medium to generate site-specific information (TM 3).

Additional considerations for the modification of the ISO test medium are further outlined.

The ISO procedure uses ammonium as the sole nitrogen source in order to counteract increases in pH, but in a natural environment, algae would obtain nitrogen from both nitrate and ammonium ions. In the present tests nitrate and ammonium at equivalent nitrogen concentrations were added to the media (Tab. 4). Some algal species prefer one nitrogen source to another and supplementing both nitrate and ammonium avoids unintentional interference by nitrogen sources when testing environmental samples.

Silicate is essential for the growth of diatoms and is a component of most freshwater systems and it was therefore added to all test and culture media. A concentration of 28.4 mg/L of $NaSiO_3 \cdot 9H_2O$ (2.8 mg/L as Si) in the test medium was sufficient for the growth of the tested diatom strain (*Nitzschia* sp.). Preliminary experiments showed that the addition of silicate did not alter the growth patterns of *Selenastrum capricornutum*.

Iron and trace amounts of various metals are required for the growth of phytoplankton. An organic chelator is required to maintain these in bioavailable forms. Modifications to the iron and chelating agent were made to bring them into stoichiometric balance, leaving no organic chelating agent (EDTA) available to complex with metals in the environmental sample. In TM 2, the concentration of these components is reduced from the ISO level to concentrations sufficient to maintain log-phase growth for most test species for a 72-hour period.

Due to specific nutrient requirements of some of the test species, modifications have to be made to the culture and test medium. These are:
- Nitrogen fixing species (*Anabaena flos-aquae* UTCC64) was not provided with a nitrogen source during culture or testing.
- Nitrogen fixing species (*Anabaena flos-aquae* UTCC64) had additional iron and trace element requirements, therefore, Stock solutions 2 and 3 are doubled in all culture and test media (TM 2) used for this species.
- Diatom species (*Nitzschia* sp.) are provided with vitamins during culturing only.

Table 4. Chemical composition of test media.

Nutrients	(mg/L) in TM 1	(mg/L) in TM 2	(mg/L) in TM 3
Stock solution 1: macro-nutrients*		Add 10 mL stock/L	
$NaNO_3$	23.8	23.8	23.8
NH_4Cl	15	15	15
$MgCl_2 \cdot 6H_2O$	12	12	12
$CaCl_2 \cdot 2H_2O$	18	18	18
$MgSO_4 \cdot 7H_2O$	15	15	15
KH_2PO_4	1.6	1.6	1.6
Stock solution 2: Fe-chelator **		Add 1 mL stock/L	
$FeCl_3 \cdot 6H_2O$		20 µg/L	
$Na_2EDTA \cdot 2H_2O$		31 µg/L	
Stock solution 3: trace elements**		Add 1 mL stock/L	
H_3BO_3		61.6 µg/L	
$MnCl_2 \cdot 4H_2O$		138.3 µg/L	
$ZnCl_2$		1 µg/L	
$CoCl_2 \cdot 6H_2O$		0.5 µg/L	
$CuCl_2 \cdot 2H_2O$		0.0033 µg/L	
$Na_2MoO_4 \cdot 2H_2O$		2.3 µg/L	
Stock solution 4: Sodium bicarbonate		Add 1 mL stock/L	
$NaHCO_3$	50	50	50
Stock solution 5: Sodium silicate		Add 2.5 mL stock/L	
$Na_2SiO_3 \cdot 9H_2O$	24.8	24.8	24.8

* A separate Stock solution 1 is prepared containing no nitrogen source to be used for culturing and testing filamentous nitrogen fixing cyanobacteria.
** Media prepared for nitrogen fixing species (*Anabaena flos-aquae*) have double the amount of Stock solutions 2 and 3 added to culture and TM 2.

3.5.1 Preparation of media stock solutions

All chemicals (analytical grade) are weighed using a spatula and disposable weighing dish on an analytical balance. Stock solutions, as outlined below, are prepared using Milli-Q water in volumetric flasks, which have been segregated for the purpose of each solution. Once all components have dissolved the solution is transferred to Pyrex bottles and stored at 4°C.

Stock solution 1: Macronutrients (per L)

$NaNO_3$	2.38 g
NH_4Cl	1.5 g
$MgCl_2 \cdot 6H_2O$	1.2 g
$CaCl_2 \cdot 2H_2O$	1.8 g
$MgSO_4 \cdot 7H_2O$	1.5 g
KH_2PO_4	0.16 g

Stock solution 2A: Fe-chelator (per L)

$FeCl_3 \cdot 6H_2O$	0.05 g
Na_2NTA	0.126 g

Stock solution 2B: Fe-chelator (per L)

$FeCl_3 \cdot 6H_2O$	0.02 g
$Na_2EDTA \cdot 2H_2O$	0.031 g

Stock solution 3A: Trace elements (per L)

H_3BO_3	0.185 g
$MnCl_2 \cdot 4H_2O$	0.415 g
$ZnCl_2$	0.003 g
$CoCl_2 \cdot 6H_2O$	0.0015 g
$CuCl_2 \cdot 2H_2O$	0.0002 g
$Na_2MoO_4 \cdot 2H_2O$	0.007 g

Stock solution 3B: Trace elements (per L)

H_3BO_3	0.0616 g
$MnCl_2 \cdot 4H_2O$	0.1383 g
$ZnCl_2$	0.001 g
$CoCl_2 \cdot 6H_2O$	0.005 g
$CuCl_2 \cdot 2H_2O$	0.000067 g
$Na_2MoO_4 \cdot 2H_2O$	0.0023 g

Stock solution 4: Sodium bicarbonate (per L)

$NaHCO_3$	50 g

Stock solution 5: Sodium silicate (per L)

$Na_2SiO_3 \cdot 9H_2O$	11.36 g

Stock solution 6: Vitamins (per L)

Thiamine	0.1 g
Biotin	0.0005 g
Cyancobalamin	0.0005 g

3.5.2 Media preparation

All synthetic testing media (TM 1 and TM 2) are prepared as outlined in Table 4. Specified volumes of stock solutions are added to Milli-Q water in the order listed in Table 4, resulting in a total volume of 1 litre of medium. The sodium bicarbonate additions are close to those required for inorganic carbon equilibrium with air, but to achieve such equilibrium all test media were, in addition, aerated (bubbled with filtered air) for two hours to overnight. When the inorganic carbon is in equilibrium with air the pH is stable. After the aeration only small pH adjustments were needed to achieve a pH of 8.00 ± 0.02.

For scenario 3 (site-specific toxicity), a natural water (receiving water) local to the environmental sample, is used to prepare the test medium (TM 3). Upon receipt this water should be stored at 4°C in the dark. If the water is highly turbid (visibly cloudy) or contains high indigenous phytoplankton populations (visibly green), it may have to be filtered through a Whatman GF/C filter (approx 1µm pore size) prior to use.

One day prior to testing, a 400 mL aliquot of receiving water is enriched with Stock solution 1 (4 mL), Stock solution 4 (0.4 mL) and Stock solution 5 (1 mL). It is presumed that the natural water source contains adequate trace elements and iron for testing purposes. This medium is prepared in inert plastic bottles, which are considered disposable. The solution is aerated for two hours and the pH measured and recorded. The pH is adjusted with acid/base additions only if it falls outside the pH range of 7.50 to 8.50. Anything outside this range is considered non-optimum for algal growth and significantly changes the speciation of the metal constituents being tested. A pH below 7.5 is adjusted up to 7.5 and the pH above 8.5 is adjusted down to 8.5. The adjustment is carried out with 0.5 N NaOH or 0.5 N HCl and adjustment volumes are recorded. The solution is aerated overnight, the pH is measured, and a final adjustment is made if necessary.

4. Reference toxicant

Reference toxicants are used to assess the reproducibility and reliability of results using the test organisms, test procedure and/or laboratory over a specific period of time. Results for a reference toxicant are compared with historical test results to identify whether they fall within an acceptable range of variability. Results, which do not fall within the acceptable range, indicate a change in test organism health or genetic sensitivity, a procedural inconsistency, or a combination of these factors. A reference toxicant may be used to confirm the acceptability of the concurrent test results and demonstrate satisfactory laboratory performance.

The reference toxicant should be included with every test for every species tested. This section outlines the procedure used to prepare reference toxicant concentrations used during testing.

4.1 POTASSIUM DICHROMATE STOCK SOLUTION (60 mM)

Potassium dichromate ($K_2Cr_2O_7$) in the purest form commercially available should be obtained from a reputable company supplying laboratory chemicals. The potassium dichromate (1.7651g) is weighed using a Teflon coated spatula and a disposable weigh dish. It is prepared to a 100 mL volume in a volumetric flask segregated for this use. Once the potassium dichromate is dissolved, the solution is transferred to a 100 mL Pyrex bottle and stored at 4°C for up to 6 months.

The dichromate ion is the toxic component in potassium dichromate and the concentrations are listed as molar weights of $K_2Cr_2O_7$. The weight unit equivalent of 60 mM is 6239 mg/L chromium.

4.2 REFERENCE TOXICANT DILUTIONS

Before each experiment (same day), the potassium dichromate stock solution is used to prepare a 3000 μM and a 12 μM dichromate working solution. These solutions are prepared in 50 mL Falcon brand centrifuge tubes as shown in Table 5.

Dilution schemes for the reference toxicant are outlined in Table 6. An example would be to make a 1.25 μM final concentration, where 0.025 mL of the 3000 μM stock would be added to 9.975 mL of Milli-Q water. Dilutions of the reference toxicant are prepared in 16 x 125 mm borosilicate glass acid-washed test tubes. Water is added to each dilution tube first and then followed by the appropriate concentration of working solution. All dilutions are mixed by vortex and covered with plastic wrap until used.

Table 5. Dilution scheme for preparation of reference toxicant dilutions.

Stock concentration (μM)	Concentration used for dilution	Volume used in dilution (mL)	Volume Milli-Q water (mL)
3000	60 (mM)	2.5	47.5
12	3000 (μM)	0.2	49.8

Table 6. Dilution scheme for preparation of reference toxicant dilutions.

Final test concentration of reference toxicant (μM)	Stock (μM)	Volume of stock (mL)	Volume of Milli-Q water (mL)
0.375	12	3.75	6.25
0.5	12	5	5
0.625	3000	0.025	9.975
0.75	3000	0.03	9.97
1	3000	0.04	9.96
1.25	3000	0.05	9.95
1.5	3000	0.06	9.94
2	3000	0.08	9.92
2.5	3000	0.1	9.9
3	3000	0.12	9.88
3.75	3000	0.15	9.85
5	3000	0.2	9.8

It should be noted that the range of dilutions listed may not cover all those necessary for testing the five species of algae. Each laboratory will have to experimentally determine the appropriate dilutions required for their test system. Five concentrations of the reference toxicant will be used for each species of algae, ideally these concentrations should bracket the IC25 concentrations and may vary among laboratories and analysis techniques.

5. Environmental samples

It is often necessary to test environmental samples at concentrations near 100%. This section outlines the procedures used for handling and preparation of environmental samples.

5.1 PREPARATION OF ENVIRONMENTAL SAMPLES

Samples must be transported and stored taking into account their individual characteristics. Generally, dark storage at 4°C is recommended for a limited period of time while non-repeatable storage by freezing (samples must not be frozen after thawing) may be necessary for longer time periods. Testing should begin as soon as practically possible, unless it can be shown that the sample is stable under the storage conditions.

One day prior to testing, 400 mL of the sample are enriched with equal concentrations of each stock solution as the selected test medium. If more than one test medium has been selected, more than one environmental sample will have to be prepared. Aerate the enriched sample for 2 hours and adjust the pH to match the selected test medium using 0.5 N NaOH or 0.5 N HCl. The sample is then aerated overnight with further pH adjustment the next morning if required. With every acid or base addition, the sample must be aerated another hour to allow for equilibration. Acid or base additions should be continued until the aeration period does not cause the pH to change more than ± 0.2. If the last aeration period has not changed the pH more than ± 0.2, adjust to the test medium pH without further aeration. The pH of the environmental sample must be equivalent to the test medium ± 0.05. Maintain a pH adjustment record for future reference.

5.2 CHOICE OF SAMPLE CONCENTRATIONS

To screen samples of unknown toxicity, a wide range of concentrations is best. A selection of nine sample concentrations ranging from undiluted to 1/2000 will (in authors' experience) provide an adequate data set for non-linear parametric statistics. The dilutions outlined in Table 7 may not be adequate for all types of samples and other ranges may need to be adapted. If time and sampling procedures permit, it is desirable to receive a sample for screening purposes and a couple of weeks later receive a second sample from the same location for more definitive testing, with a narrowed range of concentrations.

It is recommended that concentrations of environmental samples always be documented in reference to percent final test concentration. Due to dilutions (nutrient enrichment and algal additions) the highest concentration of the environmental sample reported for this test procedure is 90.6%.

5.3 SAMPLE DILUTIONS

When testing to determine the toxicity of environmental samples, the dilutions are prepared using the selected test medium. Dilutions of the sample are made in 16 x 125 mm borosilicate glass acid-washed test tubes. The selected dilutions can be prepared according to the scheme shown in Table 7. A 1 in 4 dilution of the sample is prepared with 22.5 mL of selected test medium and 7.5 mL of the environmental sample in an acid washed 25 x 150 mm borosilicate glass disposable acid-washed test tube. This 1 in 4 dilution is used to prepare further dilutions (Tab. 7). Test medium is added to the dilution tubes, followed by the pH adjusted environmental sample. After dilutions are prepared, they are mixed by vortex, covered with plastic wrap and allowed to equilibrate for a minimum of 1 hour. Using this scheme, an example would be to prepare a 1/400 dilution by adding 0.1 mL of the 1 in 4 stock to 9.9 mL of the dilution water.

Table 7. Dilution scheme for preparation of environmental samples used in the algal growth inhibition test.

Dilution	Sample dilution	Volume of sample (mL)	Volume of dilution water (mL)
1/4000	1/4	0.01	9.99
1/1600	1/4	0.025	9.975
1/800	1/4	0.05	9.95
1/400	1/4	0.1	9.9
1/160	1/4	0.25	9.75
1/80	1/4	0.5	9.5
1/16	1/4	2.5	7.5
1/8	1/4	5	5
1/3.33	Undiluted	3	7
1/2	Undiluted	5	5
1/1.33	Undiluted	7.5	2.5
Undiluted		10	0

6. Preparation of test species for toxicity testing

Culturing of test organisms is carried out in a two-step process. Organisms are aseptically transferred from stock cultures (maintenance of stock cultures will vary among laboratories) into 5 mL of membrane sterilized (0.2 µm nylon filters) culture media prepared as in Table 8. Incubation time is three to four days under the same conditions as testing. Care must be taken with inoculation volumes to ensure that the maximum biomass is not reached before the 4 days. Biomass increases may be monitored using optical density at 750 nm.

The second culture step is designed to prepare a sufficient number of cells in the log-phase of growth to complete the test set-up for each species. Cells are transferred from the first culture step into 20 mL of non-sterile culture medium prepared in 25 x 150 mm borosilicate glass test tubes and capped with culture lids. Initial cell densities of this step were aimed toward optical densities (OD_{750}, 1.6 cm pathlength) around 0.02 for slower growing species, such as *Microcystis* and *Anabaena*; and near 0.01 or less for *Selenastrum*, *Nannochloris* and *Nitzschia*. These cultures are mixed on a rotational mixer at 38 to 40 rpm under the same light conditions and temperature as used for testing. These cultures are incubated for 40 to 48 hours prior to testing with optical density readings measured at time 0, and three additional time intervals during the incubation period (these intervals must be a minimum of 8 hours apart).

Exponential growth of the algae is checked by plotting $\log(OD_{750})$ *versus* time (Fig. 2) which should produce a straight line.

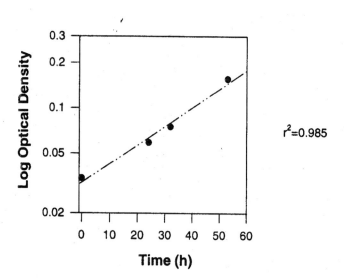

Figure 2. *Log plot of optical density (OD) versus time for assessing log phase growth of test organisms.*

Table 8. Concentration of nutrients in culture media.

Nutrients	mg/L in culture media
Stock solution 1: macro-nutrients*	**Add 10 mL stock/L**
$NaNO_3$	23.8
NH_4Cl	15
$MgCl_2 \cdot 6H_2O$	12
$CaCl_2 \cdot 2H_2O$	18
$MgSO_4 \cdot 7H_2O$	15
KH_2PO_4	1.6
Stock solution 2A: Fe-chelator **	**Add 1 mL stock/L**
$FeCl_3 \cdot 6H_2O$	50 µg/L
Na_2NTA	126 µ/L
Stock solutions 3A: trace elements**	**Add 1 mL stock/L**
H_3BO_3	185 µg/L
$MnCl_2 \cdot 4H_2O$	415 µg/L
$ZnCl_2$	3 µg/L
$CoCl_2 \cdot 6H_2O$	1.5 µg/L
$CuCl_2 \cdot 2H_2O$	0.010 µg/L
$Na_2MoO_4 \cdot 2H_2O$	7 µg/L
Stock solution 4: Sodium bicarbonate	**Add 1 mL stock/L**
$NaHCO_3$	50
Stock solution 5: Sodium silicate	**Add 2.5 mL stock/L**
$Na_2SiO_3 \cdot 9H_2O$	28.4
Stock solution 6: Vitamins***	**Add 1 mL stock/L**
Thiamine	0.1
Biotin	0.0005
Cyancobalamin	0.0005

* A separate Stock solution 1 is prepared containing no nitrogen source to be used for culturing and testing filamentous nitrogen fixing cyanobacteria.
** Media prepared for nitrogen fixing species have double the amount of stock solutions 2 and 3 added to culture medium.
*** Stock solution 6 is only used for the culturing of diatom species.

With measurements as outlined above a criterion r^2 value of a linear regression greater than 0.95 has been used to indicate exponential growth or log phase. If an alga is not in log phase at the time of the test, or if a microscopic examination shows contamination by other algae, fungi, or bacteria, then the test will not be carried out. Starting optical densities may be adjusted to ensure an adequate number of log phase cells for a given experiment.

7. Test set-up

This section outlines the preparation of the algae test inocula, the microplate configuration, and dispensing of all solutions into the microplate.

7.1 PREPARATION OF THE ALGAE TEST INOCULA

In order to prevent algal exudates and other by-products of growth being carried over into the testing, the algal test inoculum must be washed. Exponentially growing cells are poured into acid-washed 25 x 100 mm glass Corex brand high speed centrifuge tubes and centrifuged at 4000 rpm for 4 minutes. The supernatant is discarded. All washing of the algal cells is carried out with synthetic test medium (TM 1 or TM 2, based on test medium selection). The test medium is added to the tube and the pellet re-suspended by vortex. This procedure is repeated three additional times. After the fourth wash, the cells are re-suspended in synthetic test medium and counted microscopically using a haemocytometer. With the microscope, the cells are also examined to verify morphology and ensure that only one algal species is present in the culture.

Algal cells should be used for testing within 30 minutes of washing completion. Dilution to the proper cell number is carried out in the appropriate synthetic medium, in a 50 mL Falcon brand centrifuge tube and just prior to adding the cells to the test. The cells are diluted to 12 times the starting cell number to account for the dilution factor in the test set-up.

Starting cell numbers may be adjusted to ensure that the initial cell number and final cell numbers (allowing for possible stimulation) are detectable at the same sensitivity setting on the fluorometer. The following starting cell numbers have been used in our studies (per mL):

Selenastrum - 10^4
Nannochloris - 10^5
Nitzschia - 10^4
Microcystis - 10^5
Anabaena - 10^5

7.2 SELECTION OF MICROPLATE CONFIGURATION

The number of replicates for each test concentration is eight. A set of eight controls must be placed on each plate used for a particular organism. Randomization of wells in not recommended due to the excess amount of time that it would take to prepare the plates. A recommended plate layout is shown in Figure 3. The dilution water background (test medium) and the sample background are included to ensure that the background fluorescence of indigenous phytoplankton or other fluorescent compounds do not interfere with the data analysis.

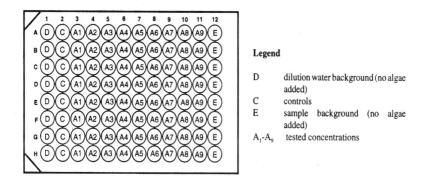

Figure 3. *Microplate layout for screening environmental samples.*

The reference toxicant is tested using five concentrations and eight replicates (Fig. 4). Controls must be included on every plate. It is best to use the remaining wells in this plate for extra control replicates, which can be pooled together at the end of the test to determine the final pH. Receiving water and environmental samples should never be placed on reference toxicant plates. The dilution water used on the reference toxicant plate will be synthetic medium (TM 1 or TM 2, depending upon test medium selected). If there are unused wells on the plate, they must be filled with 240 µL of Milli-Q water or synthetic medium to equalize evaporation when the plate is incubated.

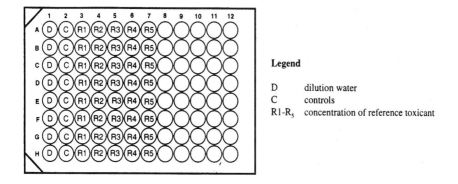

Figure 4. *Microplate layout for reference toxicant testing.*

7.3 DISPENSING TEST SOLUTIONS

The test is set up in non-tissue culture treated 96-well round bottom Corning brand microplates. Multi-channel pipettors are used for dispensing the test medium, Milli-Q water, environmental sample, reference toxicant and algae. All dilutions are mixed by vortex prior to dispensing into microplates.

Control wells for the reference toxicant plates contain:
    200 µL synthetic medium (TM 1 or TM 2)
    20 µL Milli-Q water
    20 µL algae

Reference toxicant wells contain:
    200 µL synthetic medium (same as control wells)
    20 µL potassium dichromate dilution
    20 µL algae

Control wells for environmental samples contain:
    220 µL medium (TM 1, TM 2 or TM 3)
    20 µL algae

Environmental sample test wells contain:
    220 µL environmental sample dilution
    20 µL algae

After the plates have been prepared the specially constructed lids are placed on the plates and plastic or Teflon tape is used to seal the interface of the lid with the microplate.

7.4 INCUBATION OF MICROPLATES

Microplates are placed on a microplate shaker in a controlled environment chamber/incubator (conditions outlined in Table 9). A shaking speed of 400 rpm is suggested to facilitate CO_2 mass transfer during the incubation; this helps to minimize the pH drift during testing. When TM 1 is used the duration of testing is 24 hours, while 48 hours are used for TM 2 and TM 3.

Table 9. Summary of incubation conditions.

- Temperature: 23-27°C.
- Light: Vitalite ultra high output, full spectrum, 70 to 90 $\mu E/m^2/s$ (20 to 40 $\mu E/m^2/s$ for nitrogen fixing species).
- Relative humidity: 40 to 60%.
- Agitation: 400 rpm on microplate shaker.
- Test duration: 24 hours for TM 1, 48 hours for TM 2 and TM 3.

ALGAL MICROPLATE TEST FOR HEAVY METALS

8. Observations/measurements and response evaluation

8.1 FLUORESCENCE MEASUREMENTS

In vivo fluorescence of chlorophyll *a* is the measured parameter, with growth and growth inhibition (the response) quantified from the fluorescence measurements after conversion to cell number units (surrogate biomass parameter). Plastic tape and lids are removed with no additional mixing of the wells. The fluorescence of each well is then rapidly read using a fluorescence plate reader with an excitation filter of 440 nm and an emission filter of 670 nm.

8.2 CELL ENUMERATION OF TEST WELLS

After the microplates have been read and the fluorescence values recorded, some of the wells must have the cells enumerated to establish a conversion factor from fluorescence to cell number units. Enumeration is done microscopically using a haemocytometer or an electronic particle counter. The cells are suspended by scraping the bottom and the sides of the wells with a pipette tip and drawing well contents up and down with the pipette three or four times. Microscopic observation confirms species and gives an estimate of indigenous contamination such as other algae, fungi or bacteria. The counts for each of the enumerated wells are recorded.

8.3 DETERMINATION OF FINAL PH

Contents of the replicate control wells are pooled to obtain a pH measurement using a standard laboratory pH meter. The pH of the control replicates for each algal species is used as a measure of test validity and must be recorded and reported.

8.4 DATA ANALYSIS

The response variable ("endpoint") is the average growth rate over the test period and the response is the percentage of inhibition relative to controls with no toxicant. Raw fluorescence data are converted to cell numbers, which are then used to calculate growth rates. A non-linear regression model is used to curve-fit the response *versus* concentration relationship and estimate the inhibitory concentrations with confidence intervals.

8.4.1 Estimation of biomass
For each species, fluorescence *versus* cell number plot must be generated (Fig. 5). This must consist of at least five data points and be made of counts from both test wells and controls. Counts from different tests may also be plotted. The plot should be updated periodically and monitored for consistency. Cell numbers are calculated from fluorescence results using a linear regression.

8.4.2 Calculations
Cell numbers for each test well are calculated from fluorescence readings using a linear relationship $y = mx+b$, where y is the fluorescence value, m equals the slope of

the line, x is the cell number, and b is the axis intercept. The growth rate for each well is calculated by averaging over time.

$$\text{Growth rate } (day^{-1}) = \ln(N_1/N_0)/t_{(days)} \qquad (1)$$

Where, N_1 is the number of cells at the end of the test and N_0 is the initial starting cell number.

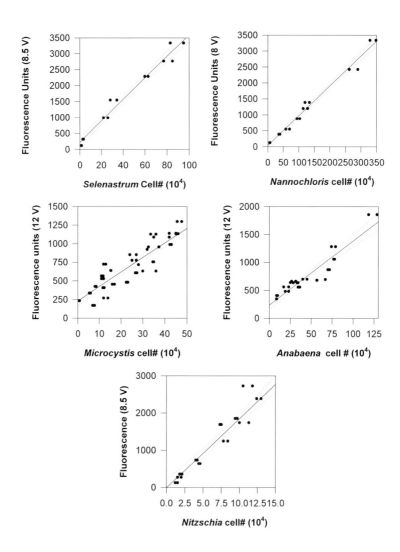

Figure 5. Fluorescence units versus cell numbers for the five test organisms.

8.4.3 Statistical analysis of responses

The growth rate from each test well is used in statistical analysis (continuous data set). It is desirable to use a non-linear regression model, which is non-symmetric around the 50% inhibitory concentration to curve-fit the data and predict the IC25 values and confidence intervals (Nyholm et al., 1992; Andersen et al., 1998).

Software is generally available to carry out non-linear regressions, but the correct inverse estimation of confidence limits needed, is unfortunately not an option in standard statistical packages and for that reason one might prefer using recently developed probabilistic boot-strap methods instead of traditional parametric regressions.

One must be sure the results of statistical analyses accurately represent the data obtained from testing. Nine concentrations are included when testing in an attempt to generate an adequate amount of partial responses so that the non-linear regression methods can accurately predict inhibitory concentrations. Not all of the nine concentrations can always be used in the regression since some of the lowest concentrations can be stimulatory and no two-parameter model, which is the type of concentration response models commonly used, can account for such non-monotonic responses.

It is suggested that only concentrations that show responses from 0 to 100% inhibition of the growth rate be included in the model. In most cases the nine concentrations in eight replicates will result in reasonably accurate IC predictions and confidence intervals. With 8 replicates elimination of outliers (except obvious extremes) is normally not necessary and statistical estimates are robust and not influenced by minor deviations from normal distribution. For the same reason the use of statistical analysis are a must and simple graphical methods should not be applied for these tests.

8.5 VALIDITY CRITERIA

The validity criteria include ensuring the quality of the test algae and maintaining the initial test conditions. At various stages in the test procedure specified criteria are stated which must be met for a test to be considered valid. Tests with different species can perform differently with respect to validity (*i.e.*, because growth rates and biomass levels differ among species). When testing is carried out to determine the toxicity of stable chemicals or compounds, the failure of one test with a certain species to meet validity criteria does not invalidate the results with the other species tested, because the full battery of tests can be obtained by repeating the failed test after corrective action. For environmental samples that are not stable over time or consistent between collections, any single test with a certain species that may have to be rejected will stand alone if repeated and cannot be claimed to be part of the five species test battery.

Testing should be carried out with organisms in exponential growth propagated under the test conditions. The test organisms are examined microscopically to verify morphology and to ensure that only one algal species is present in the culture. It is recommended that one becomes familiar with the appearance of healthy algal cultures. A culture should not be used for testing if there are any changes from normal, such as unusual cloudiness of the culture, off-color, or increased presence of bacteria. Algal cultures must be strictly unialgal, but axenic cultures are not necessary.

8.5.1 Constancy of test conditions

Growth rates are greatly affected by lighting conditions and temperature. It is important to ensure that test conditions are uniform between tests. Light and temperature levels should be recorded for each test and procedures must be in place to maintain conditions with stated levels.

Expressed toxicity as well as speciation and hence inherent toxicity of test compounds are influenced by pH and even algal growth might be affected by large pH drifts and/or concurrent carbon limitation. The pH of the test is stable for the algal cell densities used when combined with efficient and continuous shaking of the microplates. The pH of the controls is pooled and measured at the end of the test to ensure that pH has remained within the stated range. If the pH has drifted more than 0.5 units from the starting pH, the test is considered invalid for our purposes.

The growth rate of any species will vary between laboratory environments. Each laboratory must determine a range of normal growth rates for each species and medium type. A continuous plot of average control growth with standard deviation is used to produce running 95% confidence limits for a given test medium (example in Fig. 6). This plot is used to determine an acceptable range for the control growth rate. A minimum of five points must be placed on the plot before acceptable ranges can be determined. If the control growth rate is suspiciously low, the test may be considered invalid and may be repeated after corrective action has been taken.

8.5.2 Reference toxicant results

A control chart showing growth rates for unexposed algae and the predicted IC25 result with 95% confidence limits for the reference toxicant used for each test organism should be constructed to assess test reproducibility and performance (Fig. 6). Once a minimum of five data points has been plotted, the acceptable range can be determined (95% confidence limits). If the reference toxicant results do not fall within acceptable ranges, the test is considered invalid for these purposes. One such invalid result was obtained on 08-05-96 where, clearly, the IC25 for chromium lay outside acceptable limits (Fig. 6).

9. Factors capable of influencing performance of test organisms and results

No matter how stringent the controls of experimental design, there are times when organisms to do not perform as expected and test results appear atypical. This section tries to identify some of the potential problem areas in using this test method and recommendations to minimize these effects.

9.1 CULTURE OF TEST ORGANISMS

Culture media used by a laboratory for long-term storage have typically been autoclaved for sterility, however many algae show reduced growth in autoclaved media. All sterile culture media described in this protocol employ membrane filtration using 0.2 μm nylon membranes to achieve sterility. Sterile media can be stored at 4°C for up to two weeks, followed by any leftover media being discarded.

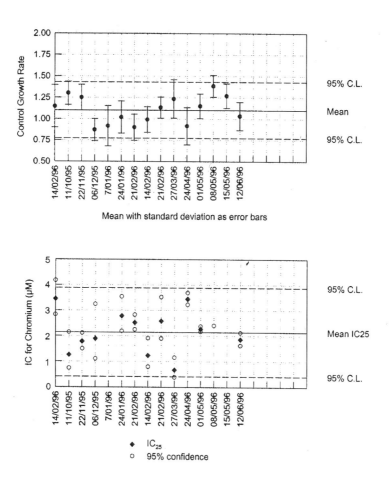

Figure 6. Quality control charts for control growth rate and reference toxicant results for Selenastrum capricornutum. C.L.: confidence limits.

Maintenance of test organisms for extended periods is time and resource consuming. By using a two-step culture process to produce organisms appropriate for testing, continual transfers can be minimized. Any organism used for testing which has come from long term storage (varies amongst laboratories) should undergo three or more transfers in the described culture media prior to testing.

There have been concerns that the chelating effects of EDTA are diminished by photo-degradation. NTA has been chosen to replace EDTA in all culture media. EDTA remains in the test medium due to the short duration of testing.

9.2 GROWTH RATES

Test medium constituents and environmental conditions affect the rate of growth of algal species. Although the conditions may not be optimal for all species tested, the outlined protocol results in the doubling of algal biomass over the first 24 hours of testing in all test media. To account for reduction in growth rates due to trace element limitation, the duration of testing for TM 1 is limited to 24 hours. (The authors are now recommending further reduction of the test duration for TM1 to 18 hours).

9.3 pH

The strict control of pH has been implemented throughout the testing protocol. The growth rates of algal species can be greatly affected by pH. Extremes of high (> 9) or low (< 6) can reduce the growth of many algal species to near zero. A pH of 8.0 is suitable (however may not be optimum) for most species as well being environmentally relevant.

Almost any result can be obtained from algal tests on materials containing heavy metals because metal speciation and expressed toxicity are greatly influenced by a number of interacting physical, chemical and biological factors. Generally, the speciation of metals determines their bioavailability, which again is a key determinant of the expressed toxicity. Free metal ions are considered most toxic, but also a number of metal complexes may be toxic. In addition to being affected directly by metal speciation, toxicity in itself may be influenced by pH, because hydrogen ions compete with heavy metals for sorptive sites on the algal surface (above issues have been reviewed by Peterson and Nyholm, 1993; Nyholm and Peterson, 1997; Campbell, 1995; Stauber and Davies, 2000).

The pH of control test wells is recorded for every test. It is assumed that the control wells contain the most rapidly growing cells and these would exert the largest pH change over the test duration.

9.4 INTERPRETATION OF RESULTS

To assess the maximum toxicity of a material, a test medium devoid of trace elements and organic chelating agents would be ideal. With this objective the goal was to find a test medium and a procedure, which would allow a minimum of a 10-fold increase in biomass before onset of deficiency. The length of the test period sustaining exponential growth on internally stored iron and trace elements will depend on initial inoculum size, growth rate, test species and test conditions. Attempts were made to pre-load the algae with trace elements including iron by growing them in a medium enriched with these constituents. This was not successful. In tests without added trace elements, algae will likely become limited by one or more elements (iron in particular) resulting in decreased growth rates within short time periods. For example, it was found that the rate of growth for *Selenastrum* decreased drastically after 24 hours and at times even sooner. It has been recommended that tests with this species be 18 hours in duration or that initial cell densities be decreased.

The rate of growth and toxicity response to the two pure metals copper and zinc were recently determined for seven different media (unpublished data). As predicted the chelator-free medium was the one revealing the highest toxicity. Data obtained with no chelator medium (TM 1), one with reduced and balanced iron and trace elements (TM 2) and ISO medium (ISO, 1989) revealed copper IC25 values of 8, 13 and 23 µg/L, respectively. Comparative data for zinc were 4, 15 and 23 µg/L, respectively. Owing to an even greater chelator excess in the U.S. Environmental Protection Agency's test protocol (Greene et al., 1989), IC25 values were double those of the complete ISO medium.

10. Application of toxicity test in a case study

In collaboration with Environment Canada the outlined test procedures were used to evaluate phytotoxicity of metal mining wastewaters. Over two years the toxicity of more than 30 mining effluents was evaluated using the three testing scenarios. It is impossible to address all the results from this testing, however, a portion of data is presented here to demonstrate the usefulness of the test protocols.

The test battery was used to examine the potential toxicity of 6 metal mining effluents. Little information regarding the mine was provided for confidentiality reasons. What was known was the type of mine the effluent came from and that the effluents were of one of three types: tailings ponds, wastewater or treated wastewater. The effluents were assigned arbitrary numbers, Mine Effluent (ME) 1 through 6 and the mine type associated is as follows:

- ME 1 – lead/zinc mine
- ME 2 – tin/cadmium mine
- ME 3 – copper/zinc mine
- ME 4 – copper/zinc mine
- ME 5 – copper/zinc mine
- ME 6 – copper/zinc mine

These particular effluents were tested under scenario 1: test medium with no chelators (TM 1) resulting in the maximum sensitivity of the phytoplankton to the toxic metals in the sample. The 25% inhibitory concentrations and confidence limits for each of the mine effluents and the reference toxicant (potassium dichromate) were estimated as described by Andersen et al. (1998). Results are summarized in Table 10.

Each of the mine effluents tested showed toxicity (growth inhibition of 25%) to at least one alga at 1% volume/volume (v/v) or less. ME 2 would be considered the most toxic inhibiting all species at a concentration of 0.03% v/v or less.

The data clearly show that no species was ever the most sensitive or the least sensitive. This underlines the value gained from extending tests done with one species to tests with several species. These sensitivity variations differed by as much as 250 times in terms of the IC25.

Table 10. Toxicity results of unknown mine effluents (ME) to the algal test battery.

Toxicant	IC25 (95% confidence limit)				
	Selenastrum	Nannochloris	Microcystis	Nitzschia	Anabaena
Dichromate (μg Cr/L)	261 (231-296)	71 (64-79)	115 (103-127)	42 (40-45)	11148 (10914-11387)
ME 1 (%v/v) Lead/zinc	0.090 (0.078-0.104)	0.017 (0.013-0.022)	0.011 (0.010-0.013)	0.032 (0.009-0.111)	0.90 (0.77-1.10)
ME 2 (%v/v) Tin/cadmium	0.018 (0.017-0.019)	0.031 (0.022-0.043)	0.007 (0.006-0.008)	0.028 (0.020-0.040)	0.013 (0.011-0.015)
ME 3 (%v/v) Copper/zinc	0.45 (0.36-0.56)	0.34 (0.25-0.47)	0.54 (0.43-0.66)	21 (18-26)	>82.2
ME 4 (%v/v) Copper/zinc	1.1 (0.50-2.2)	>82.2	0.33 (0.19-0.56)	21 (14-32)	33 (32-34)
ME 5 (%v/v) Copper/zinc	11 (9.4-13)	81 (79-84)	1.0 (0.93-1.1)	>82.2	3.2 (1.5-7.0)
ME 6 (%v/v) Copper/zinc	0.035 (0.021-0.059)	0.04 (0.028-0.058)	0.011 (0.011-0.011)	0.040 (0.031-0.051)	0.017 (0.000-0.632)

Variable toxicity profiles are demonstrated by algal species belonging to the same taxonomic grouping. Both *Selenastrum* and *Nannochloris* are green algae (*Chlorophyta*), yet they show very different toxicity responses among the six effluents tested. For ME 4, *Selenastrum* is one of the most sensitive species, while *Nannochloris* is the least sensitive species. The same is represented by *Microcystis* and *Anabaena* both being blue-green algae (*Cyanophyta*), where *Microcystis* appears to be more sensitive in all cases. At this time it is not known if morphology or nutrient requirements contribute to this difference.

As a generalization for these 6 particular mine effluents, *Nitzschia* and *Anabaena* appear to be the least sensitive species. IC25 values from the most sensitive to the least sensitive species can vary as much as 250 times as stated above. This becomes very important from an environmental perspective as the discharge of such effluents into the natural environment could potentially cause shifts in algal populations. This scenario could result in non-harmful species of algae in the environment being replaced by nuisance organisms (biofilm, taste and odor, or toxin producing species). An example of this was given by Peterson et al. (1997) who showed that hexazinone impacts on aquatic systems may lead to competitive advantages for blue-green algae, which may be undesirable. The toxicity modifying effect of the receiving water can be compared with effluent tests diluted with TM 1 and TM 2 media. Poor quality receiving waters high in organic material are more likely to reduce the toxic effects of metal containing mine effluents than those of pristine waters.

Since the development of the above test protocol (or modifications thereof), it has been used for assessing the toxicity of mine effluents, pulp mill effluents and organic compounds. The use of a battery of test organisms has proved valuable in all cases.

11. Conclusions

Dispersion of an increasing number of man-made chemicals into aquatic environments demands that effective methods be applied to assess toxicity to organisms forming the base of the food-chain, namely phytoplanktonic algae and cyanobacteria. The use of large-volume, high biomass flask testing followed by tedious microscopic examination is certainly not ideal to achieve this objective. Instead, toxic assessment of liquid samples conducted in 96-well microplates, coupled with multiple species testing and sensitive automated microplate readers, is an attractive alternative that ensures ecological relevance, high throughput and good replication.

Beyond the efficiencies of different testing platforms, it is also essential that test results give accurate estimates of what might occur in the receiving environment. The method described in this chapter provides the necessary growth requirements for test organisms without altering the toxicological properties of the samples being investigated. By introducing media with different chemical compositions, it is possible to mimic several types of natural waters, and the inclusion of receiving water testing further enhances environmental relevance. It is hoped that future developments in small-scale toxicity testing will favor the path of better and easier methods, so that the use of effects-based measurements will become more widespread.

Acknowledgements

This work was financially supported by Environment Canada's Environmental Innovation Program. The strong support of Rick Scroggins, Environment Canada, is much appreciated. The laboratory work was carried out at the Saskatchewan Research Council and the help of Ilene Cantin in the initial stages of this project is much appreciated. The Technical University of Denmark provided in-kind support.

References

Andersen, J.S., Holst, H., Spliid, H., Andersen, H., Baun A. and Nyholm, N. (1998) Continuous ecotoxicological data evaluated relative to a control response, *Journal of Agricultural Biological and Environmental Studies* **3**, 405-420.

Campbell, P.G.C. (1995) Interaction between trace metals and aquatic organisms: A critique of the free-ion-activity-model, in D.R. Turner and A. Tessier (eds.), *Metal speciation and bioavailability in aquatic systems,* J. Wiley, New York, pp. 45-102.

Greene, J.C., Bartels, C., Warren-Hicks, W.J., Parkhurst, B.R., Linder, G.L., Peterson, S.A and Miller, W.E. (1989) *Protocols for short term toxicity screening of hazardous waste sites*, U.S. EPA, EPA 600/3-78-094, 17 pp.

ISO (International Standard Organization) (1989) Water quality – fresh water algal growth inhibition test with *Scenedesmums subspicatus* and *Selenastrum capricornutum*, International Standards Organization (ISO) 8692-1989(E), Geneva, Switzerland, International Organization for Standardization, 5 pp.

Nyholm, N. and Peterson, H.G. (1997) Laboratory bioassays with microalgae, in W. Wang, J.W. Gorsuch and J.S. Hughes (eds.), *Plants for environmental studies,* Lewis Publishers, Boca Raton, Florida, pp. 226-276.

Nyholm, N., Sorensen, P.S., Kusk, K.O. and Christensen, E.R. (1992) Statistical treatment of data from Microbial toxicity tests, *Environmental Toxicology and Chemistry* **11**, 157-167.

Peterson, H.G. and Nyholm, N. (1993) Algal bioassays for metal toxicity identification, *Water Pollution Research Journal of Canada* **28**, 129-153.

Peterson, H.G., Boutin, C., Freemark, K.E. and Martin, P.A. (1997) Toxicity of hexazinone and diquat to green algae, diatoms, cyanobacteria and duckweed, *Aquatic Toxicology* **39**, 111-134.

Stauber, J.L. and Davies, C.M. (2000) Use and limitations of microbial bioassays for assessing copper bioavailability in the aquatic environment, *Environmental Reviews* **8**, 255-301.

Abbreviations

CE	controlled environment
CL	confidence limit
EDTA	ethylenediamine tetraacetic acid
FWI	Freshwater Institute
IC25	25% inhibitory concentration
ISO	International Standards Organization
ME	mine effluent
NTA	nitrilotriacetic acid
OD	optical density
r^2	coefficient of determination
RO	reverse osmosis
TM	test medium
UTCC	University of Toronto Culture Collection
UTEX	University of Texas Culture Collection.

7. *LEMNA MINOR* GROWTH INHIBITION TEST

MARY MOODY
Saskatchewan Research Council
125 - 15 Innovation Boulevard, Saskatoon
Saskatchewan S7N 2X8, Canada
moody@src.sk.ca

JENNIFER MILLER
Miller Environmental Sciences Inc.
1839 Lockhart Road, Innisfil
Ontario L9S 3E9, Canada
miller.smith@sympatico.ca

1. Objective and scope of test method

This chapter describes Environment Canada's method entitled *Biological Test Method: Test for Measuring the Inhibition of Growth Using the Freshwater Macrophyte, Lemna minor* (EC, 1999). Research has lead to minor changes in the methodology that are described where appropriate. Applications of the method are also described herein. The *L. minor* growth inhibition test is a toxicity test carried out in replicate test vessels containing ≥ 100 mL of test solution and two 3-frond *L. minor* plants, at the end of which the effects of contaminants on *Lemna* growth are determined. It can be used alone or as part of a test battery to determine and monitor the toxic effects of discrete substances or complex mixtures that could be harmful to aquatic life in freshwater ecosystems. Results of these toxicity tests can be used as part of a weight-of-evidence approach, to determine the need for control of discharges and to set effluent standards (EC, 1999). The *L. minor* growth inhibition test measures: (a) increased number of fronds; and (b) dry weight (as an indication of growth) at the end of the test. Various sample types can be evaluated using this method, including:

(1) freshwater industrial or urban effluents, elutriates, or leachates;
(2) single chemicals, commercial products, or known mixtures of chemicals; and
(3) freshwater surface or receiving waters.

The small size, structural simplicity and rapid growth of *L. minor* are some of the characteristics that make them advantageous for use in laboratory toxicity tests. In addition, the cultures are easily maintained in the laboratory, test results are reproducible, and the method is cost effective.

2. Summary of test procedure

Environment Canada's *L. minor* growth inhibition test described in this chapter is a simple, short-term test. It uses actively growing *L. minor* in a 7-day static or static-renewal assay to assess various liquid samples for their effects on *Lemna* growth. The test uses axenic stock cultures of *L. minor* which are easily maintained in the laboratory and are acclimated to test conditions prior to use in a test. The test is conducted at 25 ± 2°C in vessels containing a minimum of 100 mL of test solution. At least 3 replicates, each containing two 3-frond plants are prepared for each treatment, plus controls. Both single-concentration (*i.e.*, pass/fail test) and multi-concentration configurations can be used. Recommended light conditions include continuous full-spectrum (fluorescent or equivalent) lighting with a fluence rate of 63 to 72 $\mu mol/m^2/s$ (*i.e.*, approximately 4000 to 4500 lux).[1]

Test samples are spiked with the same nutrients (at the same concentrations) as those used in the test medium (control/dilution water). Options for test media include a modified APHA (American Public Health Association) medium (SRC, 1997) and the SIS (Swedish Standard Institute) growth medium (OECD, 2002). Observations of frond numbers and frond appearance can be easily made throughout the test, and endpoints include growth, based on increase in the number of fronds during the test, and dry weight at the end of the test. The relative sensitivity of the culture of *Lemna* being used, and the precision and reliability of data being produced, is assessed with the routine use of a reference toxicant under standardized conditions.

Table 1. Rapid summary of test procedure.

Test organism	Actively growing *L. minor*
Type of test	7-day static or static-renewal toxicity test
Test format	Glass or plastic cups or flasks
Volume of test vessels	100 to 150 mL
Lemna *inoculum*	Two 3-frond plants
Lighting	24-h continuous full spectrum light at 63-72 $\mu mol/m^2/s$
Temperature	25 ± 2°C
Experimental configuration	Controls, 5-7 dilutions of test solution in geometric series, minimum of 3 replicates per treatment
Endpoints	IC25, NOEC, LOEC, based on biomass (frond increase and dry weight)
Measurements	Frond count and dry weight at 7 days, pH at beginning and end in control, low, medium and high concentrations
Reference toxicants	KCl, Ni (NiSO$_4$), 3,5-dichlorophenol (DCP)

---

[1] The relationship between quantal flux (*i.e.*, $\mu mol/m^2/s$) and lux is highly variable and depends on the source of light, the light meter used, the geometrical arrangement, and the possibilities of reflections (ASTM, 1999). Approximate conversions between quantal flux and lux, however, for full-spectrum fluorescent light is as follows: one lux is approximately equal to 0.016 $\mu mol/m^2/s$.

3. Overview of applications reported with the toxicity test method

Duckweeds have been used as test organisms since the 1930s and were among the species first used to define the effects of the phenoxy-herbicides on plants (Blackman and Robertson-Cumminghame, 1955). Today, many important environmental legislation and guidelines developed under various authorities have included phytotoxicity testing as part of environmental monitoring and assessment (Wang and Freemark, 1995). The United States Environment Protection Agency (U.S. EPA) includes duckweed testing under the Federal Insecticide, Fungicide, and Rodenticide Act (FIFRA), Toxic Substances Control Act (TSCA), and National Pollution Discharge Elimination System (NPDES) permits under the U.S. Water Quality Act, 1987 (Wang and Freemark, 1995). *L. minor* testing is also required for herbicides under the Directive 91/414/EEC for the registration of plant protection products in the European Union (EU); and under Canada's new Metal Mining Effluent Regulations (MMER) promulgated in 2002, pursuant to the Fisheries Act (DFO, 2002).

Standardized duckweed test methods currently available and used in North America and abroad include those by: the Institute of Applied Environmental Research (ITM, 1990); the American Society for Testing and Materials (ASTM, 1991); the American Public Health Association et al. (APHA et al., 1995); the Swedish Standards Institute (SIS, 1995); the Association Française de Normalisation (AFNOR, 1996); the U.S. EPA (1996); and Environment Canada (EC, 1999). More recently, draft test methods using *L. minor* continue to be developed by the Organization for Economic Cooperation and Development (OECD, 2002), and the International Organization for Standardization (ISO, 2003).

4. Advantages of conducting the toxicity test method

Duckweed species, including *L. minor* have many attributes that make them advantageous for use in laboratory toxicity tests and assessments of freshwater systems. They are small enough that large laboratory facilities and other equipment are not necessary and only 4 L of sample is required to perform a multi-concentration test. They are large enough, however, that effects can be observed and quantified without the use of a microscope. They are relatively simple, structurally and they can grow rapidly and indefinitely under laboratory conditions that are relatively easy to maintain (Hillman, 1961; Wang, 1987; Smith and Kwan, 1989).

Lemna spp. reproduces vegetatively and their genetically homogenous populations enable specific clones to be used for all experiments. This eliminates potential effects due to genetic variability (Hillman, 1961; Bishop and Perry, 1981; Smith and Kwan, 1989). *Lemna* spp. are floating aquatic plants, inhabiting the air-water interface. They are sensitive to metals (see Section 11), surface-active substances, hydrophobic compounds, and other substances that concentrate at the surface of the water (Bishop and Perry, 1981; Jenner and Janssen-Mommen, 1989; Smith and Kwan, 1989; Taraldsen and Norberg-King, 1990; ASTM, 1991).

Environment Canada's *L. minor* Growth Inhibition Test is a short-term test that is quick to set up and take down (*i.e.*, quantify results). The use of inexpensive, disposable vessels further reduces the time and cost associated with this test. Unlike many algal toxicity tests, test solutions can be renewed, and colored or turbid wastewater or receiving water samples can be tested (Taraldsen and Norberg-King, 1990).

5. Test species

L. minor Linnaeus (Arales:Lemnaceae), commonly referred to as lesser duckweed, is the test species required for use in the Environment Canada test method (EC, 1999). It is a small vascular, aquatic macrophyte belonging to the family Lemnaceae. Members of the family Lemnaceae are structurally the simplest and smallest flowering plants in the world (Godfrey and Wooten, 1979).

The taxonomy of *Lemna* spp. is complicated by the existence of a wide range of phenotypes (OECD, 2002). *L. gibba*, another species commonly used in toxicity tests, differs from *L. minor* in that the fronds of *L. gibba* are broadly elliptic to round, its upper surface often has red blotches, and its lower surface is generally swollen (gibbous). The lack of overwintering turions (dark green or brownish daughter plants), prominent dorsal papules, and reddish anthocyanin blotches on the ventral side separate *L. minor* from another closely related species *L. turionifera* Landolt (EC, 1999).

The fronds of *L. minor* occur singly or in small clusters (3 to 5) and are flat, broadly obovate to almost ovate, ranging from 2 to 4 mm long (Hillman, 1961; Godfrey and Wooten, 1979; Newmaster et al., 1997). They are green to lime green, and glossy when fresh (Godfrey and Wooten, 1979). Reproduction in *Lemna* spp. is usually vegetative and occurs by lateral branching. The plant has a single root or rootlet that emanates from a deep root furrow in the center of the lower surface of each frond (Hillman, 1961).

L. minor is ubiquitous in nature, inhabiting relatively still fresh water (ponds, lakes, stagnant waters, and quiet streams) and estuaries ranging from tropical to temperate zones (APHA et al., 1992). Its distribution extends nearly worldwide (Godfrey and Wooten, 1979) and in North America, *L. minor* is one of the most common and widespread of the duckweed species (Arber, 1963; APHA et al., 1992). Duckweeds form an essential component of the ecosystem in shallow, stagnant waters. They are an integral portion of the food chain, providing food for waterfowl and marsh birds and occasionally small mammals. They also provide food, shelter, shade, and physical support for fish and aquatic invertebrates (Jenner and Janssen-Mommen, 1989; Taraldsen and Norberg-King, 1990; APHA et al., 1992; Newmaster et al., 1997).

Two clones (Landolt 8434 and 7730) originally collected in Canada are recommended for use in the Environment Canada method (1999) and are available from the University of Toronto Culture Collection as UTCC 490 and 492.

> University of Toronto Culture Collection
> Dept. of Botany, University of Toronto
> 25 Willcocks St., Toronto, Ontario
> Canada M5S 3B2
> Telephone: 416-978-3641
> Facsimile: 416-978-5878
> e-mail: judy.isakovic@utoronto.ca

Other clones of *L. minor* are acceptable if obtained from a culture collection, biological supply house, government laboratory or private laboratory and species is confirmed taxonomically. As clones vary in sensitivity (Cowgill and Milazzo, 1989) identification of the clone is recommended.

6. Culture/maintenance of organism in the laboratory

6.1 EQUIPMENT AND FACILITIES REQUIRED

Environment Canada's *L. minor* growth inhibition test must be conducted in a constant–temperature room, incubator, environment chamber or equivalent facility with good temperature control and acceptable lighting. The facility should be well ventilated and isolated from physical disturbances or any contaminants that could affect the test organisms. The test facility should also be isolated from the location where *Lemna* are cultured. Dust and fumes should be minimized within the test and culturing facilities. The laboratory must have instruments for measurement of the basic variables of water quality (temperature, pH) and the instruments must be well maintained and calibrated regularly (EC, 1999).

6.2 GLASSWARE

Glassware and accessories in contact with *Lemna* must be made of nontoxic, chemically inert materials and where necessary, sterile. Materials such as borosilicate glass (*e.g.* Pyrex^{TM}), porcelain, high-density polystyrene, or perfluorocarbon polyethylene plastics (e.g. Teflon^{TM}) may be used. Plastic vessels may be used for test vessels as long as *Lemna* does not adhere to the walls. To reduce the static charge that can cause *Lemna* to stick to the sides, plastic vessels may be soaked in clean distilled water before use. All vessels and accessories should be chemically cleaned (and sterilized if necessary) or disposable. Culture and test vessels should be transparent and covered to exclude dust and minimize evaporation. Test vessels must be covered with transparent covers (EC, 1999).

6.3 EQUIPMENT REQUIRED FOR CULTURE AND TESTING

Equipment recommended for *Lemna* culture maintenance includes typical items found in laboratories conducting aquatic toxicity testing such as an autoclave for sterilizing glassware and media, and a sterile laminar flow hood or similar area that can provide a sterile, draft-free workspace for culturing *Lemna*. Materials suggested for *Lemna* testing are listed below:

> Volumetric flasks (100, 500, 1000 mL).
> Erlenmeyer flasks (250, 1000, 2000 mL).
> Bunsen burner and gas source.
> Graduated cylinders (50, 100, 500 mL capacity).
> Filtration apparatus, 1 and 0.22 µm porosity filters, filter flask.
> Light meter to verify light intensity in culture and test areas.
> pH meter and probe.
> Disposable plastic cups and lids to fit, approximately 200 mL volume.
> Petri dishes (150 mm diameter), sterile, disposable.
> Magnetic stirring bars and stirring device.
> Pipetting devices and disposable tips to fit devices.
> Weighing dishes and spatula.
> Reagent-grade chemicals required for culture and test media preparation.
> Milli-Q water purification system (or equivalent such as glass distiller) for producing high quality water free of ions, organics, particles and micro-organisms greater than 0.45 µm.
> Pyrex baking dish, approximately 20 x 20 cm.

6.4 CULTURE MEDIUM

L. minor cultures are easily maintained in the laboratory by inoculation of plants in Hoagland's E+ medium (Cowgill and Milazzo, 1989). The medium originally described by Environment Canada (1999) has been modified since the publication of the method by replacing the separate $FeCl_3 \cdot 6H_2O$ and EDTA solutions with a combined solution (Stock C) containing increased amounts of $FeCl_3 \cdot 6H_2O$ and $Na_2EDTA \cdot 2H_2O$ to support healthy growth of the two clones of *L. minor* recommended (Moody, 2003). The up-dated chemical composition and method of preparation of the modified Hoagland's E+ culture medium are presented in Table 2.

6.5 STOCK CULTURE

According to the Environment Canada method (1999) stock cultures are initiated by the transfer of one or two plants from the agar slant or liquid culture obtained from the supplier into a small volume of sterile modified Hoagland's E+ medium (for example, 25 mL medium in a 25 x 150 mm test tube). Because use of aseptic technique is essential, preparation of *Lemna* cultures should be performed in a laminar flow cabinet or other pre-sterilized space with minimal airflow. Several of these stock cultures should be prepared each week to maintain the laboratory's culture in a rapidly growing state. *Lemna* that is subcultured less frequently may be viable for several weeks, but is not suitable for testing purposes. Stock cultures may be stored under reduced light and at a temperature of 4 to 10°C if necessary. To recover its fast growth rate, *Lemna* must be subcultured in fresh culture medium and grown under test conditions for at least 14 days immediately prior to use for testing.

Culture conditions recommended by Environment Canada (1999) are listed in Table 3 and described herein in Sections 6 and 7.

Table 2. Chemical composition of nutrient stock solutions and preparation of modified Hoagland's E+ culture medium.

Stock	Substance	Concentration stock solution g/L	Concentration medium mg/L	Amount of stock solution per litre medium
A[a]	$Ca(NO_3)_2 \cdot 4H_2O$	59.0	1180.0	20 mL
	KNO_3	75.76	1515.2	
	KH_2PO_4	34.0	680.0	
B[b]	Tartaric Acid	3.0	3.0	1 mL
C	$FeCl_3 \cdot 6H_2O$	1.21	24.2	20 mL
	$Na_2EDTA \cdot 2H_2O$	3.349	67.0	
D	$MgSO_4 \cdot 7H_2O$	50	500	10 mL
E	H_3BO_3	2.86	2.86	1 mL
	$ZnSO_4 \cdot 7H_2O$	0.22	0.22	
	$NaMoO_4 \cdot 2H_2O$	0.12	0.12	
	$CuSO_4 \cdot 5H_2O$	0.08	0.08	
	$MnCl_2 \cdot 4H_2O$	3.62	3.62	
	Sucrose	10 g/L		
	Yeast Extract	0.10 g/L		
	Bactotryptone[c]	0.6 g/L		
Method	Add listed ingredients to 900 mL deionized water, stir until dissolved. Adjust pH to 4.6 with NaOH or HCl and bring volume to 1 litre. Autoclave for 20 minutes at 121°C and 124.2 kPa.			
Stock solutions	Stock solutions are prepared by dissolving reagent-grade chemicals in distilled or deionized water. Stock solutions should be stored at 4°C. Protect bottles from light by covering or use amber bottles.			

[a] Add 6 mL of $6N$ HCl to stock solution A.
[b] Add 1.2 mL of $6N$ KOH to stock solution C.
[c] Peptone from casein, trypsin digested is an acceptable alternative.

6.6 CULTURE STERILIZATION

L. minor must be maintained in axenic condition. If the culture is contaminated with microorganisms such as bacteria, fungi, or algae the culture medium will become cloudy and will develop a slime layer, mould colonies, or visible contamination. The culture must be discarded and replaced or sterilized. To sterilize, several plants are treated for 30 seconds to 5 minutes by immersion in with 0.5% v/v sodium hypochlorite solution such as dilute household bleach. Plants are periodically transferred to fresh culture medium during the treatment period. Many fronds will die as a result of this treatment but some of those surviving will usually be free of contamination. These can then be used to re-inoculate new cultures (AFNOR, 1996; OECD, 2002). Plants that have been sterilized will require at least 8 weeks of routine subculture before they can be used in tests.

Table 3. Summary of culture and test conditions for L. minor growth inhibition test.

Culture	Test organism	- *L. minor*, axenic, taxonomically verified.
	Stock culture	- Axenic, actively growing *Lemna* maintained by weekly subculture in modified Hoagland's E+ medium, pH 4.6 (Tab. 2).
	Test culture	- Healthy, large-volume culture in modified Hoagland's E+ medium, inoculated 7-10 days before test. A test of this culture health in test medium must yield at least 8-fold increase in frond number.
	Acclimation culture	- Prepared by rinsing test culture twice in test medium and incubating under test conditions 18-24 hours immediately before test.
Test	Test format	- Static or static-renewal (renewed at least every three days). - 7-day test.
	Control/ Dilution medium	- Modified APHA medium, pH 8.3 for wastewaters, effluents, and leachates (Tab. 4). Nutrient enriched receiving waters (spiked with the same nutrients used in modified APHA test medium) may be used for monitoring and compliance. - SIS medium (pH 6.5) for chemicals, modified APHA for metals, nutrient-enriched receiving waters (spiked with the same nutrients used in SIS test medium) may be used to assess local toxic effects.
	Lemna inoculum	- Two 3-frond plants (total 6 fronds) per test vessel.
	Test vessel	- Disposable plastic cups, Erlenmeyer flasks or other containers that allow sufficient surface area that no overlapping of *Lemna* occurs at test end. Suitable covers include Petri dishes, transparent plastic covers made to fit plastic cups or other covers; watch glasses are not suitable covers. Test vessels are placed on dark surface to reduce reflected light.
	Test volume	- $\geq$ 100 mL, preferably 150 mL.
	Test replicates	- $\geq$ 3 replicates ($\geq$ 4 replicates if hypothesis testing to be done).
	Temperature	- Daily mean $25 \pm 2°C$, constant throughout test.
	pH	- No adjustment if pH of test solution is between 6.5 and 9.5. A parallel pH-adjusted test might be required for pH outside this range. Measure pH at start and end of test (or at times of renewal of solutions) in controls, high, medium and low test concentrations.

Table 3 (cont.). Summary of culture and test conditions for L. minor *growth inhibition test.*

Test	Aeration	- Wastewater and receiving water samples are aerated gently for 20 minutes after addition of nutrients and before test initiation or renewal of test solutions. There is no aeration of test vessels during testing.
	Light	- Continuous full-spectrum light of 63-72 µmol/m^2/s at surface of test solution, fluence rate within the entire test area (with the acceptable ± 15% variation from the selected light intensity) should be 55 to 80 µmol/m^2/s. Measure at several locations in test area once during test.
	Endpoints/ Observations	- Frond increase (final frond count minus initial frond count), in each vessel and frond appearance. Fronds may be counted on two more occasions for growth rate determination.
		- Dry weight (total dry weight per replicate after drying at either 100°C for six hours or 60°C for 24 hours).
	Calculations	- IC25 (or other ICp) derived from point-estimation techniques for frond increase and dry weight endpoints, including 95% confidence limits.
		- LOEC and NOEC derived from hypothesis testing.
Quality control	Reference toxicant	- KCl, NiSO$_4$, 3,5-dichlorophenol. Determination of IC25 (growth inhibition) is to be performed within 14 days of testing following same procedure and test medium as the definitive test. Use of control charts showing historical mean and warning limits is recommended.
	Test validity criteria	- Frond increase at least 8 fold (*i.e.,* frond number has increased from 6 to at least 48 in 7 days).
		- Recommended test conditions and procedures have been followed.

6.7 ALTERNATE CULTURE MEDIA

Other culture media for *Lemna* species are recommended in *Lemna* test methods used in Europe and the USA (U.S. EPA, 1996; OECD, 2002; ISO, 2003). However, use of modified Hoagland's E+ medium is recommended for Environment Canada's test procedure described herein, as it provides the nutrients required for high quality *Lemna* plants intended for testing with modified APHA or SIS media. A summary of conditions and procedures used in other *Lemna* test method documents, and the chemical composition of the culture and test media recommended therein is available elsewhere (Appendices C and D in EC, 1999). The International Organization for Standardization recommends Steinberg medium (ISO, 2003) for both culture and test purposes.

7. Preparation of test species for toxicity testing

7.1 STOCK CULTURE

L. minor plants suitable for use in testing are grown in modified Hoagland's E+ medium in an axenic state and subcultured on a weekly basis (EC, 1999). Plants that are healthy will have glossy, bright green fronds, with no discolored areas, chlorosis, necrosis or other damage (Fig. 1). Typical plants consist of 2 to 7 fronds. Cultures that contain mostly small colonies of one or two fronds are stressed and should not be used.

Figure 1. L. minor *grown in modified Hoagland's E+ medium culture, 5X.*

7.2 TEST CULTURE

For testing, selection of a number of plants having three fronds is required (EC, 1999). Therefore, large volume cultures in Hoagland's E+ medium must be prepared 7 to 10 days in advance of performing a toxicity test. To prepare a test culture, approximately 10 plants are aseptically transferred from a week-old stock culture (*i.e.*, test tube culture, see Section 6.5 and 7.1) into a 150 mm diameter sterile Petri dish (or other sterile container having a large surface area) filled with sterile Hoagland's E+ medium to a depth of 1 to 1.5 cm. These are incubated under test conditions for 7 to 10 days. The culture should not be crowded as this will reduce *Lemna* health; the inoculum should be limited so that the surface area at 7 to 10 days is not more than about two thirds covered in plants. Cultures that become cloudy or contain mould colonies indicate contamination through exposure to non-sterile air or equipment and must not be used for testing.

Environment Canada recommends that a test of culture health should be performed when the test culture is initially prepared. This test indicates whether the *Lemna* are suitable for use in a test. One 3-frond plant from the stock (test tube) culture is transferred into 100 mL of the test medium to be used (either modified APHA or SIS medium). After one week under test conditions, frond numbers must have increased by at least 8 fold, to 48 fronds or more, indicating that the stock culture in Hoagland's medium was healthy and that plants from the test cultures will grow well in a test (EC, 1999).

7.3 ACCLIMATION CULTURE

The day before the test is to be set up, sufficient *L. minor* (from the 7 to 10 day old uncrowded test culture in modified Hoagland's E+ medium, see Section 7.2) are rinsed twice in test medium. To rinse the culture, the spent culture medium is poured off and replaced with fresh test medium, swirling several times to remove the Hoagland's medium. The *Lemna* is then poured into a clean, shallow container (such as a Pyrex baking dish) containing at least 2-cm depth test medium, and the dish covered with plastic wrap. This is termed the acclimation culture and is incubated for 18 to 24 hours immediately prior to toxicity test preparation, under test conditions. Although it is not necessary to maintain this culture under axenic conditions, reasonable care should be taken to avoid contamination by algae or dust. Therefore handling should be done in a laminar flow cabinet or other pre-sterilized area.

8. Testing procedure

8.1 TEST FACILITY AND INSTRUMENTS

The *L. minor* growth inhibition test must be conducted in a constant-temperature room, incubator, environmental chamber, or equivalent facility with good temperature control and acceptable lighting (EC, 1999). The laboratory must have instruments to measure the basic variables of water quality (temperature, pH) and these instruments must be maintained properly and calibrated regularly. A good source of high quality water such as glass-distilled or deionized water is required to prepare culture and test media and for rinsing glassware.

A test of the facility should be carried out before toxicity testing begins. The test system (*e.g.*, vessels, covers, lighting and temperature conditions) should be assessed by conducting a non-toxicant test in which all test vessels contain test medium (ISO, 2003). The coefficient of variation of control growth (as frond increase) will indicate the variability that may be expected. ISO recommends that the coefficient of variation of growth rate of these tests be less than 10% (ISO, 2003). This test will also demonstrate whether the *L. minor* clone and culture conditions will support the increase in frond numbers (at least 8-fold increase) required for a valid test.

8.2 CHOICE OF TEST MEDIUM

The choice of control/dilution water (test medium) will depend on the test substance and objectives. The same control/dilution water must be used to prepare sample dilutions and controls. Two different media are recommended by Environment Canada (1999) for different purposes. Modified APHA growth medium (SRC, 1997) is recommended for testing of wastewaters, receiving waters, and samples containing metals (Tab. 4). Swedish Standard (SIS) growth medium (OECD, 2002) is recommended for chemicals, commercial products or known mixtures (Tab. 5).

Table 4. Chemical composition of nutrient stock solutions and preparation of modified APHA test medium.

Stock	Substance	Concentration stock solution g/L	Concentration medium mg/L	Amount of stock solution per litre medium
A	$NaNO_3$	25.5	255	10 mL
	$NaHCO_3$	15.0	150	
	K_2HPO_4	1.04	10.4	
	KCl	1.01	10.1	
B	$CaCl_2 \cdot 2H_2O$	4.41	44.1	10 mL
	$MgCl_2 \cdot 6H_2O$	12.17	121.7	
	$MnCl_2 \cdot 4H_2O$	0.4149	4.149	
	$FeCl_3 \cdot 6H_2O$	0.16	1.6	
C	$MgSO_4 \cdot 7H_2O$	14.7	147	10 mL
	H_3BO_3	0.186	1.86	
	$Na_2MoO_4 \cdot 2H_2O$	0.00726	0.0726	
	$ZnCl_2$	0.00327	0.0327	
	$CoCl_2$	0.00078	0.0078	
	$CuCl_2$	9.0×10^{-6}	9.0×10^{-5}	
Method	Add listed ingredients to 970 mL deionized water. Aerate vigorously for at least 1-2 hours to stabilize pH. If a larger volume (4 liters or more) is prepared, overnight aeration is recommended to stabilize pH. Immediately before use in test, adjust pH to 8.3 ± 0.1 using $0.5N$ NaOH or $0.5N$ HCl. The medium is not sterilized.			
Stock solutions	Stock solutions are prepared by dissolving reagent-grade chemicals in distilled or deionized water. Stock solutions are stored as separate solutions in a refrigerator at 4°C for one month.			

Table 5. Chemical composition of nutrient stock solutions and preparation of SIS test medium

Stock	Substance	Concentration stock solution g/L	Concentration medium mg/L	Amount of stock solution per litre medium
I	$NaNO_3$	8.50	85	10 mL
	KH_2PO_4	1.34	13.4	
II	$MgSO_4 \cdot 7H_2O$	15.0	75	5 mL
III	$CaCl_2 \cdot 2H_2O$	7.20	36	5 mL
IV	Na_2CO_3	4.00	20	5 mL
V	H_3BO_3	1.00	1.00	1 mL
	$MnCl_2 \cdot 4H_2O$	0.200	0.200	
	$Na_2MoO_4 \cdot 2H_2O$	0.010	0.010	
	$ZnSO_4 \cdot 7H_2O$	0.050	0.050	
	$CuSO_4 \cdot 5H_2O$	0.005	0.005	
	$Co(NO_3)_2 \cdot 6H_2O$	0.010	0.010	
VI	$FeCl_3 \cdot 6H_2O$	0.17	0.85	5 mL
	$Na_2EDTA \cdot 2H_2O$	0.28	1.40	
VII	MOPS (free acid form of this buffer)	490	490	1 mL (optional) Add if pH control of the medium is important
Method	Add listed ingredients to 900 mL deionized water. If buffer is required, add 1 mL of stock solution VII (optional). pH is adjusted to 6.5 ± 0.2 with either 0.1 or $1N$ HCl or NaOH and volume is adjusted to 1 litre. Prepare medium 1 to 2 days before use in a test to allow medium pH to stabilize, then adjust pH if necessary. The medium is not sterilized.			
Stock solutions	Stock solutions are prepared by dissolving reagent-grade chemicals in distilled or deionized water. Stock solutions I to V are sterilized by autoclaving at 120°C for 15 minutes or by membrane filtration (0.2 μm pore size). Stock solutions VI and VII are sterilized by membrane filtration (they should not be autoclaved) and then added aseptically to the medium. Stock solutions are stored as separate solutions in a refrigerator at 4°C. Stock solutions I to V have a shelf life of 6 months, but stock solutions VI and VII should be discarded after 1 month.			

8.3 TESTING USING RECEIVING WATER

Environment Canada (1999) describes the option of using receiving water as the test diluent for testing effluent in instances where site-specific information is required about the potential toxic effect of an effluent or other substance on a particular receiving water. Receiving water may also be tested at a single concentration (for example 97%) to determine the presence or extent of toxicity near an industrial site. A sample of receiving water or upstream water (collected adjacent to the source of the contamination but removed from it, or upstream of the source) is collected concurrently with the effluent (test) sample. The water is filtered through 1 μm porosity glass fibre filters, such as Whatman GF/C filters (or additionally through 0.22 μm pore size filters to prevent test invalidation by growth of algae). The water is then enriched with the same reagent-grade chemicals, at the same concentration used to make the modified APHA growth medium (10 mL/L of each stock solution A, B, and C). In instances where the toxic effect of a specific chemical in receiving water is to be tested, receiving water may be enriched with the same concentration of nutrients as those used to prepare SIS growth medium and this water used as control/dilution water. If receiving water is used as test diluent, an additional set of controls must be prepared using the appropriate test medium prepared in distilled or deionized water. These two types of controls must be tested concurrently with the test sample and compared statistically. *Lemna* must meet the required 8-fold increase of fronds in modified APHA or SIS medium for test validity.

8.4 TEST SETUP

Universal test procedures that apply to any Environment Canada *L. minor* growth inhibition test are listed in Table 3 and described herein in Sections 8 and 9. The 7-day test includes the following two options:

(1) a static test where the test solutions are not renewed during the test and
(2) a static-renewal test, where the test solutions are replaced at least every three days during the test.

The static renewal test is recommended for test solutions in which the concentration of the test substance (or a biologically active compound) can be expected to decrease significantly during the test period due to factors such as volatilization, photodegradation, precipitation, or biodegradation (ITM, 1990; OECD, 2002).

8.4.1 Preparation of test sample
The same control/dilution medium (test medium) must be used to prepare all control and test concentrations. The appropriate control/dilution medium should be freshly prepared as described in Tables 4 and 5. Both controls and test dilutions must be brought to 25 ± 2°C. If a sample requires filtration (wastewater mixed with receiving water for example), the pH of the sample is measured before and after filtration. An aliquot of each of the same nutrient stock solutions used to prepare the modified APHA growth medium (stock solutions A, B, and C) is added to the test wastewater or receiving water at a ratio of 10 mL per 1000-mL sample. This dilutes the sample to 97%, which is the maximum nominal concentration of the test sample that can be tested. The test sample is then aerated gently for 20 minutes to equilibrate the sample with the added nutrients and to stabilize pH. Oil-free compressed air should be used and provided at a rate of aeration not greater than

100 bubbles/minute, using a disposable plastic or glass tube with small interior aperture (*e.g.*, 0.5 mm).

For tests using SIS medium as control/dilution water (*i.e.*, for testing of chemicals), the test chemical is dissolved and diluted in the medium, in a geometric series of concentrations. In some instances, such as testing of pesticides, the chemical may be sprayed as a foliar spray. Regardless of how test solutions are prepared, the concentration, solubility and stability of the chemical in the test medium under test conditions should be determined before the test is initiated. If a solvent, emulsifier or dispersant is used to assist solubility of the chemical to be tested, additional control solutions containing the highest concentration of the agent must be prepared for comparison purposes. Analysis of the test chemical during a static test may indicate that the concentration of the chemical or one of its ingredients has decreased to less that 80% of the nominal concentration. In this case, the static-renewal test option must be followed, in which *Lemna* are transferred to fresh test solutions on at least two occasions during the test (EC, 1999).

8.4.2 pH

According to Environment Canada (1999), toxicity tests should normally be conducted without adjustment of pH. However, if the sample of test substance causes the pH of any test solution to be outside the range of 6.5 to 9.5 and the toxicity of the test substance rather than the effect of pH is being assessed, the pH of the test solutions or enriched test sample may be adjusted (EC, 1999). Depending on the objectives of testing, the sample may be neutralized (adjusted to pH 7.0), or adjusted to within ± 0.5 pH units of the test medium (control/dilution water). Another option is to adjust the pH of the test sample or dilution to within the range of 6.5 to 9.5 using appropriate solutions at ≤ $1N$ HCl or NaOH. If adjustment of pH by more than 0.5 pH units is required, a further 30 minutes of aeration followed by another pH measurement is recommended. Once a test is initiated, pH is monitored but not adjusted. Volumes of HCl and NaOH must be recorded and used to calculate the nominal initial concentration of the test substance. It may be desirable to conduct a parallel test of the sample that is not pH adjusted. For static and static-renewal tests, pH must be measured at the beginning (before *Lemna* plants are added) and end of the test in at least the highest, medium and lowest test concentration. For static-renewal tests, pH must also be measured immediately before and after each test solution renewal.

8.4.3 Single-concentration tests

Single-concentration tests (*e.g.*, those used for assessing whether an effluent "passes" or "fails" in Canadian regulatory frameworks) would only use one test concentration, normally full-strength (97% in the case of this method) effluent, leachate, receiving water, elutriate or an arbitrary or a single prescribed concentration of chemical (EC, 1999). Controls would be used, as described earlier for multi-concentration tests, and all procedures previously described would be employed.

8.4.4 Test sample dilutions

Dilutions of the test sample are prepared in the control/dilution medium following a geometric dilution scheme (EC, 1999). For any test that is intended to estimate the

ICp, NOEC/LOEC, or both, at least five concentrations plus a control solution (consisting of 100% test medium) are recommended. The goal is to bracket the endpoint with test concentrations that have partial inhibition of growth at 7 days. An appropriate geometric series of concentrations is one in which each successive concentration is a multiple of 0.5 of the previous concentration. For example, concentrations of wastewater and receiving water samples could be 97, 48.5, 24.3, 12.1, 6.1, 3.0, 1.5% v/v.

Test dilutions may be prepared directly in test vessels, adding appropriate amounts of control/dilution water and test sample. Alternatively, test dilutions may be prepared in volumetric flasks at the desired concentrations and then dispensed into test vessels. The volume of the test solutions must be equal in any test and in a volume of 100 to 150 mL. Test dilutions are allowed to equilibrate for one hour before addition of *Lemna* plants. The test must begin with the same number of replicates for each treatment, and at least three replicates are prepared for each test concentration and the control. If endpoints are to be calculated using hypothesis tests (*i.e.*, NOEC/LOEC), a minimum of four replicates must be used.

8.4.5 Addition of Lemna

Lemna plants used must be from cultures grown (as described above) that meet the minimum criterion for health (*i.e.*, 8-fold frond increase in 7 days) (EC, 1999). Plants having three fronds of identical size and condition are selected from the acclimation culture and transferred to a container of test medium. Two plants (a total of 6 fronds) are randomly assigned to each chamber and immersed briefly in the test solution. A disposable plastic inoculating loop is convenient for this. Care must be taken to not contaminate the three frond *Lemna* plants that have been set aside for setting up the test. Therefore rinsing or replacing the inoculating loop after introducing plants into each test vessel is recommended. Test vessels are covered and transferred to the incubator. All vessels must be exposed to the same intensity of light and should be placed on a dark background (such as black paper) to reduce reflection of light. Care must be taken to ensure that plants do not adhere to the walls of the test vessels. The day the *Lemna* are initially exposed to solutions of test substance is designated Day 0. Day 7 is the day the test is terminated.

9. Post-exposure observations/measurements and endpoint determinations

9.1 FROND INCREASE

The adverse effects of toxic substances will reduce the increase in frond number compared to controls and may also affect frond appearance. According to Environment Canada (1999), frond count at 7 days must include every frond and every visible protruding bud, living or dead. Plants may be observed using a dissecting microscope or other magnifying device with a light directed into the side or bottom of the cup. Vessels that have *Lemna* accidentally stuck or dried to their sides during the test are removed from the test and those replicates are eliminated from endpoint calculations. Observations of the following should also be made and recorded for each test vessel: chlorosis (loss of pigment); necrosis (localized dead tissue on fronds, which appears brown or white); yellow or abnormally sized fronds;

gibbosity (humped or swollen appearance); colony destruction (single fronds); root destruction; and loss of buoyancy. If the test design requires calculation of growth rates, two additional frond counts will be required, for example at days 3 and 5. Under toxic stress, *Lemna* may grow many tiny buds that must be included in the total frond count, and may result in underestimation of toxicity (Wang, 1990). Frond area, which may be determined electronically through image analysis (ISO, 2003) and dry weight are endpoints not subject to this effect. For data analysis, mean frond increase (frond count at 7 days minus initial frond inoculum) is calculated for each test vessel.

Appearance of fronds in test cups must be compared to the appearance of fronds in controls. The appearance of fronds in control vessels will vary depending on the test medium used. *Lemna* plants grown in modified APHA, a lean medium containing no chelator (no EDTA), will be paler in colour than when grown in Hoagland's E+ medium. In SIS medium, plants may be larger and slightly paler green compared to plants grown in Hoagland's E+ medium.

9.2 DRY WEIGHT

After counting, all *Lemna* fronds are dried and weighed. All plants (including roots) in each test vessel are collected, blotted and placed in a dried and pre-weighed weighing boat. Plants are dried for 6 hours at 100°C or for 24 hours at 60°C. Upon removal from the oven, the weighing boats are placed in a desiccator, cooled, and weighed on a balance that measures consistently to 0.01 g. Delay in weighing may allow absorption of water vapor by the *Lemna* indicated by an increase in weight of a boat that has been replaced in the desiccator after initial weighing. If the increase in weight is greater than 5%, the boats should be re-dried for 1 to 2 hours. The total weight of fronds in each test vessel (each replicate) must be determined.

9.3 ADDITIONAL MEASUREMENTS

Environment Canada (1999) requires that temperature, pH, and light conditions be measured during a test.

Temperature must be monitored throughout the test. As a minimum, it must be measured daily in representative test vessels (*i.e.*, at least in the high, medium and low concentration test solutions). Continuous recordings or daily measurements of maximum and minimum temperatures are acceptable.

The pH must be measured both before the *Lemna* plants are added and at the end of the test in controls and in at least the high, medium and lowest test concentrations. For static-renewal exposures, the pH must also be similarly measured immediately before and after each renewal of test solutions.

Light fluence must be measured at least once during the test period at several points in the test area and at the same distance from light source as the *Lemna* plants.

For testing chemicals, Environment Canada recommends that test solutions be analyzed to determine the chemical concentrations to which the *Lemna* are exposed. In a static test, sample aliquots are taken from all replicates in at least the high, medium, and low test concentrations, and the controls before the *Lemna* are exposed and at the end of the test, as a minimum. In a static-renewal test, sample aliquots are taken from at least the high, medium, and low test concentrations, and the controls, at the beginning and end of each renewal period, and on the first and last days of the test (EC, 1999). All samples are analyzed according to proven methods with acceptable detection limits. Toxicity results for any tests in which the concentrations of the test solutions are measured should be based on those measured concentrations unless there is justification otherwise. Each test solution should be characterized by the geometric average of the measured concentrations to which the organisms were exposed (EC, 1999).

9.4 ENDPOINT DETERMINATION

The endpoints of the test are based on the adverse effects of test substances on the growth of *L. minor*, assessed by comparison with the controls. They should be calculated for both the reduction in frond increase and decrease in final dry weight of fronds compared to the control. Various statistical endpoints can be calculated from these data. The rationale and methods of calculation follow and are discussed in the *Guidance Document on Statistical Methods for Environmental Toxicity Tests* (EC, 2004). The inhibiting concentration for a specified percent effect (ICp) is recommended as the primary endpoint for this test, and may calculated on the basis of biomass (frond increase or dry weight) or on the basis of growth rate of a measured endpoint, such as frond count, frond area or dry weight. For regulatory purposes in Canada, *Lemna* endpoints are based on biomass (DFO, 2002). Other jurisdictions may differ. Statistical endpoints based on average specific growth rate are recommended in recently published *Lemna* test methods (OECD, 2002 and ISO, 2003). The ICp can be derived statistically using point-estimation techniques. The 95% confidence limits must be given for any ICp reported (EC, 1999).

9.4.1 Determination of IC25

The following methods for calculating IC25 values based on biomass endpoints are described in full by Environment Canada (1999, 2004). Determination of the IC25 is derived from data that shows a reduction in performance (frond increase or dry weight) of the plants in test dilutions compared to controls. The ICp should not be derived from an extrapolation (*i.e.* the data should extend above and below the percent effect of interest). To estimate the IC25, there should be at least one concentration causing more than 25% effect relative to the control and at least one concentration causing less than 25% effect relative to the control (but greater than 0% effect).

Calculate percent inhibition of frond increase for each test replicate:

$$\% I = \frac{M - X}{M} \times 100 \qquad (1)$$

where: $\% I$ = percent inhibition,
  M = average increase in frond number; or average total dry weight of fronds at test end in the control test chambers,
  X = increase in frond number; or dry weight of fronds at test end in the test vessel.

The increase in frond number is calculated by subtracting the initial number of fronds in a given test vessel from the final number of fronds in the same vessel. No initial dry weight measurement is made. The percent inhibition (% I) of each test endpoint is then plotted separately against the logarithm of test concentration. The IC25 should be read from an eye-fitted line and any major disparity between this approximate graphic IC25 and the subsequent computer-derived IC25 must be resolved.

A computer-generated IC25, NOEC (no observed effect concentration) and a LOEC (lowest observed effect concentration) may be estimated by a number of commercially available statistical programs. Methods that can be used include the program *ICPIN* (referred to in Chapter 3 of this volume) based on smoothing and interpolation (Norberg-King, 1993), as well as programs using linear regression or non-linear regression techniques (EC, 2004). Consultation with a statistician is recommended to ensure that valid and relevant statistical procedures are followed. IC25 data reported in Section 11 were determined by linear interpolation using ToxCalc (Tidepool Scientific Software), version 5.0 (1994).

9.5 VALIDITY CRITERIA AND QUALITY CONTROL

9.5.1 Test validity criteria
The mean number of fronds in the control vessels must have increased from the initial inoculum of 6 (two 3-frond plants) by at least 8 fold by the end of the 7-day test. That is, the mean number of fronds in all control vessels must be 48 or more for the test to be considered valid. There is no validity criterion for dry weight. All recommended procedures and conditions shall have been followed.

9.5.2 Quality control
Laboratory procedures must be in place to ensure that environmental conditions for *Lemna* culture and testing are adequate. Charting of control performance will alert the experimenter to changes in physical or chemical conditions that may influence the sensitivity of the *Lemna*. For example, a gradual decrease in light output from fluorescent bulbs in the culture area may be reflected in decreasing frond production. Regular performance of reference toxicant tests, within 14 days of any testing or monthly should be performed. Descriptions of reference toxicant chart preparation are provided in Environment Canada (1990; 2004). Environment Canada's *L. minor* Biological Test Method (1999) recommends potassium chromate or dichromate (Cr VI), potassium chloride (KCl), and/or 3,5-dichlorophenol (DCP) as reference toxicants, however, subsequent research (Moody, 2003) has shown that Cr VI can be problematic, and Ni ($NiSO_4 \cdot 6H_2O$), is preferred. Control and reference toxicant charts for *L. minor* clone UTCC 492 are plotted with mean frond count in controls and IC25 for frond increase using Ni. In controls, the cumulative mean frond increase (95% confidence limits) is 66.2 (41.2 – 91.1) fronds (Fig. 2). Reference toxicant tests have a cumulative mean (and 95% confidence limits) of 13.0 (4.15 – 41.0) µg/L (Fig. 3).

10. Factors capable of influencing performance of test organism and testing results

Consistent and predictable performance of a test organism is desirable in toxicity testing so that toxicant effects can be reliably attributed to the toxicant alone. Culture health is a primary factor that will affect success or failure in testing. Use of high quality chemicals and deionized water and careful attention to culture handling is vital.

Selection of test and culture media must be appropriate to the objectives of the test. Use of test media containing large amounts of chelator will decrease the effect of metals. Modified APHA medium is recommended for use with metals and wastewaters from metal mines because it contains no chelator. Control of pH within narrow limits is also desirable for testing these substances. Modified APHA has been specifically designed for use in testing of mine effluents, which under the current Canadian Metal Mining Effluent Regulations (DFO, 2002) require the pH of discharged effluents to be between 6.0 and 9.5. Aeration of modified APHA test medium before use is carried out to equilibrate the bicarbonate and carbon dioxide in the medium with air. Final pH of controls in 18 tests carried out in 2002 and 2003 (after 7 days growth in modified APHA) had increased from 8.3 to 8.55 (Moody, unpublished data).

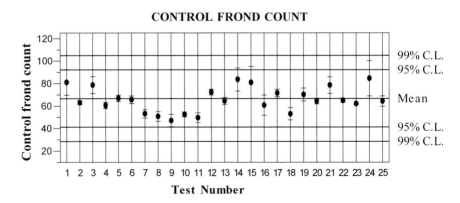

Figure 2. Control chart for Lemna minor *UTCC 492 showing mean frond count and 95% confidence limits in controls and 95% and 99% confidence limits of the cumulative mean of all tests.*

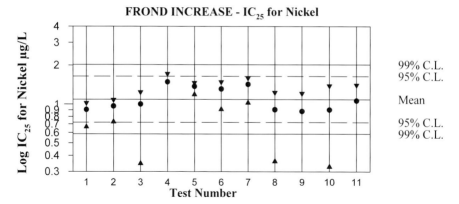

Figure 3. Control and reference toxicant charts for Lemna minor *UTCC 492 showing IC25 with 95% confidence limits and the mean IC25 with 95% and 99% confidence limits of the mean.*

11. Application (of the toxicity test) in a case study

Regulated mines in Canada are required to perform sublethal toxicity tests twice each year as part of an Environmental Effects Monitoring program under the Fisheries Act (DFO, 2002). This program specifies *L. minor* as one of four organisms (a fish, a plant, an invertebrate and an alga) to be tested by mines discharging effluent to the aquatic environment. The test protocol required is Environment Canada's *Biological Test Method: Test for Measuring the Inhibition of Growth Using the Freshwater Macrophyte, Lemna minor* (EC, 1999).

Testing of four metals was carried out at Saskatchewan Research Council (Moody, 2003) to provide baseline toxicity data for interpretation of *L. minor* tests. The toxicity of Cu, Cd, Ni and Zn to two *L. minor* clones obtained from the University of Toronto culture collection (UTCC 490 and 492) was compared. Stock and test cultures of *Lemna* were grown in modified Hoagland's E+ medium. All tests were preceded by range finding tests and met the validity criterion of mean control frond count of at least 48 (8-fold increase in fronds) after 7 days growth in modified APHA test medium. Metal toxicants solutions [zinc sulphate ($ZnSO_4 \cdot 7H_2O$); cadmium sulphate ($3CdSO_4 \cdot 8H_2O$); copper sulphate ($CuSO_4 \cdot 5H_2O$); nickel sulphate ($NiSO_4 \cdot 6H_2O$)] were prepared at 120 mM, diluted in deionized water (ASTM type 1) and added to test medium in appropriate concentrations using the same volume for each dilution. Controls containing an equivalent amount of deionized water were prepared. Frond counts and dry weights for each replicate were determined after 7 days. IC25 was calculated by linear interpolation (as described in section 9.4.1 of the protocol) using ToxCalc (Tidepool Scientific Software), version 5.0 (1994). Although at least three definitive tests were performed for each toxicant, one representative test result is reported for each *L. minor* clone (Tab. 6). Mean frond count in controls for each test is reported to demonstrate that each test met the frond increase validity criterion of at least 48 fronds in 7 days growth. In some cases, 95% confidence limits could not be calculated, presumably because toxicant concentrations did not elicit high enough % inhibition values.

Table 6. Metal tests of L. minor UTCC strains 490 and 492 (µg/L elemental metal).

Metal	Strain	Mean control fronds (7 d)	IC25 (95% confidence limits)	
			Frond Increase	Dry Weight
Zn	490	56.9	**63.3** (26.0 – 117)	**255** (0 - >981)
	492	83.5	**60.8** (38.2 – 112)	**188** (n/c)[a]
Cd	490	58.8	**16.7** (8.40 – 23.1)	**163** (81.2 – 228)
	492	72.3	**19.5** (0 – 34.4)	**198** (147 – 316)
Cu	490	58.8	**10.8** (3.50 – 30.7)	**415** (n/c)
	492	83.5	**10.8** (2.93 – 19.0)	**75.1** (0 – 370)
Ni	490	56.9	**13.6** (1.98 – 20.7)	**30.4** (8.63 – 52.0)
	492	83.5	**23.5** (15.4 – 29.2)	**50.9** (n/c)

a) not calculable

Growth of *Lemna* strain UTCC 492 was consistently greater than that of UTCC 490 at the end of 7 days. However, the toxicity of metals as IC25 was not significantly different between the two strains as demonstrated by overlapping 95% confidence limits. Dry weight is consistently a less sensitive endpoint than frond increase for these metals; however, this cannot be expected to be the case for all toxicants. Further applications of the test for metal mine effluents, herbicides, and other substances entering the aquatic environment will demonstrate the use of this test for hazard assessment.

12. Accessory/miscellaneous test information

Handling of test samples for *L. minor* testing should be standardized to ensure that collection, transportation, storage, preparation and testing are consistent with objectives of the test. Generally, a 4-litre sample of effluent or leachate is sufficient for a static non-renewal Environment Canada *Lemna* test that includes test concentrations of 97, 48.5, 24.3, 12.1, 6.1, 3.0, and 1.5%. Renewal of test solutions will require greater volumes of test sample and may require several daily collections. Testing of effluent and leachate samples should commence as soon as possible after collection, within 1 day whenever possible and no later than 3 days after sampling (EC, 1999). Samples should be kept cool (1 to 7°C, preferably 4 ± 2°C) during transport but must not freeze. Upon arrival, the temperature of the sample must be measured, an aliquot adjusted to test temperature, and the remaining sample stored without headspace at 4 ± 2°C.

The properties of a test chemical or chemical mixture should be obtained. These may include the concentration of the major constituents and impurities, water solubility, vapor pressure, chemical stability, dissociation constants, toxicity to humans and aquatic organisms and biodegradability. Where aqueous solubility is in doubt or problematic, acceptable procedures for the chemical preparation and solubility in test water may be obtained or determined experimentally (EC, 1999). Analysis of the test chemical in controls, high, medium and low test concentrations at the beginning and end of the test is recommended if an analytical procedure is available (see Section 9.3).

As mentioned in Section 3, a number of *Lemna* test methods are in use internationally for regulatory and research purposes. These methods are compared in detail in the appendix of Environment Canada (1999). For further detail refer to the specific methods, as updated. Details of three methods are compared (Tab. 7) with *Biological Test Method: Test for Measuring the Inhibition of Growth Using the Freshwater Macrophyte, Lemna minor* (EC, 1999, with minor modifications as described in this chapter). Included are the Standard Guide for Conducting Static Toxicity Tests with *Lemna gibba* G2, E1415-91 (ASTM, 1991), the Draft International Standard ISO/DIS 20079 Water quality - Determination of toxic effect of water constituents and waste water to duckweed (*L. minor*) – Duckweed growth inhibition test (ISO, 2003) and Guidelines for the Testing of Chemicals, Revised Proposal for a New Guideline 221, *Lemna* sp. Growth Inhibition Test (OECD, 2002).

13. Conclusions/prospects

Culture and testing with *L. minor* is cost-effective and simple to establish in a laboratory. Although regular testing of mine wastewaters and effluents is a recent application of the test in Canada, *Lemna* species have been used in environmental protection applications and by many international organizations. With ISO and OECD in the process of finalizing standardized tests using *Lemna* spp. (OECD, 2002; ISO, 2003), this test will continue to be used internationally for monitoring and protecting aquatic habitats. The surface-dwelling habit of *Lemna* is unique from the perspective that they are sensitive to aquatic as well as foliar exposures.

Physiological endpoints such as nutrient uptake (N, P, and C^{14}) may demonstrate sensitive toxicity responses in a shorter test. Advances in instrumentation such as the use of image analysis (ISO, 2003) make endpoints such as frond number, area and color convenient to determine, particularly for large-scale applications.

Acknowledgements

The assistance of Charlene Hudym and Leanne Crone of the Saskatchewan Research Council for document preparation and editing is appreciated.

Table 7 Comparison of Lemna test methods

	Environment Canada	ASTM	ISO	OECD
Test organism	Lemna minor	Lemna gibba G3	Lemna minor	Lemna gibba, Lemna minor
Type of test	7-day, static or static-renewal toxicity test	7-day, static	7-day, static	7-day, static, semi-static and flow-through
Test substances	Freshwater industrial or urban effluents, elutriates or leachates, chemicals, commercial products, mixtures of chemicals, freshwater surface or receiving waters	Chemicals, commercial products, known mixtures, aqueous effluents, leachates, oils, particulate matter, sediments, surface waters.	Substances and mixtures containing water, treated municipal waste water and industrial effluents	Chemicals
Test format	Glass or plastic cups or flasks, volume at least 100 mL, covers	Glass or plastic vessels such as beakers, Erlenmeyer flasks, jars, covers	Cylindrical vessels such as beakers, crystallizing dishes, Petri dishes, minimum volume 150 mL, covers	Glass (or other chemically inert material) vessels of minimum volume 100 mL and minimum depth of 20 mm, covers
Test medium	Modified APHA or SIS	Hoagland's (E+ or M-Hoagland's without sucrose or EDTA), 20X-AAP, other media capable of supporting an increase in biomass of at least 5-fold within 7 days in controls	Steinberg, SIS, modified APHA	Modified SIS for $L.\ minor$, 20X-AAP medium for $L.\ gibba$, Steinberg
Volume of test vessels	100 to 150 mL	200 to 500 mL, ratio of test chamber size to volume of test solution should be 5 to 2.	Minimum volume 100 mL	Minimum volume 100 mL
Lemna inoculum	Two 3-frond plants, axenic culture	Three to five plants totaling 12 to 16 fronds, axenic culture	Ten to sixteen fronds, from plants having 2 or 3 fronds, axenic culture	Nine to twelve fronds from plants consisting of 2 to 4 fronds, axenic culture
Lighting	24-h continuous full spectrum light at 63-72 $\mu mol/m^2/s$	24-h continuous warm white fluorescent light at 6200 to 6700 lux	24-h continuous full spectrum light at 85-125 $\mu mol/m^2/s$	24-h continuous warm or cool white light at 6500-10000 lux and photosynthetically-active radiation (400-700 nm) of 85-125 $\mu E/m^2/s$
Temperature	$25 \pm 2°C$	$25 \pm 2°C$	$24 \pm 2°C$	$24 \pm 2°C$

	Environment Canada	ASTM	ISO	OECD
Experimental configuration	Controls, 5-7 dilutions of test solution in geometric series, minimum of 3 replicates per treatment	Controls, at least five concentrations of test material, each at least 60% of the next higher one, minimum of 3 replicates per treatment	Controls, at least five concentrations in a geometric series, at least 3 replicates for test concentrations, six replicates for controls	Controls, at least five concentrations in a geometric series, at least 3 replicates
Endpoints	IC25, NOEC, LOEC, based on biomass (frond increase and dry weight)	IC50 (7 days), NOEC based on biomass (frond or plant number, dry weight, root length, fresh biomass, C-14 uptake, chlorophyll, total Kjeldahl nitrogen)	E_rC50 based on average specific growth rate, measuring frond number and at least one additional parameter (frond area, dry weight or chlorophyll)	EC_x, based on average specific growth rate, final biomass or area under the growth curve for the endpoint: frond number and at least one other parameter of biomass (total frond area, dry weight or fresh weight), LOEC, NOEC
Measurements	Frond count and dry weight at 7 days, pH at beginning and end in control, low, medium and high concentrations	One or more measurement of biomass at 7 days, pH at beginning and end in control, low, medium and high concentrations	Endpoints (frond number and one other) measured at 7 days, control frond number at least every 48 to 72 hours.	Frond and colony numbers at test start, frond numbers at least every 3 days for growth rate determination, and one or more endpoint (total frond area, dry weight, fresh weight), pH at beginning and end in each treatment, or upon renewal of test solutions
Reference toxicants	KCl, Ni (NiSO$_4$), 3,5-dichlorophenol (DCP)	Not described in method	3,5-dichlorophenol, KCl	3,5-dichlorophenol

References

AFNOR (Association Française de Normalisation) (1996) Determination of the inhibitory effect on the growth of *L. minor*, XP T 90-337, 10 pp.

APHA (American Public Health Association), AWWA (American Water Works Association), and WEF (Water Environment Federation) (1992) Toxicity, *Standard Methods for the Examination of Water and Wastewater*, 18th ed., APHA, AWWA, and WEF, Washington, DC, Part 8000, pp. 8-29 to 8-32.

APHA (American Public Health Association), AWWA (American Water Works Association), and WEF (Water Environment Federation) (1995) Toxicity, *Standard Methods for the Examination of Water and Wastewater*, 18th ed., APHA, AWWA, and WEF, Washington, DC, Part 8000, pp. 8-40 to 8-42.

ASTM (American Society for Testing and Materials) (1991) Standard Guide for Conducting Static Toxicity Tests with *Lemna gibba* G3, *Book of ASTM Standards*, E-1415-91, ASTM, Philadelphia, PA, pp. 1-10.

ASTM (American Society for testing and Materials) (1999) Standard Guide for Use of Lighting in Laboratory Testing, *Book of ASTM Standards,* E-1733-95, Vol. 11.05, *Biological Effects and Environmental Fate; Biotechnology; Pesticides*, ASTM, Philadelphia, PA, pp. 1279 – 1289.

Arber, A. (1963) Water Plants: A Study of Aquatic Angiosperms, Wheldon and Wesley Ltd. and Hafner Publishing Co., New York, NY.

Bishop, W.E. and Perry, R.L. (1981) Development and evaluation of a flow-through growth inhibition test with duckweed (*Lemna minor*), in D.R. Branson and K.L. Dickson (eds.), *Aquatic Toxicology and Hazard Assessment: Fourth Conference*, ASTM STP 737, American Society for Testing and Materials, pp. 421-435.

Blackman, G.E. and Roberson-Cuningham, R.C. (1955) Interrelationships between light intensity, temperature, and the physiological effects of 2:4-dichlorophenoxyacetic acid on the growth of *Lemna mino, Journal of Experimental Botany* **6**, 156-176.

Cowgill, U.M. and Milazzo, D.P. (1989) The culturing and testing of two species of duckweed, in U.M. Cowgill and L.R. Williams (eds.), *Aquatic Toxicology and Hazard Assessment: 12th Volume*, ASTM STP 1027, American Society for Testing and Materials, Philadelphia, PA, pp. 379-391.

DFO (Department of Fisheries and Oceans) (2002) Metal Mining Effluent Regulations, *Extract Canada Gazette*, Part II, Vol. 136, No. 13, June 19, 2002.

EC (Environment Canada) (1990) Guidance Document on the Control of Toxicity Test Precision Using Reference Toxicants, Environmental Protection Service, Report EPS 1/RM/12, Ottawa, ON, 85 pp.

EC (Environment Canada) (1999) Biological Test Method: Test for Measuring the Inhibition of Growth Using the Freshwater Macrophyte*, Lemna minor*, Environmental Protection Service, Report EPS 1/RM/37, Ottawa, ON, 98 pp.

EC (Environment Canada) (2004) Guidance Document on Statistical Methods to Determine Endpoints of Toxicity Tests, Environmental Protection Service, Report EPS 1/RM/xx, Ottawa, ON, in preparation.

Godfrey, R.K. and Wooten, J.W. (1979) *Aquatic and Wetland Plants of Southeastern United States*, University of Georgia Press, Athens, Georgia.

Hillman, W.S. (1961) The Lemnacea, or duckweeds. A review of the descriptive and experimental literature, *Botany Revue* **27**, 221-287.

ISO (International Organization for Standardization) (2003) Water quality – Determination of toxic effect of water constituents and waste water to duckweed (*Lemna minor*) – Duckweed growth inhibition test, Draft International Standard ISO/DIS 20079, Geneva, Switzerland.

ITM (Institute of Applied Environmental Research). (1990) Method for Toxicity Test with the Floating Plant *Lemna minor, Duckweed*, ITM report 7.

Jenner, H.A. and Janssen-Mommen, J.P.M. (1989) Phytomonitoring of Pulverized Fuel Ash Leachates by the Duckweed *Lemna minor*, *Hydrobiologia* **188/189,** 361-366.

Moody, M. (Unpublished) Mean pH after 7 days growth in 18 tests carried out in 2002 to 2003.

Moody, M. (2003) Research to assess potential improvements to Environment Canada's *Lemna minor* test method, Saskatchewan Research Council Publication No. 11545-1C03.

Newmaster, S.G., Harris, A.G. and Kershaw, L.J. (1997) *Wetland Plants of Ontario*, Lone Pines Printing and Queen's Printer for Ontario, Edmonton, AB.

Norberg-King, T.J. (1993) A Linear Interpolation Method for Sublethal Toxicity: the Inhibition Concentration (ICp) Approach (Version 2.0), U.S. Environmental Protection Agency, Environ. Res. Lab.-Duluth, Duluth, NM, Tech. Report 03-93 of National Effluent Toxicity Assessment Centre.

OECD (Organization for Economic Cooperation and Development) (2002) OECD Guidelines for the testing of chemicals, revised proposal for a new guideline 221*, Lemna* sp*.* Growth Inhibition Test (draft).

Smith, S. and Kwan, M.K.H. (1989) Use of aquatic macrophytes as a bioassay method to assess relative toxicity, uptake kinetics and accumulated forms of trace metals, *Hydrobiologia* **188/189**, 345-351.
SRC (Saskatchewan Research Council) (1997) The *Lemna minor* Growth Inhibition Test, Saskatchewan Research Council, Saskatoon, Saskatchewan.
SIS (Swedish Standards Institute) (1995) Water quality – determination of growth inhibition (7-d) *L. minor*, duckweed, Svensk Standard SS 02 82 13, 15 pp.
Taraldsen, J.E. and Norberg-King, T.J. (1990) New method for determining effluent toxicity using duckweed (*Lemna minor*), *Environmental Toxicology and Chemistry* **9**, 761-767.
Tidepool Scientific Software (1994) ToxCalc User's Guide: Comprehensive Toxicity Data Analysis and Database Software, Tidepool Scientific Software, 80 pp.
U.S. EPA (United States Environmental Protection Agency). (1996) Aquatic plant toxicity test using *Lemna* spp., Tiers I and II "Public Draft", *Ecological Effects Test Guidelines* OPPTS 850.4400, United States Environmental Protection Agency Prevention, Pesticides and Toxic Substances (7107) EPA 712-C-96-156, 5 pp.
Wang, W. (1987) Toxicity of nickel to common duckweed (*Lemna minor*), *Environmental Toxicology and Chemistry* **6**, 961-967.
Wang, W. (1990) Literature review of duckweed toxicity testing, *Environmental Research* **51**, 7-22.
Wang, W. and Freemark, F. (1995) The use of Plants for Environmental Monitoring and Assessment, *Ecotoxicology and Environmental Safety* **30**, 289-301.

Abbreviations

AAP	Algal Assay Procedure
AFNOR	Association Française de Normalisation
APHA	American Public Health Association
ASTM	American Society for Testing and Material
C^{14}	radioactive carbon 14
$CaCl_2$	calcium chloride
$Ca(NO_3)_2$	calcium nitrate
$CdSO_4$ (or $3CdSO_4 \cdot 8H_2O$)	cadmium sulphate
C.L.	control limits
$CoCl_2$	cobalt chloride
$Co(NO_3)_2$	cobalt nitrate
Cr	chromium
$CuCl_2$	copper chloride
$CuSO_4$	copper sulphate
DCP	3,5-dichlorophenol
DFO	Department of Fisheries and Oceans
EC	Environment Canada
ECx	Effective concentration for a (specified) percent effect
EEC	European Economic Community
EDTA	ethylenediamine tetraacetic acid ($C_{10}H_{16}O_8N_2$)
$FeCl_3$	ferric chloride
FIFRA	Federal Insecticide, Fungicide, and Rodenticide Act
H_3BO_3	boric acid
HCl	hydrochloric acid
%I	percent growth inhibition
ICp	Inhibiting Concentration for a (specified) percent effect

ICPIN	computer program developed by the U.S. EPA to estimate ICp by applying smoothing and interpolation (Norberg-King, 1993).
ISO	International Organization for Standardization
ITM	Institute of Applied Environmental Research
KCl	potassium chloride
kPa	kilopascal(s)
KH_2PO_4	potassium dihydrogen phosphate anhydride
K_2HPO_4	potassium phosphate
KNO_3	potassium nitrate
KOH	potassium hydroxide
LOEC	lowest-observed-effect concentration
M	average increase in frond number in the control test chambers
$MgSO_4$	magnesium sulphate
$MgCl_2$	magnesium chloride
MMER	Metal Mining Effluent Regulations
$MnCl_2$	manganese chloride
MOPS	4-morpholinepropane sulphonic acid
N	Normal
N	Nitrogen
NaCl	sodium chloride
Na_2CO_3	sodium carbonate
Na_2EDTA	disodium ethylenediamine tetraacetic acid ($C_{10}H_{14}N_2O_8 \cdot 2H_2O$)
Na_2MoO_4	sodium molybdate
$NaNO_3$	sodium nitrate
NaOH	sodium hydroxide
$NaHCO_3$	sodium bicarbonate
$NiSO_4$	nickel sulphate
NOEC	no-observed-effect concentration
NPDES	National Pollution Discharge Elimination System
OECD	Organization for Economic Development and Growth
SIS	Swedish Standards Institute
SRC	Saskatchewan Research Council
TSCA	Toxic Substances Control Act
UTCC	University of Toronto Culture Collection
USEPA	United States Environmental Protection Agency
X	increase in frond number in the test vessel
$ZnCl_2$	zinc chloride
$ZnSO_4$	zinc sulphate
$\mu mol/m^2/s$ (or $\mu E/m^2/s$)	micromole (micro Einsteins) per metre squared per second.

8. SPIROTOX TEST – *SPIROSTOMUM AMBIGUUM* ACUTE TOXICITY TEST

GRZEGORZ NAŁĘCZ-JAWECKI
Department of Environmental Health Sciences
Medical University of Warsaw
Banacha 1 str. 02-097 Warsaw, Poland
grzes@farm.amwaw.edu.pl

1. Objective and scope of the test method

Protozoa play an important role in the environment as primary consumers. With bacteria they are major organisms in water self-purification systems. They are attractive in ecotoxicology due to their short life cycle, ease of culturing and high susceptibility to toxicants. The Spirotox test can be used as a screening tool for toxicity assessment of various kinds of environmental samples such as freshwaters and drinking waters. It can also be used for monitoring the toxicity of effluents before and after purification steps, leachates and sediment pore water. In a battery of toxicity tests it can be used for evaluating the toxicity of pure chemicals including volatile compounds.

The Spirotox test is carried out in 24-well disposable microplate. The technique is very simple. It can be performed with conventional laboratory equipment on little bench space and at low cost. As *Spirostomum ambiguum* has impressive dimensions, the observed effects can be seen even without a microscope.

2. Summary of test procedure

Spirotox is a 24-hour microplate test undertaken with a very large ciliated protozoan *Spirostomum ambiguum*. The test is carried out in 24-well microplate (6 x 4 wells). In a single microplate 5 samples can be tested in a screening assay or one sample with 5 dilutions in a definitive assay. Two endpoints can be observed with the use of a dissection microscope: sublethal effects such as deformations, shortening and immobilisation of the cell of the protozoan and lethality. The Spirotox test is usually conducted for 24 hours, however in some cases prolongation of the test to 48 hours may significantly increase its sensitivity. For some samples (organic compounds), however, a short 2-hour test may be sufficient, as 2h-EC50 values can match those of 24h-LC50's.

Table 1. Rapid summary of the test procedure.

Test organism	Ciliated protozoan *Spirostomum ambiguum*
Type of test	acute, static
Test format	24-well disposable microplate (6 x 4 wells)
Volume contents of wells	1 ml of test solution
Test organism numbers	10 organisms per well
Incubation time	24 hours (optionally also 2 and 48 h may be used)
Incubation temperature	$25 \pm 2°C$ in darkness
Experimental configuration	a) screening test: 5 samples + control, each with 3 replicates
	b) definitive test: 5 dilutions of a sample + control, each with 3 replicates
Endpoints	a) sublethal deformations: EC50
	b) lethality: LC50
Reference toxicants	Cd^{2+} as $Cd(NO_3)_2$; Zn^{2+} as $ZnSO_4$; SDS (sodium dodecyl sulphate)

3. Overview of applications reported with the toxicity test method

Spirostomum ambiguum has been used as a test organism in Municipal Waterworks in Warsaw (Poland) for routine monitoring of the Vistula river since 1988. The test was first carried out in glass bakers, at room temperature and a light:dark photoperiod of 16:8 h. Since 1990 intensive studies on this protozoan have been undertaken in the Department of Environmental Health Sciences, Medical University of Warsaw. A simple microplate bioassay technique with the protozoan *Spirostomum ambiguum* was first presented during the 6th International Symposium on Toxicity Assessment and On-line Monitoring in Berlin (Germany) in 1993. Since that time the methodology has been improved and the test has been applied in several fields. Initially, growth and maintenance requirements were estimated. *S. ambiguum* was found to survive more than 48 hours in a wide range of pH and total hardness. One very important finding was that the protozoa were able to live in media containing non-detectable levels of dissolved oxygen. Additionally, *S. ambiguum* can be stored for several weeks at a wide range of temperatures from 5 to 28°C.

The evaluation of the sensitivity of *S. ambiguum* began with inorganic compounds. Spirotox uses an inorganic medium as a diluent and it does not require any food during the test. In comparison with other tests Spirotox was the most sensitive to heavy metals with 24h-EC50 as low as 4, 8 and 20 µg/L for copper, silver and mercury, respectively (Nałęcz-Jawecki and Sawicki, 1998). Additionally, 1h-EC50 results were comparable to 24h-LC50's (Nałęcz-Jawecki et al., 1995). In contrast to cations, *S. ambiguum* in a short test was insensitive to inorganic anions (Nałęcz-Jawecki and Sawicki, 1998). However, extending the test from 24 h to 96 h increased the toxicity of potassium dichromate, for example, more than 10 times.

Volatile compounds are an important class of environmental pollutants. Estimation of their toxicity in standard, multi-well test plates is not simple, due to some loss of the compounds during exposure and cross contamination of wells. The special Spirotox-volatile procedure was developed for evaluating the toxicity of organic compounds (Nałęcz-Jawecki and Sawicki, 1999). A short time after introducing the protozoa into the wells of the microplate, each well was impregnated with silicone fat, and the microplate was tightly closed with a polyethylene film. Since 2003 silicone fat and polyethylene film were replaced by an adhesive film. Following intensive studies on the toxicity of organics towards Spirotox, a database comprising more than 150 compounds was created (Nałęcz-Jawecki and Sawicki, 1999; 2002a; 2002b). The results of the Spirotox test were compared to 4 bioassays: Microtox®, *Tetrahymena pyriformis, Daphnia magna* and *Pimephales promelas*. The sensitivity of the ciliated protozoa *S. ambiguum* and *T. pyriformis* was similar for most of the tested compounds. For non-polar narcotics good correlation was found between the tests and 48h-EC50 values generated with the Spirotox test were generally 3-4 times higher than those of Microtox®, *D. magna* and *P. promelas*. In contrast, no correlation was found between the Spirotox assay and other tests for polar narcotics, electrophiles and weak acid respiratory uncouplers. In these groups some compounds proved to be much more toxic to the protozoan than to the other test organisms.

S. ambiguum can survive in a broad range of pH from 5 to 8. In this sense, it can be a valuable tool in evaluating toxicity/pH relationships. Our first paper in this field studied the relationship for nitrophenols (Nałęcz-Jawecki and Sawicki, 2003a). Such evaluations are helpful to estimate more adequate QSAR equations.

Apart from simple organics the Spirotox toxicity database contains the results for selected pesticides (Nałęcz-Jawecki et al., 2002a), drugs (Nałęcz-Jawecki and Sawicki, 2003b) and cationic surfactants (Nałęcz-Jawecki et al., 2003). In the group of 24 pesticides tested *S. ambiguum* was extremely sensitive to fungicides with 24h-EC50's of 2, 4 and 6 µg/L for dichlorofluanid, captan and thiram, respectively (Nałęcz-Jawecki et al., 2002a). The first report on the toxicity of pharmaceuticals towards freshwater protozoa showed that drugs used in the treatment of the human nervous system caused toxic effects towards *S. ambiguum* at concentrations lower than 1 mg/L (Nałęcz-Jawecki and Sawicki, 2003b). The Spirotox test was used as one element of a battery of four bioassays to investigate the biological activity of a new group of cationic surfactants (Nałęcz-Jawecki et al., 2003). The toxicity of tested compounds varied from 0.2 to 1 mg/L. Structure analysis studies showed that the presence of a long hydrophobic chain lowered the toxicity for the Microtox® test, but did not alter that in the protozoan test.

Traditionally, pulp and paper processing activities have been considered as a serious source of environmental pollution. Resin acids and phenolic compounds are two main classes of toxic compounds that were identified in effluents of this industrial sector. A battery of bioassays comprising Spirotox, Microtox® and Thamnotoxkit F™ was used for evaluating the biological activity of effluents from pulp and paper mills (Michniewicz et al., 2000; Nałęcz-Jawecki et al., 2000). The Toxicity Equivalency Unit (TEU) approach was applied for estimating the potential toxicity of individual compounds towards aquatic organisms. Spirotox was the most sensitive to resin acids, with the exception of 12,14-dichlorodehydroabietic acid, which was most toxic to Thamnotoxkit F™. Microtox® was much more sensitive than the crustacean and protozoan to phenols and chlorophenols. TEU's for 7 effluents were calculated based on the concentrations of the 17 major toxicants and the EC(LC)50 values for individual compounds. The effluents were toxic in all bioassays conducted, especially Microtox®. However, real toxicity expressed in toxicity units (TU) was much higher than the predicted TEU values suggesting that part of the toxicants remained undetected by chemical analysis.

Cyanobacteria can produce a wide range of toxins (Namikoschi and Rinehart, 1996). Toxic and non-toxic strains can be found together. There is no simple method to distinguish toxic cyanobacterial blooms from non-toxic ones. Historically, the mouse bioassay was used to evaluate the biological activity of cyanobacteria. A battery of six bioassays was used to examine the toxicity of cyanobacterial blooms from Central Poland (Nałęcz-Jawecki et al., 2002b; Tarczyńska et al., 2000; 2001). Spirotox and Thamnotoxkit F™ were the most sensitive bioassays and their toxicity results were correlated with the microcystin LR concentration. However, following a 2-year monitoring study (Tarczyńska et al., 2001), this correlation was not confirmed, though these bioassays were also the most sensitive.

Apart from environmental applications Spirotox was used for quality control of medical devices (Nałęcz-Jawecki et al., 1997). Protozoa are unique in that they are both eucaryotic cells as well as complete, unicellular, self-sufficient organisms (Ricci, 1990). Our preliminary results (data not published) showed that *S. ambiguum* were comparably sensitive to medical device extracts as the legal tissue culture tests with mammalian lymphocytes (method performed according to the Polish Pharmacopeia). There were no false positive samples for Spirotox, while 10% of samples proved toxic in the Microtox® test and non-toxic to mouse lymphocytes used in the legal test. Clearly, the protozoan test is not meant to replace legal tests, but it can be used as a screening tool for monitoring the production of medical devices.

4. Advantages of conducting the Spirotox test

The Spirotox test is a very simple acute bioassay conducted with a ciliated protozoan. Protozoa play an important role in natural and artificial ecosystems as primary consumers and are main components of water self-purification systems. Thus, they should be incorporated in a battery of bioassays. In the Spirotox test a simple mineral medium is used as a diluent and no food is added during the test. It minimises the influence of complexation and sorption of toxicant(s) to the components of medium and food particles. *Spirostomum ambiguum* is a very large protozoan 2-3 mm long, hence, scoring of test results is very simple even with the naked eye. Additionally, due to this "convenient size" of the protozoan, manipulation during the test is simple and transferring organisms to test vessels takes only a few minutes. The Spirotox test is carried out in disposable, standard multiwells. Dilutions of the sample are performed directly in the multiwell plate. In addition to a potentially short exposure time, testing in multiwells reduces the "consumption" of glassware and minimizes sample contamination and/or sorption to laboratory materials. Its short and simple procedure enables the initiation of more than 5 tests per hour.

S. ambiguum has simple environmental requirements (Tab. 2). It is not an anaerobic organism, but it can survive at a very low level of oxygen. Hence, different kinds of samples can be investigated including leachates and effluents with high TOC.

Table 2. Environmental requirements of S. ambiguum.

Parameters	
pH	5.0 – 8.0
Dissolved oxygen (% saturation at 25°C)	0 – 100
Total hardness ($CaCO_3$ mg/L)	0.3 – 250
Salinity (NaCl mg/L)	3 – 1100
Temperature (°C)	5 – 28

5. Test species

Spirostomum ambiguum is a very large ciliated protozoan, 2 –3 mm long, easily seen with the naked eye (Fig. 1). It has been used as a test organism for nearly one hundred years (Czerniewski et al., 1935; Seyd, 1936). Due to its size it has been a very useful organism in studies on regeneration (Seyd, 1936) and cytological and microscopic observations (Finley et al., 1964). Due to its high sensitivity to chemical, mechanical or electrical stimulation, it was a valuable tool in physiological studies (Rostkowska and Moskwa, 1968; Ettienne, 1970; Applewhite, 1972; Jones, 1966). *S. ambiguum* can be easily cultured in laboratory at low cost and with modest bench space.

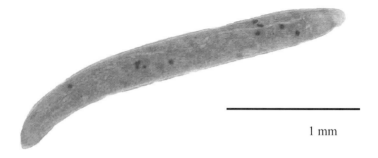

Figure 1. Spirostomum ambiguum.

5.1 TAXONOMY

Phylum	Protozoa
Class	Ciliata
Order	*Spirotricha Bütschli*
Suborder	*Heterotricha Stein.*
Species	*Spirostomum ambiguum Ehrbg.*

S. ambiguum lives in small forest ponds. It has been observed in Kampinos National Park in Central Poland and in Central France (Grolière and Njine, 1973). The strain described in this paper has been cultured in Municipal Waterworks in Warsaw for more than 25 years and in our department since 1990. The strain can be obtained from the Department of Environmental Health Sciences, Medical University of Warsaw, Banacha 1, str., 02-097 Warsaw, Poland. It can survive without food up to 6 weeks within a 5 – 28°C range.

6. Culture / maintenance of organisms in the laboratory

All culturing, maintenance and toxicity testing areas should be free of potential toxicant input. Culture of the protozoa must be separated from the toxicity testing area. *Spirostomum ambiguum* is very sensitive to heavy metals especially silver and copper. Special attention should be given to water systems in terms of tap water filters and tap water installation. Do not use silver filters! If the tap water installation is made of copper, carefully rinse all materials and equipment with glass distilled or de-ionized water.

Spirostomum ambiguum can be cultured in a 5 L aquarium containing 4 L of natural, unpolluted water. Cultures should be maintained at room temperature (15 – 25°C) in darkness or in a dim light.

6.1 MATERIALS

Materials required for culturing and testing *S. ambiguum* are listed below.

Box 1. *Materials required for the culturing and testing of* S. ambiguum.

Glass aquaria or beakers: 5 to 10 L capacity.
Food: Flaked oats + dried alder leaves (50:1). Alder leaves are used to prevent the development of fungi. Other leaves with a high level of tannin could be used.
Graduated glass pipettes: 1 and 10 mL.
Adjustable automatic pipette (1 mL).
Automatic pipette (1 mL).
Polyethylene or glass micropipettes (0.5 mL) for transferring the test organisms.
Polystyrene or glass Petri dishes (10 x 50 mm and 15 x 100 mm).
Disposable, rigid, polystyrene microplates 24-well (6 x 4). Capacity of well ~3.5 mL.
Adhesive sealing film for multiwell plates should be used only for testing volatile compounds.
Beakers: 100, 250 and 1000 mL capacity.
Graduated cylinders: 25 and 100 mL capacity.
Volumetric flasks: 100, 200 and 1000 mL capacity.

6.2 EQUIPMENT

All equipment should be adequately maintained and regularly calibrated. Any equipment in contact with protozoa, reagents, test samples, etc. must be made of chemically inert material: glass, stainless steel and plastic, and of course, clean and free of substances which could interfere with testing.

Box 2. Equipment required for the culturing and testing of S. ambiguum.

Laboratory incubator set at $25 \pm 2.0°C$. Do not use incubators made of copper.
Refrigerator.
Dissection microscope with a magnification of 8 x.
pH-meter.
Millipore Super – Q™ water purification system or equivalent (glass distilled).
Analytical balance for weighing chemicals.
Magnetic stirrer.
Vortex.
Thermometer.

6.3 WASHING OF GLASSWARE

All reusable glassware must be cleaned carefully. Any glassware used in culture/ maintenance of the organism should be washed without cationic detergents. In our department an acid detergent is used as cleaning medium. Then, glassware is rinsed twice with tap water (from a non copper installation!) and rinsed with de-ionized water. Finally, the glassware is oven dried at 105°C. New glassware should be rinsed with acid (5% HNO_3) and then treated as above.

<u>Avoid copper and silver!</u> Use only glass or stainless steel materials. Do not use water filters made of silver. If the tap water installation in your lab is made of copper, use distilled water to rinse the glassware.

For final rinsing, de-ionized water should be used. It may be replaced with glass distilled water, but not with distilled or double distilled water, where metal containers are used.

6.4 PREPARATION OF REAGENTS AND CULTURE MEDIA

6.4.1 Culture medium

The protozoa are cultured in natural water. Different sources of water were evaluated, but no good artificial water has been found so far. Each laboratory should therefore choose a suitable water source. The chemical composition of water used in our laboratory is shown in Table 3.

Protozoa are grazed on bacteria that are fed with a mixture of flaked oats and dried alder leaves. Flaked oats can be bought in health food markets (use oats with no preservatives!). Alder leaves are added in order to prevent the development of fungi. Other leaves with a high level of tannin could be used. In our laboratory alder leaves are collected in natural forests in spring. Then, they are dried at 60°C and comminuted. The food is prepared by mixing 1 g of leaves with 50 g of flaked oats.

Table 3. Physico-chemical composition of water sources used for culturing S. ambiguum.

Parameters	Source 1: 230 m deep	Source 2: 330 m deep
pH	7.0	6.7
Hardness ($CaCO_3$ mg/L)	104	89
Conductivity (mS)	0.76	0.66
Cl^- (mg/L)	102	94

6.4.2. Diluent

All chemicals must be of analytical grade quality. Millipore Super Q™ or glass-distilled water must be used for preparation of reagents.

As a diluent, a diluted Tyrod solution is used. Per L, it comprises: 125 mg NaCl, 3.13 mg KCl, 3.13 mg $CaCl_2$, 1.56 mg $MgCl_2$, 15.63 mg $NaHCO_3$ and 0.78 mg NaH_2PO_4. Two 100-fold stock solutions are prepared with the reagents listed below (all salts anhydrous). To start off, label two 200 mL volumetric flasks: Tyrod 1 and Tyrod 2, then add 150 mL of water to each.

Weigh each chemical and add individually to each solution. Ensure that each chemical is dissolved prior to adding the next chemical. Then, adjust the volume of solutions to 200 mL with water. Stock solutions can be stored at 4°C up to 2 months.

Box 3. Ingredients of stock solution Tyrod 1.

Sodium chloride	NaCl	2.50 g
Potassium chloride	KCl	62.6 mg
Calcium chloride	$CaCl_2$	62.6 mg
Magnesium chloride	$MgCl_2$	31.2 mg

Box 4. Ingredients of stock solution Tyrod 2.

Sodium bicarbonate	$NaHCO_3$	312.6 mg
Sodium phosphate	NaH_2PO_4	15.6 mg

6.4.3 Diluent – Tyrod solution (Tyrod)

Normal strength Tyrod solution is prepared by adding 10 mL of each stock Tyrod solutions (1 and 2) to a 1 L beaker filled with about 900 mL of water. Mix well between each addition. Use moderate mixing with the magnetic stirrer and adjust the final pH to 7.4 ± 0.2 with 1N HCl or 1N NaOH. Then the volume of the solution should be adjusted to 1 L in a volumetric flask.

6.4.4 Procedure of culturing *Spirostomum ambiguum*

Spirostomum ambiguum is not cultured under axenic conditions. Hence, no sterilisation of test containers and media is required. Culture should be carried out at room temperature (15-25°C), in darkness or in a dim light.

New cultures of *S. ambiguum* should be started in small, 250 mL beakers, then transferred to 1 L beakers and finally to 5 L aquaria.
- Pour 200 mL of culture water into a 250 mL beaker and add 200 mg of food.
- Cover the beaker with Petri dish and leave it at room temperature for 2 days.
- Inoculate the culture with the protozoa. The organisms are sent to the laboratory in ampoules in culturing medium, which also contain an inoculum of bacteria. Whole contents of the ampoule should be poured into the beaker.
- Once a week add 100 mg of food to the beaker.

After 4 weeks transfer the culture to a 1 L beaker, prepared as follows:
- Pour 600 mL of culture water into the beaker and add 400 mg of food.
- Cover the beaker and leave it at room temperature for 2 days.
- Inoculate the protozoa by pouring the contents of the small beaker to the 1 L beaker.
- Once a week add 100 mg of food to the beaker.

After 4 weeks transfer the culture to a 5 L aquarium, prepared as follows:
- Pour 3.5 L of culture water into a 5 L aquarium and add 1 g of food.
- Cover the aquarium and leave for 2 days.
- Inoculate the protozoa by pouring ½ of the culture from the 1 L beaker into the aquarium.

6.4.5 Maintenance of the culture

A minimum of three aquaria should be operating at the same time. Every working day the culture should be observed carefully. The culture is healthy if the protozoa are swimming in the whole volume of medium or if they are "grazing" on the flaked oats. Twice a week the protozoa should be observed under the dissection microscope and pH of the medium should be measured. The change of pH must not be greater than 0.5. Every week ½ of the water should be replaced with a fresh supply and 0.5 g of food should be added. Every month a new aquarium should be prepared and inoculated with 500 mL of the old culture.

7. Preparation of protozoa for testing

Before testing the protozoa should be separated from the culture medium. Using a 10 mL glass pipette transfer the dense culture of protozoa from the bottom of the aquarium to a 25 mL capacity graduated cylinder. Fill the cylinder with Tyrod solution. Wait a few minutes until the protozoa drop to the bottom of the cylinder. (Do not wait too long! After an additional few minutes the protozoa will start swimming in the whole medium and you will have to start from the beginning). Then carefully pour out as much medium as you can. Fill the cylinder with the Tyrod solution and repeat the rinsing three times. Finally transfer the suspension of cells in the Tyrod solution to a small Petri dish (50 mm of diameter).

8. Testing procedure

8.1 INFORMATION/GUIDANCE REGARDING TEST SAMPLES PRIOR TO CONDUCTING BIOASSAYS

8.1.1 Chemicals

For the health and safety of laboratory personnel, physical, chemical and toxicological (if available) properties of the substance(s) to be tested should be obtained. Stock solutions of each substance should be prepared in MilliQ water. Stock solutions of chemicals not readily soluble in water may be prepared by using organic solvents, *e.g.,* methanol, acetone and DMSO (dimethylsulphoxide). If solvent is used to dissolve a chemical in preparation for testing, an additional solvent control must be incorporated into the experiment at the highest concentration used.

If the pH of a stock solution is outside the 5-8 pH range, it must be adjusted to the nearest border (*i.e.*, samples with an initial pH lower than 5 should be adjusted to 5 and samples with an initial pH exceeding 8 should be adjusted to 8).

For volatile compounds a special procedure should be performed.

8.1.2 Environmental samples

Environmental samples should be collected according to standard procedures. Samples are usually placed in clean, labelled containers of inert material, filled to the brim (minimal headspace) and transported in the dark on ice. Twenty mL of sample are sufficient for conducting the Spirotox test from the range-finding to the definitive test, although it is recommended to collect 0.5-1 L of sample if physico-chemical measurements are also to be made. Environmental samples should be tested as soon as possible but no longer than 3 days after collection. Prior to testing, samples should be stored in a refrigerator at 4°C.

As the Spirotox test is based on visual observations of protozoa under the dissection microscope, suspended solids and coloured samples are not sources of interference. Hence, there is no necessity to filter the sample. If the pH of a sample lies outside a 5-8 pH range, it must be adjusted to the nearest border (*i.e.*, samples with an initial pH lower than 5 should be adjusted to 5 and samples with an initial pH exceeding 8 should be adjusted to 8).

8.2 SELECTING A TESTING PROCEDURE

The experimental procedure depends on the type of sample and objective of the assay. Two main procedures can be performed utilizing a screening assay and one requiring dilutions. A screening test (see Section 8.4) helps to identify toxic samples from a large number of (possibly non toxic) samples for further definitive assays. For example, a screening test should be carried out for evaluating the toxicity of drinking water sources and/or low-contamination freshwaters. If the toxicity of a sample is unknown and unpredictable, a range-finding dilution assay (see Section 8.5) should be performed with 1 L dilutions ranging from 100 to 0.1%. If the approximate toxicity of a sample is known, a definitive dilution test (see Section 8.6) can be carried out and EC(LC)50 values estimated.

8.3 SELECTING TEST CONCENTRATIONS

If the approximate toxicity of a sample to *S. ambiguum* is known after a screening test, test concentrations are prepared that will encompass a range of responses from 0% to 100%.

For chemicals, whose toxicity is unknown, stock solutions should be prepared at the following concentrations.
- For water-soluble compounds the highest tested concentration is usually 100 mg/L. Only for special purposes (*e.g*, solvents tests) will concentrations exceed 1000 mg/L.
- For compounds of low solubility in water the concentrations should be close to their solubility limit.
- If an organic solvent is used, the stock solution should be at least 50 times more concentrated that the highest test concentrations. The concentration of the solvent in the test should not be higher than the NOEC (no observed effect concentration). The NOEC for methanol, ethanol and acetone for Spirotox is 2%.

In the range-finding test, numerous concentrations are assayed (*e.g.,*: 100; 50; 25; 12.5; 6.25; 3.12; 1.56; 0.78; 0.39; 0.20; 0.10%). From the results of the range-finding test, the concentrations for the definitive test should be chosen. Under ideal conditions the test should include at least one concentration that will have no effect and at least one concentration that will kill all the protozoa. However, EC(LC)50 values can be estimated if at least one concentration causes a toxic effect below 50% and at least one above 50%.

8.4 SCREENING TEST

The test design incorporates one concentration of the undiluted sample in 3 replicates. Five samples (and control) can be assayed in one multiplate (Fig. 2). A rinsing row serves to prevent dilution of the toxicant during the transfer of test organisms from a Petri dish to the test wells.

	1	2	3	4	5	6
A	Sample 1	Sample 2	Sample 3	Sample 4	Sample 5	Control
B	Sample 1	Sample 2	Sample 3	Sample 4	Sample 5	Control
C	Sample 1	Sample 2	Sample 3	Sample 4	Sample 5	Control
D	Sample 1	Sample 2	Sample 3	Sample 4	Sample 5	Control

R I N S I N G   R O W

Figure 2. Configuration of a microplate in the screening test.

The screening test procedure is explained in Box 5.

Box 5. Screening test.

Prepare a sufficient number of multiwells (one per 5 samples).

Dispense 1 mL of each tested sample into all 4 wells of one column.

Dispense 1 mL of Tyrod solution into the wells of column No 6 in each multiplate.

Transfer the protozoa into the wells with a micropipette. A glass or a plastic micropipette may be used. Transfer is usually carried out under a dissection microscope at a magnification of ~8 x. It is also possible to use a magnifying lens.

Place the Petri dish with the rinsed protozoa (see Section 7) under the dissection microscope. While looking at the end of the micropipette, catch approximately 40 protozoa. Dispense this number into each well of rinsing row (D). Do not touch the sample with the pipette to avoid contamination! If that happens change or rinse the pipette.

Transfer exactly 10 protozoa from each rinsing well to the corresponding wells in each column of the microplate. Change micropipette or rinse it with Tyrod after each sample.

8.5 RANGE-FINDING TEST

The range-finding test is a preliminary assay designed to establish the approximate toxicity of an unknown sample. Test design incorporates a control and eleven 2-fold dilutions of tested sample in two replicates. Figure 3 illustrates the experimental disposition of a microplate.

	1	2	3	4	5	6
A	Sample 100 %	Sample 50%	Sample 25%	Sample 12.5%	Sample 6.25%	Sample 3.12%
B	**Control**	Sample 0.10%	Sample 0.20%	Sample 0.39%	Sample 0.78%	Sample 1.56%
C	Sample 100 %	Sample 50%	Sample 25%	Sample 12.5%	Sample 6.25%	Sample 3.12%
D	**Control**	Sample 0.10%	Sample 0.20%	Sample 0.39%	Sample 0.78%	Sample 1.56%

Figure 3. Configuration of microplate in the range-finding test.

The insertion of the control between the highest concentration of sample is meant to check for the presence of volatile compounds. If this occurs, control mortality is observed. In this case a special Spirotox-volatile procedure should be performed.

The range-finding test procedure is explained in Box 6 and schematised in Figure 4. Sample dilutions are prepared directly in the multiplate.

Box 6. Range-finding test.

Dispense 1 mL of Tyrod into all wells of the microplate with the exception of A1 and C1.
Dispense 1 mL of a tested sample into A1, A2, C1 and C2.
Using the same pipette and pipette tip mix the contents of A2 by withdrawing and dispensing the sample 5 consecutive times.
Transfer 1 mL from A2 to A3 and mix contents with the pipette. Continue this process until A6, then transfer 1 mL from A6 to B6, and continue this process until B2. Discard 1 mL of sample from B2 to a waste container. B1 is the control well.
Repeat this dilution procedure in rows C and D.
Transfer protozoa into the wells with a micropipette. A glass or a plastic micropipette may be used. The transfer is usually carried out under a dissection microscope at a magnification of ~8 x. It is also possible to use a magnifying lens.
Place the Petri dish with the rinsed protozoa (see Section 7) under the dissection microscope. While looking at the end of the micropipette, catch exactly 10 protozoa. Drop them into each well starting from the controls. Do not touch the sample with the pipette to avoid contamination. If that happens change or rinse the pipette.
Verify that 10 organisms are actually in each well of the microplate.

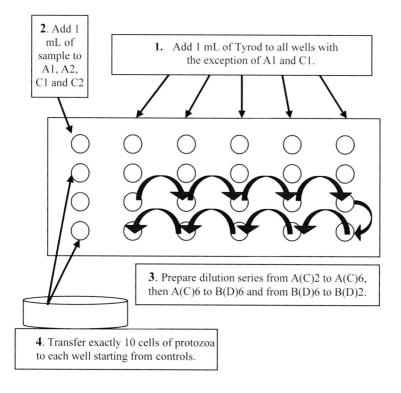

Figure 4. Range-finding test.

8.6 DEFINITIVE TEST

The definitive test is an assay designed to establish the precise toxicity of a sample. Test design incorporates a control and five 2-fold dilutions of tested sample with three replicates. Figure 5 displays the experimental disposition of a microplate.

	1	2	3	4	5	6
A	Sample 100 %	Sample 50%	Sample 25%	Sample 12.5%	Sample 6.25%	Control
B	Sample 100 %	Sample 50%	Sample 25%	Sample 12.5%	Sample 6.25%	Control
C	Sample 100 %	Sample 50%	Sample 25%	Sample 12.5%	Sample 6.25%	Control
D	Sample 100 %	Sample 50%	Sample 25%	Sample 12.5%	Sample 6.25%	Control

R I N S I N G   R O W

Figure 5. Configuration of microplate in the definitive test.

A rinsing row serves to prevent dilution of the toxicant during the transfer of the test organisms from a Petri dish to the test wells. The procedure of the definitive test is explained in Box 7 and schematised in Figure 6. Sample dilutions are prepared directly in the multiplate.

Box 7. Definitive test.

Dispense 1 mL of Tyrod into all wells of the microplate with the exception of column 1.

Dispense 1 mL of a tested sample into wells in the columns 1 and 2.

Using the same pipette and pipette tip, mix the contents of A2 by withdrawing and dispensing sample 5 consecutive times.

Transfer 1 mL from A2 to A3 and mix contents with the pipette. Continue this process until A5. Discard 1 mL of sample from A5 to a waste container. A6 is the control well.

Repeat this dilution procedure in rows B, C and D.

Transfer protozoa into the wells with a micropipette. A glass or a plastic micropipette may be used. The transfer is usually carried out under a dissection microscope at a magnification of ~8 x. Yet it is also possible to use a magnifier.

Place the Petri dish with the rinsed protozoa (see Section 7) under the dissection microscope. Looking at the end of the micropipette, catch approximately 40 protozoans. Dispense them into each well in the rinsing row (D). Do not touch the sample with the pipette to avoid contamination! If that happens change or rinse the pipette.

Transfer exactly 10 protozoa from the rinsing wells to the corresponding wells in the column of the multiwell starting from the control.

8.7 EXPOSURE CONDITIONS

Experimental microplates are placed in an incubator set at 25 ± 2°C, without illumination. Total exposure time is 48 h.

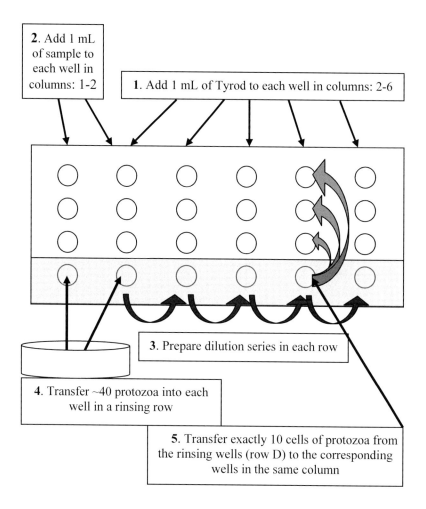

Figure 6. Definitive test.

9. Post-exposure observations/measurements and endpoint determinations

9.1 ENDPOINTS

Two toxicity effects can be observed with the use of a dissection microscope (magnification of 8 x).

- Sublethal responses: such as bending, shortening of the cell and immobilisation of the protozoan. Some deformations of the protozoan are presented in Figure 7.
- Lethal response: spherical deformation and autolysis. After autolysis the protozoa disappear, so one must be sure that they were added into the well of the microplate! A dead *S. ambiguum* is shown in Figure 8.

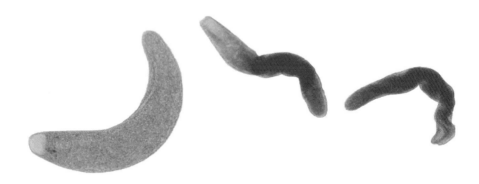

Figure 7. Sublethal deformations of Spirostomum ambiguum.

Figure 8. Autolysis of Spirostomum ambiguum.

9.2 SCORING THE RESULTS

- Place the multiwell under the dissection microscope.
- Check all the wells of row A, B and C and record the number of living normal and deformed (def) protozoa in each well. Subtract the number of living (normal and deformed) cells from 10 = the number of dead organisms (let). Report the number of dead and deformed protozoa on the data sheet. Keep in

mind that if the protozoan is scored as dead, it was first deformed and should also be considered as deformed (see example below).
− Calculate the % of sublethal responses (% def) and lethal responses (% let) in each column of the microplate.

Table 4. Scoring of the results – an example.

Row	Normal protozoa	Deformed protozoa (def)	Dead protozoa (let)
A	3	3	10 − (3+3) = 4
B	4	2	10 − (4+2) = 4
C	3	4	10 − (3+4) = 3
Effects (Σ)	-	def = 9 + 11	let = 11
Effects (%)	-	% def = (9+11)/30 x 100 = 67%	% let = 11/30 x 100 = 37%

Table 5. Spirotox data sheet.

24 h	100%		50%		25%		12.5%		6.25%		Control	
	def	let	def	let	def	let	def	let	def	let	def	let
A	-	10	5	5	2	1	1	-	-	-	-	-
B	-	10	2	8	5	2	-	1	-	-	-	1
C	-	10	7	3	3	-	1	-	2	-	1	-
Effects (Σ)	0+30	30	14+16	16	10+3	3	2+1	1	2+0	0	1+1	1
Effects (%)	100	100	**100**	**53**	**43**	**10**	10	3	7	0	7	3

9.3 ENDPOINT DETERMINATION

The endpoint reported depends on the type of test. In the screening test percent of effects (% let and % def) caused by the sample are presented. In the definitive test EC50 and LC50 values are typically calculated. An EC50 is a sample concentration causing 50% of sublethal effects. An LC50 is a sample concentration causing 50% mortality of exposed organisms.

9.3.1 Screening test
− If the % def value is lower than 20% the sample is considered non toxic.
− If the % def value is between 20 and 50% the sample is considered somewhat toxic.
− If the % def value is greater than 50% the sample is considered toxic and a

range-finding and/or definitive test should be performed to estimate the EC(LC)50 value.

9.3.2 Definitive test
Calculate the % def and % let values and report results on the data sheet.

There are several procedures for calculating EC(LC)50 values. Methods used to estimate the EC(LC)50 from multi-concentration tests depend on the number of partial deformities (sublethal effects) and mortalities (lethal effects) observed. *S. ambiguum* rarely gives partial mortality in more than one concentration. A simple, graphical procedure described below is sufficient in most cases. It is based on the U.S. EPA method (Weber, 1993). The procedure is described below in Box 8 and in Figure 9. If the results are scored not only after 24 h but also after 48 h and/or 2 h, the calculation should be made for each time period.

Box 8. Graphical procedure for estimating EC(LC)50 values.

Choose two % def values: one lower than 50% and the other greater than 50%. See bolded values in Table 5.

Indicate the concentrations on the Y-axis and corresponding % def values on the X-axis.

Connect the plotted points with a straight line.

Read the EC50 value at the intersection of the plotted line and the vertical 50% effect line.

Estimate the LC50 value in the same way.

EC50 and LC50 values from data presented in Table 5 are 27% and 48%, respectively.

This graphical procedure can be carried out with any computer programme that allows calculation of log values (*e.g.*, MS Excel). The macro can be obtained from the author of this chapter (grzes@farm.amwaw.edu.pl).

9.4 CONDITIONS FOR VALIDITY AND BUILT-IN QUALITY CONTROL

9.4.1 Conditions for test validity
The test is valid if toxic effects (both deformations and lethal effects) observed in control wells do not exceed 10%.

9.4.2 Built in quality control
Deviation from normalcy (in the case of a test result with a reference toxicant) may indicate a change in laboratory performance (health of test organisms, culture, contamination, faulty diluent, improper washing of glass or procedural error). Three reference toxicants were chosen for the Spirotox test:

Cd^{2+} as $Cd(NO_3)_2$; Zn^{2+} as $ZnSO_4$; SDS (sodium dodecyl sulphate)

The reference tests should be performed with the same batch of protozoan culture and the same batch of Tyrod stock solutions.

Reference toxicant data should be within ±2 standard deviations of values obtained in previous tests. Data based on 50 experiments performed by the Department of Environmental Health Sciences, Medical University in Warsaw, Poland, yielded the average values shown in Table 6.

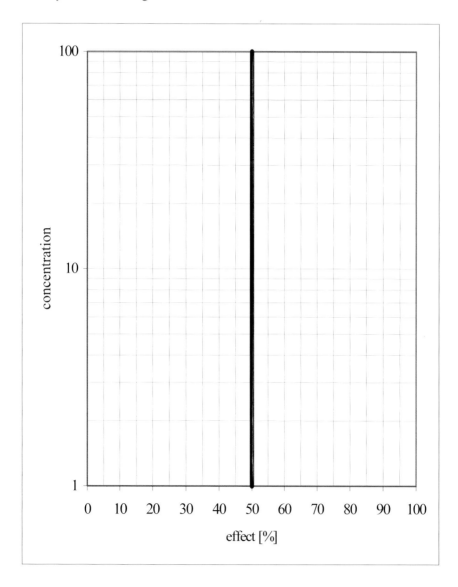

Figure 9. Graphical interpolation sheet.

Table 6. Reference toxicant data for 24h LC50's.

Reference toxicant	Mean (mg/L)	± SD	Range (± 2 SD)
Zn^{2+}	0.472	0.101	0.270 – 0.674
Cd^{2+}	0.224	0.050	0.124 – 0.324
SDS	6.92	1.83	3.26 – 10.58

10. Factors capable of influencing performance of test organism and testing results

Toxicity tests, regardless of their degree of standardisation, can be subject to various factors linked to procedure that will limit their applications if they are not properly addressed. Some such factors may be excluded by application of special procedures (volatile compounds). Factors having the potential to adversely influence testing results are briefly recalled and commented below.

10.1 ADHERENCE OF CHEMICALS TO WELLS

Because small 1 mL wells are used, high area to volume ratio can cause great toxicant adhesion on the test containers. The problem is most crucial, when very low concentrations of toxicants are tested. Some researchers do not use plastic microplates due to potential toxicant adhesion on wells. Similarly, toxicants can adhere to glass beakers (*e.g.*, metals), especially when beakers are used repeatedly. Since affinity for plastic or glass is chemical-dependent, no material is always the most convenient.

10.2 VOLATILE COMPOUNDS

For testing volatile compounds the special Spirotox-volatile procedure should be applied, in which each well is tightly closed with a plastic film.

10.3 LOW WATER-SOLUBLE COMPOUNDS

In testing low water-soluble compounds some precipitation of the tested substance can occur. This can lead to two specific problems: increased toxicity due to suspension uptake by protozoa or lower toxicity due to a decreasing concentration of the substance. If precipitation is linked to evaporation of an applied organic solvent, the Spirotox-volatile procedure can be employed. If no obvious reason can be accounted for, any noted precipitation effect should be reported.

10.4 MISCELLANEOUS FACTORS

All types and brands of 24-well polystyrene microplates may not be adequate for the test. Some microplates may be toxic to the protozoa. It is highly recommended to

11. Application examples (case studies) with the protozoan toxicity test

An application of the Spirotox test described in this chapter was first presented during the 7th Meeting of the Central and Eastern European Regional Section of SECOTOX (Society of Ecotoxicology and Environmental Safety) in Brno in the Czech Republic (Nałęcz-Jawecki et al. 2002c).

During the 1960's and 1970's unwanted pesticides were deposited in several hundreds of tombs all over Poland. After a few years the tombs started to leach out chemicals into ground water. During liquidation and remediation works 30 ground water samples were collected in the vicinity of 8 tombs. Their toxicity was evaluated with the following battery of bioassays: Microtox®, Spirotox, Protoxkit F™, Thamnotoxkit F™ and Daphnia test. First, screening tests were performed (Tab. 7) followed by definitive tests with the toxic samples (Fig. 10). EC(LC)50 results were then transformed into toxic units [TU = 100%/EC(LC)50].

Fifty seven percent of samples were not toxic according to the Spirotox test. Four samples (13%) caused toxicity effects between 20 and 50%, and were considered somewhat toxic. Nine samples (30%) were toxic in the Spirotox test and a definitive test was then performed (Fig. 10).

Table 7. Evaluation of ground water sample toxicity in the Spirotox screening test.

Sample	% def	Sample	% def	Sample	% def
M1	10	R1	13	S1	**100 (T)**
M2	**100 (T)**	R2	0	S2	**100 (T)**
M3	13	R3	3	S3	**100 (T)**
B1	**100 (T)**	P1	0	W1	**100 (T)**
B2	7	P2	7	W2	0
B3	10	P3	0	W3	10
B4	17	P4	**100 (T)**	K1	**100 (T)**
B5	**90 (T)**	D1	30 (ST)	K2	45 (ST)
B6	0	D2	16	K3	30 (ST)
B7	3	D3	13	K4	33 (ST)

T – toxic sample
ST – somewhat toxic sample
def - deformities

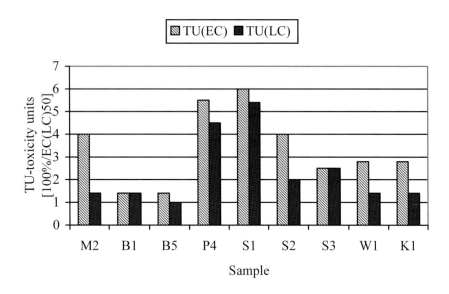

Figure 10. Toxicity of ground water samples in the definitive Spirotox test.

12. Accessory/miscellaneous test information

Laboratory personnel need not be specifically educated in biological/microbiological techniques, but must be trained in analytical techniques such as weighing, pipetting etc. Considerations for safety must be in place prior to carrying out tests with toxic substances and effluents, which may be not only toxic, but also infectious.

In terms of applicability, the Spirotox test can be carried out on any liquid medium, including coloured samples, suspensions and samples with low dissolved oxygen levels. Its use was reported for effluents, ground and surface waters, solid waste leachates and different extracts. At present, most of the published data concern the sensitivity of the test towards inorganic and organic compounds.

An experienced operator can easily initiate 5 microplate tests per hour. Twenty samples can be diluted and microplates filled in one-half of a work day. Post exposure counting is simple and takes only a few minutes per microplate, so scoring results and data reduction can be done during another one-half day. Hence, a batch of 20 microplates could be processed daily, depending on the time necessary for sample preparation (*e.g.*, adjusting of pH).

Costs to process 20 samples depend on the purchase price of microplates (the most expensive material) and wages of laboratory personnel.

The Spirotox test is currently performed in four scientific and university laboratories in Poland and in Municipal Waterworks in Warsaw (Poland).

13. Conclusions/prospects

The Spirotox test was introduced as a simple and low-cost toxicity test with ciliated protozoans. Protozoa play an important role in the environment as primary consumers. With bacteria they are major organisms in water self-purification systems. Hence, they should be incorporated into a battery of bioassays.

In order to explore new fields of application for this test, a sediment/soil direct contact procedure is now under investigation. By way of this direct contact test, cells are also exposed to particle-bound substances of low water solubility. Direct contact tests are of interest because they have higher ecological relevance than tests performed on pore waters or solvent extracts.

New endpoints allowing shorter exposure times (1-2 hours) are also being explored. They include physiological observations of food uptake and biochemical techniques with fluorescent dyes. Assuming that such tests are found to be sufficiently sensitive, they would then have useful applications as screening tools in the assessment of waterworks.

References

Applewhite, P.B. (1972) Drugs affecting sensitivity to stimuli in the plant *Mimosa* and the protozoan *Spirostomum*, *Physiology & Behavior* **9**, 869-871.

Czerniewski, Z. (1935) Działanie niektórych środków nasennych i alkaloidów na *Spirostomum ambiguum* Ehrbg., *Acta Biologiae Experimentalis* **9**, 91-110.

Ettienne, E.M. (1970) Control of contractility in *Spirostomum* by dissociated calcium ions, *Journal of General Physiology* **56**, 168-179.

Finley, H.E., Brown, C.A. and Daniel, W.A. (1964) Electron microscopy of the ectoplasm and infraciliature of *Spirostomum ambiguum*, *Journal of Protozoology* **11**, 264-280.

Grolière, C-A. and Njine, T. (1973) Etude comparée de la dynamique des populations de ciliés dans différents biotopes d'une mare de forêt pendant une année, *Protistologica* **9**, 5-16.

Jones, A.R. (1966) Uptake of 45-Calcium by *Spirostomum ambiguum*, *Journal of Protozoology* **13**, 422-428.

Michniewicz, M., Nałęcz-Jawecki, G., Stufka-Olczyk, J. and Sawicki, J. (2000) Comparison of chemical composition and toxicity of wastewaters from pulp industry, in G. Persoone, C. Janssen and W. De Coen (eds.), *New Microbiotests for Routine Toxicity Screening and Biomonitoring*, Kluwer Academic/Plenum Publishers, pp. 401-412.

Nałęcz-Jawecki, G. and Sawicki, J. (1998) Toxicity of inorganic compounds in the Spirotox test - a miniaturized version of the *Spirostomum ambiguum* test, *Archives of Environmental Contamination and Toxicology* **34**, 1-5.

Nałęcz-Jawecki, G. and Sawicki, J. (1999) Spirotox – a new tool for testing the toxicity of volatile compounds, *Chemosphere* **38**, 3211-3218.

Nałęcz-Jawecki, G. and Sawicki, J. (2002a) The toxicity of tri-substituted benzenes to the protozoan ciliate *Spirostomum ambiguum*, *Chemosphere* **46**, 333-337.

Nałęcz-Jawecki, G. and Sawicki, J. (2002b) A comparison of sensitivity of Spirotox biotest with standard toxicity tests, *Archives of Environmental Contamination and Toxicology* **42**, 389-395.

Nałęcz-Jawecki, G. and Sawicki, J. (2003a) Influence of pH on the toxicity of nitrophenols to Microtox® and Spirotox tests, *Chemosphere* **52**, 249-252.

Nałęcz-Jawecki, G. and Sawicki, J. (2003b) The toxicity of selected pharmaceuticals to the protozoa *Spirostomum ambiguum* and *Tetrahymena termophila*, *Fresenius Environmental Bulletin* **12**, 840-3.

Nałęcz-Jawecki, G., Demkowicz-Dobrzański, K., Sawicki, J. and Oleszczuk, K. (1995) Application of a simple microplate bioassay technique with protozoan *Spirostomum ambiguum* and the Microtox test system to examination of wastes, *Arch Environ Protection* **21**, 205-12.

Nałęcz-Jawecki, G., Rudź, B. and Sawicki, J. (1997) Evaluation of toxicity of medical devices using Spirotox and Microtox tests. Part 1. Toxicity of selected toxicants in various diluents, *Journal of Biomedical Materials Research* **35**, 101-105.

Nałęcz-Jawecki, G., Michniewicz, M., Stufka-Olczyk, J. and Sawicki, J. (2000) Toxicity of major pollutants occurring in the pulp and paper mill effluents, *Acta Poloniae Toxicol* **8**, 263-9.

Nałęcz-Jawecki, G., Kucharczyk, E. and Sawicki, J. (2002a) The sensitivity of protozoan *Spirostomum ambiguum* to selected pesticides, *Fresenius Environmental Bulletin* **11**, 98-101.
Nałęcz-Jawecki, G., Tarczyńska, M. and Sawicki, J. (2002b) Evaluation of the toxicity of cyanobacterial blooms in drinking water reservoirs with microbiotests, *Fresenius Environmental Bulletin* **11**, 347-351.
Nałęcz-Jawecki, G., Choromański, D., Wołkowicz, S. and Sawicki, J. (2002c) The assessment of ground water contamination around pesticide wastes deposited in tombs in Northwestern Poland. 7th Meeting of the Central and Eastern European Regional Section SECOTOX, Brno, Czech Republic, 14-16 October, 2002.
Nałęcz-Jawecki, G., Grabińska-Sota, E. and Narkiewicz, P. (2003) The toxicity of cationic surfactants in four bioassays, *Ecotoxicology and Environmental Safety* **54**, 87-91.
Namikoschi, M. and Rinehart, K.L. (1996) Bioactive compounds produced by cyanobacteria, *Journal of Industrial Microbiology* **17**, 373-384.
Ricci, N. (1990) The behavior of ciliated protozoa, *Animal Behaviour* **40**, 1048-69.
Rostkowska, J. and Moskwa, W. (1968) The influence of magnetic field on susceptibility for toxic compounds in *Spirostomum ambiguum* Ehrb., *Acta Protozoologica* **5**, 305-13.
Seyd, E.L. (1936) Studies on the regulation of *Spirostomum ambiguum* Ehrbg., *Archiv Fur Protistenkunde* **86**, 454-469.
Tarczyńska, M., Nałęcz-Jawecki, G., Brzychcy, M., Zalewski, M. and Sawicki, J. (2000) The toxicity of cyanobacterial blooms as determined by microbiotests and mouse assays, in G. Persoone, C. Janssen, W. De Coen (eds.), *New Microbiotests for Routine Toxicity Screening and Biomonitoring*, Kluwer Academic/Plenum Publishers, pp. 527-533.
Tarczyńska, M., Nałęcz-Jawecki, G., Romanowska-Duda, Z., Sawicki, J., Beattie, K., Codd, G. and Zalewski, M. (2001) Tests for the toxicity assessment of cyanobacterial bloom samples, *Environmental Toxicology* **16**, 383-390.
Weber, C.I. (1993) Methods for measuring the acute toxicity of effluents and receiving waters to freshwater and marine organisms, IV Edition. US EPA/600/4-90/027F Cincinnati, Ohio, p. 80.

Abbreviations

def	deformed
% def	% of sublethal deformity responses
EC50	sample concentration causing a 50% sublethal effect
DMSO	dimethylsulfoxide
LC50	sample concentration killing 50% of exposed organisms
let	lethal
% let	% of lethal response
NOEC	no observed effect concentration
QSAR	Quantitative structure-activity relationship
SDS	sodium dodecyl sulfate
ST	somewhat toxic sample
T	toxic (sample)
TEU	Toxicity Equivalency Unit
TOC	Total Organic Carbon
TU	Toxicity Units.

9. ROTIFER INGESTION TEST FOR RAPID ASSESSMENT OF TOXICITY

TERRY W. SNELL
School of Biology
Georgia Institute of Technology
Atlanta, GA 30332-0230, USA
terry.snell@biology.gatech.edu

1. Objective and scope of test method

This method is intended as a screening tool for rapid toxicity assessment. The test is designed for use with fresh or marine waters, to evaluate chemicals, surface waters, effluents, pore waters, drinking waters, and contamination emergencies. Because of its speed and simplicity, this test could easily be integrated into a battery of tests representing several species. Rotifers are generally responsive to a wide variety of toxicants, including metals, organics, pesticides, and endocrine disruptors. The ingestion test is performed in 1 hour in 24-well microplates in volumes of 750 µL.

2. Summary of test procedure

Test animals are obtained by hatching resting eggs (cysts), encysted dormant embryos that remain viable for years when kept cold, dark, and dry. Resting eggs enable researchers to eliminate the pre-test culture that is required to obtain most test animals. No pre-culture eliminates a major source of variability in toxicity tests, reduces cost, and the expertise required of personnel to perform the test (Persoone, 1991). Since rotifer resting eggs hatch synchronously, physiologically uniform animals of similar age can be used for the test. Approximately 15 newborn rotifers are placed into each well containing 750 µL of test solution. The format of a 24-well plate allows for a control and five test concentrations, each with four replicates. Animals are exposed to the test solutions for 45 minutes and then 5 µm red microspheres are introduced into each well for 15 minutes. The rotifers readily ingest these microspheres in the absence of toxicant stress. Rotifer ingestion rate is a dose-dependent function of toxicant concentration, as toxicity increases rotifers feed less (Juchelka and Snell, 1994). The red microspheres accumulate in rotifer stomachs so that after 15 minutes, their guts appear bright red. This can be easily seen under a dissecting microscope at 25X magnification. Rotifers with red guts are scored as feeding and those with no visible red as non-feeding. The number of red microspheres in the gut is not counted, only the presence or absence of red color. This experimental design allows for the calculation of percent feeding in each of the four replicates. Statistical analysis can be performed with the same procedures as used in analyzing percent survival data in acute toxicity tests.

Table 1. Summary of the rotifer ingestion toxicity test.

Test animal	Freshwater: *Brachionus calyciflorus* Marine: *Brachionus plicatilis*
Test type	Rapid screening
Test format	24-well plate
Test volume	750 µL per well
Test duration	1 hour
Source of test animals	Hatching cysts, commercially available
Rotifers per replicate	10-15
Temperature	20-30°C
Salinity	*Brachionus calyciflorus* : 0-5 ppt *B. plicatilis* : 3-40 ppt
Light	No specific requirements
Dilution water	Artificial freshwater or seawater, natural surface water
Endpoint	Percent feeding
Reference toxicants	Copper ($CuSO_4$), pentachlorophenol
Ingestion in controls	Should exceed 80%

3. Overview of applications reported with rotifer toxicity tests

Rotifer cysts were introduced to ecotoxicology by Snell and Persoone (1989a,b) who described a 24-hour acute toxicity test conducted with hatchlings from cysts. This test has been validated and adopted as a standard method (ASTM 1440) and is commercially available as a test kit (Rotoxkit F) from Microbiotests, Inc. (see below). A method to estimate chronic toxicity using asexual reproduction has been developed (Snell and Moffat, 1992) and published as standard method 8420 in Standard Methods for the Examination of Water and Wastewater (2001). It also is commercially available as a test kit (Rotoxkit F chronic) from Microbiotests, Inc. (see below). A number of other endpoints have been developed using cyst hatchlings as starting material such as resting egg production (Preston et al., 2000; Preston and Snell, 2001), swimming (Charoy et al., 1995), enzyme activity (Burbank and Snell, 1994) and stress protein gene expression (Cochrane et al., 1994). Tests to estimate toxicity based on rotifer ingestion rate were developed by Fernandez-Casalderry et al., (1992, 1993a and b) and Juchelka and Snell (1994), then expanded to cladocerans and ciliates (Juchelka and Snell, 1995). This latter work employed fluorescent microspheres and quantified fluorescence in rotifer guts using epifluorescent microscopy. Although useful in research, this method requires expensive equipment that is not widely available to quantify rotifer ingestion rate. The method described here simplifies the estimation of rotifer ingestion as an endpoint for toxicity tests. The use of rotifers in ecotoxicology has been reviewed by Snell and Janssen (1995; 1998).

The rotifer *Brachionus calyciflorus* was chosen for this test because it is an herbivore with a broad diet, feeding non-selectively on particles in the size range of 2-15 µm (Starkweather, 1987). Ingestion is an ecologically important process which is incorporated into most bioenergetics models (Starkweather, 1987). Energy ingested is directly linked to reproductive output, a key element of fitness and long-term survival of a population. Ingestion rate, therefore, should be a good estimator of chronic toxicity. Ingestion tests to estimate toxicity in other zooplankters have been described. CerioFAST is a method to measure ingestion rate of *Ceriodaphnia dubia* which is based in ingestion of fluorescently labeled yeast (Jung and Bitton, 1997).

4. Advantages of conducting the rotifer ingestion test

One of the main advantages of the rotifer ingestion test is its speed. The test can be conducted in one hour on five test solutions plus a control, each with four replicates. The test also requires little technical expertise since it is initiated with rotifers hatched from cysts and no difficult manipulations are required. After a few practice sessions, even inexperienced people should be able to conduct the rotifer ingestion test and produce useable data. Small volumes of test material are required, so this test is well suited for testing pore waters, incorporating into a battery of tests, or guiding the bioassay-directed fractionations of toxicity identification evaluations. No expensive equipment is required to perform the test, so it is of particular interest for performance in the field or in developing countries. The cost per sample for estimating toxicity is attractive compared to other toxicity tests. The sensitivity of the rotifer ingestion test compares favorably to other endpoints (Juchelka and Snell, 1994; 1995; Preston and Snell, 2001) and other species (Snell and Janssen, 1998). The disadvantages of this test include the short exposure time which may not be long enough for slow acting toxicants to have an effect, the small size of rotifers which requires a good quality microscope to clearly see them, the small exposure chamber which may increase sorption of test compounds, and the fact that this test is currently not approved as a standard method.

5. Test species

Rotifers are classified in the phylum Rotifera, one of several phyla of lower invertebrates. There are approximately 2000 rotifer species named; they are divided into two classes, Digononta and Monogononta (Nogrady et al., 1993). Monogononts reproduce parthenogenetically, but in response to specific environmental cues, they reproduce sexually yielding dormant embryos called cysts (resting eggs) which have been used in toxicity testing (Snell and Janssen, 1995). Most rotifer species inhabit fresh and brackish waters (Wallace and Snell, 2001), but there are some genera, like *Synchaeta*, where the majority of species are marine (Nogrady, 1982). In coastal marine habitats, rotifers sometimes comprise the dominant portion of the zooplankton biomass (Egloff, 1988). They are also abundant in marine interstitial habitats, the interstitial water of soils (Pourriot, 1979), and in water clinging to mosses, liverworts and lichens (Ricci, 1983). In freshwater lake plankton (Stemberger, 1990) and in river sediments (Schimd-Araya, 1995), rotifers often are abundant with high species diversity.

Rotifers play an important role in the ecological processes of many aquatic communities (Pace and Orcutt, 1981). As suspension feeders, planktonic rotifers influence algal species composition through selective grazing (Bogdan and Gilbert 1987; Starkweather, 1987; Arndt, 1993). Rotifers often compete with cladocera and copepods for phytoplankton in the 2 to 18 µm size range. Along with crustaceans, rotifers contribute substantially to nutrient recycling (Esjmont-Karabin, 1983). Rotifers are food for many fish larvae (Lubzens et al., 1997).

The genus *Brachionus* is large with over 25 species distributed in marine and freshwater habitats all over the world (Nogrady et al., 1993). The species *Brachionus calyciflorus* and *B. plicatilis* are a complex of cryptic species with many distinct populations (Gomez et al., 2002). The geographical strain of *B. calyciflorus* typically used in toxicity testing was collected in Gainesville, Florida, in 1983 (Snell et al., 1991) and has been used to produce cysts in the laboratory ever since. The *B. plicatilis* strain was originally collected in the Azov Sea, Russia, in 1983 (Snell and Persoone, 1989b) and likewise has been a source of cysts. These strains were selected because of their ability to produce cysts, not because of their extraordinary sensitivity to toxicants. Rotifer cysts for toxicity testing can be purchased from Microbiotests, Inc., Venecoweg 19, 9810 Nazareth, Belgium, tel. 3293808545, fax 3293808546, e-mail microbiotests@skynet.be (contact the company for distributors in various countries). Rotifer cysts should be stored in a freezer (-20°C).

6. Culture/maintenance of rotifers in the laboratory

There is no culture required for the rotifer ingestion test. Test animals are obtained by hatching cryptobiotic stages (cysts) that are commercially available (see Section 5). Because the duration of the test is only one hour, there also is no food required to feed test animals. Disposable plastic 24-well plates are used, so there is no glassware to wash.

Water to dilute test solutions may be prepared from high quality deionized or distilled water. Artificial freshwater may be used for *Brachionus calyciflorus* and artificial seawater for *B. plicatilis*.

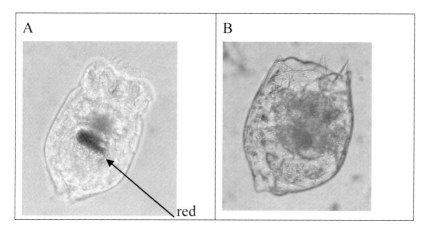

Figure 1. A photomicrograph at 400X magnification of a B. calyciflorus *which has ingested red carmine particles (A) and one which has not ingested (B). The key trait is the dark gut which would appear red in color.*

6.1 ARTIFICIAL FRESHWATER

Prepare standard synthetic freshwater by adding 96 mg $NaHCO_3$, 60 mg $CaSO_4 \cdot 2H_2O$, 60 mg $MgSO_4$, and 4 mg KCl to 1 L of deionized or distilled water. Mix well on a magnetic stirrer and adjust pH to 7.5 with 10 M KOH or HCl. Use within one week. This is a moderately hard standard freshwater, with hardness of 80-100 mg $CaCO_3$ per liter and alkalinity of 60-70 mg per liter.

6.2 ARTIFICIAL SEAWATER

Prepare standard synthetic seawater with a salinity of 15 parts per thousand (ppt) by adding: 11.31 g NaCl, 0.36 g KCl, 0.54 g $CaCl_2$, 1.97 g $MgCl_2 \cdot 6H_2O$, 2.39 g $MgSO_4 \cdot 7H_2O$, 0.17 g $NaHCO_3$ to 1 L of deionized or distilled water. Mix well on a magnetic stirrer and adjust pH to 8.0 with 10 M KOH or HCl. Use within one week.

Other waters: bottled mineral water (no gas), dechlorinated tap water, surface water, well water, natural seawater, and waters from other sources can be used as dilution water in rotifer toxicity tests. Prior to their use, ingestion studies should be conducted to ensure acceptable levels of feeding can be obtained in the negative control (*e.g.*, $\geq 80\%$ feeding).

7. Preparation of rotifers for toxicity testing

Rotifers are prepared for the ingestion test by hatching cysts. Hatchlings are collected within a few hours of their birth so they all are of similar age and physiological condition. They need no feeding in their first day since they are well provisioned with energy by their mothers. There is no need for acclimation and there are no nutritional issues since there is no feeding during the test. Ingestion in the test is quantified by uptake of inert microspheres.

7.1 ROTIFER CYST HATCHING

Hatching should be initiated 18 hours before the start of a test for *B. calyciflorus*. Place about 30 ml of artificial freshwater or mineral water into a clean Petri dish, empty the contents of one vial of rotifer cysts (a few thousand cysts) into the water and rinse the vial to remove all cysts (Snell et al., 1991). Incubate the Petri dish at 25°C in the light of one or two 20 Watt fluorescent tubes (1000-4000 lux) for 16-18 hours. Make sure that the cysts are submerged during the incubation by rinsing the sides of the hatching dish using a pipette. Hatching should start after about 15 hours and 1-2 hours later the rotifers can be transferred to the 24-well test plate. Cooler temperatures, low or high pH, elevated hardness and alkalinity can delay hatching. When hatching is delayed the cause often is low temperature or poor water quality. The problem is usually corrected by bringing temperature to 25°C or switching to a different water source. If hatching is delayed, check cysts hourly to insure collecting test animals within a few hours of hatching. Hatching of *B. plicatilis* cysts should be initiated 24 hours before a test in 15 ppt seawater in conditions as described above. Unused cysts should be stored in a freezer (-20°C).

8. Testing procedure

8.1 HANDLING SAMPLES

Water samples should be collected and handled according to standard procedures. Surface waters, pore waters, and effluents should be transported at cool temperatures in containers that protect them from light. The rotifer test should be performed within 24 hours of sample collection. Because the rotifer ingestion test requires only 750 µL per test well, usually 50 mL per sample is plenty to perform the test. When testing pure chemicals that have low solubility in water, a carrier solvent such as acetone can be used. This requires a solvent control to be included in the experimental design.

8.2 PREPARING A DILUTION SERIES

As the rotifers are hatching, prepare a dilution series of the test compound or effluent according to standard methods. If the sample contains debris or large floating and/or suspended solids it may be necessary to first coarse-filter it through a sieve that has 2-4 mm mesh openings. If the sample contains organisms, it should be filtered through a sieve with 60 µm mesh openings. Centrifugation (2000-4000 rpm for 3 minutes) is effective for removing small suspended particles. CAUTION: filtration or centrifugation may remove some toxicants if they are bound to particles.

It is advisable to measure pH, conductivity or salinity, total alkalinity, total hardness, and total residual chlorine in the undiluted effluent or surface water. If these water chemistry parameters are very different from the dilution water, this can reduce rotifer ingestion in the absence of toxicity.

A concentration-response test on effluent or pore water consists of a control and a minimum of five concentrations commonly selected to approximate a geometric series, such as 100%, 50%, 25%, 12.5% and 6.25%. One method of preparing a dilution series is as follows: pipette 10 mL of effluent or pore water sample into a test tube (**NEVER** pipette by mouth). Label this as the 100% test solution. Pipette 5 mL of the 100% sample into a second test tube and add 5 mL of dilution water, mix thoroughly and label this tube as the 50% test solution. Pipette 2.5 mL of the 50% solution into a third test tube and add 7.5 mL of dilution water, mix and label as the 25% test solution. Repeat this procedure for the 12.5%, and 6.25% test solutions. If 100% mortality has occurred in the higher concentrations after the 45 minute exposure, lower concentrations should be tested such as 3.1%, 1.6%, and 0.8%.

When testing a single chemical of unknown toxicity, it is best to do a range-finding test first. This is accomplished by creating a log series (0.01, 0.1, 1, 10, 100 mg/L) and identifying the lowest concentration where effects are observed. This is used as the highest concentration in a second, definitive test with a log concentration series spanning the two log concentrations. For example, if effects were observed at 10 mg/L in the range-finding test, a concentration series of 1.6, 2.5, 4.0, 6.3, 10 mg/L could be used in the definitive test. This series was calculated by subtracting 0.2 from the log 10 and calculating 10^x (antilog) to give the five test concentrations.

8.3 FILLING THE TEST WELLS

The rotifer ingestion test is conducted in 24-well polystyrene plates (Corning 25820 or equivalent) and consists of a negative control and five test concentrations. Notice that these plates are labeled as columns 1-6 across and rows A-D down. Pipette 0.75 mL of dilution water into well A in column 1 (A1, the upper left most well) of the test plate. This well is dilution water without toxicant and will serve as the negative control. Fill wells B1, C1, and D1 with dilution water in a similar fashion. This experimental design provides four replicates for each treatment. Working from the lowest concentration, pipette 0.75 mL of the first test concentration into wells A2, B2, C2, and D2 of column 2 of the test plate. Repeat this procedure for the wells in columns 3-6.

8.4 ADDING THE ROTIFERS

Beginning with the control, use a small bore micropipette to transfer about 15 rotifers from the hatching dish into well A1 of the test plate. Rotifers can be concentrated in the hatching dish by shining a light from one side. Repeat this transfer for the remaining wells, adding about 15 rotifers to each well. The exact number of rotifers added is not important at this point because they will be counted at the end of the test. Minimize the transfer of water along with the rotifers. For best results, rotifers should be 2-6 hours old. Rotifers 0-1 hour old may not feed.

8.5 INCUBATION AND SCORING OF THE TEST PLATE

Incubate the test plate at about 25°C in darkness for 45 minutes. After incubation, place the plate under a dissecting microscope and observe the rotifers at about 10X magnification, recording whether most rotifers are swimming in each well. If the sample is so toxic that it has killed the rotifers, there will obviously be no ingestion. In this case, lower concentrations should be tested. Pipette 0.01 mL of a concentrated suspension of 5 μm diameter red microspheres (Bangs Laboratories, www.bangslabs.com; similar products may be available from other suppliers[*]) into 6 mL of dilution water and shake to mix well. Add 0.03 mL of this microsphere suspension to each test well. The final microsphere concentration in the test wells should be about 250,000/mL. Allow the rotifers to feed for 15 minutes. At the end of the feeding period, animals should be killed by adding one drop (~50 μL) of 10% formalin solution to each well. This does not affect the red color of the beads, so the test can be scored at a later time. This allows, for example, the test to be conducted in the field and scored back in the lab. The rotifers in each well should be observed under the microscope at 25X magnification and the number of feeding and non-feeding in each concentration should be counted. Your data can be recorded in a table (Box 1) that looks like this:

Box 1. Example of a table used to report test data.

Test concentration	Rep.	Well	# feeding	# not feeding	Percent feeding	Swimming after 45 minutes?
0 (control)	1	A1	10	0	100	yes
	2	B1	12	1	92	yes
	3	C1	13	1	93	yes
	4	D1	11	0	100	yes
1	1	A2	12	2	86	yes
	2	B2	14	1	93	yes
	3	C2	11	3	79	yes
	4	D2	10	2	83	yes
2	1	A3	13	4	76	yes
	2	B3	11	3	79	yes
	3	C3	9	2	82	yes
	4	D3	14	5	74	yes

---

[*] Carmine can be substituted for the microspheres and is cheaper. A carmine suspension is a mixture of particles of various sizes, and can be prepared by adding 1 mg carmine to 2 mL dilution water. Mix well to suspend the fine particles. Add 10 μL of this suspension to each well after the 45 minute exposure. Rotifers will accumulate red color in their guts after a few minutes of feeding. Carmine can be obtained from several suppliers. We have purchased it from Fisher Scientific (https://www.fishersci.com/index.jsp, product number AC19020-0050).

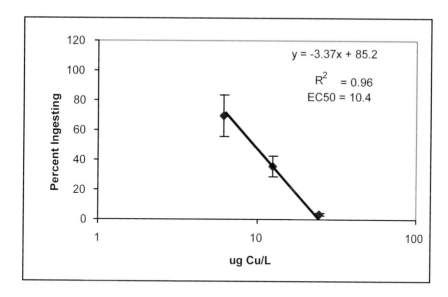

Figure 2. Example of an ingestion concentration-response curve for B. calyciflorus *exposed to copper. Vertical lines equal one standard deviation. The 0 concentration (control) ingestion was 90.1%, but cannot be plotted on this log scale.*

9. Post-exposure observations/measurements and endpoint determinations

9.1 CRITERIA FOR TEST VALIDITY

For this test to be valid, a red color should be observed in > 80% of the guts of control animals. Since this is a sublethal assay, test concentrations should not immobilize (kill) the rotifers after 45 minutes exposure. Ideally, only the highest test concentration should have 0% ingestion and there should be intermediate values between this and the negative control.

9.2 DATA ANALYSIS

Percent feeding may be arcsine transformed and then a one-way analysis of variance can be performed on the data according to standard methods. A Dunnett's test can be used to compare all treatments to the control. This analysis will produce a no observed effect concentration (NOEC) and a lowest observed effect concentration (LOEC). Alternatively, a probit or trimmed Spearman-Karber test may be performed to estimate NOECs. An EC50 can be estimated from percent feeding data by calculating a linear regression of log concentration versus percent ingesting (Fig. 2). It is advisable to maintain a cumulative record of control performance so that the range of results expected under your conditions can be characterized.

10. Factors influencing the rotifer ingestion test

Most problems with the rotifer ingestion test center around the dilution water. A typical symptom is markedly reduced ingestion in the controls. Even carefully prepared dilution water can be unusable if it is too old or if deionization was inadequate. Rather that performing experiments to determine the cause of these problems, it is usually more prudent to simply change water sources. For example, bottled mineral water (no gas) is usually a good source. Often problems with dilution water quality are seasonal, disappearing as mysteriously as they arrived.

A second source of problems could involve rotifer cyst hatching. This often occurs due to poor dilution water quality. Changing water sources usually alleviates the problem. A second possibility is storage of cysts in poor conditions. It is recommended that cysts be stored in a freezer at $-20°C$. Cysts will age more rapidly at room temperature and at high humidity. Hatching also can be delayed by low temperatures, low or high pH, or inadequate lighting. This can be avoided by following the guidelines provided.

A third source of problems could be using rotifers very soon after hatching. For about the first hour after birth rotifers do not feed, so it is important to collect hatchlings that are at least two hours old. Hatchlings older than about six hours may have reduced feeding due to starvation. Rotifer ingestion may also be suppressed by a heavy load of suspended particles in the test medium. These samples should be filtered or centrifuged to reduce this effect.

11. Applications of the rotifer ingestion test

The sensitivity of the rotifer ingestion test has been compared to reproduction and mortality endpoints for five organics, three metals, and two insecticides (Juchelka and Snell, 1994). The 48-h reproduction NOEC was more than two-fold lower than the 1-h ingestion NOEC for phenol, dimethylphenol, cadmium, copper, mercury and diazinon. Similar reproduction and ingestion NOECs were observed for pentachlorophenol, naphthol, and chlorpyrifos. Rotifer ingestion NOEC was a more sensitive endpoint than 24-h mortality by at least two-fold for all ten toxicants tested except copper. Rotifer ingestion as an endpoint can be more or less sensitive than mortality, depending on the toxicant (Fernandez-Casalderry et al., 1992; 1993a and b). The response of *Brachionus calyciflorus* ingestion rate to toxicants has been compared to that of *B. plicatilis*, *Ceriodaphnia dubia*, and *Paramecium aurelia* by Juchelka and Snell (1995). Ingestion rate was used to investigate the effects of UV-B exposure on *B. calyciflorus* (Preston et al., 1999).

The rotifer ingestion test has been used to assess the toxicity of pore waters collected from 13 urban creeks in the Atlanta area (Juchelka and Snell, 1995). The *B. calyciflorus* ingestion test was compared to ingestion by *Ceriodaphnia dubia*, and *Paramecium aurelia* and esterase enzyme activity by the yeast *Candida tropicalis* and the bacterium *Bacillus subtilis*. The *B. calyciflorus* test detected toxicity in pore water from 9 of the 13 sites. The *P. aurelia* test detected toxicity at 7 sites and the *C. dubia* and *B. subtilis* tests detected toxicity at 3 sites. No toxicity was detected by

C. tropicalis at any site. The rotifer test found two sites to be toxic that were not toxic in any of the other tests and one site was non-toxic in all five of the tests.

12. Conclusions/Prospects

The rotifer ingestion test allows investigators to estimate toxicity in surface water, effluent, and pore water samples as well as to characterize toxicity in solutions of pure chemicals. With this test, toxicity can be quantified rapidly, inexpensively, and with minimal training of personnel. The method would benefit from application to a wide variety of environmental problems so that its usefulness and limitations can be more fully understood.

Acknowledgements

The author is grateful for the insightful comments of Don Versteeg, Guido Persoone, and Roberto Rico-Martinez which have improved this paper.

References

ASTM (American Society for testing and Materials) (2002) Standard Guide for Acute Toxicity Test with the Rotifer *Brachionus*, ASTM, Vol. 11.05, method E 1440, pp. 806-813.
Arndt, H. (1993) Rotifers as predators on components of the microbial web (bacteria, heterotrophic flagellates, ciliates)-a review, *Hydrobiologia* **255/256**, 231-246.
Bogdan, K.G. and Gilbert, J.J. (1987) Quantitative comparison of food niches in some freshwater zooplankton, *Oecologia* **72**, 331-340.
Burbank, S.E. and Snell, T.W. (1994) Rapid toxicity assessment using esterase biomarkers in *Brachionus calyciflorus* (Rotifera), *Environmental Toxicology Water Quality* **9**, 171-178.
Charoy, C.P., Janssen, C.R., Persoone, G. and Clement, P. (1995) The swimming behavior of *Brachionus calyciflorus* (Rotifera) under toxic stress: I. The use of automated trajectometry for determining sublethal effects of chemicals, *Aquatic Toxicology* **32**, 271-282.
Cochrane, B.J., De Lama, Y.D. and Snell, T.W. (1994) Polymerase chain reaction as a tool for developing stress protein probes, *Environmental Toxicology and Chemistry* **13**, 1221- 1229.
Egloff, D.A. (1988) Food and growth relations of the marine zooplankter, *Synchaeta cecelia* (Rotifera), *Hydrobiologia* **157**, 129-141.
Ejsmont-Karabin, J. (1983) Ammonia nitrogen and inorganic phosphorus excretion by the planktonic rotifers, *Hydrobiologia* **104**, 231-236.
Fernandez-Casalderry, A., Ferrando, M.D. and Andreu-Moliner, E. (1992) Filtration and ingestion rates of *Brachionus calyciflorus* after exposure to endosulfan and diazinon, *Comparative Biochemistry Physiology* **103C**, 357-361.
Fernandez-Casalderry, A., Ferrando, M.D. and Andreu-Moliner, E. (1993a) Effect of the insecticide methylparathion on filtration and ingestion rates of *Brachionus calyciflorus* and *Daphnia magna*, *Science of the Total Environment* **suppl**, 867-876.
Fernandez-Casalderry, A., Ferrando, M.D. and Andreu-Moliner, E. (1993b) Feeding behavior as an index of copper stress in *Daphnia magna* and *Brachionus calyciflorus*, *Comparative Biochemistry and Physiology* **106C**, 327-331.
Gomez, A., Serra, M., Carvalho, G.R. and Lunt, D.H. (2002) Speciation in ancient cryptic species complexes: Evidence from the molecular phylogeny of *Brachionus plicatilis* (Rotifera), *Evolution* **56**, 1431-1444.

Juchelka, C.M. and Snell, T.W. (1994) Rapid toxicity assessment using rotifer ingestion rate, *Archives* of *Environmental Contamination and Toxicology* **26**, 549-554.

Juchelka, C.M. and Snell, T.W. (1995) Rapid toxicity assessment using ingestion rate of cladocerans and ciliates, *Archives* of *Environmental Contamination and Toxicology* **28**, 508-512.

Jung, K. and Bitton, G. (1997) Use of Ceriofast^{tm} for monitoring the toxicity of industrial effluents – comparison with the 48-h acute *Ceriodaphnia* toxicity test and Microtox^R, *Environmental Toxicology and Chemistry* **16**, 2264-2267.

Lubzens, E., Minkoff, G., Barr, Y. and Zmora, O. (1997) Mariculture in Israel - past achievements and future directions in raising rotifers as food for marine fish larvae, *Hydrobiologia* **358**, 13-20.

Nogrady, T. (1982) Rotifera, in S.P. Parker (ed.), *Synopsis and Classification of Living Organisms*, McGraw-Hill, pp. 865-872.

Nogrady, T., Wallace, R.L. and Snell, T.W. (1993) Rotifera, *Biology, Ecology and Systematics*, Vol. 1, The Hague, Belgium, SPB Academica Publishing bv.

Pace, M.L. and Orcutt, J.D. (1981) The relative importance of protozoans, rotifers and crustaceans in freshwater zooplankton communities, *Limnology and Oceanography* **26**, 822-830.

Persoone, G. (1991) Cyst-based toxicity tests: I. A promising new tool for rapid and cost-effective screening of chemicals and effluents, *Zeitschrift fur Angewandte Zoologie* **78**, 235-241.

Pourriot, R. (1979) Rotifères du sol, *Revue d'Écologie et de Biologie du Sol* **16**, 279-312.

Preston, B.L. and Snell, T.W. (2001) Full life-cycle toxicity assessment using rotifer resting egg production: implications for ecological risk assessment, *Environmental Pollution* **114**, 399-406.

Preston, B.L., Snell, T.W. and Kneisel, R. (1999) UV-B exposure increases acute toxicity of pentachlorophenol and mercury to the rotifer *Brachionus calyciflorus*, *Environmental Pollution* **106**, 23-31.

Preston, B.L., Snell, T.W., Robinson, T.L. and Dingmann B.J. (2000) Use of the freshwater rotifer *Brachionus calyciflorus* in a screening assay for potential endocrine disruptors, *Environmental Toxicology and Chemistry* **19**, 2923-2928.

Ricci, C. (1983) Life histories of some species of Rotifera Bdelloidea, *Hydrobiologia* **104**, 175-180.

Schmid-Araya, J.M. (1995) Disturbance and population dynamics of rotifers in bed sediments, *Hydrobiologia* **313/314**, 279-301.

Snell, T.W. and Persoone G. (1989a) Acute toxicity bioassays using rotifers. II. A freshwater test with *Brachionus rubens*, *Aquatic Toxicology* **14**, 81-92.

Snell, T.W. and Moffat, B.D. (1992) A two day life cycle test with the rotifer *Brachionus calyciflorus*, *Environmental Toxicology and Chemistry* **11**, 1249-1257.

Snell, T.W. and Janssen, C.R. (1995) Rotifers in Ecotoxicology: A review, *Hydrobiologia* **313/314**, 231-247.

Snell, T.W. and Janssen, C.R. (1998) Microscale toxicity testing with rotifers, in P.G. Wells, K. Lee and C. Blaise (eds.), *Microscale Aquatic Toxicology - Advances, Techniques and Practice*, CRC Lewis Publishers, Florida, pp. 409-422.

Snell, T.W. and Persoone, G. (1989b) Acute toxicity bioassays using rotifers. I. A test for marine and brackish water with *Brachionus plicatilis*, *Aquatic Toxicology* **14**, 65-80.

Snell, T.W., Moffat, B.D., Janssen, C. and Persoone, G. (1991) Acute toxicity tests using rotifers: IV. Effects of cyst age, temperature, and salinity on the sensitivity of *Brachionus calyciflorus*, *Ecotoxicology and Environmental Safety* **21**, 308-317.

Standard Method 8420 (2001) Rotifers, in *Standard Methods for the Examination of Water and Wastewater*, pp. 8-62 – 8-65.

Starkweather, P.L. (1987) Rotifera, in T.J. Pandian and F.J. Vernberg, (eds.) *Animal Energetics,*Vol. 1, Protozoa through Insecta, Academic Press, Orlando, FL.

Stemberger, R.S. (1990) An inventory of rotifer species diversity of northern Michigan inland lakes, *Archives fur Hydrobiologia* **118**, 283-302.

Wallace, R.L. and Snell, T.W. (2001) Rotifera, in J.H. Thorp and A.P. Covich (eds.) *Ecology and Classification of North American Freshwater Invertebrates*, Academic Press, New York, pp. 187-248.

Abbreviations

$CaCl_2$	calcium chloride
$CaCO_3$	calcium carbonate
$CaSO_4 \cdot 2H_2O$	calcium sulfate
HCl	hydrogen chloride
IC50	interference concentration where 50% of individuals are affected
KCl	potassium chloride
KOH	potassium hydroxide
LOEC	lowest observed effect concentration
M	molar
mM	millimolar
$MgCl_2 \cdot 6H_2O$	magnesium chloride
$MgSO_4$	magnesium sulfate
NaCl	sodium chloride
$NaHCO_3$	sodium bicarbonate
NOEC	no observed effect concentration
ppt	parts per thousand
rpm	revolutions per minute.

10. ACUTE AND CHRONIC TOXICITY TESTING WITH *DAPHNIA* SP.

EMILIA JONCZYK
Stantec Consulting Ltd.
R.R.2, Nicholas Beaver Road, Guelph
ONTARIO N1H 6H9, Canada
ejonczyk@stantec.com

GUY GILRON
Golder Associates Ltd.
2390 Argentia Road, Mississauga
Ontario L5N 5Z7, Canada
ggilron@golder.com

1. Objective and scope of the test method

1.1 *DAPHNIA* SP. 48-HOUR ACUTE LETHALITY TEST

The primary objective of the 48-hour toxicity test using *Daphnia* sp. is to evaluate the acute toxicity of effluents, chemicals, and elutriates on freshwater crustaceans (in particular, cladocerans, which are common zooplanktonic organisms in freshwater ecosystems). This test has two predominant uses: (a) regulatory compliance testing and monitoring of industrial effluents (*e.g.*, pulp and paper and metal mining sectors); and, (b) one of several toxicity tests used in regulatory test batteries for ecological risk assessment of commercial chemicals (*e.g.*, OECD Chemicals Programme, New Substances Notification Regulation (NSNR), Canada, U.S. EPA Pesticide Registration, United States). Moreover, due to its widespread use as a regulatory compliance test for aquatic toxicity to invertebrates in many countries (*e.g.*, Canada, U.S.A., European Union countries), it is also commonly used as a representative pelagic invertebrate test organism in freshwater toxicity test batteries supporting environmental monitoring, assessment and ecological risk assessments where fresh surface waters may be influenced by a variety of organic and/or inorganic contaminants.

The types of samples tested with this method include: industrial wastewaters, receiving waters, produced waters, and (pure, both organic and inorganic) commercial chemicals. The test design, however, is also appropriate for the testing of soil elutriates, where this application is appropriate (*e.g.*, Environment Canada, 1990a). The test is conducted in small beakers or standard test tubes, and often incorporates replication (*e.g.*, 5 organisms in each of 4 replicates, 10 organisms in each of 2 replicates, or 3 replicates with a minimum of 10 organisms) and basic water quality monitoring (*i.e.*, temperature, dissolved oxygen, pH, hardness, and conductivity). The measurement endpoints generally evaluated are the 48-hour LC50 (for survival), and the 48-hour EC50 (for immobility). Test organisms are generally assessed for the two assessment endpoints after each 24-hour interval.

The main attractive features of this test are its simplicity, short exposure duration, sensitivity, ease with which cultures are maintained, and level of standardization (these are discussed in more detail below). As a result, it is a relatively small-scale and highly cost-effective test. When species such as *Daphnia magna* are used, another attractive feature is the size of test organisms. Their relatively large body size, as compared to other cladocerans (such as *Ceriodaphnia dubia*), make them highly visible to the naked eye and therefore easier to monitor.

1.2 *DAPHNIA* SP. 21-DAY CHRONIC REPRODUCTION TEST

The primary objective of the 21-day toxicity test using *Daphnia* sp. is to evaluate the reproductive (multi-generational) toxicity of commercial chemicals on freshwater crustaceans. This test, in contrast to the acute test described above, is predominantly used as part of regulatory test batteries for the ecological risk assessment of commercial chemicals (*e.g.*, OECD Chemicals Programme, Toxic Substances Control Act, United States, Pest Management Regulatory Agency - PMRA, Canada). However, the test design has also been used in some innovative applications relating to effluents and receiving waters (*e.g.*, Moran et al., 1994; see Section 10.2), as well as long-term assessment of discharge wastewater monitoring (BEAK, 1985).

The samples tested using this method mainly include commercial chemicals (both organic and inorganic). The test is conducted in small beakers, plastic tubes with a mesh bottom, and incorporates replication (*i.e.*, 1 organism per replicate with a minimum of 10 replicates in a static-renewal test design; or 2 replicates of 10 organisms, or 4 replicates of 5 organisms, each for the flow-through test design) and water quality monitoring (*i.e.*, temperature, pH, hardness, and conductivity). There are several biological endpoints that can potentially be used. These are: survival (21-day LC50), immobility (21-day EC50), young production/fecundity (21-day ICx/IC50), growth rate (21-day ICx/IC50), biomass (21-day ICx/IC50), and indicators of a stressed population (*e.g.*, production of males, occurrence of ephippia; time to first brood).

In addition to those characteristics listed above for the 48-hour acute test design (Section 1.1), the main attractive feature of the chronic test is the relatively large number of generations produced by these organisms in a relatively short period (*e.g.*, 5 batches or 3 generations of young in 3 weeks).

2. Summary of the test procedure

The *Daphnia* sp. acute and chronic tests are used to assess the toxicity of liquid wastewaters, commercial chemicals, elutriates, or chemical mixtures. A rapid summary of the test procedures is provided below in Table 1.

Table 1. Rapid summary of test procedures.

Test organism	- *Daphnia* sp. (*e.g.*, *D. magna* or *D. pulex*). - Preferably obtained from established in-house culture, commercial suppliers or university, government or private laboratories.
Type of test	Acute, chronic (life cycle).
Test format	- Static (acute). - Static-renewal, flow-through (chronic).
Volume contents of test vessels	- 150-200 mL (acute, static). - 50-100 mL (chronic, static-renewal). - 30-40 mL (chronic, flow-through).
Organism numbers per test vessel	- 10 per test vessel (acute). - 1 in static-renewal design; 5 in flow-through design (chronic).
Test replicates	- Minimum of 3 in a single-concentration test, at least one or more in an LC50 test; for chemical product testing: preferably 2 in LC50 test (acute). - Minimum 3, preferably 4 (chronic, static-renewal); minimum 2, preferably 4 (chronic, flow-through).
Lighting	Fluorescent "cool-white" with 16 hours light, 8 hour dark photoperiod with 400 to 800 lux intensity at water surface.
Temperature	$20 \pm 2°C$.
Design/configuration of test vessel(s)	Plastic or glass made of non-toxic material.

Table 1 (continued). Rapid summary of test procedures.

Measurements of test parameters	• Observations of organisms for mortality and/or immobility at the beginning, middle (*i.e.*, 24 hours) and end of test (acute). • Observations of organisms for mortality, immobility daily and young production at least 3 times per week and preferably once daily (chronic).
Endpoints	• Mortality and immobility (acute). • Mortality, immobility, young production, time to first brood, growth (dry weight or body length) (chronic). • LC50, EC50, ICx/IC50, NOEC, LOEC.
Reference toxicants	Sodium chloride (NaCl), potassium chloride (KCl), zinc sulphate (ZnSO$_4$ · 7 H$_2$O), hexavalent chromium, cadmium chloride (CdCl$_2$), and sodium pentachlorophenol (NaPCP).

In the acute test design, *Daphnia* sp. neonates ($\leq$ 24 hours old) are exposed to a range of test concentrations (a minimum of five exposure concentrations and a control) or in a pass/fail design (*i.e.*, 100% test substance and control) for 48 hours, to determine mortality and/or immobility in the test sample. The test can be conducted under static or static-renewal conditions, and is conducted at 20 ± 2 °C, under a photoperiod regime of 16 hours light to 8 hours dark. The photoperiod regime used is supported by a body of previous research (Buikema et al., 1980). Complete darkness is also acceptable (and is recommended if photodegradable substances are being tested). The test is invalid if more than 10% of control organisms die or are immobile during the exposure period; some standards provide other validity criteria (see Tab. 5). Subsequent to test termination, an LC50 (based on the mortality) and/or EC50 (based on immobility) may be calculated and reported.

In the chronic (reproduction potential) test design, *Daphnia* sp. neonates ($\leq$ 24 hours old) are exposed to a range of test concentrations (as above) for 21 days, to determine mortality, immobility and young production per live adult female at test termination. The test is conducted either under static-renewal (*i.e.*, renewal of test solutions daily or thrice weekly at a minimum) or flow-through conditions. As with the acute test design, chronic testing is conducted at 20 ± 2°C, under the 16 h:8 h light:dark photoperiod regime. Moreover, at least five concentrations of the test substance and a control are tested. In the static-renewal test design, at least 10 replicates, each containing one test organism, are exposed (see Fig. 1), while the flow-through design is conducted in duplicate (at a minimum) with ten organisms introduced to the vessels at test initiation or in 4 replicates, each containing 5 test organisms (see Fig. 2). The test is invalid if more than 20% of control organisms die, and the mean number of live offspring produced per live adult female at test termination is ≤ 60, in the control. Test endpoints include: mortality at test termination, fecundity (*i.e.,* number of young produced per live adult at test termination in each concentration), and time to first brood. An optional endpoint is growth (change in biomass) of adults at test termination (*i.e.,* expressed as dry weight or body length). Based on data generated from the test, the following statistical endpoints may be calculated: 21-day LC50 (survival), 21-day EC50 (immobility), Lowest-Observed-Effect-Concentration (LOEC)[1], No-Observed-Effect-Concentration (NOEC)[1], IC25 and IC50 for reproduction (*i.e.,* concentrations that will results in a 25% or 50 % reduction in reproductive output) and (optionally) IC25 and IC50 for growth (*i.e.,* concentrations that will results in a 25% or 50 % reduction in first generation growth based on average dry weight).

Figure 1. Daphnia *sp. chronic test, static-renewal design (1 daphnid in each of 10 exposure vessels) (Photo courtesy of J.-F. Férard).*

3. Overview of applications with the method

Daphnia sp. have been applied widely in aquatic toxicity studies since the early 1970s. Comprehensive reviews of *Daphnia* sp. toxicity studies were undertaken and test method standardization research was developed in the 1980s (Buikema et al., 1980; Lee et al., 1986; Poirier et al. 1986; Greene et al., 1988; Baird et al., 1989). To date, the acute test design has been applied internationally for the screening of potentially hazardous chemicals and the monitoring of industrial effluents. As indicated earlier, the acute test design is widely used in many countries for regulatory purposes. Chronic tests, on the other hand, have a cosmopolitan application for ecological risk and hazard assessments of chemicals (Thurston et al., 1985; Ferrando et al., 1999) and herbicides (Klapes, 1990). Moreover, the test is often applied in

---

[1] LOEC, NOEC are to be phased out as a summary of toxicity (conclusion of OECD meeting in Brunschweig, 1996); determinations of ICx should be favoured.

environmental assessments with contaminated media and field studies. In North America, the *Ceriodaphnia dubia* test (U.S. EPA, 2002a) is a more common chronic/sublethal test used for environmental monitoring purposes.

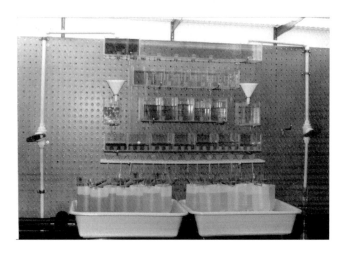

Figure 2. Daphnia *sp. chronic test, flow-through design*
(Note: Daphnids are placed in the plastic vessels located in the water baths).

3.1 ACUTE TEST DESIGN

In Canada, in the mid- to late-1980s, the Ontario Ministry of Environment developed test methods for evaluating acute toxicity to *Daphnia magna* (Poirier et al., 1988). In the mid-1990s, acute toxicity limits were set and implemented under the Municipal-Industrial Strategy for Abatement (MISA) program. The development of effluent discharge regulations for the protection of aquatic life, such as the federal Pulp and Paper Effluent Regulations (PPER) and most recently, the Metal Mining Effluent Regulations (MMER), catalyzed the development and establishment of standard methods for the evaluation of acute toxicity of effluent discharges. This phenomenon was also paralleled in many other developed countries, where a similar evolution in aquatic toxicology was occurring. For example, the U.S. EPA incorporated *Daphnia* sp. acute testing in their National Discharge Permit Effluent System (NPDES), and Whole Effluent Toxicity (WET) studies (Heber et al., 1995; Dorn, 1996). Test methods used by environmental laboratories in Canada to evaluate the ecotoxicity of environmental media (*i.e.*, water, sediment and soil), are based on a series of standardized biological test methods (*i.e.*, toxicity tests) which were initially published in 1990. The acute *Daphnia* sp. test method was one of the first of these to have been developed, reviewed, published (and amended three times since its initial publication) by the Method Development and Applications Section of Environment Canada. These methods provide test-specific guidance on how to conduct the toxicity tests, with full descriptions of culture and test conditions (for different test media), quality assurance and quality control requirements, and reporting requirements. Despite efforts to improve standardization, the acute *Daphnia* sp. test has been described in many studies with more or less diverse modifications (*e.g.*, whether test organisms are fed during test exposure, selection of neonates by age and/or size, exposure duration, choice of reference toxicants, use of parthenogenetic eggs (ephippia) for starting the test). Researchers have also recently focused on the influence of genotype, improvement and standardization of culture conditions, effects of culture medium composition, bioavailability of contaminants, and other abiotic interactions.

3.2 CHRONIC TEST DESIGN

The chronic test design evolved predominantly in Europe and grew out of the development of the acute test. From 1984, the European members of the Organization for Economic Cooperation and Development (OECD), and the International Organization for Standards (ISO) began to undertake critical validation and standardization studies of the chronic test design. The aim was to develop this design for use as a chronic sublethal invertebrate (multi-generational) reproduction test for commercial chemical testing. The results of a number of successful interlaboratory (*i.e.*, round-robin, ring test) standardization exercises helped to promote the use of this test design.

4. Advantages of conducting the test method

As discussed above, there are a number of significant advantages of using this test method for numerous applications. The most important of these advantages are listed in Table 2 below.

Table 2. Advantages of using Daphnia *sp. toxicity test methods (acute and chronic test designs).*

Aspect	*Details*
Ecologically relevant	• Cladocerans are ubiquitous pelagic crustaceans in freshwater ecosystems, particularly in North America and Europe; they are key organisms in the aquatic food chain (*i.e.*, they are primary consumers).
Endpoint measurements simple to assess	• For the acute test, survival and immobility are endpoints that are easy to determine (*i.e.*, microscope or by eye). • For the 21-day chronic test, reproduction is a basic biological endpoint, but the test design (*i.e.*, multi-generational) allows for the measurement of other relevant endpoints (*e.g.*, mortality, immobility, young production/fecundity, growth rate, biomass, stressed population (*e.g.*, production of males, ephippia), not requiring any special equipment.

Table 2 (continued). Advantages of using Daphnia *sp. toxicity test methods (acute and chronic test designs).*

Aspect	Details
Easy to use in a laboratory context	• Daphnids reproduce parthenogenetically. The lab culture therefore is 'clonal' in nature; therefore, genetic makeup generally does not confound test results; nevertheless, clonal variation has been observed among laboratories (Baird et al., 1990; Barber et al., 1990). Moreover, sex ratio does not require monitoring. • It is relatively easy to obtain, acclimate, culture and handle test organisms in the laboratory. The test requires relatively little bench space. • Standard aquatic toxicity laboratory equipment (*e.g.*, glassware (pipettes, beakers, and test tubes), water baths, microscopes, etc.) is used to culture test organisms and to conduct the test. • It is relatively easy to determine culture health from several simple biological indicators (*e.g.*, fecundity, presence of males and ephippia in cultures). • Only a light and pipette are required in order to handle test organisms; a dissecting scope is required for final endpoint measurement.
Well-established culturing and test methodology	• The history of the test method's development, and the fact that its general design has been independently applied and validated by many users worldwide increases confidence in its use. In addition, nutritional and chronic requirements have been well studied.
Strong toxicity database	• Due to its long historical use, large toxicity databases for numerous contaminants and contaminant types (*e.g.*, metals, organics) have been developed using the test method.
Economical observation regime	• Test parameters are monitored every 24 hours, which makes the test convenient logistically; for example, if the test is set up at 10:00 in the morning, subsequent monitoring is conducted at the same time the next day, and so on. In laboratories that work on regular workday schedules, this is advantageous, as technicians do not need to attend the laboratory at odd hours.
Small sample volume	• For acute effluent testing, the sample volume requirement for the *Daphnia* sp. Test is relatively low in comparison with the fish (*e.g.*, rainbow trout) tests (*i.e.*, 1 L *versus* 10-20 L).

Table 2 (continued). Advantages of using Daphnia *sp.* toxicity test methods (acute and chronic test designs).

Aspect	Details
Quality Assurance/Quality Control	• Good reliability (standardized, well-developed methodology). • Good repeatability (*e.g.*, intra-laboratory precision: CV has been reported to range between 3 and 6.4% for 3 laboratories; U.S. EPA, 1993). • Good reproducibility (*e.g.*, inter-laboratory precision: CV has been reported to range between 32 and 40% among 20 laboratories; U.S. EPA, 1993). • Robust (*i.e.*, parthenogenetic reproduction of cultures assures that genetic stock of test organisms is uniform).

5. Test species

5.1 DISTRIBUTION

Small freshwater cladocerans (a Suborder of the invertebrate Subphylum Crustacea), commonly known as 'water fleas', are found in stagnant waters in many parts of the world. While several different daphnid species have been used as test organisms in aquatic toxicology research studies, *Daphnia magna* and/or *Daphnia pulex* are the most common species used in routine testing. While the two species are morphologically and ecologically similar, *D. magna* usually lives in hard to moderately hard water (*i.e.*, water hardness of > 80 mg/L) in lakes and ponds, while *D. pulex* lives in ponds, and quiescent sections of streams and rivers; this latter species, in contrast with *D. magna*, can tolerate a wider range of water hardness from soft to hard. Both species can be found in the same water bodies, however, by mid-summer, *D. pulex* populations usually outcompete *D. magna* and dominate the community (Lynch, 1983; Pennak, 1989).

5.2 TAXONOMY AND MORPHOLOGY

Daphnia sp. are freshwater pelagic crustaceans whose current taxonomic position is as follows:

Phylum:	Arthropoda
Subphylum:	Crustacea
Class:	Branchiopoda
Subclass:	Phyllopoda
Order:	Diplostraca
Suborder:	Cladocera
Family:	Daphniidae
Genus:	*Daphnia*

A light micrograph of *Daphnia magna*, the most common daphnid used in testing, is presented below in Figure 3. A schematic diagram, illustrating daphnid anatomy, is also provided in Figure 4. Visually, the two species can generally be distinguished according to size; mature female *D. magna* can attain a length of between 5 and 6 mm, while mature female *D. pulex* can only attain a length of up to 3.5 mm. Aside from size, the two species can be morphologically differentiated with certainty only by examining the post-abdominal claws for two characteristics, size and number of spines, using a light microscope. *D. magna* has a uniform row of 20 or more of small uniform teeth, while *D. pulex* possesses 5-7 stout teeth on the middle pectin (see Fig. 4). A more detailed taxonomic and morphological discussion of the two species can be found in Brooks (1957), or Amoros (1984).

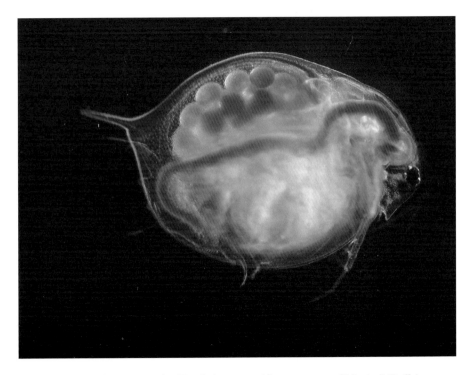

Figure 3. Light micrograph of Daphnia magna *(photo courtesy of Martin O'Reilly).*

5.3 ECOLOGY

Daphnia sp. populations are usually small to non-existent during the winter and early spring, but with temperature increases above the range 6°-12°C, they increase in abundance and can reach up to as high as 200-500 organisms/L (Pennak, 1989). Populations usually decrease during the summer months in response to environmental conditions, and often increase in autumn, followed by a decline again in winter. Throughout most of the year, daphnid populations consist mostly of females, while males are abundant in spring or late autumn. Production of males in the population appears to be induced by low temperatures or high densities, and subsequent accumulation of excretory products, and/or a decrease in available food. Males are substantially smaller than are females, have larger antennae, a modified abdomen, and front legs fitted with a stout hook used in clasping the female during copulation.

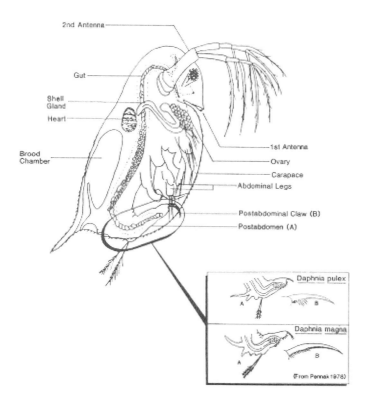

Figure 4. Anatomy of female Daphnia *sp., illustrating differences between* D. pulex *and* D. magna *(from Environment Canada, 1990b) (Note: A= Postabdomen; B=Postabdominal claw)*.

5.4 REPRODUCTION AND LIFE HISTORY

D. magna reproduce by cyclic parthenogenesis, in which males contribute to the genetic makeup of the young during the sexual stage of reproduction, while *D. pulex* may reproduce either by cyclic or obligate parthenogenesis in which zygotes develop within the ephippium by ameioitic parthenogenesis, with no genetic contribution from males. Therefore, both ephippial and live-born neonates are genetically identical to their mothers (U.S. EPA, 2002b). Under standard optimal culturing conditions in the laboratory, both species of *Daphnia* reproduce by parthenogenesis. When cultures undergo physical stress (*e.g.*, overcrowding, low food supply, temperature outside the range of $20 \pm 5°C$), production of males is common, with the daphnids reproducing sexually, and produce ephippia.

There are four distinct life stages in the daphnid life history, specifically: (1) egg; (2) juvenile; (3) adolescent; and, (4) adult. The life span of *Daphnia* species depends mainly on environmental conditions, and usually increases as the temperature decreases. *D. magna* can generally live up to 40 days at $25°C$, and about 56 days at $20°C$, while *D. pulex* may live up to 50 days at $20°C$ (Pennak, 1989), however, some researchers report that mean lifespan can be as high as 90 days (Prof. J.-F. Férard, University of Metz, pers. comm.). Typically, a clutch of 6 to 10 eggs is released into the brood chamber, however as many as 57 have been reported (Pennak, 1989), and up to 90 have been may be produced in laboratory cultures (D. Poirier, Ontario Ministry of the Environment, pers. comm.). After the eggs have hatched into the brood chamber, the young are born after approximately 2-3 days during the female's molting (*i.e.*, shedding of the external carapace). Daphnids mature within 6-14 days (average = 7-10 days) and their time-to-maturity depends primarily on body size. The highest rate of growth occurs during the early life stages (*i.e.*, instars). There are 3-4 and 3-5 instars in the *D. pulex* and *D. magna* life cycles, respectively. Each instar is terminated by a molt, and growth occurs immediately thereafter while the new carapace is still elastic. Subsequent to the juvenile stages, the short adolescent period begins, and consists of one instar. During this stage, the first batch of eggs develops fully in the ovary. Usually, eggs are deposited into the brood chamber minutes after molting, and the developing young are released just prior to the next molt. During the whole life cycle, *D. magna* have approximately 6-22 (or more) adult instars, while *D. pulex* have approximately 18-25 instars. The duration of instars increases with age until the first brood, but again, depends upon environmental conditions. A given instar lasts approximately three days under optimal culturing conditions, but may last up to one week when conditions deteriorate. At the end of each adult instar, the female bears young, molts, increases in size, and releases a new clutch of eggs into the brood chamber; all of this can occur within several minutes. The number of young born depends greatly on food availability and other environmental conditions, mainly temperature. The neonates from the two first broods are less resistant, and have less reserve substances than the subsequent broods. The highest number of *D. magna* neonates born occurs during the fifth adult instar, and during the tenth adult instar for *D. pulex*. Subsequently, the number of young released decreases. On average, 40-50 young may be born during each instar. However, as many as 78 neonates have been reported for one brood (Dr. Sylvie Cotelle, University of Metz, pers. comm.).

5.5 FEEDING

Daphnids, under natural conditions, feed predominantly on algae and bacteria. The highest density of daphnids occurs during algal blooms, during which time proteins and carbohydrates are plentiful in their ambient environment. In the laboratory, *D. magna* prefer bacteria to algae. For example, some laboratories, including our own, have found that, when fed a standard YCT invertebrate formula (*i.e.*, yeast, Cerophyll® and fermented trout chow; see Appendix 1 for details of preparation), reproduction increases in comparison with an exclusive algal diet. Other laboratories, however, have observed that daphnid reproduction increases when they are fed a mixed algal diet (*e.g.*, *Selenastrum capricornutum*, *Chlorella vulgaris*, and diatoms). *D. pulex* use bacteria as food only when algae are unavailable. Food source and availability affects *Daphnia* reproduction and its sensitivity to toxicants. It has been demonstrated that both *D. magna* and *D. pulex* fed diatoms were more tolerant to pollutants than those fed an exclusively green algal diet (U.S. EPA, 2002b).

5.6 CHOICE OF TEST ORGANISMS

Either *D. magna* or *D. pulex* may be used in the aquatic toxicity testing described in this chapter. The selection of test species should be made based upon the objectives of the study and the characteristics of the receiving water body in question, in particular, water hardness. *D. magna* should be used in testing where water is moderately hard to very hard (Cowgill, 1991). Use of this species is recommended in water of hardness > 80 mg/L. In the natural environment, *D. magna* is found in waters where the hardness is > 150 mg/L (Pennak, 1989). Moreover, *Daphnia magna* is most often used due to its larger size and ease of culturing. It is recommended that *D. pulex* be used in studies requiring very soft to moderately hard water (*i.e.*, if the water hardness is < 80 mg/L). Since younger organisms are generally more sensitive to toxicants than are their older counterparts, only neonates ≤ 24 hours old, and born after the second brood, are used in aquatic toxicity tests/studies.

6. Culture and maintenance of test organisms

A list of recommended culturing and holding conditions is presented below in Table 3.

6.1 EQUIPMENT AND CULTURING FACILITIES

Appropriate laboratory facilities and equipment are required in order to establish in-house cultures of daphnids. Prior to establishing in-house *Daphnia* sp. cultures, the minimum requirements of temperature and light control, standard laboratory equipment, and good quality water and food need to be met.

Although *Daphnia* sp. can survive at temperatures between 2 - 30°C, temperature should be controlled by physical means. Environmental chambers, water baths, aquarium heaters, or a temperature-controlled room may be used to accomplish this. Moreover, a source of "cool white" fluorescent light having the required intensity, together with appropriate timing apparatus, should also be established. Finally, an ample supply of good quality food and water must be available prior to establishing a permanent *Daphnia* sp. culture.

Table 3. Recommended conditions for culturing and holding Daphnia sp.

Aspect	Details
Source of *Daphnia* sp. culture	Commercial suppliers, university, governmental or private laboratories (see Tab. 10).
Culture vessels	Plastic or glass made of non-toxic material.
Culture water	Reconstituted water (with the addition of vitamin B_{12} and selenium) or good quality natural waters (*e.g.*, groundwater, surface water, or dechlorinated tap water), or well-defined culture media, such as M4 or M7 (OECD, 1998).
Temperature	$20 \pm 2°C$.
Dissolved oxygen	> 60 and $< 100\%$ saturation.
pH	6.0 to 8.5.
Total hardness	• As close as possible to control dilution water hardness to avoid osmotic stress (within 20% of control/dilution water; Environment Canada, 1990b; 2000). • Species-dependent: *D. magna* > 80 mg/L; and *D. pulex* < 80 mg/L.
Water quality monitoring	Temperature, dissolved oxygen, and pH, all monitored on a weekly basis (minimum).
Lighting	Avoid natural lighting. Fluorescent "cool-white" with 16 hours light, 8 hour dark photoperiod; 400 to 800 lux intensity at water surface.
Feeding regime	Daily (if possible); green algae (a mixture of 2 or more algal species; 4-7 day old stock is recommended) or a mixture of green algae (as above) and YCT (Yeast, Cerophyll™, Trout chow).
Test organism handling	Minimal, using a wide-bore pipette or siphon.
Culture health criteria	Brood stock with no ephippia, $\leq 25\%$ weekly cumulative mortality, time to first brood ≤ 12 days, 2-5 week-old females producing ≥ 15 young per brood, on average.

For daphnids to be successfully cultured in the laboratory, the facility must be established in a location isolated from any disturbances, dust, fumes and/or odours. Culture vessels and all required supplies must be made of non-toxic material (preferably plastic or glass) and must not come into contact with copper, lead, brass, galvanized material, rubber or any other known toxic material. Any materials and equipment (*i.e.*, glassware, plastic ware, pipettes, and/or nets) used in the facility should be handled separately; specifically, equipment should be labeled 'culturing' or 'testing'. In order to eliminate the potential of cross-contamination or possible contamination due to improper cleaning or re-use, it is strongly recommended that some equipment be used exclusively for culturing.

A partial list of equipment recommended for culturing is provided below:

- plastic or glass vessels of various sizes, preferably clear for light filtering;
- Pasteur pipettes (disposable, glass);
- Petri dishes;
- dissecting microscope(s);
- standard laboratory equipment (*i.e.,* thermometers, probes, etc.) to monitor temperature, dissolved oxygen, pH, conductivity, hardness, and light intensity;
- analytical balance;
- light table/box; and,
- standard laboratory chemicals used for synthetic water preparation.

6.2 EXTERNAL PROCUREMENT OF TEST ORGANISMS

Stock cultures of daphnids may be obtained from a number of sources, including commercial suppliers, university or government research laboratories, and/or private consulting laboratories. A list of some North American suppliers is provided in Section 12. Prior to, or during the process of ordering a new culture, it is recommended that a reliable courier (*i.e.*, one that is able to guarantee delivery of organisms within 24 hours) be chosen to transport the new culture to the laboratory. In order to establish an in-house culture, a small number of organisms (*i.e.*, 10-20) is required.

Upon arrival, all organisms should be carefully observed for any signs of stress or disease. All organisms used to establish in-house cultures should be in good physical condition and should swim freely. Moreover, new cultures should be quarantined upon arrival until the "founder" (culture representative) meets the health criteria outlined below in Section 6.6. Until that time, the newly arrived *Daphnia* sp. stock must not be mixed with any existing cultures. The quarantined organisms should be held separately in clearly-labeled containers, preferably in another culturing room or in an environmental chamber dedicated to quarantine. Usually, the quarantine period should last approximately one month, after which time an evaluation of the new *Daphnia* sp. stock can be conducted.

6.3 CULTURE WATER

Water used for culturing *Daphnia* sp. should be of the highest quality. Although all *Daphnia* sp. can be cultured in "natural" water (*i.e.*, groundwater, surface water,

Table 4. Preparation of reconstituted water using reagent-grade chemicals (from U.S. EPA, 2002b,c).

Hardness/quality	Reagent added (mg/L)[1]				Approximate final water		
	$NaHCO_3$	$CaSO_4 \cdot 2H_2O$	$MgSO_4$	KCl	pH[2]	$Hardness$[3]	$Alkalinity$[3]
Very soft	12.0	7.5	7.5	0.5	6.4-6.8	10-13	10-13
Soft	48.0	30.0	30.0	2.0	7.2-7.6	40-48	30-35
Moderately hard	96.0	60.0	60.0	4.0	7.4-7.8	80-100	57-64
Hard	192.0	120.0	120.0	8.0	7.6-8.0	160-180	110-120
Very hard	384	240.0	240.0	16.0	8.0-8.4	280-320	225-245

[1] Analytical-grade chemicals added to de-ionized water.
[2] Approximate equilibrium of pH post-aeration.
[3] Expressed as mg $CaCO_3$/L

dechlorinated tap water), the use of reconstituted water (containing vitamin B_{12} and selenium) is highly recommended, for the following reasons:

- it can be relatively easily prepared in the laboratory;
- is of known quality;
- it will support organism survival, growth and reproduction;
- it will generate reproducible results; and,
- it has been demonstrated to yield results comparable across laboratories.

Reconstituted hard water (*i.e.*, total hardness of 120-180 mg/L as $CaCO_3$), mixed with natural water and supplemented with vitamin B_{12} and selenium, should be used for culturing *D. magna*, and synthetic moderately hard water (*i.e.*, total hardness of 60-100 mg/L as $CaCO_3$) is recommended for culturing *D. pulex*. Only de-ionized water, such as Millipore MilliQ® or Super Q® water should be used for the preparation of reconstituted water. Also, only analytical grade chemicals should be used for the preparation of reconstituted water. The chemicals should be carefully weighed using an analytical balance, dissolved in water, and vigorously aerated overnight in order to buffer the pH. Physico-chemical parameters of the water should always be monitored and recorded prior to use. The initial pH of culturing water should be between 6.0 and 8.5 (preferably 7.0 to 8.0). Table 4 provides guidelines for the preparation of reconstituted water of desired hardness.

But reconstituted water alone is generally unable to sustain active and healthy *D. magna* neonates for a long duration. For long-term sustainability of the cultures, it is recommended that a mixture of reconstituted and "natural" waters is used. For example, our laboratory has successfully cultured *Daphnia magna* for over 15 years, using a mixture of reconstituted and dechlorinated tap water in a 3:1 ratio, with a vitamin B_{12} supplement (at a concentration of 50 µg/L). To ensure high quality of the "natural" water used, OECD (2004) recommends measuring some chemical parameters at least twice a year (or when it is suspected that water quality has changed significantly) (Tab. 5).

Table 5. Some chemical parameters of acceptable natural waters (adapted from OECD, 1998).

Parameters	**Concentrations**
Particulate matter	< 20 mg/L
TOC	< 2 mg/L
Un-ionized ammonia	< 1 µg/L
Residual chlorine	< 10 µg/L
Total organophosphorus pesticides	< 50 ng/L
Total organochloride pesticides plus PCBs	< 50 ng/L
Total organochlorine	< 25 ng/L
Cu, Pb, Zn, Hg, Cd, Ni	< 1 µg/L

If dechlorinated tap water is used, it is highly desirable to conduct chlorine analysis on a regular basis. If the culture water is from a surface or ground water source, conductivity and TOC (or COD) should be measured daily. Dissolved oxygen levels should be maintained at 60 to 100% saturation. It is recommended that culture water be monitored on a regular basis (*i.e.*, preferably daily, but at a minimum, weekly); all records of this monitoring activity should be logged, and kept on file.

6.4 CULTURE VESSELS

Any clear glass or plastic vessels, made of non-toxic material may be used for culturing, as long as they contain an adequate volume of the culture medium. All new glassware should be pre-washed and soaked in de-ionized water prior to use. After the culture is established, all culturing vessels should be scrubbed in hot chlorinated water and left to dry prior to its further use on a weekly basis. This is done in order to remove accumulated food, metabolic wastes, slime and/or other debris. If calcium build-up is noted, vessels should be rinsed with 10% nitric acid, followed by multiple rinses in tap water and soaked for 24 hours in de-ionized water, then dried prior to use. One to four litre glass vessels are recommended, depending on organism loading in each vessel. For optimal growth and reproduction of daphnids, a minimum of 15 to 25 mL of water is required for each individual female adult to promote rapid growth and high reproduction; a higher water volume (per individual) results in faster growth and higher reproduction.

6.5 LIGHT, TEMPERATURE AND FEEDING REGIMES

Standard ambient laboratory light is sufficient for culturing *Daphnia* sp. The health and survival of daphnids does not appear to be influenced by either very low or very high light intensities. However, for successful culture maintenance, a minimum photoperiod of 16 hours light is required. The rationale for this requirement relates to the fact that less than 8 hours of light will induce the production of males in the culture, which is undesirable, while 16 hours of light will stimulate asexual reproduction, which is preferable (Buikema et al., 1980). For standardization purposes, 400-800 lux light intensity at the water surface should be applied.

With respect to temperature, all cultures should be maintained at $20 \pm 2°C$ at all times. If temperatures are outside this range, they should be gradually adjusted at a rate not exceeding $3°C$ per day until it reaches $20 \pm 2°C$ and held at that temperature for a minimum of two weeks. As outlined above, temperature may be controlled by the use of environmental chambers, water baths, aquarium heaters, or a temperature-controlled room.

For normal daphnid survival, growth, and reproduction, proper feeding is essential. Ideally, a mixed diet of at least two green algal species (*e.g.*, *Selenastrum capricornutum, Chlorella vulgaris*) supplemented with YCT (this is optional) yield the best results in *Daphnia* sp. survival and reproduction. The addition of 5 to 10 mL of concentrated algal stock (ideally containing $30\text{-}35 \cdot 10^6$ algal cells/mL) and 0.5 mL of YCT per litre of culture water is recommended. OECD (2004) recommends using an algal quantity per daphnid per day equivalent to 0.2 mg/L TOC. (Note: Details of algal culture are described in Chapter 3 of this volume. Details of YCT preparation are presented in Appendix 1).

6.6 ACCLIMATION OF NEWLY ACQUIRED STOCK

Upon arrival, all stock cultures received at the laboratory should be carefully inspected, and the number of live and dead organisms, water temperature and hardness should be recorded. A culturing log book/sheet should be established in order to document all relevant details pertaining to the source and date of shipment arrival, water quality, and other required culturing information (*i.e.*, species, receiver, stock number, etc.).

If the water temperature upon arrival is within 3°C of the laboratory culture water, organism acclimation can commence immediately. However, if water temperature differs substantially (*i.e.*, > 3°C difference), acclimation should occur over a longer period, within which temperature changes (*i.e.*, increase or decrease) should not exceed more than 3°C per day. Acclimation should commence by preparing a culture vessel (*e.g.*, 2 L glass jar, plastic beaker or bowl) filled with laboratory culture water, as described above. Culture water parameters should be measured and recorded, as follows: dissolved oxygen, pH, conductivity, hardness, alkalinity and temperature. The culture water should be equilibrated to match the water temperature upon arrival; approximately 25% of the daphnids should then be transferred to one of the culture vessels. The remaining *Daphnia* sp. stock should be kept in the shipping container, which can be gently aerated at this time. The organisms should be fed approximately 10 mL of green algae and 500 µL of YCT (the culture water should remain a slight green color). The animals should be observed during the first 24 hours, to confirm that they appear healthy. Subsequently, the remaining individuals should be transferred to other culture vessels at various intervals over the next two days. It is strongly recommended that new cultures be established in several containers, in the event of unforeseen problems with the culture media, the newly-arrived organisms, or the prepared food. All organisms to be used in the culture should be handled carefully and gently, but as little as possible. Any organisms that are dropped, exposed to air, or injured during handling should be discarded. Wide-bore disposable glass pipettes fitted with a bulb should be used for transferring organisms. All organisms to be used in testing must meet the culture health criteria specified below. After the organisms have been properly acclimated, the *Daphnia* sp. cultures should be initiated using only one organism. A 'clonal' culture may be established after three or four days of acclimation and careful observation of the original organisms. This is done by selecting one healthy individual to initiate a quality control (QC) culture. One neonate (referred to as the 'founder') is transferred to a freshly-prepared 200 mL culture vessel. Care should be exercised when transferring daphnids with the blunt end of a pipette and releasing it underneath the water surface. If a daphnid is exposed to air, the air bubbles may get trapped within the carapace, preventing it from swimming freely (*i.e.*, the daphnid may initially float at the water surface; the surface tension may immobilize it and subsequently cause death). This culture is monitored through three broods, as described further in Section 6.8. It usually takes approximately 7-10 days for the *Daphnia* sp. to produce its first brood, and approximately 13 to16 days to produce three broods. Careful monitoring, enumeration of neonate production, and proper record keeping in a culture logbook will help determine the health status of the culture. It is essential that the 'founder' meet the following health criteria:

- no more than 25% mortality should occur in the brood stock within 7 days prior to testing;
- no ephippia must be present in the parental stock;
- the time-to-first-brood must occur within 12 days of culture initiation; and,
- brood stock females 2-8 weeks old must produce an average of 15 or more young per brood.

It is critical to monitor the culture for signs of stress. Signs of culture stress include: excessive mortality in the culture containers, overcrowding, food deprivation, production of males, and visual occurrence of ephippial eggs. If the 'founder' does not meet the health criteria as described above, then the 'founder' and its neonates are not to be used for testing, and should be discarded. A culture should then be re-initiated by selecting another neonate from the mass culture, and repeating the procedure described above. If the 'founder' meets the health criteria, then it should be transferred to a 2 L jar of culture water. This organism should be allowed to produce one more brood; this brood will now serve as the laboratory's culture source. The original culture 'founder' should be saved, preserved, and mounted on a slide for taxonomic identification; this slide should be kept on file for verification of species identification. The recommended procedure for slide preparation is provided in Section 12.2.

6.7 ESTABLISHMENT OF A MASS CULTURE

The use of a mass culture is recommended for laboratories that routinely conduct *Daphnia* sp. toxicity testing. The mass culture serves as a backup brood source in case of a massive population failure. Mass cultures contain organisms of various sizes and ages, and must not be (directly) used for testing. Depending on the laboratory's needs, the mass culture should be reset once every other month or more frequently, if required. A mass culture may be initiated in a small aquarium (approximately 8 to 10 L), a large glass jar (3 L), or any other suitable container filled with culture water and fed approximately 20 mL of green algal culture. Healthy young neonates obtained from culture vessels or even those produced by a culture representative (ensuring that it is from a third or subsequent brood) should be transferred by pipette into mass culture containers to establish and/or reset the mass culture. Pertinent information, including the date, the founder's batch number, and the number of young used to start the culture, should be written on the side of the aquarium or jar, and in the culture logbook. *Daphnia* sp. mass cultures should be fed at least twice weekly with approximately 10 mL of concentrated algal stock (Note: feeding mass cultures with YCT is not recommended, due to the potential bacterial accumulation, and subsequent dissolved oxygen decrease). Contents of the mass culture vessel should be stirred at least twice weekly to re-suspend algae. In addition, the mass culture should be 'thinned' every so often (weekly) in order to decrease the rapidly-growing population and eliminate culture stress due to overcrowding, lack of food, and/or other factors. At the same time, half of the culture water should be replenished to reduce metabolic waste build-up. 'Thinning' can be accomplished simply by drawing a small aquarium net two to three times through the culture and disposing of the captured daphnids. Monitoring of the mass culture for signs of stress (*e.g.*, occurrence of ephippia, loss of coloration, lack of eggs in mature females) is recommended; these observations should be recorded in the culture logbook. If any signs of culture stress are observed, the entire mass culture should be discarded and another one initiated immediately. It is essential to keep detailed records of culture initiation, health monitoring, physico-chemical parameter measurements of culture water, culture disposal, other treatments, and any other critical information.

6.8 CULTURE QUALITY CONTROL (QC)

Health of the culture should be monitored closely by frequent checks of time of first brood release, average young production, and mortality. The easiest way to accomplish this is to set up a representative jar cultured under conditions identical to the main culture, but different in that only a single organism is loaded into a smaller (*i.e.,* 200 mL beaker) culture vessel (*i.e.*, this is a microcosm of a large vessel where the loading density (*i.e.*, mL of culture water per organism) and culture conditions (including feeding regime) are identical. For example, if culturing is conducted in a 1.5 L vessel containing 15 adult daphnids and fed 5 mL of algae (and optionally 2 mL of YCT), then a single culture representative should be held in a vessel containing 150 mL of culture water and fed 0.33 mL algae with 0.13 mL YCT and held under the same culturing conditions (*e.g.*, light, photoperiod, temperature, etc.) as the main stock.

The QC vessel and the container with the representative neonate should be fed with algae twice weekly or more frequently if required, so that the jar always has a light green appearance. The feeding regime will depend upon the algal density (*i.e.*, the more dense the culture, the less food will need to be added). A recommended feeding regime is 5 mL of green algae (*e.g., Selenastrum capricornutum;* concentrated at 30 to $35 \cdot 10^6$ cell/mL), and 500 µL of a YCT suspension (at 1.6 to 1.9 g/L as dry solids) for each litre of culture water. Each culture vessel should be carefully inspected daily for neonate production. When the representative neonate has reproduced its first brood (usually within 7-10 days), the maturing daphnid from the small representative jar should be transferred to a clean beaker filled with culture water and food; the newly-born neonates are then enumerated. At the same time, physico-chemical parameters of the culture water are measured and recorded. In addition, the date and number of neonates released per brood, time-to-first-brood, or average number of young per brood should be recorded in the culture logbook. When the representative daphnids produce young, the larger vessel containing 40 adults may produce young as well (there is high possibility that most, but not all, will produce young). Therefore, the large culture vessel should be carefully monitored for young production. If the daphnids in this jar have also reproduced, a small volume of the culture water is sub-sampled to measure and record physico-chemical parameters (under the appropriate culture batch number). The water should be replaced, maturing daphnids should be enumerated and transferred, and the newly-born neonates should be discarded. If the brood in the jar has produced its 3rd brood and if the neonates are $\leq$ 24-hour old, these young may be collected and used in toxicity testing. Regardless of neonate production, the large QC containers should be cleaned at a minimum of once per week to remove accumulated build-up of metabolic wastes, slime and shed carapaces of growing organisms.

Generally, it takes between 12 and 16 days for daphnids to 'pass' the *Daphnia* sp. culture health criteria (as indicated in Section 6.6 above).

When the QC representative in the 200 mL beaker has successfully produced three healthy broods according to the criteria listed above, the adults should be transferred from large 3 L vessels to smaller 1 L (or similar) culturing containers, in which they are held for the next three to five weeks. It is advisable to split the main culture into sub-cultures to facilitate simpler handling, and to sustain a reasonable loading density. The representative adult held in the 200 mL beaker (containing 100 mL culture media) is kept separate to continue to monitor the average number of neonates produced by each batch.

If the representative neonate takes longer than 12 days to produce its first brood, if ephippia are present in the culture vessel, or if mortality exceeds 25% within the culture vessel, and/or if young production is ≤ 15 young/brood, then daphnids produced from that batch must not be used for testing. Therefore, all daphnids from that batch must be discarded; these details should be recorded in a logbook. A new culture should be initiated as soon as possible.

6.9 MAIN CULTURE VESSELS

Main culture vessels, also referred to as 'overnight containers', are established to hold daphnids that have met culture health criteria, and are ready to produce organisms to be used in testing. Usually, females 2-8 weeks old are held in overnight containers. Once the daphnids in the QC jar(s) have met the health criteria, organisms held in the large vessels should be transferred to overnight containers according to the procedure described below in Box 1. The large vessel should then be split into a few smaller containers for easy handling and monitoring, and to protect against loss of the entire population (if held in one container only). For example, the large 3 L container holding 45 adult daphnids would be transferred into three 1 L jars, each holding 15 organisms. In this case, the loading density will be approximately 67 mL of culture water per individual.

Box 1. Transfer of Daphnids to overnight containers.

- Fill two or more 1 L overnight containers with culture water, and label each with the batch number (assigned from the passing QC vessel) and assign the vessel number. Label two or more other overnight containers with the same numbers and set them aside for the next day's transfer. Remember that all culturing containers must be of the same make (glass or plastic), holding capacity and shape.

- Transfer the adults from the QC vessel to the first set of overnight culturing containers. Do not transfer more than 15 adults to each container. There should be a minimum of 25 mL and preferably 100 mL of culture water per each adult daphnid, otherwise the container will be overcrowded; this may affect young production.

Box 1 (continued). Transfer of Daphnids to overnight containers.

- The water in overnight containers must be renewed after the young are born, but it is not necessary to renew water in bowls where no neonates were produced. Regardless of no young production, it is recommended that a daily refreshing of ¼ of the overlying water in every container be implemented, in order to re-suspend settled algae. Moreover, water exchange is necessary to maintain dissolved oxygen levels between 90 and 100% saturation. If this is not done within a few days, waste by-products will accumulate, detritus and slime will build up on the sides of the vessels, bacteria and other micro-organisms will grow, and an accumulation of unconsumed algae will diminish the available oxygen. To avoid a possible population crash due to any of these factors, it is recommended that the culture water be fully refreshed every 3rd day, at a minimum.
- Remove the second set of the previously-prepared empty overnight containers, labeled to match the vessels housing daphnids and line them on the bench. Fill the containers with culture water and equilibrate temperature to match the culture.
- Using a pipette fitted with a rubber bulb, transfer and enumerate the adults from each overnight container to the corresponding refreshed vessel, making sure that the temperature difference between bowls is $\leq 1°C$. Record the time of transfer; because you will need to track the birth time of neonates (remember that you must use ≤ 24 hours old for testing, so the time of young production will be important).
- Sub-sample 50 mL from one of the overnight containers, and monitor DO, pH, conductivity and temperature. Record all parameters in the logbook.
Feed each container housing adult *Daphnia* sp. with approximately 5 mL of algae and 500 µL of YCT suspension. OECD (2004) recommends using an algal quantity per daphnid per day equivalent to 0.2 mg/L TOC.
- Estimate the number of neonates produced in each remaining container and record the overnight production in the culture logbook. Set aside the ≤ 24 hrs old neonates for use in testing. - Record any observations of stress or mortality.
- Inspect the single representative daphnid in the 200 mL beaker. If neonates are present, transfer the adult to a beaker containing 100 mL fresh culture water and add 0.5 mL of algae and 50 µL of YCT suspension. As above, OECD (2004) recommends using an algal quantity per daphnid per day equivalent to 0.2 mg/L TOC. Record the number of neonates in the logbook and discard. If there are no young produced in the QC jar, the algae in the jar should be re-suspended daily, the jar must be cleaned thoroughly at least once per week and fed, as described as above.
- Observe and record any abnormal swimming behaviour, mortality or ephippia present in the overnight containers or the 200 mL beaker. Discard the entire vessel containing *Daphnia* sp., if there is excessive mortality or ephippia present.
- If neonates are not required for testing, or after-test neonates have been transferred, clean bowls in the culture area using a scrub pad and hot tap water (no soap or other disinfectants should be used since they may be toxic to daphnids). Turn the containers upside down on the counter, and air dry overnight. Always keep one bowl of neonates on the culturing table, in case a sample arrives late, or in case a test or mass culture needs resetting.
- Discard adult daphnids in overnight containers 8 weeks subsequent to their initiation date, since young production will gradually decrease in senescent females.

7. Testing procedures

While the acute test design is usually conducted as a static test, both test designs (*i.e.*, 48-hour acute, 21-day chronic life cycle/reproduction) may be conducted as static-renewal or flow-through. Most often, acute tests are conducted under static conditions (*i.e.*, the test solutions are prepared only once). The chronic test design is most often conducted as a static-renewal test, in which the test solutions are prepared and exchanged daily (or at least thrice weekly; *e.g.*, Monday, Wednesday and Friday), or as a flow-through test, in which the test solutions are continuously renewed at a rate of one full volume exchange per day. Flow-through tests are usually conducted from preparation of stock solution delivered with a peristaltic pump, mixing it with control/dilution water in a dilutor system, and distributed to test vessels through stainless steel or Tygon™ tubing (see Fig. 2).

Prior to test initiation, the sample being tested should be handled in an appropriate manner, all test solutions should be prepared and monitored, organism availability should be confirmed, and all other test logistics should be met. Recommended test conditions for acute testing with *Daphnia* sp. are presented in Table 6, while conditions for chronic testing are outlined in Table 7.

A partial list of equipment recommended for conducting *Daphnia* sp. testing includes:

- graduated cylinder(s);
- volumetric flask(s);
- mixing containers;
- testing containers;
- volumetric pipette(s), repetition pipette(s);
- waterproof marker, labels and colored tape;
- standard laboratory equipment to monitor test solutions: DO, pH, and conductivity probes, light meter(s), thermometer(s), amperometric titrator to measure chlorine, Hach® kit to measure ammonia, probe or titrator to measure hardness;
- analytical balance (with 10 µg accuracy);
- environmental chamber, incubator, water bath or equivalent for temperature control;
- light table/box;
- dissecting microscope;
- squirt bottle(s);
- Petri dishes of various sizes;
- disposable glass pipettes fitted with rubber bulbs;
- magnetic stirrer, stir bars, bar removal magnet;
- small aquarium nets to capture adult daphnids;
- aluminum dishes;
- dessicator; and,
- standard laboratory chemicals for reconstituted water preparation and reference toxicant testing.

Table 6. *Recommended test conditions for acute testing with* Daphnia *sp.*

Source of test organisms	Preferably obtained from established in-house cultures that have met culture health conditions; commercial suppliers or university, government or private laboratories.
Organism age	≤ 24 hours old.
Brood number	3rd or subsequent brood.
Loading density	≥ 15 mL per each organism (Environment Canada method; Environment Canada, 2000). ≥ 2 mL per organism (ISO method; ISO, 1996).
Number of replicates	Minimum of 3 in a single-concentration test, at least one or more in an LC50 test; for chemical product testing: preferably 2 in acute LC50 test.
Number of organisms per vessel	10
Culture vessels	Plastic or glass made of non-toxic material.
Control/dilution water	Preferably reconstituted water (see Tab. 4); good quality natural waters (groundwater, surface water, or dechlorinated tap water) may be used alone or in combination with reconstituted water.
Temperature	$20 \pm 2°C$.
Dissolved oxygen	> 40 and < 100% saturation.
pH	6.0 to 9.0 (see Tab. 4).
Hardness	• As close as possible to control dilution water hardness to avoid osmotic stress (within 20% of control/dilution water; Environment Canada, 1990b; 2000). • Species-dependent: *D. magna* > 80 mg/L; and *D. pulex* < 80 mg/L.
Lighting	Avoid natural lighting. Fluorescent "cool-white" with 16 hours light, 8 hours dark photoperiod with 400 to 800 lux intensity at water surface.
Feeding during the test	None.
Test solution renewal	None.

Table 6 (continued). Recommended test conditions for acute testing with Daphnia *sp.*

Monitoring physico-chemical parameters	Temperature, DO, pH, conductivity, and hardness monitored at the minimum at the beginning and end of test. Observations of organisms for mortality and/or immobility at the beginning and end of test.
Handling	Minimal, using wide-bore pipette or siphon.
Assessment endpoints	Mortality and/or immobility.
Test validity criteria	• Control organisms must have ≤ 10% mortality and ≤ 10% immobile (Environment Canada, 2000). • DO ≥ 2 mg/L (~20% saturation); sensitivity to potassium dichromate > 0.6 and < 1.7 mg/L (ISO, 1996).

Table 7. Recommended test conditions for chronic testing with Daphnia *sp.*

Source of test organisms	Preferably obtained from established in-house culture; commercial suppliers or university, government or private laboratories.
Organism age	≤ 24 hours old.
Brood number	3rd or more subsequent brood.
Loading density	≥ 50 mL per each organism (OECD, 1998).
Number of replicates	≥ 10 in static-renewal design; > 2 (4 recommended) in flow-through design (OECD, 1998).
Number of organisms per vessel	1 in static-renewal design; 10 in flow-through design (OECD, 1998).
Culture vessels	Plastic or glass made of non-toxic material; may be fitted with mesh on the bottom.
Control/dilution water	Reconstituted water (with the addition of vitamin B_{12} and selenium) or good quality natural waters (*e.g.*, groundwater, surface water, or dechlorinated tap water), or well-defined media, such as M4 or M7 (OECD, 1998).
Temperature	20 ± 2°C.
Dissolved oxygen	> 40 and < 100% saturation.
pH	6.0 to 8.5.

Table 7 (continued). Recommended test conditions for chronic testing with Daphnia *sp.*

Hardness	Within 20% of control/dilution water. Species-dependent: *D. magna* > 80 mg/L; and *D. pulex* < 80 mg/L.
Lighting	Avoid natural lighting. Fluorescent "cool-white" with 16 hours light, 8 hour dark photoperiod with 400 to 800 lux intensity at water surface.
Feeding during the test	Once daily in a static-renewal test; recommended three times daily in flow-through test (OECD, 1998).
Feeding regime	3.3 mL concentrated (3.0 to 3.5 $\cdot$ 10^6 cell/mL) green algae and 1.3 mL YCT daily per each adult in each test vessel; or, one or more green algal species using a regime of 0.1– 0.2 mg C/daphnid/day; if YCT is added, it should be less than 2 mg TOC/L (OECD, 1998).
Test solution renewal	Once daily in static-renewal design; three times daily in semi-static-renewal design, or continuously in flow-through design.
Monitoring physico-chemical parameters	• Temperature, DO, pH, conductivity, hardness monitored at test initiation and in the new solution at each renewal and end of static-renewal test design. • Temperature, DO, pH, conductivity, hardness monitored at a minimum at the beginning, weekly thereafter, and at the end of the test (flow-through test design). • Observations of organisms for mortality, immobility daily and young production at least 3 times per week and preferably once daily (at least three times per week).
Handling	Minimal, using wide-bore pipette or siphon.
Endpoints	Mortality, immobility, young production, time to first brood, growth (dry weight or body length).
Test validity criteria	Control organisms must have ≤ 20% mortality, on average ≥ 60 young produced by each surviving control adult female, no ephippia must be present in control.

7.1 HEALTH AND SAFETY CONCERNS

Safety precautions and proper personal protective equipment (PPE) should be available to technicians when they are testing samples of unknown properties. In addition, the laboratory should be equipped with a fully-stocked first aid kit, a fire extinguisher, an emergency shower, an eye-wash station, and a phone, at a minimum.

Some wastewater samples contain live bacteria and pathogenic protozoans, some may be corrosive or alkaline, and others may contain noxious gases. Therefore, technical staff undertaking testing should be careful when handling such samples and should wear PPE, as appropriate. For example, a lab coat, safety glasses, vinyl gloves, and close-toed shoes are often adequate to protect lab staff while testing routine wastewater samples. Since the composition of most wastewater and new commercial chemical samples is often unknown or poorly characterized, they should initially be considered potentially hazardous to human health. Whenever dealing with samples potentially containing untreated sludge or human waste, staff should take all health and safety precautions prior to handling them. Moreover, it is recommended that personnel handling environmental samples immunize themselves for Tetanus, Hepatitis B, Polio and Typhoid.

All chemical product samples should arrive for testing accompanied by a Material Safety Data Sheet (MSDS), with information pertaining to the physical and chemical properties of the substance, including (but not limited to): water solubility, vapour pressure, chemical stability, mammalian toxicity, potential carcinogenicity, purity, lot number and storage and handling requirements. At all times, samples should be clearly labeled (or coded) and stored as recommended by the MSDS or testing procedures. Strong acids and oxidants, and volatile and flammable substances should be stored appropriately, and always handled under a fume hood.

All wastes generated during testing should be handled and disposed of safely, and in an appropriate manner. It is recommended that each testing laboratory have its own disposal requirements based on guidelines and recommendations set by the appropriate regulatory jurisdiction (*e.g.,* municipal, provincial/state, or federal).

7.2 SAMPLE PREPARATION

For sample preparation procedures, slightly different approaches are applied for handling wastewater (*e.g.,* effluents, elutriates, leachates, or pore waters) and chemical or chemical product samples. Handling of both sample types is described separately below. Regardless of sample type, each sample should be accompanied by sample submission and/or chain-of-custody records.

7.2.1 *Wastewater samples*

When a wastewater sample arrives at the laboratory for testing, the sample temperature must be measured immediately (*i.e.,* no later than within 1 hour of sample arrival at the laboratory). A sample arriving in several containers must be composited prior to testing to ensure its homogeneity. When compositing these sub-samples, the collection containers must be agitated thoroughly prior to pouring to ensure re-suspension of any settleable solids in the samples. Samples may be composited by pouring the contents of all containers into a clean plastic or glass container large enough to hold the contents; the sample must be well mixed. Relevant information pertaining to the sample (*e.g.,* colour, turbidity, odour, presence of particulates, flocculants, etc.) should be observed and recorded on the sample submission and/or chain-of-custody form. If necessary (*i.e.,* when tests are initiated on the day they arrive), the temperature of samples should be adjusted by heating or cooling using a water bath or other method (*e.g.,* environmental chamber). Samples or test solutions must not be heated by immersion heaters since this may alter chemical constituents in the sample, and potentially its toxicity.

Any unused portions of the sample(s) should be returned to their original container(s), appropriately labeled, and stored without air space (e.g., best accomplished using supple plastic bags or flexible carboys), in darkness, at $4 \pm 2°C$. If the sample is stored in multiple containers, each container should be labelled with a waterproof marker with the number (i.e., 1 of 3, 2 of 3, and 3 of 3), sample code (e.g., station number or location, etc.), sampling date, and date of receipt. If the test cannot be initiated on the day of arrival to the laboratory, the sample should be stored as indicated above (i.e., without air space, in darkness, at 4 ± 2 °C).

Wastewater sample test solutions are prepared by diluting the sample with control/dilution water to generate a range of different test concentrations. Prior to test solution preparation, the initial physical and chemical parameters should be established. To prepare test solutions of wastewater samples the procedures outlined below (Box 2) should be followed.

Box 2. Wastewater sample test solution preparation.

- To measure initial parameters pour a small amount (approximately 50 mL) of sample into a small cup or beaker, and measure (at a minimum): DO, pH, conductivity, temperature, and hardness. Other chemical parameters, such as total ammonia, total residual chlorine, and alkalinity may also be required, depending on the testing objectives. If the latter are required, a larger sample volume may be needed to conduct monitoring (e.g., to measure total residual chlorine using an amperometric titrator, a 200 mL minimum sample volume is needed, or more if results are duplicated). For total ammonia measurements using a Hach kit and the Nessler method, at least 25 mL of sample are required (more if measurement is to be replicated). When testing with wastewater samples containing ammonia or chlorine, duplicate measurements are recommended for quality control purposes, and to ensure the accuracy of measurements.
- If (and only if) the DO of wastewater samples is < 40% or > 100% of air saturation, moderately pre-aerate the sample at an aeration rate of 25-50 mL/min/L using a new, disposable Pasteur pipette for no longer than 30 minutes, regardless of whether or not 40-100% air saturation is achieved; excessive aeration may strip the test sample of volatile substances, thereby altering its toxicity.
- If the pH is outside the range 6.0 - 9.0, one may consideration additional parallel testing with pH adjustment. If pH is outside this range, mortality may occur due to pH alone. Sample pH is not be adjusted unless specifically requested by the client, or if outlined specifically in the study objectives. Samples tested in regulatory monitoring programs must never be tested with pH adjustment.
- All tests conducted using *Daphnia magna* as the test species should be tested with samples yielding a hardness of at least 25 mg/L as $CaCO_3$. If the sample hardness is < 25 mg/L, *D. magna* mortality may occur due to low hardness; therefore, sample hardness may be adjusted (see next paragraph) to a minimum of 25 mg/L (or higher depending on study objectives) prior to preparing dilutions. In general, hardness must be measured prior to test initiation if its conductivity is < 100 µmhos/cm. When testing with *D. pulex*, sample hardness adjustment is not required.

Box 2 (continued). Wastewater sample test solution preparation.

- If sample hardness adjustment is required, it should be done prior to test concentration preparation. To adjust sample hardness, the following steps should be followed: if required, the hardness of each sample may be adjusted to particular hardness using reagent-grade chemicals in the ratio: 1.6 $NaHCO_3$ to 1.0 $CaSO_4 \cdot 2H_2O$ to 1.0 $MgSO_4$ to 0.067 KCl. For each desired incremental increase of 5 mg/L hardness, add to each 1 litre of sample the following chemicals: 6.0 mg of $NaHCO_3$ plus 3.75 mg of $CaSO_4 \cdot 2H_2O$ plus 3.75 mg of $MgSO_4$ plus 0.25 mg of KCl (Environment Canada, 2000). Determine the volume of hardness-adjusted sample required for testing/monitoring and calculate the quantity of chemicals required to obtain the necessary hardness. Ensure that calculations are checked. Measure the required sample volume and pour into a clean container, add a clean magnetic stir bar and transfer onto magnetic stirrer. Carefully weigh out the four chemicals, one at a time, using an appropriate analytical balance. Record the appropriate lot numbers of the chemicals used and all other required information on the bench sheet, including the preparation date.
- Dissolve the chemicals, one at a time, in the sample, using the magnetic stirrer. Ensure that all chemicals are completely dissolved before they are added to the barrel. Follow the same steps for all four chemicals. (Note: to completely solubilize the $CaSO_4 \cdot 2H_2O$, it must be divided into three or four separate aliquots, to be dissolvable).
- It should be noted that there are other simpler techniques for hardness adjustment. For example, preparation of a stock of hardness-adjusted solution, where 1 mL = 1 mg/L hardness, and add to the test solution.
- Wastewater samples often contain other microorganisms or zooplankton. Therefore, the sample should be examined for the presence of indigenous zooplankton species, since the presence of other crustacean species may confound test results, and require additional unnecessary work. To check for the presence of other zooplankton in the sample, simply pour a small amount of the sample into clear jar or plastic disposable cup or beaker, hold it against light (or above a light box) and carefully examine it for any live or moving organisms. If indigenous species are observed, remove them using a pipette. If there are too many organisms to effectively remove with a pipette, the sample may be screened using a 60 μm Nitex mesh or fine aquarium net. A sample should never be filtered through glass or paper filters, since this filtering may remove toxicity associated with the presence of toxicants, fibres, flocculants and other particulate matter.
- All sample pre-treatments such as sample pre-aeration, hardness adjustment, and removal of microorganisms, should be recorded on the bench sheet.

Depending upon the objectives of the study, testing may be conducted with a multi-concentration set-up to determine a 48-hour LC50, or with a single-concentration test set-up (usually replicated) to determine the toxicity of one sample (usually 100% wastewater or receiving water; or a pre-determined chemical concentration also known as load testing). Generally, the standard LC50/EC50 test includes a minimum of five test concentrations of the wastewater sample, and a control (dilution water only). Test concentrations may be selected based on either an arithmetic or geometric series. OECD (2004) recommends using a geometric series only. A dilution (or separation) factor of ≥ 0.5 is commonly used for testing. Increasing a dilution/separation factor beyond 0.5 may not necessarily increase accuracy and precision of the test endpoint. However, if based on historical data, or it is suspected that the toxicity may fall between 100% and 50% dilution, an increase in the dilution/separation factor to 0.6 or 0.7 (or similar) can be used (see examples provided in Table 8). If the toxicity of the sample is unknown, range-finding (also known as preliminary) testing is recommended, and is standard practice. Range-finding tests generally consist of fewer, widely-spaced sets of concentrations, usually prepared using 0.1 (*i.e.*, 100%, 10%, 1 %) or 0.3 (*i.e.*, 100%, 30%, 9 %) dilution/separation factors, and conducted with fewer numbers of organisms (*i.e.*, usually 5) over a short duration.

Table 8. Examples of concentration series for various dilution separation factors.

Separation factor	Examples of concentration series
1:10	100%, 10%, 1%, 0.1%...
1:3	100%, 33%, 11%, 3%...
1:2	100%, 50%, 25%, 12.5%...
1:1.28	100%, 78%, 61%, 48%...

Prior to test solution preparation, an adequate supply of laboratory control/dilution water, and all required materials, such as clean glassware, cylinder, test vessels, should be available and ready for use. Generally, for a single LC50/EC50 acute test, a 250 mL volume of each test solution is adequate, where 200 mL is used for testing, with 50 mL being used for routine physico-chemical monitoring. Therefore, although a 1 L sample is usually requested by the laboratory, a minimum of 500 mL of the wastewater sample is actually used for testing. If the test is being conducted in duplicate or triplicate, the sample volume would be increased appropriately. In addition, a much higher test solution volume will be required when conducting chronic testing, and generally 2 to 10 L of sample per renewal treatment may be required. It is recommended that the test vessel be rinsed with the prepared solution immediately subsequent to solution preparation. All test vessels should be pre-labeled with known test concentrations using a waterproof marker and/or coloured tape. Prior to test dilution preparation, each vessel should be marked with a unique sample code, date and time of test initiation, and the exposure concentration (*e.g.*, 100%, 50%, 25%, 12.5%, 6.25% and control). In addition small cups or beakers (for physico-chemical parameter monitoring) should be prepared and labeled in advance, to match the test vessels. To prepare test concentrations, the procedures in Box 3 below should be followed.

Box 3. Preparation of test exposure concentrations (multi-concentration test).

- Take a graduated cylinder of appropriate size, rinse it out with a small amount of sample, and discard the rinse.
- Pour twice the required test volume of the 100% sample (*e.g.*, 500 mL) into the graduated cylinder.
- Cover the top of the cylinder with Parafilm or a clean, gloved hand, and mix the sample by inverting it twice, and pour a small volume (*i.e.*, 50 mL) into the vessel labelled "100%".
- Swirl the sample in the vessel to rinse it out and pour it into the monitoring cup/beaker.
- Pour the required volume (*e.g.*, 200 mL of the 100% sample) into the 100% vessel and set it aside for testing. There should be 250 mL of sample left in the cylinder.
- Dilute the 250 mL of sample left in the cylinder with 250 mL of dilution water.

Box 3 (continued). Preparation of test exposure concentrations (multi-concentration test).

- Mix as above and pour out 50 mL of the 50% dilution into the test vessel labelled "50%".
- Swirl, pour into the monitoring cup/beaker, pour 200 mL into the 50% vessel and set it aside for testing.
- Continue to dilute the remaining solution in the cylinder until the lowest concentration of sample has been prepared and dispensed.
- Pour 200 mL of dilution water into the vessel marked "control". It is imperative that each test has its own separate control. Moreover, each test vessel (including the control) must contain an identical volume of solution.

A single-concentration test (also known as a "load test") is usually replicated (a minimum of three, preferably four). This design will only require 100% (or the highest pre-determined test concentration), and the same number of control replicates. For a single-concentration test, prepare and label each test vessel with the sample code, testing date and time, and replicate number. Also, prepare and label two small (50 mL) monitoring cups/beakers as 'control' and '100%'. Prepare 3 or more if desired, replicates of control water and 100% sample as follows:

Box 4. Preparation of single-concentration test vessels.

- Homogenize the 100% sample in the container in which it was sampled.
- Pour approximately 50 mL into one of the test vessels labelled 100%.
- Swirl and transfer to the next 100% vessel.
- Swirl and transfer to the last 100% vessel.
- To discard the sample, rinse it down the drain.
- Pour 200 mL of the 100% sample into each of the three vessels marked 100% (*i.e.*, replicates A, B, and C).
- Pour 50 mL of the 100% sample into a 50 mL monitoring cup/beaker; the same procedure should be conducted using dilution water, and three replicate vessels marked control (*i.e.*, replicates A, B and C).
- Again, each test must have its own separate control. Each test vessel, including the control, must contain an identical volume.

7.2.2 Chemical product samples

A range-finding (*i.e.*, preliminary) test is commonly conducted with chemical products, where the outcome of test results are completely unknown. The range-finding test is conducted prior to definitive testing in order to determine the toxicity of the test item and to narrow the set of exposure concentrations prior to definitive testing. The range-finding test consists of a widely-separated range of concentrations. Commonly, a dilution factor of 10 is used during test concentration preparation (*e.g.*, 10,000, 1,000, 100, 10, 1 and a control). The preparation of test solutions and test conditions are exactly the same as that applied in the definitive test, however, the number of replicates, test organisms, and the test duration are generally reduced. Most range-finding tests are conducted for 48 hours. Caution should always be exercised when attempting to interpret data resulting from range-finding tests, particularly when the data are required for definitive testing of a chemical for which aquatic toxicity is unknown. It is recommended that range-finding tests be conducted according to a similar design and test conditions as the planned definitive test, using, whenever possible, at least 5, but generally 10 organisms per test vessel, and conducting the test for the same duration planned for the definitive test.

Chemical product test solutions may be prepared using: 1) pure, undiluted chemical (used "as is" or solubilized in a carrier when dealing with sparingly-soluble substances); or, 2) a stock solution. On occasion, working with chemical products that are sparingly soluble in water, a "carrier" may be required in order to solubilize the chemical, but, in any case, the test substance in the test solutions should not exceed its limit of solubility in the dilution water (OECD, 1998). Depending on the nature of the test chemical, various carriers may be used. Most often, organic solvents such as methanol, acetone, and dichloromethane are selected. The concentration of the carrier should not exceed 100 mg/L (OECD, 1998), and an additional control exposure containing the carrier should be tested concurrently to eliminate any doubt regarding the impact of the carrier's toxicity on the test organisms. Most importantly, the toxicity of the chosen carrier should be known prior to its use, in order to eliminate any unnecessary re-testing.

Prior to test solution preparation, the amount of stock or "pure" chemical product required for attaining the highest concentration in the total volume required for the test, should be calculated. Acute testing will generally require a volume of at least 1 L for single replicate testing or 2 L for duplicate testing. Chronic testing may often require greater volumes, depending on the test system design. Generally, in a static-renewal test, 1 L to 2 L of each test concentration is prepared for testing and monitoring, however, a greater volume may be required, depending upon the test design. In order to eliminate possible error and to ensure that testing logistics are in order, the calculation should be re-checked by another person. Measure the appropriate volume of stock, or weigh using an analytical balance, the calculated amount of "pure" chemical and transfer it into a graduated cylinder and top up to desired volume – in this example, 0.5 L for single-replicate tests and up to 1 L for duplicate testing with dilution water (preferably reconstituted water). Ensure that water used for dilutions is at 20 ± 2°C and that it is "acceptable for use" according to the procedures described under reconstituted water preparation.

Confirmation of whether the sample should be pH- or hardness-adjusted, or manipulated in other ways should be made by consulting with a person experienced in aquatic toxicity testing, depending upon specific study objectives. If pH adjustment is necessary, it is generally implemented either by adjusting the pH of the stock or adjusting the pH of the highest test concentration (and prior to final test solution preparation). OECD (2004) indicates that if there is a marked change in the pH of the dilution water subsequent to the addition of the test substance, the test should be repeated, adjusting the pH of the stock solution to that of the dilution water prior to addition of the test substance. The pH adjustment should be made in such a way that the stock solution concentration is not changed to any significant extent, and that no chemical reaction or precipitation of the test substance is caused. These pH adjustments are made using sodium hydroxide (NaOH) and/or hydrochloric acid (HCL) stock solutions.

The procedures outlined below (Box 5) describe the preparation of test solutions for a single-replicate test (total required volume is 250 mL of each test solution). If duplicate acute testing is planned, then the volume must be doubled, and 1 L of the highest test concentration is prepared. If chronic testing is conducted, the required volume of test concentrations must be calculated and adjusted accordingly, prior to making solutions.

Box 5. Preparing test dilutions using "pure" chemical product and stock solution.

1) Prepare dilutions using the "pure" chemical product.
- Calculate the amount of chemical required to prepare the highest test concentration and request verification by another staff member.
- Weigh the appropriate amount of sample using an analytical balance.
- Rinse and fill the control vessel with 200 mL of dilution water and fill the matching 50 mL monitoring cup/beaker, and set aside.
- Remove a clean graduated cylinder and pour in approximately 100 mL of dilution water.
- Add the appropriate amount of the previously-weighed chemical sample and top it up with control/dilution water to the required volume (in this example, 500 mL).
- Cover with Parafilm or a clean, gloved hand. Mix the sample in the graduated cylinder by inverting it twice and pour 50 mL into the vessel labeled as the highest concentration. Ensure that the sample is completely solubilized and homogeneous.
- Swirl the sample in the vessel to rinse, and pour it out into the 50 mL monitoring cup/beaker.
- Pour 200 mL of the sample into the vessel labeled as the highest test concentration and set it aside. There should be 250 mL left in the cylinder. For a 0.5 dilution series, dilute the 250 mL of sample left in the cylinder with 250 mL of dilution water.
- Mix by inverting twice and pour out 50 mL of the 50% dilution into the container labeled as the next highest concentration. Swirl the 50 mL of dilution, pour into the matching monitoring cup/beaker, and pour 200 mL from the graduated cylinder into the test vessel. Set aside.
- Continue to dilute the remaining 250 mL in the cylinder until the lowest concentration of sample has been prepared and dispensed.

2) To prepare test dilutions using stock solution:
- Rinse and fill the control vessel with 200 mL of dilution water and put aside.
- Remove a clean graduated cylinder and pour approximately 100 mL of dilution water into it.
- Add the appropriate volume of stock solution and top up with control/dilution water to the required volume (500 mL).
- Cover the cylinder with Parafilm or clean gloved hand.
- Mix the sample in the graduated cylinder by inverting twice and pour 50 mL into the vessel labeled as the highest concentration to rinse the vessel.
- Swirl the sample in the vessel and pour out into the 50 mL monitoring cup/beaker, and pour 200 mL of the sample into the vessel labeled as the highest concentration and set aside. There should be 250 mL left in the cylinder.
- For a 0.5 dilution series, dilute the 250 mL of sample left in the cylinder with 250 mL of dilution water.
- Mix by inverting twice and pour out 50 mL of the 50% dilution into the container labeled as the next highest concentration.
- Swirl the 50 mL of dilution water, pour into the matching monitoring cup/beaker, and pour 200 mL from the graduated cylinder into the test vessel. Set aside.
- Continue to dilute the remaining 250 mL in the cylinder until the lowest concentration of sample has been prepared and dispensed.

Sample monitoring is conducted as described above.

7.3 TEST SET-UP

After all of the test solutions have been prepared, sample parameters can then be monitored; pH, conductivity, dissolved oxygen and temperature are monitored in each monitoring cup/beaker and recorded on the bench sheet. Dissolved oxygen in the solutions should be within 40%-100% saturation. If any solution is outside this range, and if the sample has not yet been aerated, then all solutions, including the control, should be aerated, for no more than 30 minutes at a rate of > 25, but < 50 mL/min/L. If the dissolved oxygen is very low (*i.e.,* < 40% saturation) then aeration should be applied at the higher rate (*i.e.,* 50 mL/min/L).

Once sample chemistry has been monitored, and samples are at 20 ± 2°C, daphnid neonates (one to five daphnids should be introduced sequentially to each test solution, including the control until five to ten daphnids are exposed in each vessel) are randomly transferred from the overnight containers into the test vessels. The temperature difference between the culture vessel and the test solution should be ≤ 2°C. It is necessary to transfer daphnid neonates in a vessel containing a large volume (*e.g.,* 0.5 L) of dilution water. The culture water is filtered with a 60 µm Nitex screen. This transfer is commonly used to isolate neonates from the culture medium and to avoid the presence of algae, mould, or other suspended matter[2]. Proper handling and transferring of the daphnids (*e.g.,* underneath the water surface, to prevent cross-contamination and adherence of the daphnids to the sides of vessels) should be practiced, as per the procedures described above. For acute testing, at least 5, but generally 10 organisms (≤ 24 hours old) are exposed in each test vessel. For chemical testing, it is strongly recommended that test exposures (*i.e.,* concentrations) be replicated, in order to obtain greater precision. For the chronic (*i.e.,* 21-day) test design, one organism is exposed in each test vessel (in the static-renewal test design) consisting of a minimum of 10 replicates, while 5 organisms are exposed in each test vessel (in the flow-through system), consisting of at least 4 replicates.

To eliminate bias in organism response during exposure and any possible minute differences between light and temperature surrounding the vessels, it is recommended, where appropriate, that randomization (out of sequential order) of test vessel position be applied. (Note: it is more difficult to apply randomization in flow-through testing due to the overall system design and set-up).

The date and time of test initiation (*i.e.,* the time when the first organism is placed in the test solution in the first vessel), and the name of the person conducting the test, should be recorded on the bench sheet. In addition, the batch number of neonates used in the test, as well as the average number of neonates produced by the QC replicate corresponding to that batch, should also be recorded. Test jars should be randomly placed in an environmental chamber or water bath set at 20 ± 2°C. Light intensity should be measured at the surface of the test solutions and recorded on the bench sheet. (Note: light intensity should be maintained within the range 400 - 800 lux).

---

[2] All these foreign elements may decrease the bioavailability of the test substance.

7.4 TEST MAINTENANCE

7.4.1 Acute test design

Standard acute testing with *Daphnia* sp. is conducted over a 48-hour period. The temperature of the water bath (or environmental chamber) should be monitored and recorded daily. Test temperature must be maintained at 20 ± 2°C during the test exposure. When conducting a test using a water bath to control temperature, the level of water in the bath should be topped up if necessary. Daphnids should not be fed during the test.

In the static acute test design, mortality/immobility may be evaluated after 24 hours (OECD, 1998), or unless specifically outlined in the study objectives, or requested by the client. When making observations, the test vessels should not be disturbed in any way. The number of daphnids exhibiting abnormal swimming behavior or immobility should be enumerated and recorded on the bench sheet. During the 24-hour observation time, it is sometimes difficult to determine organism mortality with 100% certainty. Some chemicals (*e.g.*, benzene) can impair the nervous system of daphnids, thereby causing immobility. Therefore, without careful observation of the heartbeat under a dissecting microscope, determination of mortality is uncertain. Even if they appear to be dead after 24 hours, daphnids should not be removed from the test vessel.

If a static-renewal test design is used, a set of fresh solutions should be prepared and physico-chemical parameters of both the freshly-prepared, and the "old" solutions should be monitored. Careful observations of all test organisms should be made at this time and live daphnids should be transferred to freshly-prepared solutions. Also at this time, all dead organisms should be discarded; all observations should be recorded on the bench sheet.

7.4.2 Chronic test design

The standard chronic test with *Daphnia* sp. is conducted over a period of 21 days. Because the test duration is much longer than the acute test design, this test is usually conducted using either a static-renewal or flow-through design (to eliminate dissolved oxygen depletion and to expose the organisms to the toxicant in a manner that simulates environmental conditions). The static-renewal design is more commonly conducted, due to cost constraints and an overall simpler system design. Usually, daily static-renewal is chosen for testing, since daphnid response to the test material is often unknown. If organism response is known or anticipated, and there is an additional incentive to reduce costs, a thrice-weekly static-renewal system may be used (*e.g.*, for a test initiated on a Monday, fresh solutions are prepared and monitored on Monday, Wednesday and Friday). Observations of mortality and young production, and the transfer of parental organisms to fresh solutions, are also conducted on those days.

Regardless of the test design chosen, the temperature must be maintained at 20 ± 2°C, and should be monitored and recorded at least once daily, and preferably continuously. Use of temperature loggers (*e.g.*, StowAway™, TidbiT™, Hobo®Temp, or other less expensive models) which can easily be put into one additional test vessel subjected to the same conditions as all other test treatments but required only for temperature monitoring is recommended.

Fresh test solutions should be prepared in the same manner as they were at test initiation, and monitored accordingly. Parameters for both initial (*i.e.*, newly-prepared) and final (*i.e.*, old solutions to which organisms were exposed) solutions should be measured daily or at the intervals established in the study objectives (*e.g.*, daily or thrice-weekly).

All first-generation organisms should be transferred to freshly-prepared solutions, if conducting a static-renewal test design. This task is not required when using the flow-through test design. However, if food, metabolic wastes and/or other particulate matter is observed settling from the test solution, and is accumulating on the bottom of the test vessel, or if there is slime build-up on the sides of test vessels, gentle cleaning of the vessels with a small brush is recommended (Note: there should be one brush dedicated to each test treatment including control). Moreover, if food has accumulated on the bottom, the test container should be swirled to re-suspend the food. Stress or any unnecessary disturbances to test organisms should be avoided.

Test organism mortality and young production should be observed and recorded concurrently with test solution renewal. It is expected that the first brood of control organisms will be released within 12 days of test initiation. Frequently, the first brood in the *D. magna* 21-day test is released between days 7 and 10. After all of the first-generation daphnids have been transferred, all neonates (live or dead) should be enumerated and recorded. A convenient way to count the young is to pour the old solution into a large glass Pyrex dish, set the dish on a light box and remove the young one by one (while counting) using a Pasteur pipette. Transfer of the young to small beakers should be done in case recounts are needed. Another convenient way to count the young is to pour the old solution through a small 60 μm Nitex screen placed into dilution water (to avoid stress), then rinsing the young into a small Petri dish, removing with a Pasteur pipette, and counting all neonates. Detailed records of mortality, young production, abnormal behavior (*e.g.*, erratic swimming), etc., should be kept for each test vessel.

Test organisms should be fed a mixture of *Selenastrum* and YCT daily. The recommended feeding regime is: 3.3 mL concentrated *Selenastrum* (*i.e.*, 3.0 to $3.5 \cdot 10^6$ cell/mL) and 1.3 mL YCT daily per parent, in each test vessel. OECD (2004) recommends using an algal quantity per daphnid per day equivalent to 0.2 mg/L TOC. Daphnids are fed subsequent to the transfer of organisms into fresh test solutions. Photoperiod should be maintained at 16 hours light and 8 hours darkness with light intensity at the surface between 400 - 800 lux.

8. Post-exposure observations/measurements, endpoint determinations, and quality control issues

8.1 POST-EXPOSURE OBSERVATIONS/MEASUREMENTS

The acute test is terminated after 48 hours. At this time, all test vessels should be removed from the environmental chamber, incubator or water bath. Using a light box, test organisms should be carefully observed in each test vessel. The chronic test is terminated 21 days after test initiation. As with acute testing, all vessels are removed and organized by replicates and treatments. Careful observations of mortality, young production, and growth determination (optional) are made; all signs of stress (*e.g.*, swimming abnormalities, floating, coloration changes, etc.) should also be noted. All observations are conducted in the same manner as described in the test monitoring section above.

Different post-exposure observations can be made. According to standards and guidelines related to the acute test design, mortality is preferred over immobility (representative of mortality). For mortality, all live and/or dead daphnids should be enumerated. When uncertain whether the organism is dead, remove it with a Pasteur pipette, transfer it onto small Petri dish, and carefully observe it under a dissecting microscope. Death is confirmed when there is a lack of body and appendage movement, and the absence of a heart beat, when observed through a dissecting microscope. Immobilization has been defined as "the inability to swim during a 15-second time period following gentle agitation of test solution, even if the antennae are still moving (OECD, 1998; Environment Canada, 1990b). Generally, the existence of a heart beat distinguishes a live (and immobile) from a dead organism (*i.e.*, when the heart is still beating, but the organism does not move, it is defined as immobile; when the heart does not beat, the organism is dead). The number of dead organisms, and any swimming abnormalities or immobility, is recorded on the bench sheet. If the sample is very dark, the sample can be poured into a larger flat glass dish and held over a light box to observe the organisms. The sample should be returned to the container after making these observations. [Note: with some narcotic toxicants, daphnids may become completely immobile and the heart rate may slow to 1 to 2 beats per minute. In such cases, the beating of the heart becomes the final criterion for death. If such narcosis is suspected, but careful and prolonged observation of the heart cannot be made, the number of immobile daphnids at 48 hours should be recorded]. In addition to mortality and immobility, observations in the chronic test should include details relating to young production (as described above).

An optional endpoint is determination of test organism growth, based on average dry weight or body length. In this case, the size of first-generation adult daphnids alive at the end of the test should be determined using dry weight (wet weight is not acceptable) or body length. Normally, a mean dry weight is determined for pooled adults from each test vessel. Dry weight is determined by drying daphnids on small, pre-weighed aluminum dishes to a constant weight at 100°C for 24 hours (or at 60°C for 48 hours), followed by cooling in a dessicator and weighing on an analytical balance, to the nearest 10 µg. If body length is used for growth determination, the length of each individual is measured using the distance from the apex of the helmet to the base of the posterior spine. Dry weight is the preferable method of growth determination because it provides an indication of effect of the test material on biomass production, and therefore energy transfer from one trophic level to the next (ASTM, 1998), and is less labor intensive than the body length determination.

After making observations of test organisms, the following physico-chemical parameters are measured: dissolved oxygen, temperature, conductivity and pH. All measurements should be recorded on the bench sheet.

8.2 ENDPOINT DETERMINATIONS

The statistical endpoints for toxicity testing using daphnids are based on adverse effects of the test substance on organism survival, reproduction and growth. For analysts who are not familiar with the statistical analyses used, or who are not familiar with statistical computer software used to calculate test endpoints, should solicit the assistance and/or advice of a biostatistician, where appropriate.

The endpoints of the acute test design are based on daphnid mortality and/or immobility. For a single-concentration (load) test, the endpoint is reported as a single mean value derived from the percent mortality and/or immobility, as determined in each of the test treatment replicates, and compared to the control. In the multi-concentration test, a 48-hour LC50 (with 95% confidence intervals) is derived from the test data, and is based on percent mortality. Moreover, the EC50 (with 95% confidence intervals) is calculated based on percent immobility. If the test is conducted with replication, it is recommended that data be pooled from replicates prior to the derivation of an LC50.

There are six methods for estimating an LC50/EC50, as follows: graphical, Binomial, Spearman-Karber, Moving Average, trimmed Spearman-Karber (TSK), and Probit. LC50/EC50s are routinely calculated (by most aquatic toxicology laboratories) using various statistical computer software. The recommended program for use is the standard program originally developed by Stephan (1977), which provides LC50 estimates (and confidence intervals) using three methods (*i.e.*, Probit, trimmed Spearman-Karber, and Moving Average). For the most accurate outcome, data should have at least one test concentration with no mortality, two concentrations with partial mortalities, and one concentration with complete mortality. It is recommended that any computer-derived LC50s be checked using graphical extrapolation by plotting percent mortality against test concentrations on the logarithmic-probability (or Log-Probit) graph paper. In addition, and if required by the study objectives, the NOEC and the LOEC for survival from the multi-concentration test may also be calculated using hypothesis testing (U.S. EPA, 2002b).

The endpoints in the chronic test design may be based on test organism survival, fecundity (*i.e.*, young production) and growth. The growth endpoint is optional and should be outlined in the study design.

The 21-day LC50 is calculated in the same manner as described above. If the objective of the chronic test conducted with either a liquid wastewater sample or a commercial chemical is to estimate the "safe or no-effect" concentration, it is critical to understand how the statistical test endpoints relate to those concentrations. The NOEC and LOEC values are calculated by hypothesis testing. Various statistical methods such as Fisher's Exact Test, Dunnet's Procedure, Steel's Many One-Rank Test or Wilcoxon Rank Sum Test (with the Bonferroni adjustment) may be applied (U.S. EPA, 2002a). Commonly, statistical analyses of variance and hypothesis testing and are conducted using the TOXDAT program developed by Gulley (1988, updated 1989).

The reproduction (and if desired, growth) endpoints of the test are calculated using point estimation techniques, by deriving either IC25 or IC50 values. Similar to the above-mentioned endpoints, the ICx values are generally calculated using commercially-available statistical computer software. Bootstrap techniques and interpolation estimates for evaluating chronic toxicity endpoints from sublethal tests are well documented (Norberg-King 1988; 1993). The most commonly used program for determination of ICx values is the ICPIN program developed by Norberg-King (1993).

Separate analyses are conducted for the LC50, EC50, IC25, IC50 endpoints and the NOEC and LOEC values. Concentrations at which there is complete mortality or immobility in any of the test concentrations are included in estimation of the LC50, EC50, IC25, IC50 but are excluded from the analyses of NOEC and LOEC for reproduction and growth.

8.3 QUALITY CONTROL AND REFERENCE TOXICANT TESTING

To assure and control the quality of the test data generated, mandatory quality control requirements should be routinely conducted by the laboratory (Environment Canada, 1990c). At a minimum, the following aspects should be considered when developing a quality control program for the *Daphnia* sp. testing:

- proper and essential record keeping;
- all lab equipment used for monitoring should be checked and calibrated at regular intervals;
- control/dilution water should be monitored regularly to assure its quality and consistency;
- the quality of test organisms should be assessed by their performance in cultures, reference toxicant tests, and the negative controls of tests;
- the sensitivity of test organisms should be monitored regularly using reference toxicant tests;
- the quality of food used in culturing and testing should be assessed and checked prior to use;
- the adequacy of the test system design (*e.g.*, test type: static, static-renewal, flow-through; test vessels; test conditions: temperature, pH, light, and other test conditions) should be periodically checked and evaluated; and,
- the performance and capability of personnel conducting testing should be considered and assessed.

Performance evaluation (PE) and/or other interlaboratory ('round-robin') testing should also be implemented, where possible. Moreover, laboratory accreditation according to appropriate national/international standardization programs/agencies (*e.g.*, ISO 17025) should also be sought, whenever possible.

At a minimum, reference toxicant testing should be conducted on a monthly basis (and preferably parallel to tests using the same batch of organisms), and at least within two weeks of testing. Reference toxicant testing will not only assess the sensitivity of test organisms, but will also evaluate the proficiency of technical staff, and overall test system design. While many chemicals are suitable as reference toxicants for *Daphnia* sp. (see Environment Canada, 1990c), any risks to human health should be considered prior to selecting them. For *Daphnia* sp. testing (both acute and chronic designs) the most common reference toxicants are: sodium chloride (NaCl), potassium chloride (KCl), and zinc sulphate ($ZnSO_4 \cdot 7H_2O$). The use of hexavalent chromium (Jop et al., 1986), cadmium chloride ($CdCl_2$), and sodium pentachlorophenol (NaPCP) have yielded excellent results as reference toxicants. However, these reference chemicals should be evaluated on a lab-specific basis, due to human health concerns. A reference toxicant test is conducted in the same manner and under the same conditions as the definitive tests described in this chapter. To demonstrate acceptable performance, reference toxicant data should be analyzed for trends using control charting, where consecutive LC50 or EC50 values with their warning limits (±2 standard deviations) and control limits (±3 standard deviations) are plotted on a regular basis. Details pertaining to general procedures for reference toxicant testing are described in Environment Canada (1990a,b,c, 2000) and U.S. EPA (1993; 2002a,b).

9. Factors capable of influencing performance of test organism and test results

Several factors can influence the outcome of a test. The most important of these are as follows: experience and skill of the personnel conducting testing, test organism health, age, and sensitivity (Meyer et al., 1987), temperature control; and, quality and quantity of food used in culturing and testing. Other physico-chemical factors which should be considered are: dilution water hardness and pH, Ca/Mg ratio, alkalinity, sample storage conditions (*e.g.*, temperature, air space), and storage time when effluents are tested. In particular, temperature, water hardness and pH can significantly influence the toxicity of some chemicals. Moreover, Ca/Mg ratio and alkalinity were tested by Muller (1982) using potassium dichromate as a reference toxicant. With respect to the Ca/Mg ratio, Muller (1982) observed a lower 24h EC50 for Ca only and Mg only, and at least a 7-fold higher 24h EC50 for a Ca/Mg molar ratio. In the same study, Muller (1982) also observed, in experiments with an alkalinity range between 0 to 250 mg/L (expressed as $CaCO_3$), that low alkalinity produces lower LC/EC50s than high alkalinity.

The results will also depend upon the species used for testing (*Daphnia magna versus Daphnia pulex*), and test conditions such as: dissolved oxygen, food, and water quality. Clonal variation has also been considered as a factor in test response. The precision and repeatability of the test will also depend on the number of organisms used for testing, as well as the overall test design system (*i.e.*, replication) test vessel composition (*e.g.*, plastic, glass, Teflon), test system design (*e.g.*, static, static-renewal, flow-through). Details related to these factors are provided below:

- Dilution water hardness can have a significant influence on test outcome. For example, when soft *versus* hard control/dilution water is used during preparation of test solutions and the main toxicant of concern is a metal (*e.g.*, copper, zinc, lead), it is expected that test results will differ (*e.g.*, for most metals, soft water will produce lower LC50s than hard water).

- The toxicity of almost every known toxicant (organic or inorganic) will be influenced by pH. Extreme pH values (*i.e.*, acidic or alkaline) outside the recommended range of 6.0 to 9.0 can produce adverse effects (mortality) in *Daphnia* sp. Even small changes in sample pH may affect solubility and/or precipitation of the toxicant from aqueous solution. For example, copper is generally more toxic at lower pH, which can usually be attributed to a greater concentration of free copper ions. Other metals, such as aluminum, are known to be more toxic at low (*i.e.*, < 6.0) and high (> 8.0) pH values.

- The sample storage conditions that are most significant are temperature, darkness, and air space. If effluent samples are not stored under proper storage conditions, such as temperature (*i.e.*, 4 ± 2 °C), darkness, and no airspace, chemical interactions among sample constituents may be modified, potentially confounding test results. Storage time should be less than 2 days (ISO, 1996) and 5 days (Environment Canada, 1990a) for effluent, especially those containing predominantly organic constituents, due to their degradability.

- The experience and skill of laboratory staff conducting the test can have a significant impact on test results (U.S. EPA, 2002a,b,c). Better precision, hence lower variability, can be expected from tests conducted by more experienced staff. In this regard, a strong training program, analyst proficiency monitoring (by senior laboratory staff), and reference toxicant testing and trend monitoring, can all be used to increase precision, thereby decreasing impact on test results.

- The quality and quantity of food used in culturing and testing (chronic test only) is not standardized, and because of the importance of nutrition for growth and reproduction of daphnids, may also have a significant influence on test results. Moreover, oxygen demand and pH level may shift with increased food in the test vessels, thereby affecting the response of daphnids (*e.g.*, too much food may increase pH and deplete oxygen).

- Although parameters such as test organism age, health and sensitivity are mainly standardized using culture health criteria (described above), minute differences in these parameters with certain contaminants can confound test results. In general, newly-released neonates (< 4 hours old) will be more sensitive to toxicants than older ones born 20-24 hours old.

- Temperature control is paramount to ensuring that results are accurate and precise. While temperature is standardized within a certain range, daphnids will respond to higher temperatures by producing more young more rapidly than those tested at lower temperatures.

- Clonal variation has also been considered to be a factor in test response. Large differences in interclonal responses were observed in acute tolerance to cadmium (EC50 range from 0.06 to 100 µg/L). Interclonal variation in chronic stress tolerance was also observed, although in this case differences were relatively small (EC50 range from 25 to 50 µg/L for DCA and 0.6 to 6 µg/L for cadmium) (Baird et al., 1990). Moreover, Soares et al. (1992)

reported that genotype-environment interactions played a key role in determining chronic responses of *Daphnia magna* to two toxicants (sodium bromide and 3,4-dichloroaniline); in this latter study, differences between genotypes, although significant, were not large.

Finally, outliers may confound test results if not taken into consideration prior to final analysis and reporting. An outlier is an inconsistent or questionable data point that appears not to represent a general and/or a toxicant concentration-response trend. Outliers are usually detected during the data compilation process, followed by statistical analysis. Whenever possible, an explanation should be sought for identification of an outlier. When there are no reasonable support for an outlier, the data point should be discarded and the test should be repeated. Alternatively, a statistical analysis may be conducted with and without the outlier and both test results reported.

10. Application of *Daphnia* sp. testing in case studies

10.1 ACUTE TEST DESIGN

Since daphnids are predominantly used to evaluate industrial effluent toxicity, *Daphnia* sp. are also routinely used in toxicant identification evaluation and toxicity reduction evaluation (TIE/TRE) studies with industrial effluents (*e.g.*, U.S. EPA, 1991; Chapter 5, Volume 2 of this book). These studies aim to identify the culprit toxicant(s) in effluent toxicity. Subsequent to manipulation of the effluent, each fraction is tested using the 48-hour acute test with *Daphnia* sp. While these fractions can be tested with other organisms (*e.g.*, rainbow trout, fathead minnow), the advantage of using *Daphnia* sp. tests are three-fold. Firstly, the sample volume required is minimal (*i.e.*, a minimum of 10 mL are required for a test). Secondly, the test duration is much shorter than a 96-hour fish test (*e.g.*, rainbow trout); this will also allow one to run several tests in a fraction of the time required to obtain test results and reach a conclusion, and move to the next tier of TIE testing. Finally, the sensitivity of these organisms is specific to certain causative toxicants in comparison to other organisms (*e.g.*, generally, *D. magna* will respond more sensitively to metals or a wide range of other toxicants than will fish (Hamelink et al., 1986; Hartwell et al., 1995; Karen et al., 1999; Meyerhoff et al., 1985).

Discharge limits set by regulators in Ontario (Canada) often include a "no acute toxicity" criteria (*i.e.*, LC50 > 100%) for rainbow trout and *Daphnia magna* tests. These limits are often a component of Ontario *Environmental Protection Act* effluent limit regulations.

In one case study, monitoring of mining effluent using rainbow trout and *Daphnia magna* showed no acute toxicity to rainbow trout (96-hour tests), however, there was 73% mortality and 27% immobility observed in *Daphnia magna* single-concentration (triplicate) testing. As a result, an investigation of effluent acute toxicity was undertaken. A TIE was conducted where various sample manipulations and fractionations were performed to determine groups of toxicants or a specific toxicant responsible for *Daphnia magna* toxicity and/or contributing to it.

A scaled-down LC50 multi-concentration test (*i.e.*, only 4 test concentrations) was conducted on the various effluent fractions using *Daphnia magna*. Based on test results, several TIE treatments proved successful in removing or significantly reducing toxicity such as: filtration at initial pH, adjustment to pH 11, followed by readjustment to initial pH, as well as EDTA and thiosulphate additions. Further comparisons of measured metal concentrations to LC50 values gleaned from the primary literature indicated that copper was present in high enough concentrations to account for the observed *Daphnia magna* toxicity. To facilitate the interpretation of the toxicity data, test results were converted to toxic units (TU) in order to express the toxic potency of the effluent as a fraction or proportion of the lethal threshold concentration. Higher TUs signify greater toxicity (whereas LC50 are inversely proportional to toxicity). Toxicity to *Daphnia magna* observed in the original test was compared to the expected toxicity, based on literature LC50 values for copper (*i.e.*, 0.04 mg/L) and the actual concentrations of copper in the sample. Both the effluent TU and the copper TU were calculated as 1.6, indicating that copper may have accounted for all of the measured toxicity of the effluent. Because filtration successfully removed acute toxicity of the untreated effluent, that treatment was repeated and sampled/analyzed for chemical confirmation. Specifically, an ICP-MS metal scan was conducted on the pre- and post- filtered sample and a toxicity test was also conducted to confirm that filtration removed toxicity to *Daphnia magna*. Chemical analysis confirmed copper removal by 50 % subsequent to filtration, coincident with the loss of toxicity. Finally, during the follow-up phase of the study, results of chemical and toxicity analyses from historical data were also reviewed. It was determined that periodically, total copper concentrations in the effluent were high enough to account for observed toxicity, based on comparisons to data found in the literature. The *Daphnia* sp. acute test proved to be a valuable tool in estimating and predicting the toxicity of this mining effluent.

10.2 CHRONIC TEST DESIGN

In the early 1990s, Moran et al. (1994) utilized an *in situ* flow-through toxicity testing to evaluate the impact of ongoing industrial effluent discharges to the St. Clair River, near Sarnia, Ontario. The chronic test design was implemented with *Daphnia magna* and *D. pulex* using growth, reproduction and survival endpoints as biological indicators of impact on these common pelagic invertebrates. The toxicity testing system used in the study was designed according to a modification of the 21-day chronic test for *D. magna* survival and reproduction (ASTM, 1990). The experimental design included a control/reference station located upstream of all industrial and urban inputs, and a comparable (in terms of habitat) downstream station. A series of toxicity tests were conducted using both flow-through and static renewal apparati. This innovative modification of the 21-day chronic test allowed for *in situ* monitoring of the St. Clair River water quality, due to routine industrial discharges and spill events. The conclusions from the results of the study indicated that the health of pelagic invertebrates, exposed to ongoing discharges within the study area had no detrimental short- or long-term effects on daphnids.

In another case study conducted by Beak Consultants in 1984, the *Daphnia* sp. chronic test was used to evaluate the long-term impacts of steel plant effluent discharges on freshwater invertebrates. *D. magna* was selected as the test organism since this species was found in local water bodies and in Lake Erie (the receiving environment to which the treated effluent was to be discharged). The daphnid reproduction studies were conducted according to several different methods in two separate, but concurrent, experiments. In the first experiment, a funnel apparatus was used for incubation of reproducing *D. magna* and enumeration of their young. In the second experiment, individual testing "baskets" were used as incubation chambers, eliminating the potential for crowding effects. Both experiments utilized identical effluent and dilution water treatments, facilitating a comparison between the two methods. Each experimental method was applied over two generations in six different test solutions. With the funnel apparatus, second-generation daphnids in 5 out of 6 test solutions were more productive than those in the first generation. With the testing "baskets", the second generation was more productive in 4 out of 6 test solutions. These results suggested a general increase in productivity during the second generation, which was corroborated by literature data, and reflected natural cycles in laboratory populations. Based on statistical analyses applied to the test data, it was clear that neither the funnel nor basket experiments indicated any general reproductive impairment attributable to discharge water.

In both case studies, the *Daphnia* sp. chronic test demonstrated its usefulness as a testing tool for *in situ* and laboratory studies where chronic outcomes of toxicity on invertebrate populations was investigated.

11. Accessory/miscellaneous test information

11.1 EXPERTISE REQUIRED TO CONDUCT THE TEST

While *Daphnia* acute (*i.e.*, 48-hour) testing is relatively simple as compared to a number of other chronic invertebrate tests (*e.g., Ceriodaphnia dubia* 7-day reproduction and survival, *Hyalella azteca* 10-day growth and survival tests), it still requires testing personnel with basic training in an aquatic toxicity laboratory, that have abilities in the care and handling of small aquatic organisms (*e.g.*, use of pipettes, light microscopes, handling test organisms, etc.). Moreover, the chronic testing with *Daphnia* sp. requires a higher level of mastery, and only those personnel that have previously conducted several to many years of culturing and acute testing should implement this latter test system. Prior to the application of the method, it is recommended that personnel be supervised by analysts experienced in the use of conduct of, and interpretation of data from aquatic toxicity tests (U.S. EPA, 2002a,b). Also, proficiency with the conduct of reference toxicant tests is also required prior to undertaking testing.

11.2 MEDIA TESTED

The predominant use of the *Daphnia* sp. acute test (especially in Canada) is for regulatory effluent testing, aquatic effects monitoring, aquatic ecological risk assessment of surface waters, elutriates (*i.e.*, leaching of contaminants from soils, sediments or wastes) and chemical substances. This comprises liquid effluent testing of industrial wastewaters, such as: pulp and paper, mining, iron and steel, chemical and sewage treatment mills and plants. For example, in the province of Ontario

(Canada), the test is mandatory for evaluation of effluents as part of the Municipal Industrial Strategy for Abatement (MISA); federally, the test is used in the Pulp and Paper Effluent Regulation (PPER), and the Metal Mining Effluent Regulation (MMER). For the latter program, the test is used for monitoring, but not compliance testing. Moreover, when effluents are deemed to be toxic, these species can further be used in TIE/TREs (see case study above, Chapter 5, volume 2 of this book).

The use of *Daphnia* sp. acute and chronic tests have also been used to evaluate pesticides and other commercial chemicals (*e.g.*, Ferrando et al., 1999). They are also a part of different test batteries used for evaluating the toxicity of wastes (see Chapter 11, volume 2 of this book).

The acute and chronic test designs are also used, together with an acute fish test (*e.g.*, rainbow trout, bluegill sunfish) for the hazard classification of commercial chemicals; in this regard, the results of the test with commercial chemicals are reported to regulatory agencies (*e.g.*, Toxic Substances Control Act in the U.S.; the New Substances Notification Regulation (NSNR) in Canada for registration of the chemical in question, and are often included in the "Ecological Effects" sections of Material Safety Data Sheets (MSDS) for commercial chemicals.

11.3 SENSITIVITY

It is generally known that the *Daphnia* sp. test is more sensitive to some metals than are fish tests, and less sensitive to other toxicants, such as ammonia and organics (Karen et al., 1999; Hamelink et al., 1986; Hartwell et al., 1995; Meyerhoff et al., 1985). In comparison to the *Ceriodaphnia dubia* test, *Daphnia* sp. tests are generally less sensitive, however, in comparison with tests using sediment-dwelling invertebrates (*e.g., Hyalella azteca* and *Chironomus tentans*), the test is more sensitive. Copper is an example of a contaminant with this profile (Suedel et al., 1995).

11.4 ALTERNATIVE SPECIES

While most of the acute and chronic testing utilizes *D. magna* or *D. pulex*, the U.S. EPA acute test method allows for either *Daphnia* sp. or *Ceriodaphnia dubia* to be used in testing (U.S. EPA, 2002b). Moreover, the *Ceriodaphnia dubia* chronic reproduction test is similar to the *Daphnia* sp. chronic test, in its scope, where test duration is shorter. However, *Ceriodaphnia dubia* is not always ecologically relevant.

11.5 AVAILABLE TEST METHODS

As the *Daphnia* sp. acute test is one of the better established tests world-wide, many different regulatory and standardization agencies have developed national or international test method protocols or guidance documents. The *Daphnia* sp. chronic test, on the other hand, is less well established, due to its use in specialized chemical product testing. It has, however, been embraced by a number of the same regulatory and standardization agencies. Table 9 outlines the most up-to-date method descriptions available internationally.

Table 9. Available test methods.

Test method design	Agency (reference)	Application of test
Acute test method	Organization for Economic Cooperation and Development (OECD, 2004)	Predominantly used for chemical testing
	International Organization for Standards (EN ISO 6341, 1996)	Predominantly used for chemical testing
	United States Environmental Protection Agency (U.S. EPA, 2002b)	Predominantly used for Whole Effluent Testing (WET)
	Office of Prevention, Pesticides and Toxic Substances (OPPTS) (U.S. EPA, 1996a,b,c)	Predominantly used for pesticides and toxic substances
	American Society for Testing and Materials (ASTM, 1984)	Predominantly used for pesticides and toxic substances
	Association Français de Normalisation (AFNOR, XP T90-380, 2003)	Predominantly used for pesticides and toxic substances in the presence of humic acids
	Environment Canada (1990a; revised 2000)	Predominantly used for effluent testing (reference method), but also other applications (*e.g.*, elutriate, leachate, chemicals)
	Environment Canada (1990b)	
Chronic test method	OECD (1998)	Predominantly used for chemical testing
	ISO (1987)	Predominantly used for chemical testing
	OPPTS (U.S. EPA, 1996,a,b,c)	Predominantly used for chemical testing
	ASTM (1990)	Predominantly used for chemical testing

11.6 ALTERNATIVES FOR ENDPOINT DETERMINATION

In addition to the various endpoints discussed above, the only alternative for endpoint determination would be the time to lethality of 50% of exposed organisms. This is expressed as the LT50. However, this endpoint is not routinely used in regular applications, but may be of interest in research studies, or specific environmental applications, such as the impact of chemical spills on aquatic invertebrates.

11.7 SAMPLE PROCESSING

The numbers of tests processed by an experienced technical staff will depend upon many factors, such as:
- type of sample being tested;
- availability of test organisms;
- whether samples need to be aerated prior to testing;
- whether temperature needs to be adjusted; and
- whether pH is within the required range.

For chemical samples, the exposure concentrations will need to be known in advance, otherwise, a preliminary or 'range-finding' test will need to be carried out. In general, for example, for effluents, an experienced lab technician may be able to process an estimated 12-15 effluent samples, however, no more than 5 chemical samples can be processed by a single technician in a single working day. Due to the complexity of test design, the chronic test would require at least a full day to initiate or terminate the test.

11.8 LEVEL OF STANDARDIZATION

As discussed above, the currently-available *Daphnia* methods have been highly standardized over the evolution of the tests. In this regard, the methods have been applied to many samples tested in many different laboratories in many different countries world-wide. The test has also been validated through national and international interlaboratory testing studies.

For example, in the early development of the acute test, Grothe and Kimerle (1985) conducted an interlaboratory round robin study with nine industrial, government and commercial laboratories in the United States. The results of the study indicated that when well-defined test protocols are used, the test showed good reproducibility among participating laboratories (Grothe and Kimerle, 1985). During the same time period, Environment Canada (Atlantic Region) also conducted two interlaboratory comparison studies for acute toxicity of potassium dichromate to *D. magna*, and found that, despite differences in laboratory dilution water quality, there was relatively good agreement among most participating laboratories (Parker, 1983; 1985).

In the early international development of the 21-day chronic test, the Institut National de l'Environnement et des Risques Industriels (INERIS) (France) conducted a "ring test" with several laboratories from many OECD countries (INERIS, 1985). Moreover, for the acute and chronic tests, in 1988 and 1995, respectively the University of Sheffield coordinated and reported on a series of "ring tests" with almost 50 participating laboratories from different OECD countries (Sims et al., 1995). The results of these exercises indicated that, in general, the tests were highly reproducible among well-established aquatic toxicology laboratories with expertise and experience in conducting this type of testing, however, there was variation among laboratories, due to *Daphnia magna* genotype.

Other similar studies relating to test standardization have also been conducted. For example, in 1989, the Petroleum Association for Conservation of the Canadian

Environment commissioned an interlaboratory "round robin" comparison among 4 Canadian aquatic toxicology laboratories and found that toxicity could only reliably be reported in a 'window' of concentration, rather than a single value (PACE, 1989). A recent evaluation of interlaboratory data (Novak et al., 2004) reported on the among- and within-laboratory variability of acute *Daphnia magna* tests conducted over a six-year period from 1994 to 2000. Among-laboratory coefficients of variation (CV), estimated from performance evaluation samples with reference toxicants ranged from 7.5 to 53% with a median of 12.9%. Moreover, the within- and among- laboratory CVs for reference toxicity tests using sodium chloride were 4.6 and 8.7%, respectively, and the within- and among-laboratory CVs for reference toxicity tests using zinc were 27.3 and 33.3%, respectively. The results from this study indicated that overall, the magnitude of variability observed from these tests were comparable to, or lower than, the variability associated with analytical chemistry methods.

Based on these studies, there is clear evidence that the *Daphnia* sp. acute and chronic tests yield highly reproducible results, and are therefore, highly reliable.

12. Miscellaneous information

12.1 POTENTIAL NORTH AMERICAN SUPPLIERS OF *DAPHNIA SP.*

To obtain *Daphnia* sp. starter culture, contact the following suppliers:

Table 10. Potential North American suppliers of Daphnia *sp.*

Suppliers	Telephone/e-mail	Websites
Aquatic Research Organisms, P.O. Box 1271, One Lafayette Road, Hampton, NH 03842, USA	800-927-1650 arofish@aol.com	www.holidayjunction.com/aro
Carolina Biological Supply Company, 2700 York Road, Burlington, NC 27215-3398, USA	800-334-5551 phil.owens@carolina.com	www.carolina.com
Chesapeake Cultures Inc. P.O. Box 507, Hayes, VA 23072, USA	804-693-4046 growfish@c-cultures.com	www.c-culture.com
Aquatic Biosystems Inc. 1300 Blue Spruce Drive, Suite C, Fort Collins, CO 80524, USA	800-331-5916	www.aquaticbiosystems.com

12.2 PROCEDURE FOR SLIDE PREPARATION FOR TAXONOMIC VERIFICATION OF *DAPHNIA SP.*

To prepare slides for permanent records in order to verify the taxonomy of daphnids, follow the step-by-step procedure described below.

Box 6. Procedure for Daphnia *slide preparation.*

-Pipette the organism and transfer to a small Petri dish.
-Using a Pasteur pipette, remove any excess water.
-Add a few drops of 70% ethanol to relax the animal so that the post-abdomen is extended. With enough experience or practice, extension of the post-abdomen may be accomplished by putting sufficient pressure on the cover slip.
-Put a few (one to three) drops of mounting medium on a glass microscope slide. The recommended mounting medium is CMCP-9/9AF Medium. It may be prepared by mixing one part of CMCP-9AF with two parts CMCP-9. For more viscosity and faster drying, CMC-10 stained with acid fuchsin may be used.
-Using a pipette, transfer the organism onto the microscope slide, and put it into the drop of mounting medium.
-Place a cover slip ensuring that no air bubbles are trapped underneath. Exert minimum pressure during this exercise, because more pressure will extend the post-abdomen.
-Allow the medium to dry; this will take approximately 24 hours to 48 hours, depending upon air temperature and humidity.
-After the slide is completely dry, place CMC-10 around the cover slip edges to make the slide permanent.
-Identify to species using an appropriate taxonomic key (*e.g.,* Pennak, 1989).
-Label the slide with a permanent waterproof marker.
-Store for permanent record.
---

13. Conclusions/prospects

Internationally, the *Daphnia* sp. acute and chronic tests are among the most popular, well-standardized, widely-used, reproducible and reliable tests available to aquatic toxicologists and ecological risk assessors. Due to the ubiquitous nature of these organisms, in particular in North America, and the sensitivity of the species used in testing, the acute test methods continue to be used for regulatory compliance related to effluent discharges, and as a screening tool for ecological risk assessments of surface waters in many countries. Despite the introduction and emerging use of chronic and sublethal tests for this purpose, the *Daphnia* sp. acute lethality/immobility test continues to be used as an initial indicator of aquatic toxicity.

The chronic test design has, and will continue to be, used for the evaluation and testing of commercial chemicals. While there have been other chronic invertebrate reproduction tests emerging that may eventually replace this test in some jurisdictions (*e.g., Ceriodaphnia dubia* 7-day reproduction test), the *Daphnia* sp. chronic test has, with modifications, proven to yield important information regarding potential multi-generational effects of continuous discharge of industrial effluents.

Acknowledgements

We would like to dedicate this chapter to our late colleague and dear friend, Martin O'Reilly, with whom we worked for over 10 years, and who supported us in toxicity testing. Tim Moran provided details related to the use of the 21-day *in situ* chronic study as part of our case study description. Finally, we would like to thank the manuscript reviewers, who provided many valuable comments and suggestions on an earlier draft.

References

AFNOR (2003) Water quality - Determination of the inhibition of the mobility of *Daphnia magna* Strauss by chemicals in presence of organic carbon in the form of humic acids - Acute toxicity test, Report XP T90-380 (Tentative Standard).

ASTM (American Society of Testing and Materials) (1984) Standard Practice for Conducting Static Acute Toxicity Test on Wastewaters with *Daphnia*, 1987, in *Annual Book of ASTM Standards*; *Water and Environmental Technology*, Vol. 11.01; D-4229.

ASTM (American Society of Testing and Materials) (1990) Standard Guide for Conducting Renewal Life-Cycle Toxicity Tests with *Daphnia magna*, 1987, in *Annual Book of ASTM Standards, Water and Environmental Technology*, 11.04; E-1193.

ASTM (American Society of Testing and Materials) (1998) Standard Guide for Conducting *Daphnia magna* Life-Cycle Toxicity Tests, in *Annual Book of ASTM Standards, Water and Environmental Technology*, 11.04; E 1193 - 97.

Amoros, C. (1984) Introduction pratique à la systématique des organismes des eaux continentales françaises: Crustacés Cladocères, *Extrait du Bulletin mensuel de la Société Limnéénne de Lyon* **53**, 3-4.

Barber, I., Baird, D.J. and Calow, P. (1990) Clonal variation in general responses of *Daphnia magna* Straus to toxic stress. II. Physiological effects, *Functional Ecology* **4**, 409-414.

Baird, D.J., Barber, I. and Calow, P. (1990) Clonal variation in general responses of *Daphnia magna* Straus to toxic stress. I. Chronic life-history effects, *Functional Ecology* **4**, 399-407.

Baird, D.J., Barber, I., Bradley, M., Calow, P. and Soares, A.V.M. (1989) The *Daphnia* bioassay: a critique., *Hydrobiologia* **188/189**, 403-406.

BEAK (Beak Consultants Limited) (1985) Phase II Bioassay Monitoring of a Steel Plant Process Water – Summer 1984, Report prepared for Stelco Inc., Nanticoke, ON.

Brooks, J.L. (1957) The systematics of North American *Daphnia*, *Memoirs of the Connecticut Academy of Arts & Sciences* **13**, 1-180.

Buikema, A.L., Jr., Geiger, J.G. and Lee, D.R. (1980) *Daphnia* toxicity tests, in A.L. Buikema, Jr., and J. Cairns, Jr. (eds.), *Aquatic Invertebrate Bioassays*, American Society for Testing and Materials, ASTM STP 715, pp. 48-69.

Cowgill, U.M. (1991) The Sensitivity of Two Cladocerans to Water Quality Variables: Salinity < 467 mg NaCl/L and Hardnes < 200 mg $CaCO_3$/L, *Archives of Environmental Contamination and Toxicology*, Springer-Verlag New York Inc., **21**, 218-223,

Dorn, P.B. (1996) Discussion-Initiation Paper 2.2, An industrial perspective on whole effluent toxicity testing, in D.R. Grothe, K.L. Dickson and D.K. Reed-Judkins (eds.), *Whole Effluent Toxicity Testing: An Evaluation of Methods and Prediction of Receiving System Impacts*, Society of Environmental Toxicology and Chemistry.

Environment Canada (EC) (1990a) *Biological Test Method: Reference Method for Determining Acute Lethality of Effluents to Daphnia magna*, Conservation and Protection, Ottawa, Report EPS/RM/14, Amended in May 1996 and December 2000.

Environment Canada (EC) (1990b) *Biological Test Method: Acute Lethality Test Using Daphnia spp.* Conservation and Protection, Ottawa, Report EPS/RM/11, Amended in May 1996 and December 2000.

Environment Canada (EC) (1990c) *Guidance Document on Control of Toxicity Test Precision Using Reference Toxicants*, Conservation and Protection, Ottawa, Report EPS 1/RM/12.

Environment Canada (EC) (2000) *Biological Test Method: Reference Method for Determining Acute Lethality of Effluents to Daphnia magna*, Environmental Technology Centre, Ottawa, Report EPS/RM/14 Second Edition.

Environment Canada (EC) (2002) *Guidance Document on Statistical Methods to Determine Endpoints of Toxicity Tests*, Fourth Draft, Environmental Protection Service, Ottawa.

Ferrando, M.D., Sancho, E. and Andreu-Moliner, E. (1999) Toxicity studies of tetradifon to *Daphnia magna*, *Ecotoxicology and Environmental Restoration* **2**(1), 14-18.

Greene, J.C., Bartels, C.L., Warren-Hicks, W.J., Parkhurst, B.R., Linder, G.L., Peterson, S.A. and Miller, W.E. (1988) *Protocols For Short Term Screening Of Hazardous Waste Sites*, Environmental Research Laboratory, Office of Research and Development, U.S. Environmental Protection Agency, Corvallis, OR 97333, Document EPA/600/3-88/029.

Grothe, D.R. and Kimerle, R.A. (1985) Inter- and Intralaboratory Variability in *Daphnia magna* effluent toxicity test results, *Environmental.Toxicology and Chemistry* **4**, 189-192.

Gulley, D.D., Boelter, A.M. and Bergman, H.L. (1989) TOXSTAT Release 3.0. Fish Physiology and Toxicology Laboratory, Department of Zoology and Physiology, University of Wyoming, Laramie, WY 82071, (updated from 1988 version).

Hamelink, J.L., Buckler, D.R., Mayer, F.L., Palawski, D.U. and Sanders, H.O. (1986) Toxicity of fluridone to aquatic invertebrates and fish, *Environmental Toxicology and Chemistry* **5**, 87-94.

Hartwell, S.I., D.M. Jordahl, J.E. Evans and E.B. May (1995) Toxicity of aircraft de-icer and anti-icer solutions to aquatic organisms, *Environmental Toxicology and Chemistry* **14**, 1375-1386.

Heber, M.A., Reed-Judkins, D.K. and Davies, T.T. (1995) Discussion-Initiation Paper 2.1. USEPA's whole effluent toxicity testing program: A national regulatory perspective, in D.R. Grothe, K.L. Dickson and D.K. Reed-Judkins (eds.), *Whole Effluent Toxicity Testing:An Evaluation of Methods and Prediction of Receiving System Impacts*, Society of Environmental Toxicology and Chemistry.

INERIS (formeraly Institut national de Recherche Chimique Appliquée -IRCHA) (1985) *Interlaboratory Test of the Sub-Chronic toxicity of Chemicals to Daphnids (Inhibition of Reproduction)*, Draft, March 5, 1985.

International Organization for Standardization (ISO) (1987) *Prolonged Toxicity Study with Daphnia magna*: Effects on Reproduction. Report No. ISO/TC 147/SC 5/GT 2N.

International Organization for Standardization (ISO 6341) (1996) Water Quality – Determination of the Inhibition of Mobility of *Daphnia magna* Straus (Cladocera, Crustacea), Third Edition, Geneva Switzerland, Report No. ISO 6341:1996.

Jop, K.M., Rodgers Jr., J.H., Dorn, P.B. and Dickson, K.L. (1986) Use of hexavalent chromium as a reference toxicant in aquatic toxicity tests, in T.M. Poston and R. Purdy (eds.), *Aquatic Toxicology and Environmental Fate*: Ninth Volume, ASTM STP 921, American Society for Testing and Materials, Philadelphia, pp. 390-403.

Karen, D.J., Ownby, D.R., Forsythe, B.L., Bills, T.P., La Point, T.W., Cobb, G.B. and Klaine, S.J. (1999) Influence of water quality on silver toxicity to rainbow trout (*Oncorhynchus mykiss*), fathead minnows (*Pimephales promelas*), and water fleas (*Daphnia magna*), *Environmental Toxicology and Chemistry* **18**, 63-70.

Klapes, N.A. (1990) Acute toxicity of the natural algicide, Cyanobacterin, to *Daphnia magna*, *Ecotoxicology and Environmental Safety* **20**, 167-174.

Lee, C.M., Turner, C.A. and Huntigton, E. (1986) Factors Affecting the Culture of *Daphnia magna*, in T.M. Poston and R. Purdy (eds.), *Aquatic Toxicology and Environmental Fate*: Ninth Volume, ASTM STP 921, American Society for Testing and Materials, Philadelphia, pp.357-368.

Lynch (1983) Ecological genetics of *Daphnia pulex.*, *Evolution* **37**(2), 358-374.

Meyer, J.S., Ingersoll, C.G. and McDonald, L.L. (1987) Sensitivity analysis of population growth rates estimated from cladoceran chronic toxicity tests, *Environmental Toxicology and Chemistry* **6**, 115-126.

Meyerhoff, R.D., Grothe, D.W., Sauter, S. and Dorulla, G.K. (1985) Chronic toxicity of tebuthiuron to an alga (*Selenastrum capricornutum*), a cladoceran (*Daphnia magna*), and the fathead minnow (*Pimephales promelas*), *Environmental Toxicology and Chemistry* **4**, 695-701.

Moran, T., Adams, D. and Beckwith, C. (1994) A *Daphnia magna* and *D. pulex In-situ* Assessment to Determine the Long term Biological Effect of Industrial, Municipal and Run-off Discharges into the St. Clair River, Proceedings of the 21st Annual Aquatic Toxicity Workshop, October 3-5, 1994, Sarnia, Ontario, Canadian Technical Report of Fisheries and Aquatic Sciences 2050.

Muller, H.G. (1982) Interference of insoluble particles with the reproduction toxicity test using *Daphnia magna*, *Bulletin of Environmental Contamination and Toxicology* **29**, 127-129.

Norberg-King, T. (1988) An Interpolation Estimate For Chronic Toxicity: The ICp Approach. National Effluent Toxicity Assessment Center, Technical Report 05-88, Environmental Protection Agency, Environmental Research Laboratory-Duluth, Duluth, MN 55804.

Norberg-King, T. (1993) A Linear Interpolation Method for Sublethal Toxicity: The Inhibition Concentration (ICp) Approach. National Effluent Toxicity Assessment Center, Technical Report 03-93, Environmental Protection Agency, Environmental Research Laboratory-Duluth, Duluth, MN 55804.

Novak, L., Holtze, K, Gilron, G., Wagner, R., Zajdlik, B., Scroggins, R. and Schroeder, J. (2004) Guidance for Conducting and Evaluating Rainbow Trout and *Daphnia magna* Acute Lethality Toxicity Tests using Canadian Metal Mining Effluent, *Environmental Toxicology and* Chemistry (in press).

OECD (Organization for Economic Cooperation and Development) (1984) OECD Guidelines for Testing of Chemicals. *Daphnia* sp., Acute Immobilisation Test and Reproduction Test Method 202.

OECD (Organization for Economic Cooperation and Development) (1998) OECD Guidelines for Testing of Chemicals. *Daphnia magna* Reproduction Test Method 211.

OECD (Organization for Economic Cooperation and Development) (2004) OECD Guideline for Testing of Chemicals. *Daphnia* sp., Acute Immobilisation Test, Test Method 202, adopted April 13, 2004, Paris, France.

PACE (Petroleum Associaton for Conservation of the Canadian Environment) (1989) Interlaboratory Comparison of the *Daphnia magna* Acute Lethality Bioassay. Prepared by EVS Consultants Ltd. PACE Report No. 89-4.

Parker, W.R. (1983) Results of an interlaboratory study on the toxicity of potassium dichromate to *Daphnia*. Laboratory Division, Air and Water Branch, Environmental Protection Service, Environment Canada Atlantic Region, July, 1983.

Parker, W.R. (1985) Results of an interlaboratory study to determine the acute toxicity of potassium dichromate to *Daphnia magna*. Air and Water Branch, Environmental Protection Service, Environment Canada Atlantic Region, 45 Alderney Drive, Dartmouth, Nova Scotia, May, 1985.

Pennak (1989) Freshwater Invertebrates of United States. 3rd Ed. Protozoa to Mollusca, John Wiley & Sons, New York, NY.

Poirier, S.H., Knuth, M.L., Anderson-Buchou, C.D., Brooke, L.T., Lima, A.R. and Shubat, P.J. (1986) Comparative toxicity of methanol and N,N-dimethylformamide to freshwater fish and invertebrates. *Bull. Environ. Contam. Toxicol.* **37**, 615-621.

Poirier, D.G., Westlake, G.F. and Abernethy, S.G. (1988) *Daphnia magna* Acute Lethality Toxicity Test Protocol. Aquatic Toxicity Unit, Water Resources Branch, Ontario Ministry of Environment, Rexdale, ON.

Sims, I., van Dijk, P., Gamble, J. and Grandy, N. (1995) Report of the Final Ring Tests of the *Daphnia magna* reproduction study. Draft Report, OECD Test Guideline Programme, June 1995.

Soares, A.V.M., Baird, D.J. and Calow, P. (1992) Interclonal variation in the performance of *Daphnia magna* Straus in Chronic Bioassays, *Environmental Toxicology and Chemistry* **11**, 1477-1483.

Stephan, C.E. (1977) Methods for calculating an LC50, in F.L. Meyer and Hamelink (eds.), *Aquatic Toxicology and Hazard Evaluation*, American Society for Testing and Material, ASTM STP 634, Philadelphia, PA, pp 65-84.

Suedel, B.C., Deaver, E. and Rodgers, Jr., J.H. (1995) Experimental factors that may affect toxicity of aqueous and sediment-bound copper to freshwater organisms. University of Mississippi, Biological Field Station, Department of Biology.

Thurston, R.V., Gilfoil, T.A., Meyn, E.L., Zajdel, R.K., Aoki, T.I. and Veith, G.D. (1985) Comparative toxicity of ten organic chemicals to ten common aquatic species, *Water Research* **19**, 1145-1155.

U.S. EPA (United States Environmental Protection Agency) (1991) Methods for Aquatic Toxicity Identification Evaluations, Second Edition, Office of Research and Development, Duluth, MN 55804. Document EPA/600/6-91/003.

U.S. EPA (United States Environmental Protection Agency) (1993) Methods for Measuring the Acute Toxicity of Effluents and Receiving Waters to Freshwater and Marine Organisms. Fourth Edition. Office of Research and Development, Washington, DC 20460, Document EPA/600/4-90/027F.

U.S. EPA (United States Environmental Protection Agency) (1996a) Ecological Effects Test Guidelines. OPPTS 850.1000 Special Considerations for Conducting Aquatic Laboratory Studies, Public Draft. Prevention, Pesticides and Toxic Substances, Washington, DC. , Document EPA-712-C-96-113.

U.S. EPA (United States Environmental Protection Agency) (1996b) Ecological Effects Test Guidelines. OPPTS 850.1010 Aquatic Invertebrate Acute Toxicity Test, Freshwater Daphnids, Public Draft. Prevention, Pesticides and Toxic Substances, Washington, DC., Document EPA-712-C-96-114.

U.S. EPA (United States Environmental Protection Agency) (1996c) Ecological Effects Test Guidelines. OPPTS 850.1300 *Daphnia* Chronic Toxicity Test, Public Draft. Prevention, Pesticides and Toxic Substances, Washington, DC., Document EPA-712-C-96-120.

U.S. EPA (United States Environmental Protection Agency) (2002a) Short-term Methods for Estimating the Chronic Toxicity of Effluents and Receiving Waters to Freshwater Organisms. Fourth Edition. Office of Water (4303T), Washington, DC 20460, Document EPA-821-R-02-013.

U.S. EPA (United States Environmental Protection Agency) (2002b) Methods for Measuring the Acute Toxicity of Effluents and Receiving Waters to Freshwater and Marine Organisms. Fifth Edition. Office of Water (4303T), Washington, DC 20460, Document EPA-821-R-02-012.

U.S. EPA (United States Environmental Protection Agency) (2002c) Understanding and accounting for method variability in whole effluent toxicity applications under the National Pollutant Discharge Elimination System Program, U.S. Environmental Protection Agency, Washington, D.C. EPA 833-R-00-003.

Abbreviations

CAEAL	Canadian Association for Environmental Analytical Laboratories
CCME	Canadian Council of Ministers of the Environment
COD	Chemical Oxygen Demand
CV	Coefficient of Variation
CWQG	Canadian Water Quality Guidelines
EC	Environment Canada
EC50	Median effective concentration (*i.e.*, the concentration estimated to cause specified non-lethal effect on 50% of the test organisms)
EEM	Environmental Effects Monitoring
ICp	Inhibition Concentration
ICP-MS	Inductively coupled plasma - Mass spectroscopy
ISO	International Organization for Standards
LC50	Median lethal concentration (*i.e.*, the concentration of material in water that is estimated to be lethal to 50% of the test organisms)
LOEC	Lowest Observed Effect Concentration
LT50	Time to lethality of 50% of exposed organisms
MISA	Municipal-Industrial Strategy for Abatement
MMER	Metal Mining Effluent Regulations
MSDS	Material Safety Data Sheet
NaPCP	Sodium pentachlorophenol
NOEC	No Observed Effect Concentration
NPDES	National Discharge Permit Effluent System
NSNR	New Substances Notification Regulation
OECD	Organization for Economic Cooperation and Development
OMOE	Ontario Ministry of the Environment
OPPTS	Office of Prevention, Pesticides and Toxic Substances
PAPRICAN	Pulp and Paper Research Institute of Canada
PE	Performance Evaluation
PMRA	Pest Management Regulatory Agency
PPE	Personal Protection Equipment
PPER	Pulp and Paper Effluent Regulations
PT	Proficiency Testing
QA	Quality Assurance
QC	Quality Control

QM	Quality Manual
SCC	Standards Council of Canada
SD	Standard Deviation
SETAC	Society of Environmental Toxicology and Chemistry
SOP	Standard Operating Procedure
TIE	Toxicant Identification Evaluation
TOC	Total Organic Carbon
TSK	trimmed Spearman-Karber
TRE	Toxicity Reduction Evaluation
TU	Toxic Units
U.S. EPA	United States Environmental Protection Agency
WET	Whole Effluent Toxicity
YCT	Yeast, Cerophyll, Trout chow.

Appendix 1 - YCT Preparation

1. Introduction

YCT is a laboratory culture food source for daphnids comprising three constituents: **Y**east, **C**erophyll® and **T**rout Chow. The procedure below describes how each component is prepared and mixed together to produce the final food stock. During the YCT preparation, only the Trout Chow is filtered at the end of the digestion period, while the Cerophyll® should be decanted and yeast should be used immediately upon preparation without settling. The food should contain 1.7 to 1.9 g solids/L.

2. Equipment and supplies required

- Magnetic stir plates with stir bars
- 3-2 L Erlenmeyer flasks (designated for YCT use only)
- 200 mL plastic bottles with cups
- 110 µm Nitex Mesh or Fine Unbleached pastry cloth
- Screw top bottles for storage
- Funnel
- Medium-sized fish net
- Log book
- 4 aluminium weight trays for dry weight
- Parafilm®
- De-ionized/distilled water
- Trout Chow Starter Food or No. 1 pellets
- Cerophyll®
- Dry Yeast

3. Preparation of YCT components

Trout Chow digestion (preparation of Trout Chow requires one week in advance).
- Add 10 g of Trout Chow to 2 L of de-ionized water, in the 2 L Erlenmeyer flask.
- Place an air stone and airline at the bottom of the flask and aerate continuously for seven days. Replace evaporated water as required. Make sure that the flask is covered with Parafilm at all times (to prevent fumes from escaping).
- At the end of the 7-day digestion period, turn off the air supply, remove the air stone and allow the mixture to settle for approximately 4 hours (a <u>minimum</u> of 2 hours).
- Decant and filter the supernatant through a standard pre-rinsed food grade, unbleached pastry cloth.

Cerophyll® preparation
- Place 10 g of Cerophyll® powder *(containing dehydrated cereal of grass leaves with natural vitamins A, B_2 and C)* into the 2 L Erlenmeyer flask with 2 L of de-ionized water.
- Cover with Parafilm® and stir overnight (24 hrs) at medium speed on a magnetic stir plate.
- Settle for 1 hour prior to use.
- Decant and use the supernatant.

Yeast preparation
- Add 10 g of dry yeast to 2 L of de-ionized water in the 2L Erlenmeyer flask. Make sure that the batch of yeast has not exceeded its expiry date.
- Stir for 1 hour on medium speed on a magnetic stir plate or shake vigorously by hand until the yeast is well dispersed.
- Use the yeast suspension immediately after mixing and <u>do not</u> allow settling.

4. Combining YCT components

- Before combining YCT components, prepare bottles and caps and clean using boiling water, or use new ones.
- Label approximately 24 - 200 mL plastic YCT bottles with the date (this is YCT batch #).
- Mix equal volumes (approximately 1,500 mL) of the filtered yeast, Cerophyll® supernatant, and Trout Chow supernatant.
- Measure out 200 mL of the mixture into each plastic bottle. Keep the mixture as homogeneous as possible by stirring occasionally.
- Carry out QC monitoring as described in Section 5.0.
- Cap each bottle and place in the freezer (Note: do not re-freeze food once it has been thawed).

5. Quality Control - Suspended Solids Monitoring

- Pre-weigh and label 4 aluminium foil trays.
- Pipette a 5 mL aliquot of YCT food from separate YCT bottles onto each tray with the exception of the QA/QC foil. Dry at 105°C in the oven for 24 hours then remove and cool in the dessicator for a minimum of 30 minutes, then weigh.
- Calculate the average dry weight of the food. An example calculation of YCT average dry weight:

    dry weight of the food and foil - dry weight of the foil = dry weight of the food

- Average the four dry weights average of the four weights (g/5 mL) x 200 mL = average weight (g/L)
- The food should be between 1.7-1.9 g dry solids/L. If solids are > 1.9 g/L, dilute approximately one 200 mL bottle and re-check suspended solids before diluting the whole batch. **NOTE: if a YCT batch needs to be diluted, ensure that an appropriate note is placed on the freezer, indicating that the specific batch needs to be diluted, and how much de-ionized water to add**. Likewise, if suspended solids are < 1.7 g/L, concentrate as appropriate.
- After concentrating the YCT batch, determine dry solids content once again.
- Each new YCT batch is tested prior to use. It may be determined by this food quality check that neither dilution nor concentration of the YCT batch is necessary.
- Record Total Suspended Solids Estimates in a log book. Ensure that another technician checks the calculations and initials.

11. *HYDRA* POPULATION REPRODUCTION TOXICITY TEST METHOD

DOUGLAS A. HOLDWAY
University of Ontario Institute of Technology
Faculty of Sciences
2000 Simcoe Street North, Oshawa
Ontario L1H 7K4, Canada
Douglas.Holdway@uoit.ca

1. Objective and scope of test method

Testing objectives are to determine the maximum concentration at which a chemical or wastewater has no statistically significant effect over 7 days of exposure on the population growth of 'pink *Hydra*' *Hydra vulgaris*, an aposymbiotic pink species, or 'green *Hydra*' *Hydra viridissima*, with stable algal symbiotes. The *Hydra* test has been shown to be particularly sensitive to metal exposure, both in pure chemical solutions and in complex wastewaters. Hydra do not seem to be as sensitive to organic toxicants. The test format involves immersing test animals over 7 days in a range of concentrations of the test solution to be assessed using disposable plastic Petri dishes as test containers.

2. Rapid summary of test procedure

Asexually reproducing (budding) test animals are immersed in a range of concentrations of test solution to be assessed. These animals plus all progeny are exposed to daily renewal of fresh solutions of the same concentration. Observations are made of the changes in the number of intact hydroids (one hydroid equals one animal plus any attached buds). The test is terminated after 7 days of exposure to the various test concentrations and the quantitative responses are analyzed statistically. The NOEC (No Observed Effect Concentration) and the LOEC (Lowest Observed Effect Concentration) are determined.

Table 1. Rapid summary of test procedure.

Test organisms	'pink *Hydra*' *Hydra vulgaris*, or 'green *Hydra*' *Hydra viridissima*.
Type of test	Population reproduction.
Test format	7-day continuous exposure; 100% daily solution renewal.
Test vessels	90 mm diameter disposable plastic Petri dishes for metal contaminant experiments, 90-mm diameter glass Petri dishes for organic contaminant experiments. Eighteen (18) Petri dishes, carefully washed and rinsed with distilled water, are required for each test.
Organism numbers for testing	Five budding *Hydra* are randomly assigned to each Petri dish.
Lighting	12:12 h light:dark photoperiod.
Temperature	$25 \pm 0.5°C$ – use of constant temperature room recommended.
Configuration of test vessels	Petri dishes should be randomly assigned on shelves in controlled temperature room or incubator.
Feeding	*Hydra* are fed daily with an excess of 500 µL of live brine shrimp nauplii (*Artemia salina*) suspension made up in test solution. After feeding *ad libitum* for 30 minutes, test solutions are then changed.
Observations	Hydroids are observed at 0 and 24 h, and then daily for 7 days.
Endpoints	The number of hydroids per test vessel ('one hydroid' is a single hydroid plus buds) is recorded at each observation time.
Daily test solution renewal	The majority of individual hydroids adhere to the surfaces of the Petri dish. Test solutions are swirled around the Petri dish by manually rotating the dish horizontally to dislodge any uneaten brine shrimp and regurgitated food pellets. The test solution from each Petri dish test vessel is then tipped into a second test vessel. Ten milliliters of the appropriate replacement test solution is added and the procedure repeated. Any hydroids dislodged during this procedure are carefully picked up in the liquid using a clean disposable glass micropipette and returned to the original test vessel. Thirty-five milliliters of the appropriate test solution are then added and any remaining brine shrimp or unhatched cysts are removed individually by pipette.
Recording of test conditions	pH, temperature and conductivity measured and recorded for each test solution daily.
Reference toxicant	Green *Hydra*: 4-chlorophenol 96 h LC50 (SE) = 34 mg/L (2.2); 6-day NOEC and LOEC = 10.3 mg/L and 22.3 mg/L. Pink *Hydra*: 4-chlorophenol 96 h LC50 (SE) = 32 mg/L (1.3); 6-day NOEC and LOEC = <1.1 mg/L and 1.1 mg/L.

Table 1 (continued). Rapid summary of test procedure.

Data analysis	The mean relative population growth rate (K) is calculated and is defined as: $$K = \frac{\ln(ny) - \ln(nx)}{T} \qquad (1)$$ where nx is the number of hydroids at the beginning of the first day (tx), ny is the number of hydroids after y - x days (ty) and T is the length of the test period in days (ty – tx).

3. Overview of applications reported with the toxicity test method

The rapid rate of asexual reproduction of *Hydra* (Cnidaria:Hydrozoa) by budding allows population reproduction effects of a potential toxicant to be determined in the laboratory. Such chronic toxicity bioassays provide a rapid, sensitive and precise approach to the measurement of environmental pollutant effects on freshwater invertebrates (Stebbing and Pomroy, 1978). Previous studies using *Hydra* in toxicity tests have shown them to be sensitive to various environmental pollutants including, insecticides (Kalafatic et al., 1991; Kalafatic and Kopjar, 1995), metals (Pyatt and Dodd, 1986; Hyne et al., 1992; Pollino and Holdway, 1999), and crude oil (Mitchell and Holdway, 2000).

For testing organic toxicants, both the acute and chronic toxicity endpoints for *Hydra* are higher than literature values for most species (Pollino and Holdway, 1999). Toxicity endpoints are generally out of the range that one would expect to find for most organic chemicals in the environment including toxicants such as the organochlorine pesticide lindane (Taylor et al., 1995), the chlorinated hydrocarbon insecticide mirex (Lue and De la Cruz, 1978), ethylene dibromide (Herring et al., 1987), PCBs, atrazine and DDT (Benson and Boush, 1983), and crude oil (Mitchell and Holdway, 2000). In each case, *Hydra* species were less sensitive to the toxicant compared with other invertebrate species. Therefore, *Hydra* may have only limited application for the toxicity testing of organic toxicants, and cannot be considered a sensitive model of organic toxicant exposure for invertebrates.

However, *Hydra* are highly sensitive to most metals tested including lead (Browne and Davis, 1977), copper (Stebbing and Pomroy, 1978; Pollino and Holdway, 1999), uranium (Hyne et al., 1992), cadmium and zinc (Holdway et al., 2001). All of these studies found *Hydra* were affected by levels found in contaminated waters, suggesting that such toxicity tests provide a sensitive measure of water quality. They have been utilized as part of a suite of toxicity testing protocols for assessing the toxicity of uranium mine wastewaters in Northern Australia (Holdway, 1991).

4. Advantages of conducting the toxicity test method

Both pink and green *Hydra* can be used in rapid, cost effective bioassays to assess the potential toxicity of inorganic toxicants to aquatic invertebrates. Results obtained with chronic toxicity endpoints for pink and green *Hydra* exposed to cadmium and zinc were comparable with literature values for other invertebrates and vertebrate species (Holdway et al., 2001). In other previous studies, *Hydra* were amongst the more sensitive species to metals tested (Browne and Davis, 1977; Stebbing and Pomroy, 1978; Holdway, 1991; Hyne et al., 1992; Pollino and Holdway, 1999). This test method assesses population reproduction attributes in a very timely and highly cost effective fashion which can be extremely difficult and/or expensive to do in many other tests or with many other types of test organisms.

The species is easy to maintain in the laboratory, has no license requirements nor animal ethics approvals associated with its use, and has environmental relevance as an important member of aquatic food web in many freshwater ecosystems (Pollino and Holdway, 1999). Large numbers of *Hydra* can be easily cultured due to their small size and rapid reproductive rate (Loomis, 1954). This enables chronic toxicity tests to assess the population reproductive effects of a toxicant, to be done in short time periods. Bioassays measuring *Hydra* population growth by asexual reproduction are a sensitive and precise way of measuring the effect of environmental pollutants (Stebbing and Pomroy, 1978). *Hydra* thus can be used in rapid, cost-effective bioassays to assess the toxicity of chemicals to freshwater invertebrates.

5. Test species

Hydrae (Cnidaria:Hydrozoa) are a ubiquitous part of freshwater aquatic ecosystems. They are freshwater micro-invertebrates that are roughly 2-3 mm wide and 5-20 mm in length. The two tissue layer body structure (ectoderm and endoderm only with a thin acellular mesoglea between the tissues instead of mesoderm) and simple body plan mean that almost all *Hydra* cells have direct contact with any toxicants in water. *Hydra* are multicellular organisms that reproduce asexually by budding under favorable conditions (Müller, 1996), although they will reproduce sexually if exposed to adverse conditions such as significant changes in water temperature, which often precede drying out or freezing in nature. Two species of *Hydra* which can be used in this toxicity test method are green *Hydra* (*H. viridissima*), with stable algal symbiotes, or pink *Hydra* (*H. vulgaris*) which are aposymbiotic containing no algal symbiotes. Both species are very sensitive to metals and represent different natural habitats: clear still waters for the green *Hydra*, and moving opaque waters for the pink *Hydra*. The method is also easily adapted for use with other species of *Hydra* such as *H. attenuata* (Blaise et al., 1997; Trottier et al., 1997; Pardos et al., 1999).

The general body plan of the organism consists of an upper or apical end called the hydranth, which consists of a dome-shaped structure having the mouth at its tip (the hypostome), and a whorl of tentacles. The remainder of the organism is known as the column, which contains 4 distinct regions: the gastric region between the tentacles and the first (apical) bud; the budding region or zone bearing the buds; the peduncle which is the region between the lowest bud and the basal disk; and the basal disk which is a foot-like structure secreting a sticky protein that fastens the animal to a substrate (Trottier et al., 1997). Visually, pink and green *Hydra* can be distinguished by their color due to the absence or presence of a green algal symbiote, respectively. A labeled photograph of a pink *Hydra* is shown in Figure 1.

Primary cultures of both pink and green *Hydra* are readily available from normal biological suppliers (*e.g.*, Ward's Natural Science Ltd., St. Catharines, Ontario, Canada L2S 3T5). Test animals are obtained from laboratory stock cultures which are maintained as described later in this method. All animals used are mature and reproducing asexually by budding, a characteristic of this type of organism under optimal environmental conditions. Only hydroids with a tentacled bud are selected as test organisms.

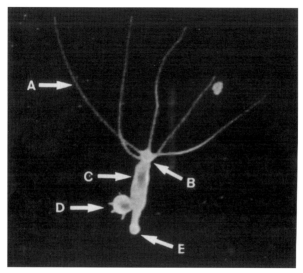

Figure 1. Labeled body parts of a pink Hydra *(A = tentacle; B = mouth; C = body column; D = bud; E = basal disc). The apical end of* Hydra *is called the hydranth and consists of the hypostome, a dome-shaped structure containing the mouth, and the tentacles. The zone between the tentacles and bud is called the gastric region while the region between the lowest bud and the basal disc is called the peduncle. The basal disc secretes a sticky substance which aids in substrate attachment. For an excellent labeled line drawing of* Hydra, *see Trottier et al. (1997).*

6. Culture/maintenance of organisms in the laboratory

6.1 FACILITIES

A room with a temperature of 25°C controlled within ± 1°C or a controlled temperature cabinet should be used for both stock cultures and for experiments. (Note: if the original stock of animals collected from the wild comes from a tropical environment, then the temperatures will be higher and should be 30°C controlled within ± 1°C). Animals should be reared and tests should be carried out with a 12-h photoperiod, the mid-point of which should coincide with solar midday. Light intensity should be of normal laboratory level (30-100 $\mu E/m^2/s$) and should not vary significantly with position within the room or controlled temperature cabinet. If light intensity does vary, then the position of the test chambers should be randomly assigned daily. *Hydra* could be sensitive to airborne contaminants and thus they should be cultured in a laboratory free from possible contaminants or kept covered.

6.2 MATERIALS AND EQUIPMENT

All materials that may come into contact with any liquid into which the animals are to be placed, or with the test animals themselves, should be chemically inert. Any pump used for the collection of water should have inert linings and impeller. Also, all tubing that is used should be of inert material. Suggested items for conducting tests are indicated below:
- Refrigerator for storage of stock solutions.
- Two or more 1 L glass bowls and lids for stock cultures.
- Reconstituted fresh water or *Hydra* medium.
- Air pump or clean and filtered compressed air supply.
- Incubator at 25 ± 0.1°C (or 30°C if *Hydra* animals are tropical).
- Materials for *Artemia* production (see below – requires sea salt and 1 L plastic conical separation funnels or equivalent).
- Eighteen (18) 45 mL disposable plastic vials with screw-capped lids.
- Eighteen (18) 90 mm diameter disposable Petri dishes with lids (total of 36 for experiment). Reusable glass Petri dishes can be substituted if desired.
- Two Perspex trays, each able to hold 10 vials.
- Screw-capped 2 L inert plastic bottles.
- pH meter.
- Thermometer (± 0.1°C).
- Conductivity meter.
- Dissolved oxygen meter.
- Automatic 0-50 mL dispensers.
- Pasteur pipettes.
- Autopipettes with plastic disposable tips.
- Binocular dissecting microscope.
- Clear trays to hold Petri dishes.

6.3 WASHING OF GLASSWARE

Washing of all glass and plastic equipment should be as follows before and after use:
(1) Equipment should be soaked in 5% laboratory detergent solution (*e.g.* Pyroneg, Extran, etc.) for 24 hours to remove any organic matter. The detergent bath should be changed after each use.
(2) All equipment should be rinsed thoroughly in de-ionized water.

(3) Equipment should then be placed in a 5% nitric acid bath (changed every 6 months) for at least 24 hours to remove any bound metals and other contaminants.
(4) All equipment should be rinsed again at least three times in de-ionized water.
(5) It is optional but desirable to then soak equipment in de-ionized water for another 24 hours before air or oven drying.

6.4 PREPARATION OF REAGENTS AND ORGANISM CULTURE MEDIA

Hydra may be cultured in an appropriate clean fresh water or reconstituted fresh water providing constant water quality and containing the essential salts in balanced quantities and hardness appropriate for the region (*e.g.*, American Public Health Association - APHA et al., 1992 - Table 8010:I Recommended composition for reconstituted fresh water). If testing in a region of soft water, use the soft water recommended recipe; if hard water, use the hard water recommended recipe (see below).

Recommended composition for reconstituted fresh water (modified from APHA et al., 1992).

Water type	Salts required mg/L				Water quality		
	$NaHCO_3$	$CaSO_4 \cdot 2H_2O$	$MgSO_4$	KCL	pH	Hardness $CaCO_3$ mg/L	Alkalinity $CaCO_3$ mg/L
Soft	48	30	30	2	7.2 - 7.6	40 - 48	30 - 35
Medium	96	60	60	4	7.4 - 7.8	80 - 100	60 - 70
Hard	192	120	120	8	7.6 - 8.0	160 - 180	110 - 120

Hydra medium can also be used if desired. A simple recipe for *Hydra* medium which contains only calcium chloride (2.94 g); N-tris [hydroxymethyl]methyl 1-2-aminoethanesulfonic acid (TES buffer) (2.2 g); ethylenediaminetetraacetic acid (EDTA, 0.080 g) dissolved in 1 L of ultrapure water, adjusted to pH of 7.0 ± 0.1 with 1 N NaOH, and then diluted with an additional 19 L of ultrapure water to make 20 L of *Hydra* medium, is provided in Trottier et al. (1997). They also provide a recipe for *Hydra* medium supplement for testing undiluted environmental samples (100% v/v) if desired.

Hydra animals are reared in the same media that would be used as dilution and control water in the toxicity tests. Test solutions (5 plus control) are made up using standard protocols such as recommended in ASTM (2002) or Cooney (1995) and will depend on the type of toxicant being tested. Generally a geometric range of solutions is used (*e.g.*, 0, 0.3, 1.0, 3.2, 10, 32 or 0, 2, 4, 8, 16, 32 respectively). A preliminary range finding experiments using an order of magnitude range (*e.g.*, 0, 10, 100, 1000, 10000, 100000) may be required to ensure exposure concentrations are below lethal levels. Use of carriers should be avoided if possible, but if required should not exceed 100 µL per liter (100 ppm) as per ASTM guidelines (ASTM,

2002). Use of a carrier control would then be required and all test concentrations and carrier control should contain the same amount of carrier (Cooney, 1995).

The brine shrimp (*Artemia salina*) culture media is a 1 L solution of salt water (approximately 30 parts per thousand) made by dissolving 30 g of course rock salt, or sea salt, in 1 L of warm (30°C) water. Brine shrimp nauplii are used as food for feeding hydra. Brine shrimp can be cultured in a variety of containers to give a continuous supply of newly hatched brine shrimp larvae (nauplii). Ideal rearing containers are conical in shape (conical 1 L plastic separation funnels are excellent) which when inverted with the neck downwards, can be bubble-aerated from the bottom with oil-free compressed air. One level teaspoon (approximately 5 g) of commercially harvested dried brine shrimp cysts (*e.g.,* Argentemia brand, Argent Chemical Laboratories, Redmond, WA 98052, USA) is added to 1 L of salt water culture solution and vigorous aeration provided from the bottom of the separation funnel to keep the eggs from settling.

Hatching time varies with temperature, which is optimal if maintained around 30°C (eggs will hatch in 18-25 hours at 30°C, while at lower temperatures hatching is delayed). To harvest the newly hatched brine shrimp nauplii, the compressed air is turned off to the individual culture funnel after 24 hours of aerated incubation to allow the nauplii to settle out at the bottom of the funnel and unhatched cysts to float to the top. These nauplii are removed by siphoning off and straining the bottom layer of water containing densely packed larvae through a fine nylon-mesh net, which is able to retain the nauplii.

The larvae are carefully washed three times with test dilution water (salt is toxic to *Hydra*) and are then transferred to a beaker containing sufficient volume of dilution water to permit equal aliquots of brine shrimp nauplii suspension to be added to each test container. The washed nauplii suspension is distributed to each tank or culture bowl using a Pasteur pipette or a syringe.

7. Preparation of test species for toxicity testing

Pink *Hydra* (*Hydra vulgaris*) and green *Hydra* (*Hydra viridissima*) are cultured in bubble-aerated dilution water (generally same source as test water) contained in covered, ventilated 1 L glass bowls. The water movement caused by the aeration results in most hydroids attaching themselves to the sides of the bowls, thus reducing the time taken to change the culture water. The *Hydra* are reared in water from the same stock that would be used as dilution and control water in the toxicity tests. Reserve stock animals should be maintained as a precaution against unknown toxic agents occurring in the dilution water.

Stock animals are fed twice a week up until one week prior to a test when they are fed daily to achieve maximal budding rates. For both the bi-weekly and daily feedings, a well-washed (rinsed 3 times with media to ensure removal of all salt) suspension of newly-hatched brine shrimp nauplii (*Artemia salina* – see above) is pipetted into each stock bowl, distributed evenly over the hydroids. The hydroids are allowed to feed for 30 minutes after which the uneaten brine shrimp and regurgitated food pellets are removed by swirling the water round each bowl and emptying it into a second 1 L glass bowl. More diluent water is added and the procedure is repeated until each bowl is free of brine shrimp. The bowls are then filled with clean water. Any hydroids removed by this process are pipetted back into their respective stock bowl. Cladocera (*e.g., Daphnia sp., Moinodaphnia sp.*) can be fed to hydroids on a weekly basis as a diet supplement if sufficient stocks are available.

Stock bowls are cleaned weekly by swirling bowls in a circular motion and gently pushing hydroids, using a finger or glass Pasteur pipette, off the side of the bowl into the water. The detached hydroids are then pipetted from the old bowl into a clean glass bowl which contains fresh water/medium to a depth of at least 2.5 cm. Because it is important to prevent bacterial buildup, glass stock bowls are cleaned as recommended previously for laboratory glassware.

Periodically, hydroids may reproduce sexually, making it difficult to maintain an isogenic population. The frequency with which this occurs can be reduced by regular cleaning of the stock cultures to avoid fouling of the water by uneaten brine shrimp, as well as by maintaining steady environmental conditions.

8. Testing procedure

Eighteen plastic disposable 45 mL vials are each filled with 35 mL of 6 concentrations of test solution (there are three vials per concentration to be tested). The vials are placed in the incubator or controlled temperature room to bring the solution to 25°C. Then, the test solutions are randomly allocated among eighteen labeled plastic or glass Petri dishes which have been placed on clear plastic trays. Each dish is first rinsed with about 5 mL of the appropriate test solution and the remaining test solution is then added to the appropriate dish.

Active asexually-reproducing hydroids (*i.e.,* one animal with one tentacled bud) from the stock culture are sequentially placed into each Petri dish until each dish contains five hydroids. The Petri dishes are covered with lids (if using plastic disposable dishes) or Parafilm (if using reusable glass dishes) to prevent evaporation and then placed in a constant temperature room or incubator held at 25°C on clear plastic trays without rearrangement. Completion of this stage constitutes the start of the test.

Observations are made on the hydroids at 0 and 24 h, and then daily for a total of 7 days by examination under a dissecting microscope. The number of hydroids per test container is recorded at each observation time.

The hydroids are fed daily after recording the data. Feeding is accomplished by individually feeding each hydroid with live brine shrimp nauplii (*Artemia salina*), made up as a suspension in the appropriate test solution, and placed into each Petri dish using a disposable glass Pasteur pipette. Feeding is allowed to proceed *ad libitum* for at least 30 minutes after which time the test solutions are changed.

The daily renewal of the test solution in each Petri dish is performed working through the test solution changes from the control to the highest test concentration, respectively, in the following manner:

- New test solutions are made up daily before renewal.
- Original test solution is swirled around the Petri dish to dislodge any uneaten brine shrimp and regurgitated food pellets.
- The test solution from each Petri dish is then tipped into a second Petri dish.
- Enough of the appropriate test solution is added to cover the bottom of the now empty dish, and the swirling procedure repeated and emptied. Any hydroids dislodged are carefully picked up with a little solution using a clean glass Pasteur pipette and returned to the original test Petri dish.
- 35 mL of new replacement test solution is then added.
- Any remaining brine shrimp or unhatched cysts are removed individually by pipette with care taken to minimize removal of the test solution.

After test solution renewal, the Petri dishes are randomly placed on clear plastic trays. The trays are then placed in a constant temperature cabinet.

Randomness is an important element of the experimental design since the statistical analyses used make the assumptions that the data are normally distributed and that the test is free from bias. Random distribution of hydroids and the random placement of Petri dishes in the controlled-temperature room or incubator can be achieved in one operation. The Petri dishes are randomly assigned to positions on trays and the hydroids are sequentially distributed.

Since the Petri dishes are randomly positioned, the distribution of hydroids, although sequential, is also random. Because the Petri dishes have a random position on the trays, they will also have a random position in the controlled-temperature room or incubator. When test solutions have to be changed, then the Petri dishes can be sorted into replicate groups for greater convenience. This avoids the continual changing of glass pipettes by working through the test solution changes from the control to the highest test concentration, respectively. At the end of the test solution change, the Petri dishes are then again randomly placed on the plastic trays.

The above procedures (*i.e.,* recording the number of hydroids per test container, replacement of test solution) are repeated daily for a total of 7 days. Occasional checks should be made on the controlled-temperature room or incubator performance with respect to constant temperature, light intensity and variation. During the test, the pH, temperature, conductivity and dissolved oxygen of each test solution should be measured and recorded at the beginning and end of each 24 h period.

9. Acceptability of test data

The test data are considered acceptable if:
- the recorded temperature of the test solution remains within the prescribed limit;
- the recorded pH is within 0.5 units of the initial pH;
- the dissolved oxygen concentration was greater than 70% of the air saturation value throughout the test at a temperature of 25°C; and
- the mean relative population growth rate (K) of the controls (replicates pooled after 7 days) is greater than 0.25 for both green and pink *Hydra*;

- the conductivity of each test solution should be within ± 5 µSi.cm⁻¹ of the values obtained on day 0.

The mean relative population growth rate (K) is calculated and is defined as:

$$K = \frac{\ln(ny) - \ln(nx)}{T} \quad (1)$$

where:
nx is the number of hydroids at the beginning of the first day (t_x),
ny is the number of hydroids after y - x days (t_y) ($n_0 = 5$), and
T is the length of the test period in days ($t_y - t_x$) which is 7 days in this case (Pollino and Holdway, 1999).

10. Analysis of test data

10.1 STEP 1. DATA PREPARATION

Prepare a data table showing the relative population growth rates (K) using the formula above for each test concentration and replicate.

10.2 STEP 2. EXAMINATION OF ASSUMPTIONS AND ERROR ANALYSIS

The replicates of the K values are used to examine the underlying assumptions of normality and equal variance by calculating 'errors' where:

$$e_1 = (rep\ 1 - rep\ 2)/\sqrt{2} \quad (2)$$

and

$$e_2 = (rep\ 1 + rep\ 2) - (2 \times rep\ 3)/\sqrt{6} \quad (3)$$

for each test concentration.
The pooled estimate of error variance (V) is given by:

$$V = \{\Sigma(e_1^2) + \Sigma(e_2^2)\}/12 \quad (4)$$

where Σ (sum) refers to the six test concentrations and the associated degrees of freedom for V is 12.
The standardized errors (e_1' or e_2' = e_1 or e_2 divided by $\sqrt{V}$) are standard normal variates if the assumptions for the analysis are met. This is confirmed by plotting the e'-values using a quantile-quantile (Q-Q) plot. An individual e'-value exceeding 2.63 in magnitude (the 1% critical value for a standardized normal deviate) is considered an 'outlier'.

If a single outlier is found it may be possible to attribute it to a single replicate observation in which case that observation should be dropped and the analysis continues. If, however, more than one outlier occurs or there is other evidence of non-normality such as non-linear plotted e-values, then the analysis is abandoned.

10.3 STEP 3. CALCULATION OF K-VALUES, DIFFERENCES FROM CONTROLS AND THEIR RATIO

Calculate: the mean K-value over replicates; the difference (d) from the control value; and the ratio d/SE where SE is the standard error given by:

$$SE = \sqrt{(v/n_c + v/n_t)} = \sqrt{(2v/3)} \tag{5}$$

where n_c and n_t are the number of replicates in the control (n_c) and treated (n_t) groups, respectively.

10.4 STEP 4. COMPARISON OF ORDERED RATIOS WITH CRITICAL VALUES

Compare the ordered ratios (smallest to largest) against the set of critical values:

Order of difference	Critical value
1	1.79
2	1.96
3	2.14
4	2.32
5	2.50

Declare any of the ordered ratios that exceed their critical values as statistically significant effects and the lowest concentration that is statistically significant (the LOEC) for this test. The no-observed-effect-concentration (NOEC) for the test is the next lowest test concentration.

11. Example of a *Hydra* reproduction test analysis

The following *Hydra* reproduction test data involved exposure of green *Hydra* to increasing concentrations of mine wastewater expressed as percentage composition. The resultant data summary is shown in the example below modified from Allison et al., 1991. Five animals were initially placed in each replicate. The high final population numbers relate to the use of tropical organisms and an experimental temperature of 30°C rather than the 25°C described in this protocol.

HYDRA REPRODUCTION TOXICITY TEST

Box 1. Data summary.

Concentration group	Offspring numbers			K-Values		
	Rep 1	Rep 2	Rep 3	Rep 1	Rep 2	Rep 3
Control	98	92	95	0.496	0.485	0.491
0.3%	83	65	93	0.468	0.427	0.487
1%	80	89	91	0.462	0.480	0.484
3.2%	65	56	87	0.427	0.403	0.476
10%	78	51	63	0.458	0.387	0.422
32%	52	46	58	0.390	0.379	0.409

Box 2. Examination of assumptions and error analysis.

Concentration group	E	e_2	Standardized errors	
			e_1'	e_2'
Control	0.0078	-0.0004	0.30	0.02
0.3%	0.0290	-0.0323	1.11	-1.24
1%	0.0127	-0.0106	-0.49	-0.41
3.2%	0.0170	-0.0498	0.65	-1.91
10%	0.0502	0.0004	1.92	0.02
32%	0.0141	-0.0237	0.54	-0.91

$\Sigma e_1^2 = 0.0041$ $\Sigma e_2^2 = 0.0042$ $V = 0.00068$ $\sqrt{V} = 0.0261$

None of the e'-values exceeds 2.63 and a quantile-quantile (Q-Q) plot of e_1' and e_2' indicates good linearity and thus this represents normally distributed data (a systematic departure from linearity indicates non-normality).

Box 3. Calculation of k-values, differences from controls and their ratio.

Concentration group	Mean K	d	d/SE	Critical value
Control	0.491	-	-	
0.3%	0.461	0.030	1.43	1.79
1%	0.475	0.016	0.76	1.96
3.2%	0.435	0.055	**2.67***	2.14
10%	0.422	0.068	**3.29***	2.32
32%	0.390	0.101	**4.81***	2.50

Where $SE = \sqrt{(v/n_c + v/n_t)} = \sqrt{(2v/3)} = \sqrt{(2 \times 0.00068/3)} = 0.021$, and *=significantly different from control at $p \leq 0.05$.

Finally, when conducting comparison of ordered ratios with critical values, three significantly different results are obtained from the data analysis (*: see Box 3). The NOEC is thus deemed to be 1% for the test while the LOEC is deemed to be 3.2%.

12. Reporting of results

The test report should include the following information:

- Identity of the test material used and how the test solutions were made up.
- Identity of the control water used.
- Date and time of start of test.
- Concentrations tested.
- Tabulated data showing daily and mean (SE) measurements of temperature, pH, dissolved oxygen concentration and conductivity for each treatment.
- Tabulated data showing the total number of animals produced at the end of the test period for each treatment.
- Definition of NOEC and LOEC for changes in population reproduction.
- Any deviation from the method described above.

13. Application examples (case studies)

The application of this recommended *Hydra* reproduction toxicity test has been demonstrated in a variety of test situations including both pure chemicals (Pollino and Holdway, 1999) and dilute mining wastewaters (Holdway, 1992) where the sensitivity of both green and pink *Hydra* were similar for metals but where pink *Hydra* were more sensitive to the organic reference toxicant, 4-chlorophenol (Tab. 2).

Other authors have successfully utilized various species of *Hydra* in acute lethal (Blaise and Kusui, 1997; Beach and Pascoe, 1998; Pardos *et al.*, 1999), sub-lethal morphology (Blaise and Kusui, 1997; Pardos *et al.*, 1999), and feeding tests (Beach and Pascoe, 1998). Such alternative tests used compartmentalized dishes or 12-well microplates which made them very convenient to run. In one study, the *Hydra* test was found to be more sensitive to metals than *Vibrio fisheri* in the Microtox test (Pardos et al., 1999).

14. Accessory/miscellaneous test information

The recommended level of expertise of testing personnel for this toxicity test is a college diploma or university first degree in biology. The toxicity testing protocol is relatively easy to master and only requires a few trial tests to become expert. The *Hydra* population reproduction toxicity test is particularly sensitive to water-borne metals and can be applied to assessing whole wastewaters such as mine run-off or effluent. There would be potential for this test to be partly automated through use of a photographic assessment system to count the numbers of animals each day.

However, to be completely automated, systems for the exchange of exposure media, feeding with live food and removal of food wastes would need to be designed.

Table 2. The lowest 6-day NOEC and LOEC values of 4-chlorophenol, endosulfan and copper ($n = 4$) (Pollino and Holdway, 1999), and for uranium mine retention pond water ($n = 1$) (Holdway, 1992) for pink and green Hydra.

Toxicants / Effluents	Species	NOEC 6-day	LOEC 6-day
4-Chlorophenol (mg/L)	Pink	< 1.1	1.1
	Green	10.3	22.3
Endosulfan (µg/L)	Pink	0.04	0.08
	Green	0.06	0.08
Copper (µg/L)	Pink	4	8
	Green	4	8
Uranium mine retention pond #2 water (%v/v)	Pink	10	32
Uranium mine retention pond #2 water (%v/v)	Green	3.2	10
Uranium mine retention pond #4 water (%v/v)	Pink	32	> 32*
Uranium mine retention pond #4 water (%v/v)	Green	32	> 32*

* Maximum waste water concentration tested was 32%.

Hydra toxicity testing can be used to measure acute lethality, sub-lethal morphological effects (*e.g.*, tentacle clubbing), sub-lethal behavior and feeding effects, chronic reproductive effects (this protocol) and regeneration effects to assess teratogenic potential of chemicals (relative toxicity of chemical to embryo relative to adult).

Costs to run this reproductive toxicity test are modest. I would estimate about 25 hours of technical time (or $500 US at $20 per hour) to assess one chemical/mixture or wastewater sample, excluding the costs of equipment and any required chemical analyses.

15. Conclusions/prospects

The *Hydra* reproduction toxicity test is a useful and relatively easy toxicity test which is able to assess the population reproductive toxicity of both pure chemicals and mixed effluents in a relatively short time of 6 to 7 days. Both species of *Hydra* discussed in this method are very sensitive to metals and represent different natural habitats: clear still waters for the green *Hydra* and moving opaque waters for the pink *Hydra*. More research is required to assess the value of *Hydra* for assessing the toxicity of organic chemicals; initial research indicates lower sensitivity to such compounds. There is also value in further development of both the morphological and behavioral assays using *Hydra* as described in Blaise and Kusui (1997) and Beach and Pascoe (1998). Perhaps these assays could be combined with the reproduction toxicity test described here.

This toxicity test could be used as part of a battery of toxicity tests to assess the impacts of industrial effluents, wastewaters or pollution-impacted river and lake waters. Other endpoints or tests that could be utilized include *Hydra* morphology, *Hydra* behavioral assays, *Hydra* regeneration assays and *Hydra* feeding tests amongst others.

Acknowledgements

I would like to thank the editor Dr. C. Blaise and three external reviewers who read and commented on initial versions of this chapter. They all provided excellent constructive criticism which greatly improved the quality of text and logic flow of the final manuscript. I would also like to thank previous technicians, graduate students and biologists who worked on developing and applying *Hydra* toxicity tests with special mention to H. Allison, K. Lok, F. Mitchell, M. Semaan, S. Templeman and Drs. R. Hyne, C. Pollino and G Rippon.

References

Allison, H.E., Holdway, D.A., Hyne, R.V. and Rippon, G.D. (1991) OSS procedures for the biological testing of waste waters for release into Magela Creek. XII. *Hydra* test *(Hydra viridissima and Hydra vulgaris)*. *Open File Record #72*, Supervising Scientist for the Alligator Rivers Region, 9 pp.

APHA (American Public Health Association), AWWA (American Water Works Association) and WEF (Water Environment Federation). (1992) *Standard Methods for the Examination of Water and Wastewater*, 18th Edition, American Public Health Association, Washington, DC, USA.

ASTM (American Society for Testing and Materials). (2002) Annual book of ASTM standards. Section Eleven Water and Environmental Technology, *Biological effects and environmental fate; biotechnology; pesticides*, Vol. 11.05, ASTM Stock Number S110502 ASTM International West Conshohocken, PA., 1716 pp.

Beach, M.J. and Pascoe, D. (1998) The role of *Hydra vulgaris* (Pallas) in assessing the toxicity of freshwater pollutants, *Water Research* **32**, 101-106.

Benson, B. and Boush, G.M. (1983) Effect of pesticides and PCBs on budding rates of green *Hydra*, *Bulletin of Environmental Contamination and Toxicology* **30**, 344-350.

Blaise, C. and Kusui, T. (1997) Acute toxicity assessment of industrial effluents with a microplate-based *Hydra attenuata* assay, *Environmental Toxicology and Water Quality* **12**, 53-60.

Browne, C.L. and Davis, L.E. (1977) Cellular mechanisms of stimulation of bud production in *Hydra* by low levels of inorganic lead compounds, *Cell Tissue Research* **177**, 555-570.

Cooney, J.D. (1995) Freshwater tests, in G.M. Rand (ed.) *Fundamentals of Aquatic Toxicology: Effects, Environmental Fate, and Risk Assessment*, Second Edition.Taylor & Francis, Washington D.C., USA, pp. 71-102.

Herring, C., Adams, J., Pollard, S. and Wilson, B. (1987) Dose response studies using ethylene dibromide in *Hydra oligactis*, *Bulletin of Environmental Contamination and Toxicology* **40**, 35-40.

Holdway, D.A. (1991) Mining and tropical freshwater environments: laboratory studies, *Australian Biologist*, **4**(5), 228-236.

Holdway, D.A. (1992) Control of metal pollution in tropical rivers in Australia, in. D.W. Connell and D.W. Hawker (eds.), *Pollution in Tropical Aquatic Systems,* CRC Press, Boca Raton, Florida, pp. 231-246.
Holdway, D.A., Lok, K. and Semaan, M. 2001 The acute and chronic toxicity of cadmium and zinc to two *Hydra* species, *Environmental Toxicology* **16**, 557-565.
Hyne, R., Rippon, G.D. and Ellender, G. (1992) pH-dependent uranium toxicity in freshwater *Hydra*, *Science of the Total Environment* **125**, 159-173.
Kalafatic, M. and Kopjar, N. (1995) Response of green *Hydra* to Pirimicarb, *Biologia* **50**, 289-292.
Kalafatic, M., Znidaric, D., Lui, A. and Wristcher, M. (1991) Effect of insecticides (Dimiline Wp 25, Torak EC 24 and Gamacide 20) on *Hydra* (*Hydra vulgaris* Pallas), *International Journal of Developmental Biology* **35**, 335-340.
Loomis, W.F. (1954) Environmental factors controlling growth in *Hydra*, *Journal of Experimental Biology* **126**, 234-236.
Lue, K.Y. and de la Cruz, A.A. (1978) Mirex incorporation in the Environment: Toxicity in *Hydra*, *Bulletin of Environmental Contamination and Toxicology* **19**, 412-416.
Mitchell, F.M. and Holdway, D.A. (2000) The acute and chronic toxicity of the dispersants Corexit 9527 and 9500, water accommodated fraction (WAF) of crude oil, and dispersant enhanced WAF (DEWAF) to *Hydra viridissima* (green *Hydra*), *Water Research* **34**, 343-348.
Muller, W.A. (1996) Pattern control in the immortal *Hydra*, *TIG* **3**, 91-96.
Pardos, M., Benninghoff, C., Guéguen, C., Thomas, R., Dobrowolski, J. and Dominik, J. (1999) Acute toxicity assessment of Polish (waste) water with a microplate-based *Hydra attenuata* assay: a comparison with the Microtox test, *Science of the Total Environment* **243/244**, 141-148.
Pollino, C.A. and Holdway, D.A. (1999) The potential of two *Hydra* species as standard toxicity test animals, *Ecotoxicology and Environmental Safety* **43**, 309-316.
Pyatt, F.B. and Dodd, N.M. (1986) Some effects of metal ions on the freshwater organisms *H. oligactis* and *Chlorohydra viridissima*, *Journal of Experimental Biology* **24**, 169-173.
Stebbing, A.R.D. and Pomroy, A.J. (1978) A sublethal technique for assessing the e!ects of contaminants using *Hydra littoralis*, *Water Research* **12**, 631-635.
Taylor, E.S., Morrison, J.E., Blockwell, S.J., Tarr, A. and Pascoe, D. (1995) Effects of lindane on the predator-prey interaction between *Hydra oligactis* Pallas and *Daphnia magna* Strauss, *Archives of Environmental Contamination and Toxicology* **29**, 291-296.
Trottier, S., Blaise, C., Kusui, T. and Johnson, E.M. (1997) Acute toxicity assessment of aqueous samples using a microplate-based *Hydra attenuata* assay, *Environmental Toxicology and Water Quality* **12**, 265-271.

Abbreviations

ASTM	American Society for Testing and Materials
APHA	American Public Health Association
d	difference from the control
DDT	dichlorodiphenyltrichloroethane
e	standardized error
EDTA	ethylenediaminetetraacetic acid
K	relative population growth rate
LC50	Lethal Concentration leading to 50% effect
LOEC	Lowest Observed Effect Concentration
NOEC	No Observed Effect Concentration
PCBs	polychlorinated-biphenyls
Q-Q	quantile-quantile
SE	Standard Error
TES buffer	N-tris [hydroxymethyl] methyl 1-2-aminoethanesulfonic acid
V	error variance.

12. AMPHIPOD (*HYALELLA AZTECA*) SOLID-PHASE TOXICITY TEST USING HIGH WATER-SEDIMENT RATIOS

UWE BORGMANN
& WARREN P. NORWOOD
National Water Research Institute - Environment Canada
867 Lakeshore Road, P.O. Box 5050
Burlington, Ontario L7R 4A6, Canada
uwe.borgmann@ec.gc.ca
warren.norwood@ec.gc.ca

M. NOWIERSKI
Department of Biology - University of Waterloo
Waterloo, Ontario N2L 3G1, Canada
monica.nowierski@ene.gov.on.ca

1. Objective and scope of test method

This chapter describes a method for testing the chronic (four week) toxicity of sediments to the freshwater amphipod *Hyalella azteca* using large water to sediment volume ratios in place of water renewal. The test can also be used to measure only contaminant bioaccumulation by initiating the experiment with adult animals instead of young, and reducing the exposure time to one week. Using a large water-sediment ratio, with sufficient equilibration time, ensures that major ions and pH in overlying water remain at acceptable levels, while still allowing contaminants to partition into the overlying water. This negates the need for water renewal, simplifies the test procedure and equipment (*e.g.*, no automated water renewal apparatus is required), ensures that contaminants leached into overlying water are not flushed out of the test system, and provides a large water volume (1 L) at the end of the test for chemical analysis. Tests are conducted in Imhoff settling cones. The "V" shape of the test vessel provides sufficient sediment depth while maintaining the large water-sediment ratio. The low sediment volumes required (15 mL) also make it possible to test small sample volumes, such as those obtained from fine sectioning of sediment cores, or when sampling only a thin top oxic layer of sediment.

2. Summary of test procedure

Tests are conducted in Imhoff settling cones using 15 mL of sediment and 1 L of overlying water, giving a water to sediment ratio of 67:1 (Borgmann and Norwood, 1999a). These are 1 L funnel shaped containers constructed of polycarbonate or glass and designed for measuring the volume of suspended solids. A number 4 silicone rubber stopper is placed in the bottom, resulting in a sediment depth of about 2.3 cm and a surface sediment diameter of 3.1 cm. Test duration is four weeks for chronic toxicity assays, but it may be shorter (*e.g.*, one week) for contaminant bioaccumulation tests. The sediment volume is sufficient for at least 15 young or adult *Hyalella* per test container. Most of the amphipods remain buried in the sediment most of the time, but emerge and become more active if food is limiting. If desired, additional amphipods can be exposed to overlying water in cages attached to the inside of the cones. This allows comparison of toxicity and/or bioaccumulation for the water-only phase as compared to combined water and sediment exposure. This is particularly useful if it is necessary to determine if water quality objectives or criteria are sufficient to protect sediment dwelling organisms. A summary of test conditions and procedures is provided in Table 1.

Table 1. Rapid summary of test procedure.

Test chamber	1 L polycarbonate or glass Imhoff settling cone fitted with a #4 silicone rubber stopper.
Test container rack	19 mm plywood stand, 40 cm high, with 8.3 cm diameter holes placed 14 cm apart (centre to centre, *e.g.*, 5 rows of 10 holes in a 148 × 70 cm rack).
Temperature and lighting	23-25°C, 16 h light: 8 h dark, fluorescent tubes placed approximately 75 cm above the top of the cones and giving a light intensity of approximately 30-40 $\mu E/m^2/s$ on the top of the cones.
Sediment volume	15 mL (smaller volumes can be used if overlying water quality is not maintained, *e.g.*, when using overlying water of very low buffering capacity).
Overlying water volume	1 L (evaporative losses are replaced with de-ionized water).
Aeration	Gentle bubbling of air through a glass rod tipped with a 250 μL polypropylene pipette tip placed 1-2 cm above the sediment surface.
Covers	Polypropylene snap lids with holes punched for insertion or glass rods for aeration.

Table 1 (continued). Rapid summary of test procedure.

Sediment addition method	Sediment should be added so as to minimize the mixing of sediment with overlying water (*e.g.*, one method is to partially fill the cones with water and then "inject" the sediment near the bottom using a large plastic syringe fitted with a large bore plastic tip, and then to add the remaining overlying water).
Equilibration time	Allow cones to equilibrate for 1 or 2 weeks before addition of test animals, unless conditions dictate otherwise (*e.g.*, when highly volatile or unstable test substances are expected in the sediment).
Test animal density	Fifteen young *Hyalella* per cone (chronic tests) or up to 15 adults (one-week bioaccumulation tests).
Feeding	Finely ground Tetra-Min® fish food flakes at a rate of 2.5 mg twice in week 1 and 2, 2.5 mg three times in week 3, and 5 mg two times in week 4. For experiments with adults, feed each cone and cage 5 mg twice a week. Reduce feeding if uneaten food accumulates and starts to rot.
Water sampling prior to and at end of test	Because overlying water volume is large, small volume (*e.g.*, 10-15 mL) water samples for pH, dissolved oxygen, ammonia and other analyses can be taken during the test. Large volumes (up to 1 L) can be collected for contaminant analysis at the end of the test and filtered and/or preserved as required for specific contaminants.
Test termination	Once the overlying water is carefully decanted for water sampling (see above) the sediment can be rinsed onto a fine mesh screen using a gentle water spray while holding the cone on its side (this is generally required due to the narrow construction of the cone). Test animals are then rinsed into a glass bowl for counting and weighing. If contaminant bioaccumulation is to be measured, survivors are put in clean water with cotton gauze and 50 μM EDTA for 24 h to clear their guts before drying or freezing.

Table 1 (continued). Rapid summary of test procedure.

Test endpoints	Survival and growth (final wet or dry weight) in chronic tests; contaminant bioaccumulation in both chronic and short-term (one-week) tests.
Caged animals	If desired, additional *Hyalella* can be exposed to overlying water in the cones inside cages in order to determine how much toxicity or contaminant bioaccumulation occurs through the dissolved phase alone.

3. Overview of applications reported with the toxicity test method

The Environment Canada *Hyalella* test method (EC, 1997) recommends the use of 300 mL high form glass beakers or jars with 100 mL of sediment and 175 mL of overlying water for conducting sediment toxicity tests, equivalent to the U.S. EPA (2000) method. There are two options for water renewal, 1) no renewal during the two-week test period, except for replacement of evaporative losses, or 2) daily renewal at the rate of two volume additions per day. An automated intermittent-renewal system is recommended if the renewal option is chosen. In contrast, chronic (four-week) sediment toxicity tests in our laboratories at the Canada Centre for Inland Waters (first in the Great Lakes Laboratory for Fisheries and Aquatic Sciences, and later in the National Water Research Institute) were initially conducted in 250 mL beakers with 40 mL of sediment and 160 mL of overlying water without water renewal (*e.g.*, Borgmann and Norwood, 1997). These static toxicity tests have worked well when testing a variety of sediments from the Great Lakes, but testing with sediments collected from Sudbury area lakes resulted in a rapid deterioration of overlying water quality. These Canadian Shield lakes contained high sulphide levels, and oxidation of sulphides resulted in very rapid acidification of the overlying water, down to pH 4 or even lower. High or complete mortality of test organisms resulted from pH stress, rather than from other contaminants. The lake waters, from which these sediments originated, however, were circum-neutral, with surface pH values above 6 and deep-water pH values always above 5.6. The laboratory tests were, therefore, not representative of natural conditions. Although a water renewal system may have overcome this problem, this would have necessitated the purchase and installation of specialized automated water renewal equipment, and it would have resulted in the potential loss of toxic substances leached from the sediments during the exposure period.

As an alternative to water renewal, Borgmann and Norwood (1999a) varied the volume ratio of sediment to overlying water. Water to sediment ratios of 40 or less resulted in rapid pH changes, but overlying water quality could be maintained by using 15 mL of sediment and 1 L of overlying water, giving a water to sediment ratio of 67:1. For comparison, a water renewal rate of two volume additions per day (350 mL) for 100 mL of sediment (EC, 1997) will result in an equivalent total water volume to sediment ratio after 19 days. In order to maintain a reasonable sediment depth, Borgmann and Norwood (1999a) conducted their tests in Imhoff settling cones. These funnel shaped containers provide a sediment depth of about 2.3 cm, in spite of the high water-sediment ratio. In 28-day tests, survival of *Hyalella* ranged from 89 to 97 % in cones at animal densities of 5 to 30 per test container. There was no apparent density effect on survival from 5 to 20 animals per cone (93-97%), and only a modest decrease in survival at 30 animals per cone (89%). By comparison, survival in beakers with 40 mL of sediment and 160 mL of water was 93% (Borgmann and Norwood, 1999a). Growth was reduced slightly at 30 amphipods per cone (mean wet weight per amphipod = 2.08, 1.61, 1.66, 1.76 and 1.44 mg at 5, 10, 15, 20 and 30 animals per cone respectively), but was always higher than in beakers (0.98 mg) in spite of the smaller sediment volume (Borgmann and Norwood, 1999a). Sediment toxicity tests with Imhoff settling cones have now been completed successfully in several studies using both field-collected and spiked sediments (Borgmann and Norwood, 1999b; Borgmann et al., 2001a; Borgmann et al., 2001b; Borgmann and Norwood, 2002; Nowierski, 2003). These included both 4-week chronic toxicity tests initiated with young amphipods, measuring growth, survival and bioaccumulation, and one-week metal bioaccumulation tests with adults at densities of 15 animals per cone. The test chambers have also worked well with Chironomids (survival and growth; 10 day test), mayflies (survival and growth; 21-day test) and Tubificid worms (survival and reproduction; 28-day test) (Borgmann and Norwood, 1999a). The Imhoff cone test method with *Hyalella* is now also being used successfully at the Pacific Environmental Science Centre, Environment Canada, North Vancouver, British Columbia, Canada (G. van Aggelen, personal communication). However, since this is a relatively new test procedure, it has not yet been used in a large number of laboratories or received round-robin testing as have some other standard test methods (*e.g.*, EC, 1977; U.S. EPA 2000; ASTM 2003).

4. Advantages of conducting the toxicity test method

Tests using Imhoff cones are easy to perform, provide large water volumes for chemical analysis, require less sediment sifting when counting surviving animals, and result in greater control survival than tests with beakers if sediments cause deterioration of overlying water quality (Borgmann et al., 2001a). They are particularly convenient when testing small sediment volumes, such as those obtained from sediment core sections. The absence of an automatic water renewal system eliminates problems with mechanical breakdowns of such systems. Analysis of contaminants in overlying water can be particularly useful when conducting sediment tests in Imhoff settling cones. The toxicity of sediments to *Hyalella* is often due to the toxicity of chemicals leached from the sediment into the water, rather than the solid phase of the sediment itself. The presence of chemicals causing toxicity can, therefore, often be determined by measuring their concentration in the overlying water, provided the concentrations in overlying water and sediment are close to equilibrium. The large water volume, and avoidance of water renewal, makes water sampling particularly appealing when conducting tests in cones. Nickel toxicity, for example, is a function of the amount of Ni bioaccumulated by *Hyalella*, and Ni bioaccumulation correlates much more closely with Ni in overlying water in tests conducted in Imhoff settling cones than in beakers (see Fig. 4 in Borgmann et al., 2001a). Concentrations of both Ni (Borgmann et al., 2001a) and Pb (Borgmann and Norwood, 1999b) in overlying water at equally toxic sediment concentrations can be much higher in tests conducted in beakers than in cones. Leaching of organic matter and other substances from sediments can reduce the bioavailability of dissolved metals, and this effect is much more pronounced at the lower water to sediment ratios obtained in beakers compared to cones. Comparison of contaminant concentrations in overlying water with criteria (*e.g.*, water quality guidelines) developed from water-only tests is, therefore, more useful in cone tests than in beaker tests.

5. Test species

Hyalella azteca is the most widely distributed amphipod in North America (Bousfield, 1958), and is common in Central and northern South America as well (Fig. 1). It is extensively used in toxicity testing and several standard sediment test methods have been produced (*e.g.*, EC, 1997; U.S. EPA, 2000; ASTM, 2003). Although native to the Americas, it is also used in toxicity testing in laboratories in Europe and Asia (*e.g.*, Blockwell et al., 1999; Othman and Pascoe, 2002). Unlike *Gammarus* sp., it is relatively easy to culture. We typically produce 500-1000 young per week in plastic beakers under static conditions on a single incubator bench of about 70 by 150 cm. It can also be cultured easily at room temperature outside an incubator. It is a detritivore/herbivore (Hargrave, 1970) and not cannibalistic. *Hyalella* used for experiments conducted in Department of Fisheries and Oceans and Environment Canada laboratories in Burlington, Ontario, were obtained from aquarium cultures originally set up from one batch of animals collected at the marshy shoreline of a small lake near Burlington (Valens Conservation Area) in 1985. Attempts at culturing *Gammarus fasciatus* and *Crangonyx* sp. using the same methods have been unsuccessful.

Although almost all studies refer to *H. azteca*, both allozyme and nucleic acid analyses strongly suggest that this is a complex of several very similar species (Hogg et al., 1998; Duan et al., 2000; Witt and Hebert, 2000). Consequently, the taxonomy needs revision and it is possible that the species name(s) of *Hyalella* populations commonly used in toxicity tests will change. This taxonomic uncertainty needs to be kept in mind when comparing results from different laboratories. If discrepancies are reported in amphipod behaviour or chemical sensitivity, these could be due to genetic/species differences.

Hyalella used in Burlington burrow readily in sediments, but only into the upper oxic layers. Most amphipods are not visible on the surface during sediment tests, but will emerge and swim above the sediments if food is limited. This intimate contact with sediments makes them a useful test organism for solid-phase toxicity tests. However, in contrast to some benthic species, *Hyalella* can also be cultured and tested in the absence of sediments if a suitable solid substrate is provided. Cotton gauze works well (Borgmann et al., 1989), and some populations of *Hyalella*, such as those originating from Redberry Lake, Saskatchewan, will actively consume it. In nature, *Hyalella* can be found above the sediment in dense vegetation. This makes culturing of *Hyalella* convenient, since young can be produced in water-only cultures and do not need to be sorted from sediment. It also allows toxicity testing in water-only tests for comparison to solid-phase sediment tests.

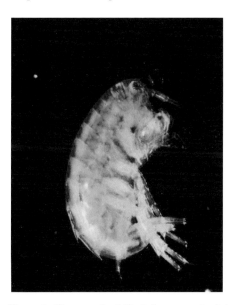

Figure 1. *Photograph of* Hyalella azteca *(male)*.

Hyalella is very sensitive to toxic chemicals. In comparison with other species, *Hyalella* is often the most sensitive, or one of the more sensitive, species. (Phipps et al., 1995; Burton et al., 1996; Suedel et al., 1997; Milani et al., 2003). Although usually used only for testing in freshwater, *Hyalella* have been cultured and used successfully in tests with waters of up to 15 g/L salinity (Ingersoll et al., 1992; McGee et al., 1993), and short terms tests with adults have been conducted at still higher salinities (Nebeker and Miller, 1988). Although *Hyalella* routinely used in Burlington (*i.e.*, originating from Valens Conservation Area) have a limited tolerance to salinity, *Hyalella* from Redberry Lake, Saskatchewan, have been cultured in 12.5 g/L Instant Ocean artificial seawater (Borgmann, 2002).

6. Culture/maintenance of organism in the laboratory

Water used for culturing and bioassays may be dechlorinated tap water, if suitable, or an artificial medium. Dechlorinated Burlington City tap water, originating from Lake Ontario, works well (hardness 126 mg/L, alkalinity 87 mg/L, Ca 36 mg/L, Mg 8.6 mg/L, Na 13 mg/L, K 1.6 mg/L, SO_4 31 mg/L, Cl 25 mg/L, pH 8.0-8.4 and dissolved organic carbon 1.6 mg/L). This is produced by filling a large polyethylene jug with tap water after allowing the tap to run for 15 minutes to flush out impurities and dissolved metals in the water line (note that copper leaching from copper pipes can be a major problem in newly constructed laboratories). A standard aquarium filter (operated by an air pump) with activated charcoal and polyethylene fibrefill is inserted and the water filtered for several days before use. If suitable tap water is not available, *Hyalella* can be cultured in artificial medium containing 1 mM $CaCl_2$, 1 mM $NaHCO_3$, 0.25 mM $MgSO_4$, 0.05 mM KCl and 0.01 mM NaBr in de-ionized water (SAM-5S medium, Borgmann, 1996). Modification of this medium is possible, as *Hyalella* tolerate a range of major ion concentrations, but bromide must be present in trace amounts or poor survival will result (Borgmann, 1996). Survival and production of young in SAM-5S has been equivalent, or slightly better, to that of dechlorinated tap water in our laboratory.

Young amphipods are obtained from adults held in polypropylene containers (2 L capacity, beaker shape) in an incubator at 25°C with a 16 h light: 8 h dark photoperiod, using fluorescent tubes placed 30 to 50 cm above the shelves and giving a light intensity of approximately 50-80 $\mu E/m^2/s$ at the level of the shelf surface, similar to de March (1977). These temperature and lighting conditions are also used for the bioassays. A temperature of 23° C is also suitable.

Each container holds 30 (± 5) adults in 1 L of water and a 5 x 10 cm piece of cotton gauze (sterile gauze bandage) as a substrate. Once a week the animals are shaken off the gauze and collected by filtration through two nested nylon mesh screens of 650 and 275 μm size and rinsed into two glass Petri dishes. The adults are retained on the larger screen and the young on the smaller screen. The Petri dish with the adults is then checked and any young present are removed with an eyedropper. The culture jars are then scrubbed clean of any accumulated algae, and the gauze and adults are returned with fresh water and food. Routine washing with soap and/or acid rinsing of the culture containers is not required or recommended, although it might be needed initially for new equipment. Incomplete rinsing can result in soap or acid residues, which are harmful. Instead, use plenty of clean rinse water.

Food consists of Tetra-Min® fish food flakes ground and sifted through a 500 μm mesh nylon screen. If other brands are used, they should be tested first and compared to Tetra-Min. Some brands will give poor reproduction of *Hyalella*. Each jar receives 5 mg of food when set up each week, and an additional 5 mg two more times per week. Food can be measured rapidly by using a "measuring spoon" consisting of a strip of plexiglass with a small depression, drilled on top at one end, which is equal to the volume of food required when levelled. Stir the water after food is added to ensure that the food settles to the bottom (a fine spray of distilled water is effective and avoids contamination between containers). Food stuck to the surface tension will not be eaten and will foul the water. Excess feeding can result in ammonia production and toxicity. Feeding rates should be adjusted as needed based on the rate of food consumption.

On average, each adult produces 1-3 young per week (*e.g.* 20 adults in each of 10 jars should provide 200-600 young per week). There will usually be a lag period of several weeks after a culture jar is set up before young are produced at these rates.

7. Preparation of test species for toxicity testing

Chronic toxicity tests in our laboratory are typically initiated with 0-1 week old *Hyalella*. Culture maintenance and separation of young from adults is performed weekly on a Monday, and tests are initiated on Tuesday to Friday. The exact age of the animals is, therefore, between 1 and 11 days when added to the test container. Comparison of the sensitivity of different age groups of young (0-2 day old up to 24-26 day old) to selected toxic substances (Cu, Cd, Zn, diazinon) suggests that there is little difference in the sensitivity of different ages of young in the first few weeks (U.S. EPA, 2000). Depending on the purpose of the test, it may be necessary to acclimate animals to the test water, if this differs significantly from the culture water.

Although animal sensitivity may not vary significantly during the first week or two of life, some researchers have expressed concern that variations in body size could potentially affect the sensitivity of growth estimates. If most animals grow at a similar rate, then a wider size range at the start of the test could result in a wider size range at test end. This could result in larger standard deviations for body size measurements requiring a larger growth reduction before statistically significant effects on growth are observed. However, an experiment conducted to test the effect of varying size ranges on growth rate estimates did not support this hypothesis (Borgmann, 2002). Sequentially screening the animals through 500, 475, 400 and 275 µm mesh screens to reduce size variation at the start of the test did not reduce the size variation at the end of the test, 4 weeks later. Furthermore, there was no significant difference in final size between groups of animals that were different in size initially. This appears to be the case because 1) the variability in growth rates among individual amphipods contributed more to variation in final size than did variation in initial sizes, and 2) the instantaneous growth rate of *Hyalella* decreases with age, allowing the younger animals to catch up partially with their older siblings, reducing the relative weight range with time (Borgmann, 2002). This suggests that no benefit is gained by size-sorting animals used to initiate toxicity tests, at least for four-week toxicity tests. Initial animal size may have a larger effect on final body size in tests of shorter duration.

In addition to chronic toxicity tests, it is sometimes desirable to conduct shorter (one-week) bioaccumulation tests with adults. These can be obtained by raising young for 4-6 weeks in the same jars used for culturing (see above).

8. Testing procedure

8.1 TEST SAMPLES

The source and method of collection of sediment samples for testing will depend on the objectives of the study and standard methods applicable to all situations cannot be given. If the purpose of the test is to estimate the toxicity of sediments in situ, then sediments should be collected using a sampler that disturbs the sediment as little as possible and allows collection of the top 1 or 2 cm of sediment. *Hyalella* only burrows into the top oxygenated layers of sediment. Sediment contaminant concentrations are often highly stratified with depth (*e.g.*, Borgmann and Norwood, 2002), and collection of bulk samples may not be representative of surface sediment conditions. Sediments should be stored refrigerated and tested as soon as possible, but never frozen (Day et al., 1995a). If sediments are largely anoxic, and if chemical changes are to be kept at a minimum, then sediments should be stored in airtight sealed containers with as little overlying water and air as possible. On the other hand, if oxygenated surface sediment are collected and these are to be tested in their original oxic condition, then it may be preferable to use larger storage containers with ample overlying water and air and relatively little sediment so as to reduce the likelihood that anoxia will develop. Tests can be conducted with unaltered field-collected sediments, or with contaminant-spiked sediments. The method of spiking will depend on the type and properties of the contaminant to be tested. One method, suitable for spiking sediments with metals, is described below under Section 8.7.

8.2 TEST EQUIPMENT

- One litre polycarbonate (or glass, if preferred) Imhoff settling cones, plugged with #4 silicone rubber stoppers (see Figs. 2 and 3).
- Plywood rack (150 x 70 cm) with five rows of 8.3 cm diameter holes (50 holes/rack) for holding the cones.
- Walk-in temperature-controlled incubator (at 23 or 25°C), or equivalent, with air supply and fluorescent lights on timers (16 h light: 8 h dark). Separate plastic airline to each cone.
- Plastic cover (snap-on polypropylene lids for plastic containers) for each cone, perforated with a hole for insertion of a glass rod. The upper end of the rod is attached to the airline and the bottom end is covered with a metal-free (natural, without colouring agents) 250 µL disposable capillary pipette tip to provide a uniform small diameter opening. Attachment of the glass rod to the airline and capillary tip is easier if an adaptor, consisting of a 1000 µL disposable pipette tip with the wide end trimmed off is used.
- Optional: cages consisting of inverted 250 mL polypropylene specimen cups with the bottom cut out (now serving as the top of the cage) and the top covered with 200 µm mesh nylon screen (now serving as the bottom of the cage). The screen is held in place by the rim of the lid, the centre of the lid having been cut out. The rim of the finished cage may be covered in plumber's Teflon tape to ensure a smooth surface which discourages test animals from clinging to the outside of the cage.

8.3 SETUP PROCEDURE

- Fill cone about one third full with test water (dechlorinated tap, artificial medium, or site water, as required). Mix the sediment to be used in its storage container to ensure homogeneity, but avoid excessive exposure to air and avoid mixing sediment immediately next to the container walls if the container is a plastic bag or thin-walled container and shows signs of oxidation of sediment at the container wall. Using a 50 mL plastic syringe with an enlarged opening (5 mm ID or more) fitted with a 10 to 20 cm plastic tube extension (10 mm ID), suck up

15 mL of sediment. Carefully extrude this into the cone below the water surface with the tip near the bottom. Keep resuspension of sediment to a minimum. Let settle while sediment is added to the next cones.

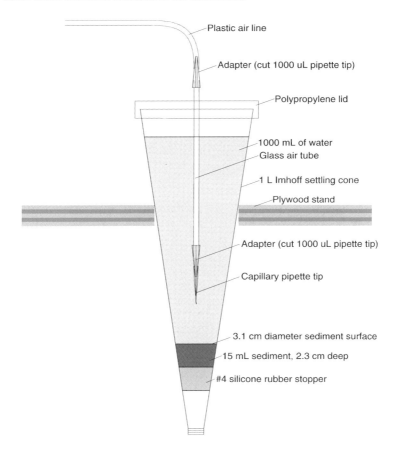

Figure 2. Schematic diagram of Imhoff cone toxicity test chamber.

- Gently fill the cone to give a total volume of 1 L of overlying water, keeping resuspension to a minimum. Place the cone in the plywood rack in the incubator, add the lid and glass rod-airline. Position the rod so the lower tip is 1-2 cm over the sediment surface. Bubble gently with air at a speed that does not cause resuspension of the sediment.
- Aerate and allow equilibration for 1-2 weeks before adding test animals when assessing impacts of metals. Shorter equilibration times should be used when assessing impacts from volatile or unstable chemicals.
- Optional: if cages are to be used in an experiment, attach them with plastic clothes-pins to the top inside of the cones. Add a 5 x 5 cm piece of pre-soaked cotton gauze inside the cage.

Figure 3. Photograph of Imhoff cone with sediment inside incubator.

8.4 TEST PROCEDURE

- After 7-14 days equilibration, remove 15-50 mL of water from each cone and measure dissolved oxygen, pH and conductivity. Remove 1 mL of water from each cone and analyse for ammonia using an ammonia probe or ammonia test kit (available from aquarium and pet stores). Collect time 0 water samples for chemical analysis, if required. Replace test water with an equivalent amount of water. Record dissolved oxygen, pH, conductivity and ammonia. If total ammonia exceeds 0.75 mM or 10 mg/L (in Lake Ontario water, Borgmann, 1994) or if oxygen is below 4 mg/L, additional aeration and/or a water change may be required. Monitor ammonia, pH, dissolved oxygen and conductivity periodically during the experiment if initial readings suggest a possible problem. Check and record temperature in several cones at opposite ends of the rack.
- Count out 15 *Hyalella* into separate 15 mL beakers with about 10 mL of water each. This is best done by adding a small number of animals (*e.g.*, 2-5) to all beakers, repeating the procedure in reverse order, and continuing until 15 have been added to all containers. This evens out operator bias in selection of larger

individuals, or other systematic variation. Use young (0-1 week old) for chronic toxicity tests (4-week exposure duration) or adults (4-6 weeks old) for short-term (1-week) bioaccumulation tests. Randomize the beakers. Add 15 amphipods to each cone, preferably using an eyedropper kept below the water surface to ensure animals do not stick to the surface tension. Add another 15 animals inside the cage, if present.

- For chronic tests initiated with young amphipods, feed each cone and each cage, if present, with finely ground Tetra-Min® fish food flakes at a rate of 2.5 mg twice in week 1 and 2, 2.5 mg three times in week 3, and 5 mg two times in week 4. For experiments with adults, feed each cone and cage 5 mg twice a week. Reduce feeding if uneaten food accumulates and starts to rot. Food added to the cone outside the cage will gradually sink due to agitation from rising air bubbles, but food inside the cage must be made to sink (e.g., by spraying with de-ionized water while replacing evaporated water). Ensure that all food has sunk after each feeding.
- At the end of the incubation period, before counting animals (this may be done the day before animals are removed and counted) measure and record ammonia, pH, dissolved oxygen and conductivity. Optional: if metal analyses are to be performed, remove 1-10 mL (depending on analytical method requirements: 1 mL for Graphite Furnace Atomic Absorption Spectrophotometry (GFAAS), 10 mL for Inductively Coupled Plasma-Mass Spectrometry (ICP-MS) of water from each cone and place in an acid washed and labelled 1 mL cryovial or larger polypropylene container. Add 10-100 µL ultra-pure concentrated (70%) nitric acid, close lid tightly and shake.
- Remove the air tube and lid. If major ion or other large-volume-requiring analyses are to be performed on overlying water, carefully decant as much of the overlying water as possible into an appropriately cleaned container, being careful not to resuspend sediment. Filter (e.g., 0.45 µm glass microfiber) water samples for major ion and dissolved organic and inorganic carbon analysis, as required. For metals analysis, filter as much water as possible, or as needed, through a 0.4 µm polycarbonate filter using an acid-washed metal-free filter head or clean disposable syringe filters. Acidify to 1% with ultra-pure concentrated nitric acid, close sample container lid tightly and shake. For other contaminants, use appropriate processing and preservation methods.
- Rinse the sediment and amphipods onto a 363 µm screen. Gently wash through as much of the sediment as possible and rinse the amphipods and remaining debris into a clean glass bowl using dechlorinated water. Record time and number of survivors.
- If metal bioaccumulation is to be measured, amphipods must have their guts cleared (Neumann et al., 1999). Place survivors in polypropylene cups with 40-60 mL of 50 µM EDTA in dechlorinated tap water, cotton gauze and 5 mg Tetra-Min food. After 24 hours gut clearance time, count and place the amphipods onto a clean Kim-wipe® paper to absorb adhering water and obtain a total wet weight. Record number of survivors and wet weight. Place in a clean cryovial and dry at 60°C for 24-72 hours.

8.5 CHEMICAL ANALYSIS

Samples for major ions and dissolved organic and inorganic carbon collected at the end of the experiment should be kept refrigerated and submitted for analysis as soon as possible. Filtered and acidified water samples are stored refrigerated and sent for metal analysis by ICP-MS (or GFAAS) as funds and analytical time are available. Samples for other analyses should be processed appropriately. Dried amphipods are stored in a desiccator until they can be digested and analysed. Subsets of the sediments used in the tests should be freeze-dried and saved for future analysis for metals, total organic carbon, loss on ignition and particle size.

Metal concentrations in *Hyalella* can be measured in individual amphipods, if desired. However, it is usually more convenient to pool 4 to 6 dried amphipods (about 0.5-4 mg dry weight total) from each container into a single sample. These are digested with 70% ultra-pure concentrated (70%) nitric acid at room temperature for 1 week, after which 30% hydrogen peroxide is added and digestion allowed to continue for another 24 h. Final sample volume can be made up to 0.5-2 mL (depending on tissue mass, approximately 1 mL per mg of mass) for analysis by GFAAS. Acid and peroxide volumes are 25 µL and 20 µL per 1 mL of total volume respectively. Alternatively, final volumes can be made up to 5 or 6 mL for analysis by ICP-MS

8.6 REPLICATION AND REFERENCE SAMPLES

Several controls and/or reference samples need to be included with each test. Toxicity controls are tests conducted with a control sediment. This is a sediment that has been tested repeatedly and is known to consistently result in good survival and growth. Inclusion of control samples demonstrates that the animals are healthy and growing at an acceptable rate under normal conditions. The origin, collection, and storage of the control sediment are not important, as long as the sediment consistently results in good survival and growth. A reference sediment is one collected from a site similar to the test site in as many aspects as possible, with the exception of the level of contamination, which should be minimal. The reference sediment is collected and processed in the same manner as the test sediments. Survival and growth may be significantly different in the reference sediment relative to the control. The reference sediment provides a benchmark against which the test sediments are compared. Separate reference and control sediments are not needed when conducting tests on spiked sediments (*i.e.*, the un-spiked sediment serves as the control). In addition to the sediment control, a gauze control may also be added. This test chamber has no sediment, but contains a 5 x 5 cm piece of cotton gauze instead. This is used solely for comparing contaminant bioaccumulation with that observed in control and reference sediments. It demonstrates if there are bioavailable contaminants present in the control sediment. Survival and growth in the gauze control is generally poorer than in sediment controls, and should not be used to compare to test samples.

A total of 15 to 20 cones is a reasonable number for a single technician to set up or take down in one day. A typical test with spiked sediments may involve duplicate cones for 6-8 test concentrations and a control. Experiments with spiked sediments should be repeated until 2 or 3 separate experiments with comparable results are obtained. Tests with field-collected sediments should include multiple sediment samples collected from each site if the objective is to compare sampling sites. A single test container is then set up with sediment from each sample. Setting up multiple test containers using sediment from the same sample represents pseudo-replication and should be avoided.

Routine reference toxicant tests for the purpose of verifying *Hyalella* health and sensitivity do not need to be performed with spiked sediments. Tests with *Hyalella* can be conducted in water-only exposures using cotton gauze as a substrate instead of sediment. It is more efficient to conduct routine reference toxicant tests in water-only exposures.

8.7 SEDIMENT SPIKING PROCEDURE

The following procedure for spiking sediments with metals has worked well in our laboratory. Other procedures may need to be used for other contaminants.
- Make up a solution of metal in de-ionized or double distilled water on a volume concentration basis (mole/L) equal to the highest desired sediment concentration. Mix equal volumes of sediment and metal-solution in polypropylene bottles and rotate the mixture for 24 h at 4 rpm on a mechanical mixer (rotating bar attached to slow speed motor). Allow to settle and decant the surface water. If the desired metal concentration is too high to dissolve fully in water initially, lower concentrations may be used and the spiking procedure repeated several times. Process a sufficient amount of control sediment in the same way, but using un-spiked de-ionized water in place of the metal spike solution.
- Make spiked sediments of lower concentration by mixing 10, 18, 32 or 56% metal-spiked sediment with control sediment and mixing thoroughly. If a wider concentration range is to be tested, then initial spiking of the sediment (see step above) may need to be conducted at several concentrations.

Further recommendations on sediment spiking can be obtained from standard reference manuals (*e.g.*, EC, 1995; U.S. EPA, 2000).

9. Post-exposure observations/measurements and endpoint determinations

Test endpoints include percent survival, final weight (total wet weight of animals per test container divided by number surviving, or individual dry weights, if desired) and contaminant bioaccumulation. As an alternate to weight, length can also be used as an endpoint (U.S. EPA, 2000). Control survival in cones is typically around 90% and final wet weight per *Hyalella* is above 1 mg after a four-week chronic test (Borgmann, 2002). An acceptability criterion of 80% survival (*e.g.*, EC, 1997, two-week test) for controls is, therefore, reasonable. The number of replicates required to detect a 25% reduction in survival or growth ($p < 0.05$) after four weeks is about 4-8 (two tailed test) or 3-7 (one tailed test) for several test types, including water-only tests and sediment toxicity tests in beakers (Borgmann, 2002). Tests conducted in Imhoff settling cones were among the more consistent ones, requiring only 2 replicates to detect a 25% drop in survival or 5 to detect a 25% drop in final weight. A 25% reduction in survival or growth after four weeks is, therefore, a convenient endpoint for chronic toxicity when using 4 or more replicates per treatment. Chronic mortality is often a more sensitive endpoint than growth in tests with *Hyalella* (Borgmann et al., 1989; 1993) and a growth reduction of 25% is often not observed at the highest test concentration with survivors. A 25% drop in final biomass (total mass of surviving animals = number surviving animals x mean mass per individual) is, therefore, sometimes a useful endpoint incorporating both survival and growth effects, especially if growth effects are minimal.

Endpoints such as EC50s (Effective Concentrations 50) and LC50s (Lethal Concentrations 50) can be computed using a variety of procedures. One of the most robust methods of computing such endpoints and their confidence limits is the Trimmed Spearman-Karber (TSK) method (Hamilton et al., 1977). This is particularly useful for tests with *Hyalella* because toxicity curves are often very steep, resulting in few partial effect concentrations. The TSK method can compute LC50s under these circumstances. A program to compute LC50s using the TSK method is available at the US EPA website http://www.epa.gov/nerleerd/stat2.htm.

Lethal and sublethal endpoints can be computed from linear interpolation of survival, growth or biomass plotted against log (concentration), after ensuring that the data are monotonic (Borgmann et al., 2001a). Alternatively, mortality or growth can be modelled using a variety of curves fitted through non-linear regression methods (*e.g.*, Borgmann and Norwood, 1999b; Borgmann et al., 2005). The latter are particularly useful if it is desirable to predict bioaccumulation or toxicity using mechanistically based models. More detailed information of endpoint determination using standard methods can be obtained in U.S. EPA (2000).

Although the test duration proposed is 28 days, other exposure periods could be used. The cost savings and time required to perform the test must to be balanced against the need for sensitivity. The Environment Canada (1997) method calls for a 2-week exposure, whereas the U.S. EPA (2000) method calls for a 10-day exposure or a 4-week exposure followed by two weeks in water-only chambers to monitor reproduction. Based on data obtained in our laboratory over a 20 year period, the median LC25 (concentration resulting in 25% mortality, corrected for control mortality) for various contaminants with *Hyalella* was equal to 162, 119, 91, 60 and 57% of the 4-week LC25 at 1, 2, 6, 8, and 10 weeks, respectively (Borgmann, 2002). Increasing exposure duration does, therefore, increase sensitivity, but will decrease the number of tests that can be run within any time period. We most often conduct chronic sediment toxicity tests using 4-week exposures, but have conducted some tests for 10 weeks and even longer (*e.g.*, multi-generation tests with tributyltin, Bartlett, 2004). Four weeks is long enough to ensure that the test represents a truly chronic exposure (*i.e.*, *Hyalella* start reproducing at 4-6 weeks of age), while avoiding excessively long exposure periods.

If a longer exposure period (*e.g.*, 10 weeks) is acceptable, then the test described here can be adapted to measure reproductive effects, if desired. Borgmann et al. (2001a) placed amphipods surviving after four weeks into fresh sediment and continued exposure for an additional six weeks. This allowed determination of total number of amphipods and biomass after 10 weeks, but first generation animals could not be distinguished from the rapidly growing young produced. Accomplishing the latter would require more frequent counting and changes of sediment. Although feasible, and of particular interest when chemicals are present which are known to affect reproduction, reproduction tests with *Hyalella* suffer from high variability. Very large numbers of replicates are needed to detect a 25% drop in reproduction in water-only tests (*e.g.*, > 40, Borgmann, 2002), although a 75% drop in reproduction (IC75) can be detected using only 5 replicates. Drastic effects on reproduction can be quantified, but subtle changes are difficult to detect. The increased effort required to measure reproduction needs to be justified in terms of increased sensitivity. However, reproduction is not usually a much more sensitive endpoint than survival in chronic tests with *Hyalella*. A review of end-point sensitivity for ammonia, Cu, Hg, Ni, Pb, Tl, Zn, tributyltin (TBT), 2,5,2,'5'-tetrachlorobiphenyl and sea salt revealed that the median of the IC75 for reproduction (*i.e.*, the effect detectable with 5 replicates) after 10 weeks was greater than the 10-week LC25, and equal to 87% of the 4-week LC25 (range 12-203%, Borgmann, 2002). Although feasible, the increased cost and time required to perform a reproduction test (e.g., a 10-week test) must to be weighed against the importance of detecting a usually modest difference between the 10-week IC75 for reproduction and the 4-week LC25. In an alternate method, Ingersoll et al. (1998) exposed *Hyalella* to sediments for four weeks and then measured reproduction in a subsequent two-week water-only exposure. They also observed that reproduction was a more variable endpoint than growth.

Contaminant bioaccumulation, either after chronic four-week exposure or in one-week tests with adults, is a useful endpoint. It is particularly useful for identifying the cause of sediment toxicity, if the relationship between bioaccumulation and toxicity is known (*e.g.*, Borgmann et al., 2001b; 2004). Only those contaminants responsible for toxicity are likely to be accumulated to concentrations above the critical body residue. Bioaccumulation is also useful for demonstrating "no-effect" due to a specific contaminant. Toxicity endpoints, unfortunately, cannot be used to rank sites when effects are below the detection limit. Toxicity responses can only be categorized as 1) significantly toxic, or 2) statistically indistinguishable from control or reference. Sites with toxicity not significantly different from control might be just under the toxic threshold and could show effects if exposure periods were longer, or contaminant bioavailability might be far below concentrations of concern. In contrast, bioaccumulation data can be categorized as 1) significantly above the toxic threshold, 2) statistically indistinguishable from the toxic threshold, or 3) significantly below the toxic threshold. The latter category can be used to rule out some substances or sites as of concern.

A one-week bioaccumulation test with adults is recommended as an alternative to four-week chronic tests when bioaccumulation is the desired endpoint because previous studies have shown that bioaccumulation reaches steady state within a week for uptake of most metals in *Hyalella* (Borgmann and Norwood, 1995; MacLean et al., 1996; Borgmann et al., 2001a). Furthermore, one- and four-week bioaccumulation tests provide equivalent results (Nowierski, 2003). The equivalence of one- and four-week bioaccumulation should, however, be confirmed when measuring bioaccumulation of other contaminants. A four-week bioaccumulation test, initiated with young *Hyalella* can be assumed to provide steady-state bioaccumulation data since the majority of the biomass of individual amphipods is produced within the exposure period (*i.e.*, ingestion, assimilation, growth, and growth-dilution effects are all incorporated in the measurement).

10. Factors capable of influencing performance of test organism and testing results

The equilibration time allotted before addition of animals to the cones needs to be considered. It is beneficial to allow the cones to sit for 1-2 weeks with sediment and water before addition of animals when testing metal-contaminated sediments. Leaching of metals from the sediments is rapid in the first few days, and then slows as equilibrium is approached (Borgmann and Norwood, 1999a). Although complete equilibrium between water and sediment is not likely to be achieved in most sediment tests, a 1-2 week equilibration time ensures that test animals are exposed to near-equilibrium conditions for a longer period. The duration of the equilibration time may need to be adjusted, however, depending on the purpose of the test and type of contaminants present. If it is desirable, for example, to detect the presence of highly unstable or volatile compounds, which may dissipate within a one-week period, a shorter equilibration period may be required. A one-day equilibration time is recommended in the Environment Canada protocol using beakers (EC, 1977), but tests at different equilibration times should be conducted to determine if this is appropriate when using cones.

Hyalella tolerate a wide range of sediment types and grain sizes (Suedel and Rodgers, 1994), but growth will vary with changes in nutritional content of the sediment. Therefore, care needs to be taken when interpreting differences in growth between test and reference sediments. This is generally not a problem when testing spiked sediments, because spiked sediment controls (*i.e.*, the un-spiked sediment) are equivalent to the test sediments nutritionally. However, reduced growth in test sediment could be the result of a lower nutritional quality relative to the reference sediment, rather than sediment toxicity. This makes mortality a more definitive endpoint for toxicity compared to growth. Fortunately, growth is often a less sensitive indicator of toxicity than is survival (*e.g.*, Borgmann et al., 1989; 1993), although Cu in sediment does appear to reduce growth at lower concentrations than survival (Borgmann and Norwood, 1997).

Data collected using metal-spiked sediments needs to be interpreted with caution when compared to metal concentrations in field-collected sediments. Metal bioavailability and toxicity is often greater in soils and sediments spiked with metal salts in the laboratory, than in field collected soils or sediments (Borgmann, 2003; Lock and Janssen, 2003; Smolders et al., 2003). This can result in over estimation of metal impacts in the field. Similarly, TBT bioavailability to *Hyalella* is also higher in laboratory-spiked sediments compared to sediments collected for TBT contaminated sites in harbours (Bartlett, 2004). Further comments on spiking procedures and the effects of storage on toxicity of spiked sediments can be found in EC (1995) and U.S. EPA (2000).

Sediment storage procedures can also affect toxicity. Autoclaving, freezing and gamma irradiation all reduced survival of *Hyalella* in sediments (Day et al., 1995a). Sediments should be stored cool, but not frozen.

Allozyme and nucleic acid analyses strongly suggest that *H. azteca* is a complex of several similar species (Hogg et al., 1998; Duan et al., 2000; Witt and Hebert, 2000) and there appears to be some association between genotype and median survival time during exposure to Cd, Zn or low pH (Duan et al., 2001). Consequently, toxicity data reported by different laboratories for "*H. azteca*" may not actually have been conducted on the same species.

11. Application (of the toxicity test) in several case studies

The Imhoff cone test method (Borgmann and Norwood, 1999a) was first developed to measure bioaccumulation and toxicity of metals in sediments from lakes near smelters at Sudbury, Ontario (Borgmann et al., 2001b). Unlike tests with sediments from the Laurentian Great Lakes, tests with sediments from Canadian shield lakes using a water-sediment ratio of 4:1 resulted in a very rapid drop in pH down to 4, and sometimes even lower. This was presumably caused by oxidation of sulfides in the water. The low pH resulted in complete mortality of test organisms and very high metal levels in overlying water. Since the pH of the water in these lakes was always above 5.6, these sediment tests represented very unnatural conditions. Using cones, with a water-sediment ratio of 67:1, alleviated this problem without the need to resort to frequent water changes. This made it possible to determine differential toxicity between sediments from the Sudbury area and reference lakes using chronic toxicity tests with *Hyalella*, *Chironomus*, *Hexagenia* and *Tubifex*. Furthermore, one-week metal bioaccumulation tests with adult *Hyalella* clearly demonstrated that Ni was the only metal accumulated to levels above the lethal body concentration. The study demonstrated how the popular "Sediment Quality Triad" approach could be extended to identify the cause of toxicity (Borgmann et al., 2001b). This would have been much more difficult, if not impossible, to do using standard water-sediment ratios and slow water renewal rates. The same approach has been used in the Rouyn-Noranda area of Quebec. In this case, Cd is the metal most likely responsible for toxic effects (Borgmann et al., 2004).

The low sediment volumes required for the cone test also made it possible to determine sediment toxicity and Ni bioavailability as a function of sediment depth and age (Borgmann and Norwood, 2002). Ten cm diameter sediment cores from Richard Lake near Sudbury, Ontario, were sectioned into slices a thin as 0.5 cm. This still provided sufficient sediment to conduct toxicity tests as well as metal analysis on the sediment. This study demonstrated that the relative bioavailability of Ni in the sediment increased with depth down to about 5 cm, and was then constant. Toxicity matched Ni bioaccumulation. From these data, which demonstrated a trend towards reducing Ni concentrations in the surface sediments, it was possible to predict that, if current trends continue, surface sediments in Richard Lake might become non-toxic in about 15 years.

The successful use of the cone method for determining not only sediment toxicity, but also metal bioavailability in sediments, has made it possible to propose a methodology for deriving true cause-effect based sediment quality guidelines (Borgmann, 2003). Unlike water quality guidelines, Canada's current interim sediment quality guidelines are based on correlations between sediment toxicity and contaminant concentrations in field-collected sediments (Smith et al., 1996; CCME,

1999). These can be used to predict possible sediment toxicity, but not the cause of that toxicity. Their purpose is to identify sites requiring sediment toxicity testing and further assessment. The difference in the method of derivation of water and interim sediment guidelines, and hence the difference in correct interpretation of sediment chemistry data, often causes confusion among those not familiar with the origin of these guidelines. Derivation of true cause-effect based guidelines, would help alleviate some of this confusion. By conducting sediment toxicity and bioaccumulation tests in cones, it is possible to determine the relationship between contaminant bioavailability and concentration in the sediment, thereby providing a basis for the derivation of cause-effect based sediment quality guidelines (Borgmann, 2003).

12. Accessory/miscellaneous test information

The temperature of 25°C is an optimum temperature for *Hyalella* culture and toxicity testing, but the test can be conducted at lower temperatures as well. For example, 23°C is commonly used for tests with other benthic species, and *Hyalella* tests can be conducted at this temperature to eliminate the need for separate temperature controlled incubators for each species (Day et al., 1995b).

The water-sediment ratio recommended in this method (1 L water over 15 mL sediment) originated from the observation that this was the minimum required to produce stable overlying water chemistry in tests with Canadian shield lakes and Lake Ontario water (Borgmann and Norwood, 1999a), but other ratios are also possible. For example, in tests designed to determine the effect of overlying water source on sediment toxicity, some tests were conducted using overlying water obtained from the same Canadian shield lakes from which the sediments were obtained (Nowierski, 2003). These lake waters, however, have extremely low buffering capacity (*e.g.*, alkalinity down to 0.054 mEq/L and Ca as low as 0.059 mM). Even a water-sediment ratio of 67:1 did not result in stable overlying water quality. For these ultra-low alkalinity waters, a water-sediment ratio of 1000:1 was used (*i.e.*, 1 mL of sediment). To provide sufficient sediment depth, the #4 silicone rubber stopper was replaced with a 1.2 cm diameter stopper, resulting in a sediment depth of 1 cm and surface sediment diameter of 1.9 cm. Such a small sediment volume still did not result in reduced growth even at densities of up to 30 *Hyalella* per cone. The very small sediment volume did seem to somewhat limit the amount of metal, which leached into the overlying water. Cadmium and nickel in overlying water in tests using laboratory water were lower (1.2 to 3.9 fold) when 1 mL of sediment was added to the test chambers compared to 15 mL, but this was proportionately much less than the 15 fold difference in sediment volume (Nowierski, 2003). While 15 mL of sediment is preferable, sediment volumes as low as 1 mL can be used for special purpose tests, if required.

Although most sediment tests with cones have been conducted using *Hyalella*, other species can be tested in these containers as well. For example, Borgmann and Norwood (1999a) also reported growth and survival of *Chironomus riparius* (10-day test) and *Hexagenia* sp. (21-day test), and reproduction in *Tubifex tubifex*. Survival and growth of *Chironomus* at densities of up to 15 animals per cone was better in cones than in beakers, similar to results observed using *Hyalella*. Tests with *Hexagenia and Tubifex* were only conducted at one density (2 animals per cone with 15 mL of sediment) and the strong burrowing activity of *Hexagenia* increased overlying water turbidity and caused a slight reduction in pH, presumably due to oxidation of buried sulfides in the sediment.

The test method described here is designed for freshwater sediments, but the relatively high salt tolerance of *Hyalella* (Ingersoll et al., 1992; McGee et al., 1993) suggests that it should also be possible to use this method to test estuarine sediments, at least with some strains of *Hyalella* (Borgmann, 2002).

The major cost involved in conducting these tests is for labour. We have set up and taken down single experiments with up to 34 cones in one day (although a long day, at 10-12 hrs). Depending on requirements, about 80 cones can be set up in one week. Maintenance is minimal once set up, until takedown. Consequently additional experiments can be set up in following weeks, before the first experiments terminate. Theoretically, up to four sets of experiments (*i.e.*, 320 cones) could be run simultaneously. Allowing seven weeks for a chronic test (set up, wait 2 weeks for equilibration and add animals, wait 4 weeks and terminate), allowing 3 more weeks for the other 3 sets of experiments running simultaneously to terminate sequentially, and allowing 2 weeks for washing and cleanup, this means that about 320 cones could be set up every 12 weeks by one technician. Compared to tests conducted in beakers, there is an increased space requirement because of the depth of Imhoff settling cones (45 cm tall by 10.6 cm wide) and an increased cost for the cones (currently polycarbonate Imhoff cones are approximately 4 times as expensive in Canada as the high form 300 mL glass beakers commonly used in sediment toxicity tests).

13. Conclusions/prospects

The use of Imhoff settling cones and large water-sediment ratios significantly simplifies toxicity tests with sediments which otherwise result in rapidly deteriorating overlying water quality, and provides a number of additional benefits. Water renewal is not necessary, even in chronic (4-week) tests, control survival is frequently improved, sorting test animals at the end of the experiment is easier due to the smaller (15 mL) sediment volume, contaminants leached into overlying water are not lost from the test system, and the final water volume (1 L) is sufficient for numerous chemical analyses. The small sediment volume required also makes it possible to conduct tests when sediment sample volume is limited (*e.g.*, samples from sediment cores). The relationship between contaminant bioaccumulation and concentration of contaminants in sediment and overlying water can be determined making derivation of cause-effect based sediment quality criteria easier. This method has been used primarily with *Hyalella* to date, but shows promise when used with other benthic species as well. While the Imhoff cone method has proven extremely useful in studies conducted at NWRI, there is a need for other laboratories to also critically test and evaluate this method in order to demonstrate its universal applicability.

Acknowledgements

M. Nowierski was supported from a Natural Sciences and Engineering Research Council grant to D.G. Dixon, as part of the Metals In The Environment Research Network (MITE-RN). The cone figure was draw by Philip McColl, Graphic Arts, NWRI. Graham van Aggelen provided useful comments on the manuscript.

References

ASTM (American Society for Testing and Materials) (2003) Test method for measuring the toxicity of sediment-associated contaminants with freshwater invertebrates, ASTM, Standard E1706-00e2.

Bartlett, A.J. (2004) Chronic effects of tributyltin in freshwater invertebrates, Ph.D. Thesis, University of Waterloo, Waterloo, Ontario, Canada.

Blockwell, S.J., Maund, S.J. and Pascoe, D. (1999) Effects of the organochlorine insecticide lindane (gamma-$C_6H_6Cl_6$) on the population responses of the freshwater amphipod *Hyalella azteca*, *Environmental Toxicology and Chemistry* **18**, 1264-1269.

Borgmann, U. (1994) Chronic toxicity of ammonia to the amphipod *Hyalella azteca*; Importance of ammonium ion and water hardness, *Environmental Pollution* **86**, 329-335.

Borgmann, U. (1996) Systematic analysis of aqueous ion requirements of *Hyalella azteca*: A standard artificial medium including the essential bromide ion, *Archives of Environmental Contamination and Toxicology* **30**, 356-363.

Borgmann, U. (2002) Toxicity test methods and observations using the freshwater amphipod, *Hyalella*, NWRI Contribution No. 02-332.

Borgmann, U. (2003) Derivation of cause-effect based sediment quality guidelines, *Canadian Journal of Fisheries and Aquatic Sciences* **60**, 352-360.

Borgmann, U. and Norwood, W.P. (1995) Kinetics of excess (above background) copper and zinc in *Hyalella azteca* and their relationship to chronic toxicity, *Canadian Journal of Fisheries and Aquatic Sciences* **52**, 864-874.

Borgmann, U. and Norwood, W.P. (1997) Toxicity and accumulation of zinc and copper in *Hyalella azteca* exposed to metal-spiked sediments, *Canadian Journal of Fisheries and Aquatic Sciences* **54**, 1046-1054.

Borgmann, U. and Norwood, W.P. (1999a) Sediment toxicity testing using large water-sediment ratios: an alternative to water renewal, *Environmental Pollution* **106**, 333-339.

Borgmann, U. and Norwood, W.P. (1999b) Assessing the toxicity of lead in sediments to *Hyalella azteca*: the significance of bioaccumulation and dissolved metal, *Canadian Journal of Fisheries and Aquatic Sciences* **56**, 1494-1503.

Borgmann, U. and Norwood, W.P. (2002) Metal bioavailability and toxicity through a sediment core, *Environmental Pollution* **116**, 159-168.

Borgmann, U., Ralph, K.M. and Norwood, W.P. (1989) Toxicity test procedures for *Hyalella azteca*, and chronic toxicity of cadmium and pentachlorophenol to *H. azteca*, *Gammarus fasciatus*, and *Daphnia magna*, *Archives of Environmental Contamination and Toxicology* **18**, 756-764.

Borgmann, U., Norwood, W.P. and Clarke, C. (1993) Accumulation, regulation and toxicity of copper, zinc, lead and mercury in *Hyalella azteca*, *Hydrobiologia* **259**, 79-89.

Borgmann, U., Néron, R. and Norwood, W.P. (2001a) Quantification of bioavailable nickel in sediments and toxic thresholds to *Hyalella azteca*, *Environmental Pollution* **111**, 189-198.

Borgmann, U., Norwood, W.P., Reynoldson, T.B. and Rosa, F. (2001b) Identifying cause in sediment assessments: bioavailability and the Sediment Quality Triad, *Canadian Journal of Fisheries and Aquatic Sciences* **58**, 950-960.

Borgmann, U., Nowierski, M., Grapentine, L.C. and Dixon, D.G. (2004) Assessing the cause of impacts on benthic organisms near Rouyn-Noranda, Quebec, *Environmental Pollution* **129**, 39-48.

Borgmann, U., Norwood, W.P. and Dixon, D.G. (2005) Re-evaluation of metal bioaccumulation and chronic toxicity in *Hyalella azteca* using saturation curves and the biotic ligand model, *Environmental Pollution* in press.

Bousfield, E.L. (1958) Fresh-water amphipod crustaceans of glaciated North America, *The Canadian Field-Naturalist* **72**, 55-113.

Burton, G.A., Jr., Ingersoll, C.G., Burnett, L.C., Henry, M., Hinman, M.L., Klaine, S.J., Landrum, P.F., Ross, P. and Tuchman, M. (1996) A comparison of sediment toxicity test methods at three Great Lakes areas of concern, *Journal of Great Lakes Research* **22**, 495-511.

CCME (1999) Canadian environmental quality guidelines, Canadian Council of Ministers of the Environment, Winnipeg, Canada.

Day, K.E., Kirby, R.S. and Reynoldson, T.B. (1995a) The effect of manipulations of freshwater sediments on responses of benthic invertebrates in whole-sediment toxicity tests, *Environmental Toxicology and Chemistry* **14**, 1333-1343.

Day, K.E., Dutka, B.J., Kwan, K.K., Batista, N., Reynoldson, T.B. and Metcalfe-Smith, J.L. (1995b) Correlations between solid-phase microbial screening assays, whole sediment toxicity tests with macroinvertebrates and *in situ* benthic community structure, *Journal of Great Lakes Research* **21**, 192-206.

de March, B.G.E. (1977) The effects of photoperiod and temperature on the induction and termination of reproductive resting stage in the freshwater amphipod *Hyalella azteca* (Saussure), *Canadian Journal of Zoology* **55**, 1595-1600.

Duan, Y., Guttman, S.I., Oris, J.T. and Bailer, A.J. (2000) Genetic structure and relationships among populations of *Hyalella azteca* and *H. montezuma* (Crustacea:Amphipoda), *Journal of the North American Benthological Society* **19**, 308-320.

Duan, Y., Guttman, S.I., Oris, J.T. and Bailer, A.J. (2001) Differential survivorship among allozyme genotypes of *Hyalella azteca* exposed to cadmium, zinc or low pH, *Aquatic Toxicology* **54**, 15-28.

EC (Environment Canada). (1995) Guidance document on measurement of toxicity test precision using control sediments spiked with a reference toxicant. Environmental Protection Service, Report EPS 1/RM/30, Ottawa, ON, 56 pp.

EC (Environment Canada) (1997) Biological test method: Test for survival and growth in sediment using the freshwater amphipod *Hyalella azteca*. Environmental Protection Service, Report EPS 1/RM/33, Ottawa, ON, 123 pp.

MacLean, R.S., Borgmann, U. and Dixon, D.G. (1996) Bioaccumulation kinetics and toxicity of lead in *Hyalella azteca* (Crustacea, Amphipoda), *Canadian Journal of Fisheries and Aquatic Sciences* **53**, 2212-2220.

Hamilton, M.A., Russo, R.C., Thurston, R.V. (1977) Trimmed Spearman-Karber method for estimating median lethal concentrations in toxicity bioassays, *Environmental Science and Technology* **11**, 714-719; Correction, *Environmental Science and Technology* **12**, 417.

Hargrave, B.T. (1970) The utilization of benthic microflora by *Hyalella azteca* (Amphipoda), *Journal of Animal Ecology* **39**, 427-437.

Hogg, I.D., Larose, C., de Lafontaine, Y. and Doe, K.G. (1998) Genetic evidence for a *Hyalella* species complex within the Great Lakes St. Lawrence River drainage basin: implications for ecotoxicology and conservation biology, *Canadian Journal of Zoology* **76**, 1134-1140.

Ingersoll, C.G., Dwyer, F.J., Burch, S.A., Nelson, M.K., Buckler, D.R. and Hunn, J.B. (1992) The use of freshwater and saltwater animals to distinguish between the toxic effects of salinity and contaminants in irrigation drain water, *Environmental Toxicology and Chemistry* **11**, 503-511.

Ingersoll, C.G., Brunson, E.L., Dwyer, F.J., Hardesty, D.K. and Kemble, N.E. (1998) Use of sublethal endpoints in sediment toxicity tests with the amphipod *Hyalella azteca*, *Environmental Toxicology and Chemistry* **17**, 1508-1523.

Lock, K. and Janssen, C.R. (2003) Influence of ageing on zinc bioavailability in soils, *Environmental Pollution* **126**, 371-374.

McGee, B.L., Schlekat, C.E. and Reinharz, E. (1993) Assessing sublethal levels of sediment contamination using the estuarine amphipod *Leptocheirus plumulosus*, *Environmental Toxicology and Chemistry* **12**, 577-587.

Milani, D., Reynoldson, T.B., Borgmann, U. and Kolasa, J. (2003) The relative sensitivity of four benthic invertebrates to metals in spiked-sediment exposures and application to contaminated field sediment, *Environmental Toxicology and Chemistry* **22**, 845-854.

Nebeker, A.V. and Miller, C.E. (1988) Use of the amphipod crustacean *Hyalella azteca* in freshwater and estuarine sediment toxicity tests, *Environmental Toxicology and Chemistry* **7**, 1027-1033.

Neumann, P.T.M., Borgmann, U. and Norwood, W. (1999) Effect of gut clearance on metal body concentrations in *Hyalella azteca*, *Environmental Toxicology and Chemistry* **18**, 976-984.

Nowierski, M.M. (2003) Sediment Metal Bioavailability and Toxicity to the Freshwater Amphipod, *Hyalella azteca*, M.Sc. Thesis, University of Waterloo, Waterloo, Ontario, Canada.

Othman, M.S. and Pascoe, D. (2002) Reduced recruitment in *Hyalella azteca* (Saussure, 1858) exposed to copper, *Ecotoxicology and Environmental Safety* **53**, 59-64.

Phipps, G.L., Mattson, V.R. and Ankley, G.T. (1995) Relative sensitivity of three freshwater benthic macroinvertebrates to ten contaminants, *Archives of Environmental Contamination and Toxicology* **28**, 281-286.

Smith, S.L., MacDonald, D.D., Keenleyside, K.A., Ingersoll, C.G. and Field, L.J. (1996) A preliminary evaluation of sediment quality assessment values for freshwater ecosystems, *Journal of Great Lakes Research* **22**, 624-638.

Smolders, E., McGrath, S.P., Lombi, E., Karman, C.C., Bernhard, R., Cools, D., Van den Brande, K., van Os, B. and Walrave, N. (2003) Comparison of toxicity of zinc for soil microbial processes between laboratory-contaminated and polluted field soils, *Environmental Toxicology and Chemistry* **22**, 2592-2598.

Suedel, B.C. and Rodgers, J.H., Jr. (1994) Responses of *Hyalella azteca* and *Chironomus tentans* to particle-size distribution and organic matter content of formulated and natural freshwater sediments, *Environmental Toxicology and Chemistry* **13**, 1639-1648.

Suedel, B.C., Rodgers, J.H., Jr., and Deaver, E. (1997) Experimental factors that may affect toxicity of cadmium to freshwater organisms, *Archives of Environmental Contamination and Toxicology* **33**, 188-193.

U.S. EPA (United States Environmental Protection Agency) (2000) Methods for measuring the toxicity and bioaccumulation of sediment-associated contaminants with freshwater invertebrates, Second Edition, EPA 600/R-99/064.

Witt, J.D.S. and Hebert, P.D.N. (2000) Cryptic species diversity and evolution in the amphipod genus *Hyalella* within central glaciated North America: a molecular phylogenetic approach, *Canadian Journal of Fisheries and Aquatic Sciences* **57**, 687-698.

Abbreviations

ASTM	American Society for Testing and Materials
CCME	Canadian Council of Ministers of the Environment
EC	Environment Canada
EC 50	Effective concentration resulting in a 50% quantal ("all or none") effect for a specified time period
EDTA	Ethylenediaminetetraacetic acid (strong metal complexing agent)
GFAAS	Graphite Furnace Atomic Absorption Spectrophotometry
LC25	Lethal concentration resulting in 25% mortality for a specified time period, corrected for control survival
IC75	Concentration resulting in a 75% inhibition of a response (*e.g.*, reproduction) for the specified time period
ICP-MS	Inductively Coupled Plasma-Mass Spectrometry, used for trace metal analysis
ID	Internal Diameter
NWRI	National Water Research Institute, Environment Canada, Burlington
TBT	Tributyltin
TSK	Trimmed Spearman-Karber
U.S. EPA	United States Environmental Protection Agency.

13. *CHIRONOMUS RIPARIUS* SOLID-PHASE ASSAY

ALEXANDRE R.R. PÉRY
& RAPHAËL MONS
& JEANNE GARRIC
Laboratoire d'écotoxicologie, CEMAGREF
3bis quai Chauveau, CP 220, 69336 Lyon, France
alexandre.pery@cemagref.fr
mons@lyon.cemagref.fr.
jeanne.garric@cemagref.fr

1. Objective and scope of the method

The objectives of the *Chironomus riparius* solid-phase assay are to assess the toxicity either of field sediments or of a chemical spiked into artificial or field reference sediments on *C. riparius* growth and survival. It can be used alone to assess whole sediment toxicity or associated with micro-scale bioassays to predict the toxic potential of freshwater sediments through a tiered risk assessment process (Côté et al., 1998). This test, usually performed in beakers (with a volume between 0.5 and 1 L), is designed for field sediments or for any chemical able to accumulate in them. The test is 7 days long, sensitive and sufficiently documented to avoid misinterpretation of the results. The culture of the organisms under laboratory conditions is easy.

2. Summary of test procedure

This test aims to assess the toxicity of sediments (spiked or field-collected) to the midge *C. riparius*. This non-biting species is widely distributed in the Northern hemisphere and plays an important role in aquatic ecosystem functioning, as prey for fish and birds. It is easy to handle for toxicity tests and easy to culture. The culture and testing do not require much space (1m² for the culture, 5 to 7 beakers per concentration of test sediment). For a benthic bioassay, the test is relatively short (7 days) but covers a significant part of the life cycle, providing information about the effects of sediments on *C. riparius* growth and survival. The measurements are easy to perform requiring only a binocular microscope. It is important to use controlled conditions of light (16:8 h light:dark photoperiod), temperature ($21 \pm 1°C$) and feeding (1.5 mg fish food/larva/day) to ensure normal growth of the organisms and to reduce the influence of confounding factors. The sensitivity of the organisms should be checked regularly using control charts, based on copper acute toxicity tests.

Table 1. Rapid summary of test procedure.

Test organism	The midge *C. riparius* Meigen
Type of test	Solid-phase assay designed to test the toxicity of spiked or field sediments
Test format	5 to 7 beakers per concentration of test sediment
Volume of test vessels	500 mL
Organism numbers for testing	50 to 70 per concentration of test sediment
Lighting	16:8 h light:dark photoperiod
Temperature	$21 \pm 1°C$
Endpoints	growth (length) and survival
Performance of measurements	At the end of the test organisms are removed from the sediments and counted, and their length is measured using a binocular microscope.
Test duration	7 days
Reference toxicant	Copper (cupric sulfate anhydrous)

3. Overview of applications reported with the toxicity test method

Fourteen-day survival and growth tests with *Chironomus* species (initially with *C. tentans* in the USA) were first proposed in the 1980s for aquatic hazard assessment of chemicals adsorbed to sediments (Adams et al., 1985). At that time toxicity tests were usually only performed with organisms from the water column, and there were concerns that such tests had only limited relevance to the sediment compartment.

At the beginning of the 1990s, chronic toxicity tests based on survival and growth measurements of the midge *C. riparius* appeared in Europe. This benthic species was chosen because of its easy culture and its relatively short life-cycle (Pascoe et al., 1989). The test was initiated with second instar larvae, a compromise between sensitivity and ease of handling. It lasted 10 days, which represents more than one third of the animal's life-cycle at 20°C; the feeding regime was 0.5 mg/larva/d. Effects on growth were assessed based on weight measurements.

Notable evolution of this test occurred over the next 12 years. First, test temperature was increased (from 20°C to 21 - 23°C) in some studies to reduce the test duration (U.S. EPA, 1994; Environment Canada, 1997; Bervoets and Blust, 2000; Péry et al., 2002). Second, an increasing number of studies focused on length rather than weight measurements (Stuijfzand et al., 2000; Péry et al., 2002; Vos et al., 2002) because gut contents can interfere with weight measures (Sibley et al., 1997). Length measurements provide more data per beaker and consequently the potential for more powerful statistical analyses (Callaghan et al., 2002). Third, a shorter test duration (7 days) has recently been proposed, based on a better understanding of *C. riparius* biology. This reduction in test duration is both a practical improvement and a theoretical necessity, because it has been shown that growth in length ceases completely after 7 days of test under *ad libitum* feeding conditions (Péry et al., 2002). A 10-day test using length as an endpoint could thus miss an effect equivalent to three days of delay. Fourth, some recent studies have focused on the influence of feeding on test results. A minimum level of 0.12 mg/larva/day is recommended to ensure good survival of the control (Ristola et al., 1999), but optimal growth is obtained only when feeding is above 0.6 mg/larva/d (Péry et al., 2002).

4. Advantages of conducting the toxicity test method

Following an evaluation of sediment toxicity studies in several parts of the United States, Burton and Scott (1992) concluded that the test with *C. tentans* was among the most efficient in assessing the toxicity of whole sediments. The same conclusions apply equally to *C. riparius*. The test is relatively brief (7 days long) but covers more than half of the larval development time, and the organisms are easy to culture and handle. In addition, because of their deep red colour, the larvae can easily be found in any sediment as soon as they have reached fourth instar.

Control survival is high (about 90%), which allows the detection of small effects on survival. Similarly, under *ad libitum* feeding conditions, the standard deviation of the length is less than 10% of the length increase during the test, thereby allowing powerful statistical analyses. Control data from one test to another are also reproducible (less than 10% variation for the mean length value).

Under controlled conditions, confounding factors such as sediment characteristics and organism density have little influence on the results of the test with *C. riparius*. When feeding *ad libitum*, there is no influence of organism density during the test (Péry et al., 2002). In addition, under these feeding conditions, the influence of sediment characteristics only accounts for 7% of the length increase (Péry et al., 2003).

5. Test species

C. riparius Meigen, from the dipteran family Chironomidae, is a non-biting midge widely distributed in the northern hemisphere at temperate latitudes. It can be found both in lentic and lotic environments, mainly in organically enriched waters. Its life-cycle includes several aquatic stages (egg, four larval instars and pupa) and an aerial adult stage.

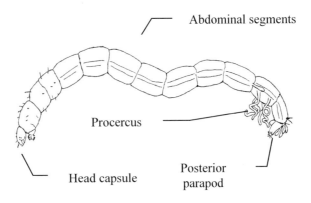

Figure 1. Diagram of C. riparius *larva.*

The larvae (Fig. 1), which are collector-gatherers, feed on sediment-deposited detritus (Rasmussen, 1984). Even if *Chironomus* species can digest bacteria, bacteria are not as quantitatively important as other components of the detrital food material (Baker and Bradnam, 1976). First instar larvae are pale white, second instar larvae pink, and third and fourth instar larvae red in colour. Larval instars can also be distinguished through head capsule width measurements (Day et al., 1994). Larvae build tubes, in which they live, mainly to protect them from predation (Baker and Ball, 1995), and also partially from pollutants present in the water column (Halpern et al., 2002). When disturbed, a larva only resumes feeding after it has built a new tube (Naylor and Rodrigues, 1995). In the field, larval growth occurs generally from April to October. Larvae overwinter, mainly as fourth instar larvae, and the majority of them reach a particular phase of this instar at the end of winter (Rasmussen, 1984; Goddeeris et al., 2001). This phenomenon is responsible for the synchronization of the life-cycles of these organisms, and allows consideration of populations as cohorts. In the laboratory, it is possible to induce diapause by short-day conditions (Goddeeris et al., 2001). As soon as larval length has reached its maximum value, growth ceases and larvae prepare for emergence and reproduction. This period can last four days under laboratory conditions (Péry et al., 2002), after which larvae become pupae and emergence occurs.

Males usually emerge earlier than females. Adults have a very short life-span (from 3 to 5 days), and females can only reproduce once, although males can mate with many females (Downe, 1973). Oviposition mainly occurs at dusk or during the night (Armitage et al., 1995), with up to 500 eggs deposited on the water surface in a mass (typically shaped like the letter C, Péry et al., 2003). The simplified life-cycle of the species is shown in Figure 2.

Chironomidae are of interest to ecologists because they form a numerically prominent part of benthic communities in nearly all freshwater habitats. For instance, Berg and Hellenthal (1992) reported annual chironomid secondary production in a stream (northern Indiana, U.S.) that accounted for 80% of the total insect secondary production.

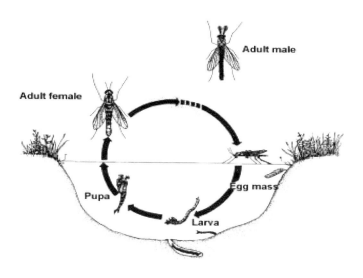

Figure 2. Diagram of C. riparius *life cycle (Ali and Morris, 1992).*

Chironomidae also possess a number of characteristics which make them valuable for sediment toxicity tests (Benoit et al., 1997; Callaghan et al., 2002). They are easy to handle and culture. They also maintain intimate contact with sediments throughout larval development (their ability to burrow into sediments makes them useful for assessing toxicants adsorbed to sediments). The life stages are easy to identify and the life history under laboratory conditions is short. Among chironomids, *C. riparius* has a particularly short development time (about 14 days at 21°C).

6. Culture/maintenance of organism in the laboratory

The culture of *C. riparius* requires little space (about 1 m²) but does require a specially-design room. First, the photoperiod must be maintained at 16:8 h light:dark with a light intensity of 500 to 1000 lux to avoid dormancy of the organisms. Second, water temperature should be maintained between 20 and 24 °C to allow relatively high production of organisms. Third, using an aeration system with water renewal (4 renewals per day) is advised to maintain the quality of the culture water.

The culture can be set up in any uncontaminated sediment. We recommend artificial silica sand, which allows easy observation and removal of the larvae. Equally, any water type (artificial or natural) can be used provided it allows normal growth and reproduction of *C. riparius*. A ratio ranging from 1:4 to 1:12

sediment:volume of water is advisable. Once the aquarium is ready, about ten hatched egg masses (five the first week, five the week after) obtained from a dedicated laboratory should be added to allow continuous organism production after about three months. Every working day, 0.5 g of fish food should be provided, ground finely using a mortar and pestle.

Equipment and facilities required (see Fig. 3):
- Aquarium: the aquarium, of about 20 L volume, must be covered, with an aperture to allow feeding of larvae and handling of the egg masses.
- Sediment: any kind of silica sand is satisfactory as sediment and should be placed in the aquarium to a depth of 2 cm.
- Aeration system: air should be gently but continuously bubbled into the water.
- Water renewal system: water should be slowly but continuously provided. An overflow system should be set up to remove any excess of water.
- Grinding device: to grind the food particles (*e.g.*, mortar and pestle).

We advise using three aquaria in parallel. To ensure genetic variability of the culture, it is wise to exchange every two months egg masses between aquaria. It may also be useful to perform similar exchanges with other laboratories (once per year, for instance). It is also necessary to remove larvae every month, from the aquaria, leaving about 200 organisms per aquarium to avoid high densities. On these occasions sediments should be replaced and the walls of the aquaria cleaned with water. Pupal exuviae should be removed twice a week.

7. Preparation of test species for toxicity testing

Tests are initiated with two day-old organisms. The preparation of test organisms should consequently be performed as follows. Four days before test initiation all available masses are taken from the culture and placed individually in small beakers with water at $22 \pm 1°C$ and a 16:8 h light:dark photoperiod. Three days before test initiation, in the afternoon, all hatched masses are removed and disposed of to ensure synchronicity of the organisms used to initiate the test. Two days before test initiation, in the morning, newly-hatched masses are placed in a container (about 200 cm^2) with silica sand (covering about half of the surface of the container), water (1.5 cm depth) and an amount of fish food corresponding to 50 mg per mass. This container is kept at $22 \pm 1°C$ under a 16:8 h light:dark photoperiod for two days. After this two-day period the test can be initiated with larvae of a length that should be between 1.5 and 2.1 mm. One egg mass is likely to provide 100 organisms.

To avoid a lack of egg masses, they can be first collected from the culture five days before test initiation. They are then placed individually in a refrigerator (at a temperature of $4 \pm 2°C$) for one day, before being added to the other egg masses at $22 \pm 1°C$. There should be no difference in response between the larvae from both treatments when initiating the test.

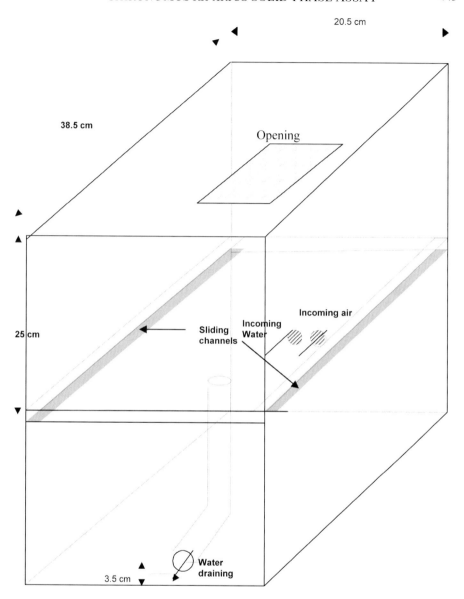

Figure 3. Diagram of an aquarium to culture Chironomus riparius.

8. Testing procedure

8.1 PREPARATION OF TEST SAMPLES

Tested sediments are either field sediments or chemical spiked sediments. When testing field sediments, we recommend the use of artificial silica sand as a control. Before using it for a test, it is necessary to prepare it by placing the silica sand in a container with water (volume ratio approximately 1:1), food (0.1 g per kg silicate) and aeration for two weeks. When testing chemical-spiked sediments, it may be more appropriate to use uncontaminated natural sediment with organic matter (> 2% organic carbon) to ensure a good spiking yield. Whatever the choice, sediment characteristics (*e.g.*, organic carbon content and particle size distribution) should be known and reported with the test results.

Immediately after collection, field sediments should be stored in the dark and kept at a temperature $\leq 7°C$ (Environment Canada, 1997). Once in the laboratory, sediments should be 2 mm sieved to remove the largest particles. They can then be stored for six weeks at a temperature $\leq 6°C$, or for six months at a temperature $\leq -17°C$. In the latter case, sediments should be reconstituted with water (volume ratio of 1:1) and aerated for two weeks prior to the toxicity test. The advantage of using frozen sediments is that indigenous organisms are killed. However, sediment structure may be substantially affected by freezing. Sediments should be stored in polyethylene, polypropylene or polycarbonate containers, with about 10% of the container volume kept empty.

When spiking poorly soluble compounds into sediment, it is advisable to use the shell coating/rolling technique, as described by Northcott and Jones (2000), to obtain an efficient and homogeneous concentration of test compound. Compound dissolved in methanol is introduced into 2 L jars, which are then rolled on a horizontal rolling mill while the solvent is removed by evaporation by blowing air into the jars. A given amount of wet sediment is then added together with water (the same as used in the test) to increase sediment fluidity. The jars are then rolled for two hours on one day, and for four hours the day after, kept at the test temperature and swirled manually each day for a number of days determined by pre-test experimentation with the test compound.

To spike soluble compounds into sediment, a given amount (0.8 L) of wet sediment is placed into 2 L jars together with the compound dissolved in water. Jars are then rolled as described previously, kept at test temperature and swirled manually each day for a number of days determined by pre-test experimentation.

In both cases, nominal concentrations can be calculated from knowledge of sediment weight, water volume and the ratio between sediment wet and dry weight. Sediments should be homogenized just prior to filling the beakers.

8.2 EXPERIMENTAL DESIGN

The test described in this chapter lasts for 7 days. On the first day of the experiment, two day-old organisms are placed in beakers, with 10 organisms per beaker. The volume of these beakers is 0.6 L and the surface area 14 cm², with sediment and water in a ratio of 1:4, and the depth of the sediment layer in the range of 1.5-3 cm.

We advise using the same type of water in the test systems as used for the culture. The beakers are placed more than 72 hours before the beginning of the test in a water bath maintained at 21± 1°C to avoid temperature variations, under a 16:8 h light:dark photoperiod and with an aeration system. This aeration should be kept as low as possible to avoid suspending food and to minimize sediment oxidation. Aeration should be stopped prior to introducing the organisms into the test beakers and resumed two hours afterwards. Conductivity, temperature, pH, and concentrations of dissolved oxygen, nitrates and ammonia should be measured on days 0, 3, 5 and 7.

We recommend the use of five to seven beakers per tested concentration or natural sediment. This permits sufficient replication for accurate parameter estimation (usually, 8% growth reduction and 20% survival reduction can be detected) and avoids waste of space. We recommend the use of a control and five concentrations spaced at geometric intervals when testing a chemical. We advise the use of a solvent control if a solvent has been used during sediment spiking. If prior knowledge is available about the toxicity of the tested compound, concentrations should be chosen in a way so that the Lowest Observed Effect Concentration (LOEC) estimated is close to the third tested concentration. For instance, if prior information is available to suggest a LOEC of 10 mg/kg, the range [control, 1, 3, 9, 27 and 81 mg/kg] could be a relevant choice of tested concentrations. If no prior knowledge exists about the chemical, it is best to carry out a range-finding test with concentrations at geometric intervals, with a ratio of 10 between each concentration.

Test vessels should be placed at random in the water bath to eliminate potential interferences due to gradients of light and temperature. This also avoids bias during the start of the test, as there is a tendency to first introduce the largest (and therefore most easily captured organisms) at test initiation.

During exposure, midges are fed daily with fish food. We use Tetramin® (Tetrawerke, Melle, Germany) but other kinds of fish food may also be used (Vos et al., 2000). Food must be placed in an appropriate volume of water, then ground to a fine powder. The daily feeding ration should be 0.6 mg/larva/day, so for 1 g of food mixed into 250 mL of water, this corresponds to 1.5 mL per beaker of this solution.

To avoid the need for technicians to work on weekends, it is possible to initiate the test on Friday with double feeding (1.2 mg/larva/day), and then to continue feeding daily from Monday to Thursday with 0.6 mg/larva/day without any adverse effects on the results of the test.

9. Post-exposure measurements and endpoint determinations

At the end of the test organisms should be killed using a solution of 20% formaldehyde and 80% water. Animals should be dipped in this solution but must be removed from it just after death to avoid distortion of their shape. No more than five organisms should be killed at any one time, to allow the operator sufficient time to place each onto a coverglass. Length (from posterior parapod to head capsule) is then measured using a binocular microscope fitted with a calibrated eye-piece micrometer. We propose the use of length instead of weight because length is easy to measure even with small organisms and also because our studies have demonstrated that variation in length measurements were lower compared to weight measurements.

The test can be considered valid if the mean control length is between 10 and 12 mm, mean control survival is above 70%, and nitrate and ammonium concentrations remained below 10 and 5 mg/L respectively.

Control charts for reference toxicants should be produced. For instance, a short-term survival test with copper (introduced as cupric sulfate anhydrous) -spiked water and a small amount of silica sand could be performed in parallel to the solid-phase assay. We recommend testing for one year with two tests a month to establish the control chart. This provides more than 20 data points, which should represent test variability throughout a year, from which a mean and standard deviation are calculated. Sediment assays are then only valid if the result of the parallel survival test, expressed as an LC50, does not differ from the control chart mean LC50 by more than three standard deviations.

To compare the results obtained for a test sediment with the control, we recommend the use of Student t-tests performed using the mean values of the replicates. To achieve this, the following parameter T should be calculated:

$$T = \frac{m1 - m2}{\sqrt{\dfrac{s1^2 + s2^2}{n-1}}} \qquad (1)$$

where $m1$ and $s1$ are the mean and standard deviation for the control, $m2$ and $s2$ are the mean and standard deviation for a test sediment, and n is the number of replicates. If $T > To$, that is the threshold for effects significance (with $To = 2.31$ if $n = 5$; $To = 2.23$ if $n = 6$ and $To = 2.16$ if $n = 7$), the tested sediment has a significant effect ($p < 0.05$) compared to the control. If the number of replicates is not the same for the control and for the tested sediments, Student t-tests can also be performed. The formula and the threshold values can be found in any book of statistics, as in the book by Spiegel (1975).

For instance, based on the survival ($T = 1.3$) and growth ($T = 6.85$) results of Table 2, we can therefore conclude that the tested sediment has no effect on survival but a significant effect on growth.

Table 2. Example of results from a solid-phase assay with C. riparius.

Replicate	1	2	3	4	5
Control growth (mm)	11.0	11.2	10.7	10.8	11.3
Test sediment growth (mm)	10.0	9.1	9.7	9.5	9.6
Control survival (%)	90	90	80	100	100
Test sediment survival (%)	80	80	80	70	90

10. Factors capable of influencing test performance and results

Several factors can influence the outcome of this assay.

(1) Temperature, light, and nitrate and ammonium concentrations must be controlled during the test as specified earlier. Temperature greatly influences the growth pattern of the organisms. If light during the test is insufficient, the phenomenon of dormancy, with growth cessation, can occur (Ineichen et al., 1979). Nitrate and ammonium are natural toxicants, which could either produce toxic effects or interact with tested toxicants. We have found sublethal effects of ammonium above 5 mg/L and sublethal effects of nitrate above 10 mg/L, but have never experienced interactions between these chemicals and the toxicants we tested.

(2) Sediment characteristics and density of organisms can influence the results of the test (Ristola et al., 1999). With the recommended feeding regime (0.6 mg/larva/day), initial density or time-varying density have no effect for up to 10 organisms per beaker (Péry et al., 2002). Sediment organic matter also has no influence at this regime (Péry et al., 2003). In contrast, even under *ad libitum* conditions, other sediment characteristics, such as particle size distribution can affect organism length at 7 days, although the extent of this influence is limited to a maximum of 7% (Péry et al., 2003).

(3) Sexual dimorphism could influence the outcome of the assay. Indeed, males weight about 40% less than females (Day et al., 1994) and their length is about 15% less than females (Péry et al., 2002). However, studies have shown that the influence of this phenomenon is very limited, with a Type I error rate of less than 3% in sediment toxicity tests owing to sexual dimorphism (Day et al., 1994; Sildanchandra and Crane, 2000).

(4) The presence of other indigenous organisms in the sediment used for the assay can influence the outcome of the assay if these organisms are food competitors or predators (Reynoldson et al., 1994). To avoid this, sediments may be frozen, dried or gamma irradiated. All treatments affect sediment structure and may influence sediment-linked contaminants. Sieving can also be used but small organisms are likely to remain in the sediment.

11. Application in a case study

In May 2003, we studied the toxicity of sediment from the River Durance near Les Mées, France. Chemical measurements of pollutants (metals and PAHs) indicated that water and sediment were of moderate but not poor quality. However, measurements of the biological community suggested poorer quality. There were consequently two hypotheses to explain these results: either the sediment was polluted by toxicants not commonly measured (*e.g.*, pesticides or other organics) or the sediment physical characteristics were responsible for poorer diversity.

The *Chironomus* solid-phase assay was performed with sediment from the River Durance, using control silica sand sediment. Results are presented in Table 3.

Table 3. Results of the solid-phase assay with C. riparius *to assess the toxicity of sediment from the River Durance.*

Replicate	1	2	3	4	5
Control mean length (mm)	11.1	10.9	11.9	10.7	11.8
Durance mean length (mm)	8.8	8.3	7.8	8.9	8.4
Control survival (%)	90	90	90	100	100
Durance survival (%)	90	90	80	100	100

Survival was not significantly affected, but effects on growth were very significant ($p < 0.01$), even taking into account possible confounding factors due to sediment characteristics (Péry et al., 2003). In conclusion, sediment from the River Durance was found to be polluted by toxicants not measured in French routine surveys of lotic waters.

12. Additional test information

(1) Organism culture and testing are not very difficult to perform. Four months of experience suffice for a technician to learn how to insure a regular output of egg masses from the stock culture and to perform tests adequately.

(2) As with other dipterans, *C. riparius* is more sensitive to organic compounds than metals (Wogram and Liess, 2001). However, some authors showed that laboratory populations can be more sensitive to toxicants than feral ones, especially when site-specific selection pressure has occurred, which can incur resistance traits (Hoffman and Fisher, 1994).

(3) The test described herein using a control and 5 different chemical concentrations or 5 different field sediments should take about 18 hours of work for a technician (4 hours to prepare the test, 4 hours on the first day, 4 hours in total for the six following days, and about 6 hours on the last day).

(4) The solid-phase assay presented herein will be soon standardized in France by AFNOR (Association Française de Normalisation) standards organization. The main difference between this test and other standardized tests (U.S. EPA, 1994; Environment Canada, 1997) is in its shorter exposure time (7 days instead of 10 days). However,

(5) Other endpoints can be studied when testing the midge *C. riparius*. First, emergence can be monitored, resulting in a 28-day test. As emergence delays are generally the consequence of effects on growth, the information gained will often confirm the results of a 7-day test. However, if pupation is likely to be targeted by the toxicants (hormones, in particular), the information from emergence tests can be relevant. Reproduction could be an interesting endpoint as well, particularly for hormonally active compounds like endocrine disruptors, but standardized test protocols are not yet available.

since a 7-day test is more meaningful and less costly than a 10-day test, we believe that international standardized growth tests with the midge *C. riparius* will soon move towards a test duration of 7 days.

13. Conclusions and prospects

The culturing and testing proposed here are easy to perform, and allow a rapid and efficient toxicity assessment for spiked and field sediments. This assay consequently has great potential to be used widely in routine tests for surveying sediments. It could also be the basis of a life cycle risk assessment for the midge *C. riparius*, which would be relevant from an ecological point of view, as chironomids are key species in many aquatic ecosystems. To achieve this, it will first be necessary to standardize a reproduction test similar to that for growth presented here.

Acknowledgements

The authors would like to thank two anonymous reviewers who helped greatly to improve the quality of the manuscript.

References

Adams, W.J., Kimerle, R.A. and Mosher R.G. (1985) Aquatic safety assessment of chemicals sorbed to sediments, in Cardwell R.D. et al. (eds.), *Aquatic Toxicology and Hazard Assessment: Seventh Symposium*, American Society for Testing and Materials, pp. 429-453.
Ali, A. and Morris, C.D. (1992) Management of non-biting aquatic midges, IFAS, University of Florida Medical Entomology Laboratory, Technical Bulletin no. 4.
Armitage, P.D., Cranston, P.S. and Pinder, L.C.V. (1995) Biology and ecology of non biting midges, Chapman & Hall, London, UK.
Baker, J.H. and Bradnam, L.A. (1976) The role of bacteria in the Nutrition of Aquatic Detritivores, *Oecologia* **24**, 95-104.
Baker, R.L. and Ball, S.L. (1995) Microhabitat selection by larval *Chironomus tentans* (Diptera: Chironomidae): effects of predators, food, cover and light, *Freshwater Biology* **34**, 101-106.
Benoit, D.A., Sibley, P.K., Juenemann, J.L. and Ankley, G.T. (1997) *Chironomus tentans* life-cycle test: design and evaluation for use in assessing toxicity of contaminated sediments, *Environmental Toxicology and Chemistry* **16**, 1165-1176.
Berg, M.B. and Hellenthal, R.A. (1992) The role of chironomidae in energy flow of a lotic system, *Netherlands Journal Aquatic Ecology* **26**, 471-476.
Bervoets, L. and Blust, R. (2000) Effects of pH on cadmium and zinc uptake by the midge larvae *Chironomus riparius*, *Aquatic Toxicology* **49**, 145-157.

Burton, G.A. and Scott, K.J. (1992) Sediment toxicity evaluations: Their niche in ecological assessments, *Environmental Science and Technology* **26**, 2068-2075.

Callaghan, A., Fisher, T.C., Grosso, A., Holloway, G.J. and Crane, M. (2002) Effect of temperature and pirimiphos methyl on biochemical biomarkers in *Chironomus riparius* Meigen, *Ecotoxicology and Environmental Safety* **52**, 128-133.

Côté, C., Blaise, C., Schroeder, J., Douville, M. and Michaud, J.R. (1998) Investigating the adequacy of selected micro-scale bioassays to predict the toxic potential of freshwater sediments through a Tier process, *Water Quality Research Journal of Canada* **33**, 253-277.

Day, K.E., Kirby, R.S. and Reynoldson, T.B. (1994) Sexual dimorphism in *Chironomus riparius* (Meigen): impact on interpretation of growth in whole-sediment toxicity tests, *Environmental Toxicology and Chemistry* **13**, 35-39.

Downe, A.E.R. (1973) Some factors influencing insemination in laboratory swarms of *Chironomus riparius* (Diptera, Chironomidae), *Canadian Entomologist* **105**, 291-298.

Environment Canada (1997) Biological test method. Test for growth and survival in sediment using larvae of freshwater midges (*Chironomus tentans* or *Chironomus riparius*), SPE 1 / RM / 32, Environment Canada, Ottawa, Ontario, Canada.

Goddeeris, B.R., Vermeulen, A.C., De Geest, E., Jacobs, H., Baert, B. and Ollevier, F. (2001) Diapause induction in the third and fourth instar of *Chironomus riparius* (Diptera) from Belgian lowland brooks, *Archives of Hydrobiology* **150**, 307-327.

Halpern, M., Gasith, A. and Broza, M. (2002) Does the tube of a benthic chironomid larva play a role in protecting its dweller against chemical toxicants, *Hydrobiologia* **470**, 49-55.

Hoffman, E.R. and Fisher, S.W. (1994) Comparison of a field and laboratory-derived population of *Chironomus riparius* (Diptera: Chironomidae): biochemical and fitness evidence for population divergence, *Journal of Economic Entomology* **87**, 318-325.

Ineichen, H., Riesen-Willi, U. and Fisher, J. (1979) Experimental contributions to the ecology of *Chironomus* (diptera) II. The influence of photoperiod on the development of *Chironomus* plumosus in the 4th larval instar, *Oecologia* **39**, 161-183.

Naylor, C. and Rodrigues, C. (1995) Development of a test method for *Chironomus riparius* using a formulated sediment, *Chemosphere* **31**, 3291-3303.

Northcott, G.L. and Jones, K.C. (2000) Spiking hydrophobic organic compounds into soil and sediment: a review and critique of adopted procedures, *Environmental Toxicology and Chemistry* **19**, 2418-2430.

Pascoe, D., Williams, K.A. and Green, D.W.J. (1989) Chronic toxicity of cadmium to *Chironomus riparius* Meigen - Effects upon larval development and adult emergence, *Hydrobiologia* **175**, 109-115.

Péry, A.R.R., Mons, R., Flammarion, P., Lagadic, L. and Garric, J. (2002) A modeling approach to link food availability, growth, emergence, and reproduction for the midge *Chironomus riparius*, *Environmental Toxicology and Chemistry* **21**, 2507-2513.

Péry, A.R.R., Sulmon, V., Mons, R., Flammarion, P., Lagadic, L. and Garric, J. (2003) A model to understand the confounding effects of natural sediments to toxicity tests with *Chironomus riparius*, *Environmental Toxicology and Chemistry* **22**, 2476-2481.

Rasmussen, J.B. (1984) Comparison of gut content and assimilation efficiency of fourth instar larvae of two coexisting chironomids, *Chironomus riparius* Meigen and *Glyptotendipes paripes* (Edwards), *Canadian Journal of Zoology* **62**, 1022-1026.

Reynoldson, T.B., Day, K.E., Clarke, C. and Milani, D. (1994) Effect of indigenous animals on chronic end-points in freshwater sediment toxicity tests, *Environmental Toxicology and Chemistry* **13**, 973-977.

Ristola, T., Pellinen, J., Ruokolainen, M., Kostamo, A. and Kukkonen, J.V.K. (1999) Effect of sediment type, feeding level, and larval density on growth and development of a midge (*Chironomus riparius*), *Environmental Toxicology and Chemistry* **18**, 756-764.

Sibley, P.K., Monson, P. D. and Ankley G.T. (1997) The effect of gut contents on dry weight estimates of *Chironomus tentans* larvae: implications for interpreting toxicity in freshwater sediment toxicity tests. *Environmental Toxicology and Chemistry* **16**, 1721-1726.

Sildanchandra, W. and Crane, M. (2000) Influence of sexual dimorphism in *Chironomus riparius* Meigen on toxic effects of cadmium, *Environmental Toxicology and Chemistry* **19**, 2309-2313.

Stuijfzand, S.C., Helms, M., Kraak, M.H.S. and Admiraal, W. (2000) Interacting effects of toxicants and organic matter on the midge *Chironomus riparius* in polluted river water, *Ecotoxicology and Environmental Safety* **46**, 351-356.

U.S. EPA (1994) Methods for measuring the toxicity and bioaccumulation of sediment-associated contaminants with freshwater invertebrates, U.S. EPA 600-R94-024, Duluth, USA.

Vos, J.H., Ooijevaar, M.A.G., Postma, J.F. and Admiraal, W. (2000) Interaction between food availability and food quality during growth of early instar chironomid larvae, *Journal of the North American Benthological Society* **19**, 158-168.

Vos, J.H., Van Den Brink, P.J., Van den Ende, F.P., Ooijevaar, M.A.G., Oosthoek, A.J.P., Postma, J.F. and Admiraal, W. (2002) Growth response of a benthic detritivore to organic matter composition in sediments, *Journal of the North American Benthological Society* **21**, 443-456.

Wogram, J. and Liess, M. (2001) Rank ordering of macroinvertebrates species sensitivity to toxic compounds by comparison with that of *Daphnia magna*, *Bulletin of Environmental Contamination and Toxicology* **67**, 360-367.

Abbreviations

AFNOR	Association Française de Normalisation
U.S. EPA	United States Environmental Protection Agency
LOEC	Low Observed Effect Concentration
LC50	Lethal Concentration leading to 50 % effect
PAH	Polycyclic Aromatic Hydrocarbon.

14. ACUTE TOXICITY ASSESSMENT OF LIQUID SAMPLES WITH PRIMARY CULTURES OF RAINBOW TROUT HEPATOCYTES

FRANÇOIS GAGNÉ
St. Lawrence Centre, Environment Canada
105 McGill Street, Montreal
Quebec H2Y 2E7, Canada
francois.gagne@ec.gc.ca

1. Objectives and scope of test method

A methodology that involves using primary cultures of rainbow trout hepatocytes for toxicity assessment of liquid samples is presented herein. This method is particularly suitable for liquids such as treated or untreated domestic and industrial wastewater; surface water, groundwater and soil leachates; sediment interstitial waters; water-soluble chemicals; and organic chemical(s) soluble in dimethylsulfoxide (DMSO).

Rainbow trout hepatocytes (RTH) have recently been proposed as an alternative to the acute lethality test, which requires the sacrifice of rainbow trout (*Oncorhynchus mykiss*). After proper validation, this microscale fish-cell assay can be used as a screening tool to assess municipal and industrial effluents likely to be toxic to rainbow trout. This alternative contributes a significant reduction in the number of fish sacrificed in applied research and monitoring studies, as well as the cost thereof. It could be used alone or as part of a battery approach to assess the ecotoxic effects of municipal and industrial effluents.

2. Summary of test procedure

Rainbow trout hepatocytes (RTH) are freshly isolated from the liver of sexually immature rainbow trouts. Hepatocytes are prepared from at least three fish to attenuate inter-individual variability. Cells are plated in sterile microplates in L-15 cell culture medium at 15°C. Then they are exposed to incremental concentrations of the test substance(s) for 24 to 48 h at 15°C in the dark. After this incubation period, cell viability is determined by either one or both of the neutral red uptake or fluorescein diacetate tests. Neutral red dye or fluorescein is readily retained in viable cells, where they are stored in the lysosomes for the former. Dead cells are unable to retain the dye, thus accumulating less of it.

Table 1. Rapid summary of the test procedure.

Characteristics of test method	Comments
Purpose	Primary cultures of rainbow trout hepatocytes are used for the evaluation of the acute cytotoxicity of miscellaneous chemicals and complex mixtures.
Principle	Hepatocytes are plated in microplates. Cells are then exposed to the sample for 24-48 h at 15°C prior to cell viability determinations.
Test organisms	Usually, hepatocytes are collected from rainbow trout (*Oncorhynchus mykiss*), which is the designated test species for regulatory purposes in Canada.
Test format	Hepatocytes are either plated in 24, 48 or 96 well microplates treated with cell cultures at cell densities ranging from 2×10^4 to 2×10^5 cells per mL, depending on the desired sensitivity and cell requirement for other cytotoxicity tests.
Cell viability assays	Cell viability is determined on the basis of loss of cell membrane permeability. Cell viability can be determined by either neutral red or fluorescein diacetate (carboxyfluorescein diacetate) retention assays.
Reference toxicant	KCl, ZnCl, DMSO or any other suitable toxicant.
Data expression	For effluents, data are usually expressed in % v/v or in toxic units (TU = 100% v/v ÷ endpoint value in % v/v). For single compounds, results are usually expressed in terms of molar concentration (µM or mM).
Toxicity endpoints	Data are either reported by a threshold concentration (TC = $[NOEC \times LOEC]^{1/2}$) or a concentration that reduces cell viability by 50 or 20% (IC50 or IC20).
Notes of interest	- This alternative can be used for screening large number of industrial wastewaters to reduce the cost and sacrifice of trout (Gagné and Blaise, 1997). - The hepatocyte test was found to be more sensitive that the fish cell line derived from rainbow trout gonads, RTG-2 (Gagné and Blaise, 1998). - Two cell viability methods are proposed which use either spectrophotometric or fluorometric instruments.

3. Overview of applications

The routine preparation of primary cultures of hepatocytes became more practical with the double-perfusion methodology developed by Seglen (1979). First, the liver is perfused with a saline solution containing a calcium chelator (*i.e.*, EGTA and EDTA) in an attempt to weaken Ca-dependent intercellular bonds. Second, the liver is perfused with collagenase (type IV) in the presence of calcium to release hepatocytes. This methodology was later optimized to extract fish hepatocytes (Klauning et al., 1985) and enabled investigators to prepare, on a routine basis, primary cell cultures. Hepatocytes are the main cell system involved in the biotransformation and elimination of nearly all xenobiotics. *In vitro* methods are very useful model systems for studying the mechanism by which toxicants produce their deleterious effects (Baksi and Frazier, 1986; Babich and Borenfreund, 1987). In the present procedure, the double-perfusion methodology was modified to reduce costs by first perfusing with a citrate saline solution and then mincing the liver tissue in albumin-citrate saline solution. Citrate is also a calcium chelator while albumin has some calcium chelation capacity and assists in hepatocyte extraction.

The routine use of *in vivo* assays, such as the rainbow trout acute lethality test, in the monitoring of wastewaters for regulatory purposes has raised some ethical issues. Indeed, this test requires the sacrifice of 120 fingerling trout per sample for the regulatory monitoring of pulp and paper, mine tailings, municipal and other effluents (Environment Canada, 1990). As a result, alternative tests have been proposed and their aims summarized under the so-called 3Rs: Reduction of the number of sacrificed animals, Refinement of tests in terms of cost-effectiveness and increased toxicological information and Replacement of conventional tests by more efficient ones (Rixon, 1995). Moreover, fish cytotoxicity data obtained from fathead minnow cells for 50 chemicals were shown to be significantly correlated with acute toxicity data obtained from the corresponding organism, indicating that fish cells could be used as surrogates for whole fish (Dierickx and Van de Vyver, 1991; Brandao et al., 1992). Primary cultures of RTH were also predictive of the acute lethality test performed with fish (Gagné and Blaise, 1997). The RTH test has been proposed as an alternative to the acute lethality test for industrial and municipal effluent monitoring (Gagné and Blaise, 1998). Moreover, this fish cell system has successfully undergone an inter-laboratory comparison evaluation (see Section 11.4).

4. Advantages of conducting the toxicity test method

The advantages and limitations of the RTH system are presented in Table 2.

Table 2. *Advantages and limitations of the rainbow trout hepatocyte system.*

Advantages	**Comments**
Control of experimental conditions	pH, temperature, nutrition components and hormone composition of the culture media; delivery of the test substance; loss of systemic influence (reduced variability in responses).
Sample volume	Fish cells require only 1 mL of effluent for testing while the fish test needs 60 to 200 L.
Fish sacrifice	The RTH test requires the sacrifice of 3 fish to run three samples, while the whole organism test requires 120 fish for just one test sample.
Ease of increasing toxicological information	Several sub-lethal biomarkers can be added to the *in vitro* tests, such as genotoxicity, endocrine disruption (estrogens), biotransformation and oxidative stress measurements, without increasing the number of sacrificed animals.
Shorter exposure times	The RTH test requires 24 to 48 h of incubation time while the fish test requires 96 h.
Limitations	
Relevance to the organism	The alternative test must be validated with its *in vivo* counterpart (whole fish) to confirm whether an *in vitro* response corresponds to a similar effect *in vivo*.
Loss of systemic effects	Some toxicants act through other cellular targets, such as nerve tissues (organochloride pesticides) or the immune system. The RTH test is sensitive to liver toxicants and those altering fundamental processes common to all cells (*e.g.*, DNA synthesis and integrity, cell division, respiration, oxidative state, excretion).
Sensitivity	Cell culture medium contains a relatively high salt concentration in the range of 7-9 g/L to maintain osmotic conditions for cells. The presence of salts could reduce the availability of metals.

5. Test species

Hepatocytes are usually prepared from salmonids (Tab. 3) such as rainbow trout (*Oncorhynchus mykiss*). This species is the designated test organism for regulatory purposes in Canada. Other fish species could also be used depending on the objectives of the study. Trout are easily obtained from commercial farms destined for fisheries and human consumption.

Table 3. Taxonomy of rainbow trout.

Kingdom	Phylum	Class	Order
ANIMALIA	CRANIATA	OSTEICHTHYES	SALMONIFORMS
Family	**Genus**		
SALMONIDAE	ONCORHYNCHUS		

Rainbow trout is a highly variable species, formerly known as *Salmo gairdneri*, but this taxon is closely related to Pacific salmon and is conspecific with the Asiatic steelhead (*Oncorhynchus mykiss*). The normal life span of rainbow trout is about 5-6 years. Resident populations inhabit small headwater streams, large rivers, lakes, or reservoirs; often found in cool clear lakes and cool swift streams with silt-free substrate. In streams, deep low velocity pools are important wintering habitats. The fish aggressively defends its feeding territories in streams. Usually requires a gravel stream riffle for successful spawning where lake populations move to tributaries to spawn. The females lay their eggs in gravel in a depression. This species survives in a wide range of temperature conditions: 4-22°C. Dissolved oxygen concentration required for survival is at least 7 ppm. Salinity of 8 ppt (parts per thousand) is the upper limit for normal development of eggs and alevins (Morgan et al. 1992). Spawning is in spring (February-June), or later depending on water temperature and location. Each trout lays between 200-9000 eggs (Wydoski and Whitney 1979), which hatch in 3-4 weeks at 10-15 °C. Fry emerge from gravel 2-3 weeks after hatching. Sexual maturity is usually achieved after 2-3 years. Sexually immature fish are selected for hepatocyte preparations which correspond to 0.5 to 2 years in age. Trout are usually commercially available from aquacultures. They could be also grown in the laboratory following established guidelines for their appropriate care (Environment Canada, 1990).

6. Laboratory preparations prior to toxicity testing

6.1 REAGENTS

All the reagents described below are of the highest quality and are readily available commercially.

6.1.1 Phosphate buffered saline (PBS)
Dissolve 7 g NaCl, 0.7 g dibasic potassium phosphate, 1.7 g sucrose, 0.75 g trisodium citrate and 1.2 g HEPES buffer in 900 mL of bidistilled water, and adjust the pH at 7.5 with NaOH 1 N. Top up the final volume to 1L with water. Filtration through a 0.22 µm pore size filter (cellulose acetate) is recommended, as is the use of sterile bottles.

6.1.2 Liebovitz L-15 medium (with glutamine and albumin)
This medium is commercially available from Sigma-Aldrich. The medium contains most of the essential amino acids, cofactors, vitamins and salts, permitting the

maintenance of viable hepatocytes without an external source of CO_2. Dissolve the powder in 95-99 mL of water. After dissolution, add 1 mL of bovine serum albumin at 0.1% (Sigma-Aldrich Company), 100 units of penicillin G, 100 µg/mL of streptomycin sulfate and 0.25 µg/mL of amphotericin B. Add 0.24 g (10 mmol) of HEPES buffer and adjust the pH at 7.5 with NaOH 1M. The medium is sterilized by filtration through a 0.22 µm pore size filter. Note that the antibiotics and antimycotic used here can be purchased as a 100-fold concentrate (Sigma-Aldrich) for safe (and ease of) handling.

6.1.3 Perfusion medium
Add 0.5 g of albumin and 0.25 g of sodium citrate to 100 mL of PBS, and adjust the pH to 7.5, if necessary. Filter through a 0.22 µm pore size membrane. The final concentration of citrate and albumin is 10 mM and 0.5 %, respectively.

6.1.4 Neutral red and fluorescein dye preparation
Neutral red (NR) is prepared by dissolving 0.04 g of neutral red in 10 mL of PBS. This stock solution is diluted 1/80 (0.1 mL to 8 mL of L-15 media) prior to the addition of hepatocytes. The revealing solution is prepared as follows: 50 mL of methanol, 2 mL of glacial acetic acid and 48 mL of bidistilled water. NR tends to agglomerate over time and agglomerates are not retained in cells. Verify proper dissolution by passing the solution through a 0.22 or 0.45 µm pore size filter several times. The filtrate should retain its reddish colour.

Carboxyfluorescein diacetate (cFDA; Molecular Probes Inc.) is prepared by dissolving 5 mg in 1 mL of DMSO (100%) to obtain a final concentration of about 10 mM. The working solution is obtained by diluting the stock solution to a final concentration of 20 µM. For example, 10 mL of working solution is prepared by mixing 20 µL of stock solution with 9.98 mL of PBS.

6.1.5 Tricaine anesthesia
Dissolve 200 mg of tricaine in 4 L of dechlorinated tap water and keep the solution at a temperature below 20°C. This step can be replaced by a quick blow on the head of the fish. These procedures are in agreement with the animal care committee of Canada.

6.1.6 Trypan blue solution
Dissolve 0.4 g of the trypan blue in 100 mL of PBS solution. Mix for 30 min and filter on a Whatman #4 paper to remove aggregates and clarify the solution.

6.1.7 Reference toxicant
Weigh 7.46 g of KCl and dissolve in 100 mL of water (1 M KCl).

6.2 MATERIALS AND APPARATUS

The following items are required to successfully prepare a primary culture of fish hepatocytes:

Bunsen burner or laminar flow hood with UV sterilization lamps.
Centrifuge with a rotor of 10-50 mL tube capacity.
Incubator saturated with humidity at 15-18°C.
Spectrophotometer for reading absorbency at 540 nm and 600 nm.
Microplate fluorometer (optional).
Optical microscope with 200 to 400 X enlargement.
Hemacytometer for cell counting (commercially available from Fisher Scientific).
pH meter.
Cell dissociation sieve kit (Sigma-Aldrich Company) or nylon sieve at 50-100 μm mesh size.

6.3 PREPARATION OF TEST SAMPLES

The liquid sample should be sterilized by filtering through a membrane of 0.22 μm porosity or centrifuged at 3000 x g for 15 min at 4°C (as in the case with sediment elutriates, for example). Toxicity is therefore associated with a particle size of less than 0.22 μm and dissolved compounds. Filtration is not required for organic extracts (*e.g.*, dichloromethane extracts of sediments in which this solvent was changed for DMSO).

6.4 PREPARATION OF TROUT HEPATOCYTES

6.4.1 Surgical operation and liver perfusion
Execute the following steps to minimize bacterial or fungal contamination. Capture a trout (10 to 20 cm long) and anaesthetize it for 3-5 min in a 4 L container containing 50 mg/L tricaine. Then place the fish on ice and open the abdominal cavity with a scalpel from the frontal fins to near the anal cavity. Locate carefully the liver from the intestinal region and nick the hepatic vein with surgical scissors. Insert the point of a 100 μL pipette tip attached to a 50 mL syringe and perfuse with about 10-25 mL of perfusion medium (PBS-citrate). The aorta can be cut to facilitate liver perfusion. The colour of the liver should change from dark red to light brown (like milky coffee). Then remove the liver from the fish and immerse it in 25 mL of perfusion medium cooled in ice. At least three livers are usually prepared to minimize inter-individual variability. Remove blood clots by washing with the perfusion media.

6.4.2 Hepatocyte dissociation and washing
Slice each liver into about 10 pieces and place them in a 50 mL centrifuge tube containing 35 mL of perfusion medium. Incubate at room temperature for 30 min with a light mixing movement (or place the sample in a sterile beaker and use a magnetic stirrer). Gently rub the liver pieces on a tissue dissociation sieve (50 to 125 μm mesh size; available at Sigma-Aldrich) to liberate hepatocytes. Centrifuge the cell suspension at 125 x g for 5 min and suspend the cell pellet with the perfusion medium. Repeat the centrifugation/resuspension steps three more times or until a clear supernatant is obtained. Under sterile conditions, suspend the cell pellet in 5 mL of L-15 medium containing 0.01 % albumin and keep it on ice.

6.4.3 Cell counting and viability estimation
Cell number and viability are determined with a hemacytometer (Fig. 1). Dilute and stain the cell suspension 1/20 in PBS solution: pipette 50 µL of cell suspension and 50 µL of trypan blue solution to 900 µL of PBS solution. Remove 10 µL of the diluted cell suspension and introduce this volume into the hemacytometer chamber. Wait 5 min at room temperature. View the slide through a microscope at 200-400 x enlargement. Count the number of cells (clear and blue cells) at each corner (where one corner is separated into 16 squares).

(Cell count per corner) x 10^4 x dilution factor (i.e., 20) = y cells / mL (1)

Blue cells are considered dead, while unstained, clear cells are considered viable. Lightly stained cells are sometimes found. In this case, they are still considered viable. The cell number yield is generally in the order of 2 x 10^6 cells/g of fish liver. This cell population consists mainly of hepatocytes (95%).

Figure 1. Counting chambers in hemacytometers. Corners A, B, C and D contain 16 squares where cells are counted. These squares (16) hold a volume of 0.1 µL. The microscopic view of the chamber is at 200 x enlargement.

7. Toxicity testing procedure

7.1 EXPOSURE OF RAINBOW TROUT HEPATOCYTES

7.1.1 Hepatocyte plating
Trout hepatocytes are usually plated in 96-well microplate(s) at a density of 5 x 10^5 cells/mL. An overview of the microplate test format is presented (Fig. 2). For example, if 20 x 10^6 hepatocytes/mL are harvested, then 5 µL of the suspension are added to 195 µL of L-15 medium to obtain a cell density of 1 x 10^5 cells/200 µL or 5 x 10^5 cells/mL (Tab. 4). Note that some cells are treated with 20% DMSO, which completely permeates cell membranes and corresponds to ~ 0% viability.

7.1.2 Exposure conditions
Trout hepatocytes are exposed in the dark at 15-18°C in an incubator saturated in humidity. Humidity ensures that no evaporation of the cell culture media occurs. The use of external gaseous CO_2 (5%) is not required with the L-15 culture media. A film permeable to gases (Sealing film for cell culture, Sigma Chemical) for covering the

microplates is sometimes required *e.g.*, if microplates are placed in large coolers for protection against bacteria and fungi that can be present in air.

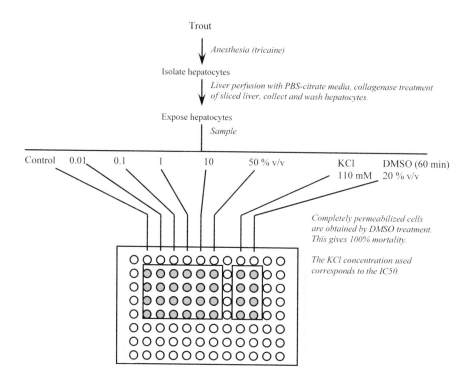

Figure 2. Overview of the rainbow trout preparation and exposure procedures.

7.2 ASSESSMENT OF CELL VIABILITY WITH THE NEUTRAL RED UPTAKE ASSAY

Viable cells accumulate the NR dye where it is preferentially absorbed in lysosomes. Less viable cells with compromised cell/lysosome membranes are not as capable of retaining the dye. The procedure is summarized in Figure 3.

7.2.1 Neutral red incubation

After the incubation period, remove the exposure medium carefully to limit cell resuspension. Centrifugation at 125 x *g* for 5 min in microplate adaptors is highly recommended. Carefully add 100 µL of the NR dilution (1/80) and incubate at room temperature for 60-90 min. After the incubation period, remove the NR solution by aspiration. If cells appear to be resuspended or lightly attached to the bottom of the wells, centrifuge the microplate at 125 x *g* for 5 min before removing the NR solution. The NR solution should be thoroughly removed at this step. An additional PBS washing step is strongly recommended; add 200 μL of PBS, centrifuge the plate, and remove the PBS solution by aspiration. Add 150 μL of revealing solution (50% methanol and 2% acetic acid) and wait 15 min for colour development.

Table 4. Preparation of serial dilutions for exposure of hepatocytes.

Effluent dilution (% v/v)	Sample volume (μL)			KCl (1M) (μL)	DMSO (100 %) (μL)	L-15 Medium (200 μL – x)[a]
	1	10	100			
0	--	--	--	--	--	200
0.01	2	--	--	--	--	198
0.1	20	--	--	--	--	180
1	--	20	--	--	--	180
10	--	--	20	--	--	180
25	--	--	50	--	--	150
KCL Concentration	--	--	--		--	
500					100	100
250					50	150
100					20	180
50					10	190
10					5	195
DMSO (20 %)					40	160

[a] L-15 medium containing 0.01% bovine albumin minus the volume (x) of the cell suspension. For example, if the volume of added cell suspension is 5 μL, then this volume should be removed from the L-15 medium (= 195 μL).
[b] A 30 min exposure period is usually sufficient for DMSO treatment.

7.2.2 Absorbency readings
Turn on the spectrophotometer at least 15 min before taking the readings. Read each well at 540 nm. If a microplate reader is not available, simply transfer the content of each well to a cuvette, fill with water to the appropriate volume (usually 1 mL total volume) and read with a tube spectrophotometer. The blank consists of the revealing solution only.

7.2.3 Fluorescence readings
Turn on the (microplate) fluorometer 15 min before analysis. Read each well at 590 nm emission with a 540 nm excitation filter. With a cuvette fluorometer, just transfer 150 μL of each microwell sample (containing the revealing solution and

stained cells) to 2.5 mL of water and read the fluorescence as described above with 10 nm bandpasses.

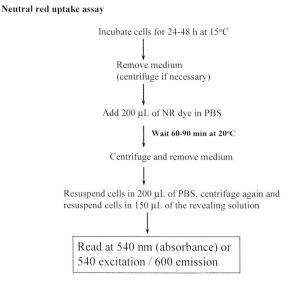

Figure 3. Overview of the neutral red uptake assay.

7.3 ASSESSMENT OF CELL VIABILITY WITH THE FLUORESCEIN DIACETATE ASSAY

Viable cells are exposed to cFDA and accumulate the dye where it is cleaved to fluorescein by the action of intracellular non-specific esterase activity. The more polar fluorescein is retained in healthy cells while compromised cells leak fluorescein in the extracellular environment. An overview of the procedure is outlined in Figure 4.

7.3.1 *Fluorescein diacetate incubation*
After the incubation period, microplates are centrifuged at 125 x *g* for 2-3 min. The exposure medium is carefully removed. Carefully add 100 µL of the 20 µM cFDA solution and incubate for 20 min at room temperature (18-20°C). After the incubation period, remove the cFDA solution by aspiration after centrifuging cells at 125 x *g* for 5 min. An additional PBS washing step can be performed by adding 200 µL of PBS, centrifuging the plate, and removing the PBS solution by aspiration. Add 150 µL of PBS solution and proceed to fluorescence reading.

7.3.2 Fluorescence readings

The formation of fluorescein in cells from cFDA by the action of non-specific intracellular esterases is measured in a fluorescence microplate reader with 485 nm excitation and 520 nm emission. The assay is sensitive with a detection limit of about 500 cells per well, depending on the microplate reader.

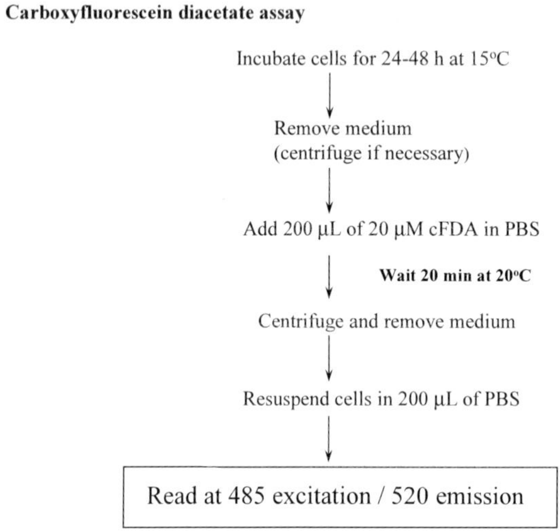

Carboxyfluorescein diacetate assay

Incubate cells for 24-48 h at 15°C
↓
Remove medium
(centrifuge if necessary)
↓
Add 200 µL of 20 µM cFDA in PBS
↓ **Wait 20 min at 20°C**
Centrifuge and remove medium
↓
Resuspend cells in 200 µL of PBS
↓
Read at 485 excitation / 520 emission

Figure 4. Overview of the carboxyfluorescein diacetate (cFDA) assay for cell viability estimation.

7.4 EXPRESSION OF RESULTS

7.4.1 Data calculation

The data (540 nm absorbency readings or relative fluorescence units) are normalized with respect to unexposed cells in terms of response ratio:

$$Response\ ratio = value\ of\ treated/mean\ value\ of\ control\ cells \quad (2)$$

Cells treated with 20% DMSO for 30 min represent 100% mortality and can be used to extrapolate cell viability with respect to unexposed cells:

$$\%\ mortality = \frac{(A540\ control - A540\ treated)}{(A540\ control - A540\ DMSO\ treated)} \times 100 \quad (3)$$

where cell viability corresponds to (100% - % mortality).

7.4.2 Measurement endpoint for toxicity

The test sample concentration that reduces by either 20% (IC20) or 50% (IC50) dye retention in cells is usually determined through linear regression of the concentration-response data. For various reasons, such as the lack of linear relationship, absence of 50% effect concentration due to low toxicity of the sample, or the desire to report a more sensitive endpoint, it may be relevant to calculate a threshold concentration (TC) where significant effects are observed. The TC value is derived with the No Observable Effect Concentration (NOEC) and the Lowest Observable Effect Concentration (LOEC): $TC = (NOEC \times LOEC)^{1/2}$. The NOEC and LOEC values are usually determined by (non-)parametric analysis of variance followed by a suitable *post hoc* test (*e.g.*, Dunnett's t-test or Mann-Whitney U test), depending on data distributions.

For complex mixtures, data are usually expressed in terms of % v/v, which is the dilution of the original effluent (100%) required to elicit a cellular response. In some cases, the dilution % is converted into toxic units following this simple relationship: Toxic unit = 100/ IC50 or TC (% v/v). An example of cell viability data obtained with the cFDA method is given in Table 5, where the threshold concentration (TC) of the effluent corresponded to 6.3 toxic units: 1) $TC = (10 \times 25)^{1/2} = 15.8\%$; 2) TU = 100%/15.8% = 6.3 toxic units.

8. Conditions of test validity and quality control

8.1 CONDITIONS FOR TEST VALIDITY

Primary cultures of rainbow trout hepatocytes should come from at least three individuals to reduce inter-individual variability in respect to sensitivity to toxicants. This condition is especially important for organisms that are genetically heterogeneous in nature (see Section 5). The initial cell viability should be at least 90% before initiating exposure to toxic substances or environmental samples. These reasons described above in addition to variations in cell culture media, laboratory environment, and the general health status of fish, justify the use of a reference toxicant. The use of a reference toxicant like KCl or ZnCl permits the comparison of toxicity responses between different cell preparations over time and laboratories. This is especially useful for monitoring programs.

8.2 QUALITY CONTROL OF TROUT HEPATOCYTE PREPARATIONS

The toxic potential of a reference toxicant (KCl) should be determined with each hepatocyte preparation to ensure the reproducibility of different cell preparations for measuring the toxic potential of environmental samples. The individual IC50s of the reference toxicant are then plotted on a quality control chart format as depicted in Figure 5. At least 8 data points are required to define the central value (mean) with boundaries (standard deviation). IC50 values that are within 1 standard deviation are considered as normal biological variation while those between 1 and 2 times the standard deviation should be treated with caution. Values outside 2 times the standard deviation are considered invalid and the results should be discarded. If an

IC50 value is close to twice the standard deviation limit, then a thorough evaluation of all culturing and tests conditions is triggered. The IC50 of the reference toxicant KCl is normally close to 100 mM with a coefficient of variation of 15%. The mean value indicates overall good reproducibility of rainbow trout hepatocyte preparations.

Table 5. Example of cell viability data obtained by the cFDA assay.

Effluent nominal concentration (% v/v)	Fluorescence units (FU)	Absorbance (600 nm)	FU/A600	Statisitical analysis (mean)
0	1802	0.135	13348	12471
	2074	0.156	13295	
	1825	0.18	10138	
	1900	0.145	13103	
0.1	1843	0.161	11447	12186
	1978	0.158	12518	
	1789	0.141	12688	
	1850	0.153	12091	
1	1938	0.154	12584	11358
	1980	0.193	10259	
	1937	0.172	11262	
	1960	0.173	11329	
10	2069	0.19	10989	11984
	2319	0.175	13251	
	2012	0.179	11240	
	2133	0.17	12547	
25	1450	0.13	11154	10351 ($p < 0.05$)[a]
	1435	0.142	10106	
	1400	0.136	10294	
	1430	0.145	9852	
KCl (100 mM)	967	0.12	8058	7701 ($p < 0.01$)[a]
	975	0.124	7863	
	987	0.132	7477	
	1000	0.135	7407	
DMSO (20 %)	839	0.115	7296	6385 ($p < 0.01$)[a]
	700	0.113	6195	
	670	0.112	5982	
	734	0.121	6066	

[a] Statistically different relative to control by Dunnett's *t*-test.

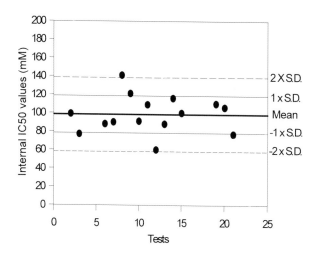

Figure 5. Quality control chart for KCl using primary cultures of rainbow trout hepatocytes. The reference toxicant (KCl) was evaluated with different preparations of three trout. S.D.: standard deviation.

9. Factors capable of influencing the results

9.1 REFERENCE TOXICANT

A reference toxicant (KCl) should always be used for screening the toxic potential of wastewater for each hepatocyte preparation. The 50% inhibitory concentration (*i.e.*, the test sample concentration reducing the intracellular level of dye by 50%) for KCl is 100 mM with a coefficient of variation of 15%. Thus, each hepatocyte preparation should be verified with this reference toxicant to ensure that RTH preparations have adequate sensitivity throughout the year.

9.2 NEUTRAL RED DYE

Neutral red dye has a tendency to precipitate during storage. Prepare a working solution daily and check precipitation by filtering the solution through a 0.45 μm pore size filter (if precipitation occurs, then nearly all the dye is retained on the filter). Repeated filtering (several times) through the same filter sometime releases the NR dye. In this event, the filtrate will take on a dark reddish colour. During the NR uptake assay for cell viability determination, a 60 min treatment time is usually sufficient. Treatment time can be increased up to 120 min if low accumulation is observed.

9.3 INITIAL VIABILITY OF HEPATOCYTE PREPARATIONS

The initial viability, which is determined at the cell-counting step, should be 90% or higher. If the initial viability is below this value, then the hepatocytes and solutions should be discarded. Repeat the extraction of hepatocytes with freshly- prepared solutions.

9.4 EXPOSURE TEMPERATURE

This variable is likely to influence the response of hepatocytes to toxicants *in vitro*. The temperature employed is usually between 12 and 20°C depending on study objectives. For regulatory purposes, the rainbow trout fingerling acute lethality test is performed at 15°C and this temperature should also be used for RTH testing.

9.5 STERILITY ISSUES

Cell culture using a complete medium, such as L-15, is subject to contamination by bacteria and fungi. This type of contamination can be controlled by adding two antibiotics and antimycotic(s) to the cell medium and filtering it through a 0.22 µm membrane in sterilized containers. All the materials should be as clean as possible and cells should be plated in a laminar-flow hood or in the vicinity of a Bunsen burner flame. Cell plates can be covered with a plastic film that allows gas exchange while excluding micro-organisms and small particles.

10. Application of the hepatocyte test in a case study

In this case study, the toxic potential of surface waters contaminated by a municipal effluent was examined to evaluate its probable zone of impact and the dilution potential of a given river system. Rainbow trout hepatocytes were used to evaluate the cytotoxic properties of a municipal effluent dispersion plume (Fig. 6). In an attempt to optimize the toxicological knowledge for these surface waters, metallothionein levels were determined by the silver saturation assay and 7-ethoxyresorufin O-deethylase (EROD) activity (a measure of CYP1A1 activity) was also followed in cells. The results show that cytotoxicity was found in surface waters collected at distances of 0.1, 1 and 4 km downstream in the municipal dispersion plume and strong responses were also observed for MT and EROD at similar distances. This case study demonstrates that primary cultures of trout hepatocytes are a simple but effective screening microbiotest for detecting both lethal and sublethal toxic effects along a chemical pollution gradient.

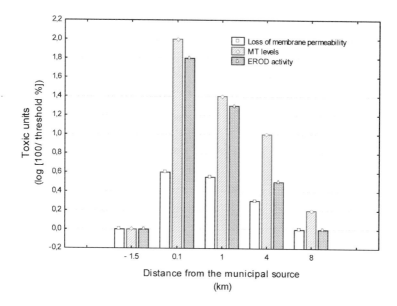

Figure 6. Cytotoxicity potential of a primary-treated municipal effluent. Trout hepatocytes were exposed to surface waters at a density of 0.5 x 10^6 cells/mL and incubated for 48 h at 15°C. Data are expressed in log of toxic units.

11. Accessory test information

11.1 STABILITY OF HEPATOCYTE PREPARATIONS

Primary cultures of rainbow trout hepatocytes can be maintained for up to six days if they are kept at 4°C. These cells were shown to be equally sensitive to KCl treatment within this time frame. Hence, the hepatocyte preparation and exposure steps should be done before the sixth day after the initial plating of cells.

11.2 DOUBLE PERFUSION METHODOLOGY

Our double perfusion methodology is cheaper to apply than the one used by Klauning et al. (1985). In our approach, the liver is perfused with saline citrate to chelate calcium, and hepatocytes are mechanically released (*i.e.*, by stirring) in the presence of citrate and albumin. Some fish species provide poor cell yields. We therefore recommend replacing the second perfusion medium with 100 units of collagenase, 0.05% albumin in L-15 medium. The washing medium should contain albumin to quench non-specific protease activity of the collagenase preparations and to remove adsorbed collagenase from the cell surface. In some cases, DNA released from lysed cells during the cell extraction procedure could form large cell aggregates (observable by the naked eye) and dramatically reduce cell yields. Adding desoxyribonuclease A (DNAse A) during the second perfusion step is recommended.

This phenomenon rarely occurs with trout but may occur with other fish species.

11.3 SENSITIVITY ISSUES

Cells are maintained in a salt-rich environment (L-15 medium), corresponding to an osmolarity of 275-320 mOsm, during the exposure period. The presence of salts could reduce the availability of metals or metalloids in the exposure medium, thereby reducing the sensitivity of the cell system towards this class of xenobiotics. We recommend validating the fish cell model with the surrogate fish assay to confirm adequate sensitivity when screening metal-rich effluents (*i.e.*, mining effluents). Assay sensitivity of stands to be increased by lowering cell density to 5,000 - 20,000 cells per mL of exposure medium. For example, plating 25,000 cells in 2 mL culture plates (24-well plates) could readily increase the sensitivity of the cell system.

11.4 STANDARDIZATION

The procedure described here underwent an inter-laboratory examination in an attempt to standardize the methodology (Gagné et al., 1999). Five laboratories were trained to prepare and expose primary cultures of trout hepatocytes to three samples and assess cell viability by the neutral red uptake assay. The IC50 values for KCl, the reference toxicant, obtained for these laboratories were in the same range of IC50 values obtained in our laboratory during an in-house intercalibration exercise (80-120 mM). These results indicate that this procedure is transferable to other laboratories with no previous experience in cell culture.

12. Conclusions and prospects

The primary cell culture system using rainbow trout hepatocytes was presented. This cell system can be undertaken as a complementary microbiotest to the rainbow trout acute (sub)lethality test, which is routinely used to screen municipal and industrial wastewater. Indeed, rainbow trout hepatocytes could be used to effectively screen large samples of municipal or industrial effluents to identify those that are most toxic and elicit sublethal effects, such as oxidative stress (lipid peroxidation) or DNA damage, from those that are less, or not at all toxic. These effluents could then be tested *in vivo* for confirmation purposes. This approach can be used provided that the cell system is properly validated (*i.e.,* shown to be as sensitive to a specific type of complex mixture as the *in vivo* test).
This cell system is also promising for detecting several sublethal effects relevant to environmental toxicology projects, such as mixed-function oxidase activity, metallothionein induction and vitellogenin gene expression. RTH testing is also relevant to toxicogenomic studies because many gene sequences are well characterized in rainbow trout. By coupling genomics and trout hepatocytes as a novel test system, effects of complex mixtures could be readily characterized at the molecular level to predict their toxic potential in chronically exposed organisms.

Acknowledgements

The author thanks the St. Lawrence Centre, Environment Canada, Quebec Region, for supporting this research initiative. The editorial comments of Jacqueline Grekin during the preparation of the manuscript are duly recognized.

References

Babich, H. and Borenfreund, E. (1987) Cultured fish cells for the ecotoxicity testing of aquatic pollutants, *Toxicity Assessment* **2**, 119-133.
Baksi, S.M. and Frazier, J.M. (1986) Isolated fish hepatocytes – model systems for toxicology research, *Aquatic Toxicology* **16**, 229-256.
Brandao, J.C., Bechets, H.H.L., Van de Vyver, I.E. and Dierickx, P.J. (1992) Correlation between the *in vitro* cytotoxicity to cultured fathead minnow fish cells and fish lethality data for 50 chemicals, *Chemosphere* **25**, 553-562.
Dierickx, P.J. and Van de Vyver, I.E. (1991) Correlation of the neutral red uptake inhibition assay of cultured fathead minnow fish cells with lethality tests, *Bulletin of Environmental Contamination and Toxicology* **46**, 649-653.
Environment Canada (1990) *Biological Test Method, Acute Lethality Test Method Using Rainbow Trout*, Conservation and Protection, Environment Canada, Report No. SPE/1/RM/9.
Gagné, F. and Blaise, C. (1997) Evaluation of cell viability, mixed function oxidase activity, metallothionein induction, and genotoxicity in rainbow trout hepatocytes exposed to industrial effluents. II. Validation of the rainbow trout hepatocyte model for ecotoxicity testing of industrial wastewater, *Environmental Toxicology and Water Quality* **12**, 304-314.
Gagné, F. and Blaise, C. (1998) Differences in the measurement of cytotoxicity of complex mixtures with rainbow trout hepatocytes and fibroblasts, *Chemosphere* **37**, 753-769.
Gagné, F., Blaise, C., vanAggelen, G., Boivin, P., Chong-Kit, R., Jonczyk, E., Marion, M., Kennedy, S., Legault R. and Goudreault, J. (1999) Intercalibration study in the evaluation of toxicity with rainbow trout hepatocytes, *Environmental Toxicology* **14**, 429-437.
Klauning, J.E., Ruch, R.J, and Goldblatt, P.J. (1985) Trout hepatocyte culture: isolation and primary culture, *In Vitro Cellular and Developmental Biology* **21**, 221-228.
Morgan, J. D., Jensen, J.O.T. and Iwama, G. K. (1992) Effects of salinity on aerobic metabolism and development of eggs and alevins of steelhead trout (*Oncorhynchus mykiss*) and fall chinook salmon (*Oncorhynchus tshawytscha*), *Canadian Journal of Zoology* **70**, 1341-1346.
Rixon, R. (1995) Validation – now or never, *Trends in Alternative Testing*, Joseph F. Morgan Research Foundation, Winter/Spring, 1-4, Ottawa.
Seglen, P.O. (1979) Hepatocyte suspensions and cultures as tools in experimental carcinogenesis, *Journal of Toxicological and Environmental Health* **5**, 551-560.
Wydoski, R. S., and Whitney, R. R. (1979) Inland fishes of Washington. The University of Washington Press, Seattle.

Abbreviations

A	Absorbance
cFDA	5,6-carboxyfluoresceine diacetate
CYP1A1	Cytochrome P4501A1
DMSO	Dimethylsulfoxide
DNAse A	Desoxyribonuclease A
IC50	Inhibitory concentration causing a 50 % effect
EDTA	Ethylenediaminetetracetic acid
EGTA	Ethylene glycol-bis(2-aminoethylether)-N, N, N',N'- tetraacetic acid
EROD	7-ethoxyresorufin O-deethylase

FU	Fluorescence units
HEPES	4-(2-hydroxyethyl)piperazine-1-ethanesulfonic acid
L-15	Liebovitz-15 cell culture media
LOEC	Lowest observable effect concentration
LC50	Lethal concentration that kills 50% of organisms
mRNA	Messenger ribonucleic acid
MT	Metallothionein
NOEC	No observable effect concentration
NR	Neutral red
PBS	phosphate buffered saline
pH	$-\log_{10}$(molar concentration of H_3O^+)
RTG-2	rainbow trout gonad cell line 2
RTH	Rainbow trout hepatocyte
S.D	Standard deviation
TC	Threshold concentration, TC= $(NOEC \times LOEC)^{1/2}$
TU	toxic unit defined as 100%v/v ÷ endpoint value (% v/v).

15. RAINBOW TROUT GILL CELL LINE MICROPLATE CYTOTOXICITY TEST

VIVIAN R. DAYEH
Department of Biology
University of Waterloo, Waterloo
Ontario N2L 3G1, Canada
vrdayeh@sciborg.uwaterloo.ca

KRISTIN SCHIRMER
Department of Cell Toxicology
UFZ-Centre for Environmental Research
Permoserstr, 15
04318, Leipzig, Germany
kristin.schirmer@ufz.de

LUCY E.J. LEE
Department of Biology
Wilfrid Laurier University, Waterloo
Ontario N2L 3C5, Canada
llee@wlu.ca

NIELS C. BOLS
Department of Biology
University of Waterloo, Waterloo
Ontario N2L 3G1, Canada
ncbols@sciborg.uwaterloo.ca

1. Objective and scope of the RTgill-W1 cytotoxicity test method

This chapter describes a rapid, inexpensive *in vitro* test for evaluating the toxicity of water samples and a potential alternative to the use of fish in routine toxicity testing. The overall objective of the procedure is to assess water samples for acute or basal cytotoxicity to fish cells. Basal cytotoxicity refers to impairment of cellular activities that are shared by all or most cells. The basic procedure can also be used to evaluate the acute cytotoxicity of individual chemicals and to understand the mechanism(s) behind their toxicity. The speed and cost-saving features arise from assessing the viability of fish cells in 96-well microplates. This reduces the cost of disposables to be used and of shipping large volumes of water samples to a central testing facility.

The described protocol is for whole-water samples, but water extracts could also be examined. Theoretically, any water could be tested, but samples containing copious amounts of microbes or particulate matter may require a filtration step, which could also lead to an inadvertent removal of some toxicants.

2. Summary of the RTgill-W1 cytotoxicity test procedure

The test procedure can be considered as the integration of three basic protocols. The first involves routinely growing a fish cell line in flasks and using cells from these culture vessels to initiate test cultures in either 48- or 96-well microplates. Each culture well receives approximately 5×10^4 or 3×10^4 cells and is confluent in approximately 3 days at which time the cells are exposed to the test solution. The second protocol describes how water samples are prepared for application to the microwell cultures of fish cells. The key preparative step is adding medium components to the water samples in order to achieve an osmolality appropriate for fish cells. The third protocol evaluates basal cytotoxicity in fish cell cultures after a 24 h exposure to the water samples. This is done with cell viability assays that utilize fluorometric dyes to monitor different cellular activities. The results are read in a fluorometric plate reader and expressed as a percentage of the control. These three basic protocols are given in detail in Sections 6 to 9 and have been described in different formats in other publications (Bols and Lee, 1994; Ganassin et al., 2000; Dayeh et al., 2003b). Table 1 summarizes the general characteristics of the test procedure.

3. Overview of applications reported with the toxicity test method

As described here, measuring basal cytotoxicity in microplate cultures of fish gill cells is a means of evaluating the toxicity of water samples. Fish cells have been used widely in microwell plates for toxicological studies (Bols et al., 2005; Castaño et al., 2003; Dayeh et al., 2002; Fent, 2001), but the procedure of this chapter has several special features. The most unique one is the addition of very few medium components to water samples as solids to make up a solution that will maintain the cells and allow the cytotoxicity of the water to be evaluated (see Section 4.2). Another is the employment of a continuous gill epithelial cell line, RTgill-W1 (Fig. 2). A third one is to monitor changes in cell viability with three fluorescent indicator dyes that can be measured with a fluorometric multiwell plate reader (Fig. 3). To date, the procedure has been applied to effluent from a paper mill (Dayeh et al., 2002) and currently is being evaluated for its usefulness in testing the toxicity of mining effluent (Dayeh et al., 2003a). In addition an early variation of the procedure was successful with oil refinery effluent (Schirmer et al., 2001). The procedure also can be used to rank the cytotoxicity of individual chemicals for the general purposes of identifying compounds that have the potential to be acutely toxic *in vivo*. Data on the basal cytotoxicity of a range of chemicals to RTgill-W1 have been obtained. This includes polycyclic aromatic hydrocarbons (PAHs) (Schirmer et al., 1998a, b), surfactants (Dayeh et al, 2002; 2004) and metals (Dayeh et al., 2003a). This information can help to interpret results obtained with whole-water samples.

Table 1. Rapid summary of RTgill-W1 cytotoxicity test.

Test organism	- Rainbow trout gill cell line, RTgill-W1
Type of test	- Acute toxicity test (24 h exposure); static
Test format	- 48 or 96-well tissue-culture treated flat bottom microwell plates
Well volume contents	- 500 µL for 48-well plates and 200 µL for 96-well plates
Initial cell plating density	- 5×10^4 for 48-well plates and 3×10^4 for 96-well plates grown until confluent monolayer has formed (~ 3 days)
Lighting	- Cell culturing in ambient lighting; dosing in reduced lighting; test exposure in darkness*
Temperature	- Cell growth and exposure at ambient room temperature ($20 \pm 2°C$)

Experimental configuration	- 48-well plates: 5 control wells, 7 serial dilutions of test solution, each with 4 replicates with cells and one no-cell control
	- 96-well plates (two configurations available)
	  1. same as above for 48-well plates (able to conduct two separate compounds)
	  2. 7 control wells, 11 serial dilutions of test solution, each with 6 replicates and one no-cell control

Measurement of cell viability	- Fluorescent indicator dyes quantified on a multiwell fluorescent plate reader - alamar Blue – metabolic activity - CFDA-AM – cell membrane integrity - Neutral red – lysosomal activity
Endpoints determined	- EC50, based on % of control cells
Reference toxicants	- Abietic acid (85% purity – Acros Organics through Fisher Scientific)

* allows additional evaluation of photo-cytotoxicity in the presence of UV irradiation.

4. Advantages of conducting the toxicity test method

Each of the three basic protocols that make up the test procedure brings advantages to the overall procedure. This begins with the use of cells in culture, the kind of cell culture, and the choice of cell type.

4.1 ADVANTAGES OF CELL CULTURES AND CELL LINES

Cell cultures in general offer several advantages over fish as assay tools for environmental samples (Bols et al., 2005; Castaño et al., 2003; Dayeh et al., 2002; Fent 2001). Commonly, results are obtained more rapidly and at less cost with the cell assays than with intact animals. The small volume of sample needed for cell assays provides convenience and saves money. For example, in Canada the pulp and paper industry has to pay for shipping large volumes of effluent, often from remote sites, to a central facility for the rainbow trout 96 h lethality test. Finally, assays with cell cultures satisfy a societal desire to reduce the use of animals in toxicology testing (see Box 1).

Two general types of cultures can be used to study animal cells *in vitro*. One is the primary culture; the other, cell lines. Primary cultures are initiated directly from the cells, tissues or organs of fish and typically last for only a few days. The two are interrelated because cell lines are developed from primary cultures. By convention (Schaeffer, 1990), the primary culture ends and the cell line begins upon subcultivation or splitting of the primary culture into new culture vessels. The cell line can continue to be propagated by repeating the cycle of allowing cell number to increase through cell proliferation followed by splitting the cell population into new culture vessels, usually flasks. This cycle of growth and splitting, which is often referred to as passaging, might be possible for only a limited number of population doublings, which is a finite cell line, or done indefinitely, which is a continuous cell line. In the case of fish, the cell lines almost always appear to be continuous or immortal (Bols et al., 2005).

Box 1. Main advantages of using RTgill-W1 cytotoxicity assay.

- Continuous supply of cells
- Low volume sample requirement
- Detection of multiple endpoints of cell viability
- Large numbers of samples can be tested
- Detection of mechanism(s) of toxicity
- Ease of sample preparation
- Rapid exposure time (24 hours)

As to the choice of cell culture type in environmental assays, cell lines have several advantages over primary cultures. Cell lines are a much more reproducible and convenient source of cells because, once established, cell lines are fairly homogeneous and can be cryopreserved indefinitely. Although a single preparation from an organ or pooled organs can yield identical primary cultures, there is the cost of maintaining the fish and of the labor involved in repeatedly initiating new primary cultures. Once a cell line is developed no further animals need to be consumed. Thus cell line assays better satisfy the desire to use fewer animals in toxicity tests.

Although much more is known about mammalian than piscine cell lines and mammalian cell lines have been used to monitor water quality (Richardson et al., 1977; Mochida, 1986), ultimately piscine cell lines should be superior in assays of water quality for several reasons. Firstly, the whole animal tests that are used to assay water employ fish, making fish cells more appropriate as alternatives. Secondly, the toxicants can be applied to fish cells at temperatures more typical of the temperatures to which fish would be exposed. A wide range of exposure temperatures can be utilized for the testing including the temperature normally used for whole fish. Thirdly, the cells of a piscine cell line should better reflect the properties of the fish from which they were derived than the cells of a mammalian cell line. Finally, fish cells tolerate being maintained in culture for a day or two in a simple exposure medium. Such a medium is L-15/ex, which was developed for studying the photocytotoxicity of polycyclic aromatic hydrocarbons (PAHs) to the rainbow trout gill cell line, RTgill-W1 (Schirmer et al., 1997). This medium has only salts, pyruvate and galactose. The simplicity favors expression of cellular responses to toxicants because protective molecules such as antioxidants are absent. As well, the medium is much less expensive than complete cell culture medium. Thus whole-water samples can be applied to fish cell cultures by being used to make up L-15/ex.

Finally, as the overall endpoint of the described procedure is basal cytotoxicity, any fish cell line might be suitable. However, the recommended cell line, RTgill-W1 has advantages besides being derived from an appropriate species, which is discussed in Section 5. RTgill-W1 cells remain attached firmly to microwell plates under a variety of culture conditions and after repetitive rinsing of the cultures and changes in solutions, which are necessary to perform the assays. RTgill-W1 is available from the American Type Culture Collection (ATCC CRL-2523), which assures quality of the line and continuity of the supply. Additionally, the cell line was derived from the gill, which is often the organ that fails during acute fish toxicity tests. However, the extent to which RTgill-W1 expresses gill epithelium properties is unknown, but future studies might identify them, which would allow the development of assays that monitor differentiated or tissue-specific functional endpoints.

4.2 ADVANTAGES OF WHOLE-WATER SAMPLES

Testing whole-water samples on cells offers several important advantages. Applying the whole sample to cultures assures that little or no toxicant is lost in any processing steps. As well, the total toxicity of the sample, encompassing all potential synergistic, antagonistic and additive interactions, is measured. The cost and time of testing is reduced because labor-intensive extraction procedures with expensive organic solvents are eliminated. As mentioned previously, the cost of shipping large volumes of effluent samples from distant sources to testing facilities is reduced because relatively small volumes are needed for testing. Finally, whole-water samples are more analogous to the protocol used to test the toxicity of water samples to fish.

Preparation of whole-water samples in L-15/ex has several advantages over the use of complete culture medium for applying whole-water samples to cells in culture. The simplicity of this medium, which contains only salts, pyruvate and galactose, favors expression of cellular responses to any toxicants that might be present because protective molecules, such as antioxidants, are absent. The medium is much less expensive than complete cell culture medium. As well, the amount of L-15/ex components can be varied easily to account for any big differences in the osmolality of whole-water samples. Finally, the simplicity of the medium reduces the growth of any microbial contamination during the 24 h of presentation of the whole-water sample to cells in culture.

4.3 ADVANTAGES OF MULTIPLE FLUORESCENT ASSAYS FOR CELL VIABILITY

Although numerous assays of cell viability have been developed, those that focus on the integrity of the plasma membrane and metabolism and utilize fluorescent dyes to indicate impairment in these cellular parameters are perhaps best. The tests can be performed on cultures after relatively brief exposure to putative toxicants. In the procedure of this chapter, the exposure is kept short (24 h) to reduce overgrowth by any microbes in the whole-water sample and to provide information about the status of the water sample as quickly as possible. One potential drawback is that toxicants inducing a particular cellular process, such as the xenobiotic metabolism, or by causing cumulative damage might be missed. More and more fluorescent dyes are becoming available commercially to evaluate different cellular parameters, including the integrity of metabolism and the plasma membrane. As well, the development of fluorometric multiwell plate readers has made the use of fluorometric dyes easy and rapid. The microwells conserve material resources by reducing the number of cells needed and increasing the number of replicates. The plate readers have the potential for high interlab reproducibility and can be coupled to computers to rapidly and easily manage data, which can allow for multiple assays. Such assays with slightly different cellular endpoints can be more sensitive than a single test and also reduce the chance of recording a false negative. Multiple cellular endpoints also have the potential of revealing the mechanisms behind the cytotoxicity of a water sample. In the procedures described in this chapter three assays are used: membrane integrity is monitored with 5-carboxyfluorescein diacetate acetoxymethyl ester (CFDA-AM); energy metabolism, with alamar Blue (AB) or resazurin; and lysosomal activity, with neutral red (NR).

4.3.1 Plasma membrane integrity and CFDA-AM
The integrity of the plasma membrane in cultures of fish cell lines has been assayed in a variety of ways, but most assays can be considered to be one of two types (Bols et al., 2005). Methods that measure the ability of the plasma membrane to exclude large bulky, charged molecules, such as dyes, constitute one type. The classic dye

exclusion technique is trypan blue, which has been applied to fish cells, but can often be tricky and tedious to use because the results must be scored under the light microscope. The alternative to dye exclusion is the capacity of the plasma membrane to retain a marker molecule. The marker can be the appearance in the medium of an intracellular molecule, such as an intracellular enzyme like lactate dehydrogenase (LDH), which can be complicated by several factors (Putnam et al., 2002).

In this chapter an esterase substrate (5-carboxyfluorescein diacetate acetoxymethyl ester, CFDA-AM) is used to measure cell membrane integrity, with the fluorescent product being the marker retained (Schirmer et al., 1997; 1998a; 1998b; 2000). CFDA-AM diffuses into cells rapidly and is converted by non-specific esterases of living cells from a nonpolar, nonfluorescent dye into a polar, fluorescent dye, 5-carboxyfluorescein (CF), which diffuses out of cells slowly. Although the CFDA-AM assay appears to monitor impairment to plasma membranes, the test as described in this chapter could result in more complex explanations. When the CFDA-AM is applied to fish cells in microwell plates after having been exposed to the test solution and read sometime later without removing the dye, the fluorescent readings or units (FU) constitute the CF both inside and outside the cells. In this case a decrease in FU with CFDA-AM actually measures a decline in the total esterase activity within a microwell cell culture (Ganassin et al., 2000; Dayeh et al., 2003b).

The decrease in esterase activity with toxicant treatment could be achieved in two general ways: interference with plasma membrane integrity or with cellular esterase activity. A loss of plasma membrane integrity would decrease esterase activity in two slightly different ways. The first of these would be the complete or partial lysis of the cells upon toxicant exposure so that the esterases are released into the medium and lost when the medium is removed and replaced with the CFDA-AM solution. Another possible cause for the diminution in esterase activity is a change in plasma membrane integrity so that cytoplasmic constituents are lost to the medium but the esterases remain contained within the cells, which are left still attached to the surface of the microwells. This change in the cytoplasmic milieu would be less able to support maximal esterase activity. Alternatively, the toxicant treatment could leave membrane integrity unimpaired but specifically interfere with cellular esterases, causing activity to decline. Examples of this would be a toxicant interfering with the uptake of the substrate, CFDA-AM, across the plasma membrane or inhibiting the catalytic activity of the esterases. These potential complexities can be overcome by carrying out the other viability assays.

4.3.2 Metabolic impairment and alamar Blue (AB)
Although metabolism by fish cell cultures has been monitored by measuring their ATP content or their ability to reduce either 3-(4,5-dimethylthiazol-2-yl)-2,5 diphenyl tetrazolium bromide (MTT) (Segner, 1998) or resazurin; the reduction of resazurin has some convenient features. Resazurin can be purchased as a commercial solution called alamar Blue (AB). AB reduction can be measured either spectrophotometrically or fluorometrically. Recovery from metabolic impairment can be evaluated by repeatedly applying the dye to the same culture over a period of days (Ganassin et al., 2000). Originally, AB was thought to be reduced by mitochondrial enzymes (De Fries and Mistuhashi, 1995), but now enzymes, such as diaphorases, with both cytoplasmic and mitochondrial locations, are thought to be responsible for

dye reduction (O'Brien et al., 2000). Thus a decline in AB reduction indicates an impairment of cellular metabolism rather than specific mitochondrial dysfunction.

4.3.3 Lysosomal activity and neutral red (NR)
Neutral red (NR) (3-amino-7-dimethylamino-2-methylphenazine hydrochloride) measures plasma membrane integrity after exposure to putative toxicants (Babich and Borenfreund, 1991; Segner, 1998), but as well, NR can detect injury specific to lysosomes. The general principle behind the use of this dye is that only viable cells accumulate NR into lysosomes (Borenfreund and Puerner, 1984). In the procedure of this chapter, NR is applied after the exposure to water samples, so the endpoint is the lysosomal accumulation of NR rather than NR retention. NR can be measured either spectrophotometrically (Borenfreund and Puerner, 1984) or fluorometrically (Essig-Marcello and van Buskirk, 1990). Although accumulating specifically in lysosomes, NR accrual and retention is dependent on an intact plasma membrane, adequate energy metabolism, and a functioning lysosome. Under most circumstances, the NR assay likely detects impairment to all three cellular parameters and the results are commonly similar to the results with other viability assays. However, hints of specific lysosomal damage have been seen. For example with the RTgill-W1 cell line, Schirmer et al. (1998b) found that immediately after UV irradiation in the presence of either acenaphthylene, acenaphthene or phenanthrene, photocytotoxicity was detected with NR but not with other indicator dyes, which suggests that lysosomes were being impaired before cell viability was lost.

5. Test species

The recommended test subject is a continuous epithelial cell line, RTgill-W1, from the gill of rainbow trout (Bols et al., 1994). Rainbow trout or *Oncorhynchus mykiss*, formerly *Salmo gairdneri*, is widely available and easily maintained. This has led to the species being used intensively in toxicology, and in some instances, to being referred to as the piscine 'white rat' (Wolf and Rumsey, 1985). As a result, more is likely known about the toxicology of rainbow trout than any other aquatic vertebrate, and rainbow trout have become incorporated into standardized toxicology tests, such as the 96 h acute lethality test (Environment Canada, 1990). In Canada, legislation requires that effluent from pulp and paper mills be assessed routinely by the 96 h rainbow trout lethality test (Environment Canada, 1989). Thus for the cell culture approach, rainbow trout is an excellent species to obtain cells from and to use in toxicity tests because the *in vitro* results can be compared to the enormous amount of *in vivo* data. In addition, as an *in vitro* alternative to rainbow trout in routine toxicity testing, cells from the same species would intuitively appear to be more suitable than cells from other species. The advantages of RTgill-W1 over other types of cell cultures and lines have been described in the previous sections.

6. Culture/maintenance of fish cell lines in the laboratory

6.1 LAB FACILITIES REQUIRED

Like any other tissue culture facility, the fish cell culture lab should also emphasize the need to maintain sterile conditions (Freshney, 2000). The ideal tissue culture facility should have an area for preparing primary cultures separate from the maintenance and experimental areas to prevent contamination. If this is not feasible a minimum of two laminar flow hoods located as far away as possible from each other is desirable (and never facing each other). One flow hood can be designated for work involving preparation of primary cultures and the other for routine maintenance and testing. Ideally, the latter should be a level 2 biosafety cabinet as opposed to the primary hood which can be a level 1 cabinet.

In addition to the working hoods, the facility must have a sink located near the entrance of the room for washing, and a low bench ideally in the middle of the room where an inverted phase contrast microscope can be placed. A centrifuge, fridge, freezer and aspirator are also needed. An incubator is desirable but not needed as most fish cells grow well at room temperature (Bols and Lee, 1991; 1994).

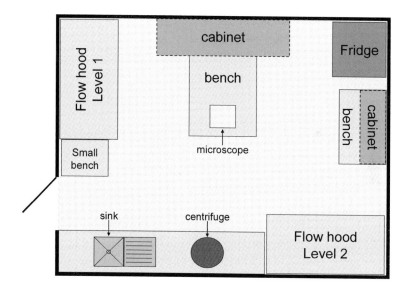

Figure 1. Minimal requirements for a small self-contained fish tissue culture lab facility.

Glassware washing and autoclaving facilities should be located nearby. Figure 1 depicts a small self-contained fish tissue culture lab. The laboratory should have restricted access and have the basic requirements to be designated Containment Level 2 as indicated in the Laboratory Biosafety Guidelines (Health Canada, 1996). These are the optimal laboratory requirements and are the same as for mammalian cell lines, but fish cell lines are easier to maintain than mammalian lines because the lower temperature requirements means incubators are not absolutely needed and microbial contamination is less frequent.

6.2 MATERIALS

Box 2. Required materials for cytotoxicity testing with RTgill-W1.

- RTgill-W1 cell culture (CRL-2523, ATCC) in a 75 cm^2 tissue culture flask.
- Leibovitz's L-15 complete medium (with FBS – see Section 6.5.1 for recipe).
- 0.53 mM Versene (EDTA; Life Technologies) diluted 1:5000 (1 x 0.2 g tetrasodium EDTA/L in PBS).
- Trypsin solution (see Section 6.5.1 for recipe).
- 75 cm^2 tissue culture flask.
- 15 mL centrifuge tubes.
- 9" Pasteur pipettes to be stored in autoclavable pipette boxes.
- 10 mL transfer pipettes (glass or sterile disposable – graduated) to be stored in autoclavable pipette boxes.
- Digital micropipette and glass capillary tubes for toxicant dispensing.
- 70% ethanol solution.
- Microwell plates (sterile, disposable): either 48- or 96-well format.

6.3 EQUIPMENT

Box 3. Equipment required to culture fish cells in the laboratory.

- Laminar flow hood, either horizontal or vertical
- Inverted phase-contrast microscope
- Vacuum aspirator
- Incubator
- Centrifuge
- Pipettor
- Hemacytometer
- Fridge

6.4 WASHING OF GLASSWARE

Glassware for cell culture must be washed in a mild laboratory detergent and should be scrubbed thoroughly and rinsed 5–7 times with hot tap water and then rinsed 3–5 times in deionized water. The glassware is then left to air dry and is sterilized with an autoclave for 30 minutes at a temperature of 121°C and pressure of 20 PSI. The autoclaved glassware is then further dried for an additional 3 hours at 90°C to remove all condensation within the glassware.

Glassware that contains toxicants and is to be reused must be washed after soaking the glassware in an acid detergent for 24 hours. This is then followed by rinses as done for cell culture glassware with autoclaving and drying as described above.

6.5 PREPARATION OF REAGENTS AND CELL CULTURE MEDIA

6.5.1 Culture maintenance

Leibovitz's L-15 complete medium containing FBS. RTgill-W1 cells are grown on the plastic surfaces of flasks and microwells in the basal medium, Leibovitz's L-15, supplemented with fetal bovine serum (FBS). This is prepared by adding aseptically 50 mL of FBS (Sigma) to a 500 mL bottle of L-15 (Sigma), which gives a solution that is commonly described as being 10% FBS. The growth medium is completed by adding 10 mL of penicillin/streptomycin (100 IU/mL penicillin, 100 μg/mL streptomycin; Sigma) to this solution and can be stored at 4°C for months.

Versene solution. A rinse with versene (0.53 mM ethylenediaminetetraacetic acid or EDTA) is used to begin the process of removing RTgill-W1 cells from the plastic growth surface, which is necessary to initiate new cultures either in flasks or microwells. Versene can be purchased as a ready to use solution (Life Technologies). Rinsing with versene chelates and removes divalent cations, allowing trypsin to function.

Trypsin solution. Trypsin detaches the RTgill-W1 cells from the growth surface. The trypsin solution is prepared by aseptically dissolving 100 mg of trypsin (Sigma) into 10 mL of Ca^{2+}- and Mg^{2+}-free Hank's balanced salt solution (Sigma) to make up a trypsin stock solution. Once dissolved, dispense 0.5 mL of the stock solution into 9.5 mL of Ca^{2+}- and Mg^{2+}-free Hank's balanced salt solution. Keep the trypsin solution sterile, store at –20°C for up to 1 year.

7. Preparation of RTgill-W1 for toxicity testing

7.1 ROUTINE CELL CULTURING

Since the fish cell lines are immortal they can be continuously cultured so that there is an endless number of cells/flasks that can be used to continue the stock culture as well as to initiate the toxicity tests.

(1) Switch on the laminar flow hood (either horizontal or vertical). All surfaces must be sprayed with a 70% ethanol solution and wiped clean. Make sure that the laminar flow hood is functioning properly with suitable air flow to ensure maximal sterility. Place all needed

(2) Under an inverted phase-contrast microscope examine the confluent flask of RTgill-W1 cells. Check that the flask is free of contamination or unexpected rounding and detachment of cells. The cells that are to be passaged should have a normal morphology and are a confluent culture (see Fig. 2).

(3) In the laminar flow hood, remove the cap of the culture flask and aspirate the old medium using a Pasteur pipette attached to a vacuum aspirator. Dispense 1.5 mL of versene to the flask and swirl around gently to cover the entire bottom of the flask. Leave on for 1 minute and aspirate off.

(4) Add 1 mL of versene and 1 mL of the trypsin solution to the flask, replace the cap and swirl around gently to cover the bottom of the flask. Observe the cells detaching under the inverted phase-contrast microscope periodically tapping on the side of the flask to assist in detachment. Do not leave the cells in the trypsin solution for greater than 5 minutes as the enzymes may cause cellular digestion resulting in cellular death.

(5) Once the cells have detached add 3 mL of complete Leibovitz's L-15 medium containing FBS. Pipette the medium up and down over the bottom of the flask ensuring that all the cells are detached and resuspended in the medium.

(6) Transfer the cell suspension to a sterile 15 mL centrifuge tube and centrifuge for 5 minutes at 200 x g.

(7) After centrifugation, aspirate the supernatant from the 15 mL centrifuge tube leaving a small amount of media ($\sim$ 0.25 mL) above the pellet, being careful not to aspirate the cell pellet. Re-suspend the pellet into the remaining media by flicking the centrifuge tube.

(8) Add 10 mL of fresh L-15 medium to the centrifuge tube and dispense 5 mL to each of two 75 cm^2 tissue culture flasks and add a further 5 mL of medium to each flask.

(9) Observe the flasks under the phase-contrast microscope to check if the culture has been divided equally and that the cells are in a single cell suspension.

(10) Place the flasks in an incubator at 18° to 22°C. When the cultures are confluent (7 to 10 days) the cells can be subcultured or harvested for use in an experiment.

7.2 PREPARATION FOR TOXICITY TESTING

Cells from a confluent flask can be used to initiate cultures for a toxicity test using multiwell tissue-culture plates. Follow the first 7 steps of the routine cell culturing protocol up to re-suspension of the cells in the 15 mL centrifuge tube and continue with the following steps.

(1) To the resuspended cells in the centrifuge tube, add 4 mL of complete L-15 medium with FBS and ensure that the cells are evenly distributed throughout the tube. Count cells using a hemacytometer to determine the density of the cells. Adjust the cell density to 1×10^5 cells/mL if using 48-well plates and to 1.5×10^5 cells/mL if using 96-well plates using fresh medium.

(2) If using a 48-well plate, add 5×10^4 cells in 500 µL of L-15 complete medium with FBS to 40 of the 48 wells, add L-15 complete medium alone to the remaining eight wells. If using a 96-well plate, add 3×10^4 cells in 200 µL of L-15 complete medium with FBS to 84 of the 96 wells, add L-15 complete medium alone to the remaining twelve wells (see Fig. 4).

(3) Once plated, allow the cells to grow for three to four days in the dark at 18° to 22°C to form a confluent cell monolayer for 48-well plates and for two to three days for 96-well plates.

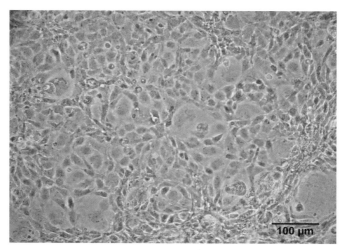

Figure 2. Confluent culture of RTgill-W1 under normal growth conditions viewed under phase contrast microscopy (100 X magnification).

8. Testing procedure

8.1 INFORMATION/GUIDANCE REGARDING TEST SAMPLES PRIOR TO CONDUCTING BIOASSAYS

8.1.1. Chemicals

All chemicals tested should be dissolved in a carrier suitable for the characteristics for that particular chemical. Dissolving chemicals in water or culture medium is ideal; however the use of ethanol or DMSO as a carrier for compounds that have low water-solubility may be necessary. When using an organic solvent, the working solutions must be at least 200 times the final concentration desired for exposure. Care must be taken when dosing cells with chemicals dissolved in DMSO/ethanol to prevent damage to the cells by the carrier alone.

8.1.2 Whole-water samples

Storage and preparation of whole-water sample. Upon receipt from the sample source, the sample should be kept at 4°C in the dark. The sample should be tested as soon as possible upon receipt due to possible degradation of potential toxicants.

The osmolality of raw whole-water samples is too low to support viability of the RTgill-W1 cell cultures, and needs to be increased to the levels of culture media (~300 mOsmkg^{-1}). In order to raise the osmolality, the salts, galactose and pyruvate of L-15 medium are added to the sample (Figure 3; Dayeh et al., 2002). This minimal medium is known as L-15/ex (Schirmer et al., 1997). At least 100 mL of sample is required because the amount of solid L-15 constituents required for this volume is the smallest that can be weighed out conveniently and accurately. See Table 2 for the amounts needed for 250 mL. Osmolality can be measured in the laboratory with an osmometer. Some work on the principle of freezing point depression; others, vapor pressure. We have routinely used the Westcor 5001B vapor pressure osmometer (Westcor, Utah, USA). Measure the raw osmolality of the whole-water sample. If the osmolality is below 90 mOsmkg^{-1}, the sample receives the normal salt concentrations of the constituents of L-15/ex (Tab. 2). Samples that have an osmolality above 90 mOsmkg^{-1} (up to a maximum of 120 mOsmkg^{-1}) receive 80% of the normal salt concentrations (Table 2).

Table 2. Measurements of L-15/ex salt constituents to add to 250 mL of raw whole-water sample.

L-15/ex salt constituents	For samples below 90 mOsmkg^{-1} (in grams)	For samples above 90 mOsmkg^{-1} (max. of 120 mOsmkg^{-1}) (in grams)
NaCl	2.0	1.6
KCl	0.1	0.08
MgSO$_4$	0.05	0.04
MgCl$_2$	0.05	0.04
CaCl$_2$	0.035	0.028
Na$_2$HPO$_4$	0.0475	0.038
KH$_2$PO$_4$	0.015	0.012
Galactose	0.225	0.18
Pyruvate	0.1375	0.11

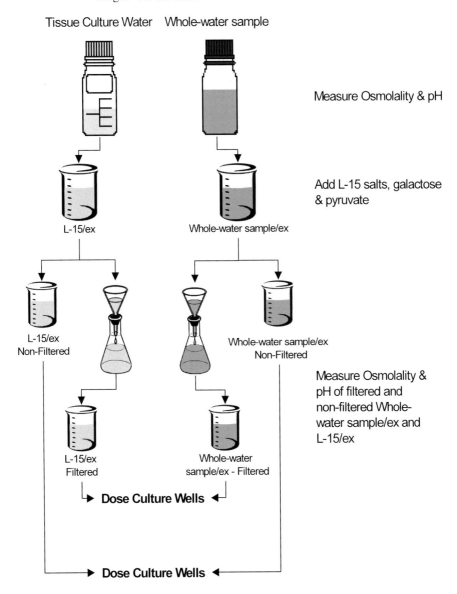

Figure 3. Scheme of L-15/ex and whole-water sample/ex preparation. Tissue culture salts are dissolved into commercial tissue culture water to give rise to L-15/ex and into whole-water samples to give rise to whole-water sample/ex. These solutions are used to dose confluent cultures of RTgill-W1 in wells of either 48-well or 96-well microplates (from Dayeh et al., 2002).

8.2 EXPERIMENTAL CONFIGURATION/DESIGN

The experimental configuration chosen for undertaking the RTgill-W1 cytotoxicity assay depends on the number of wells in the multiwell plate (see Fig. 4). Both 48- and 96-microwell plates can be used as these can be accepted by several multiwell plate readers. For both 48- or 96-well microplates a no-cell control is recommended which receives the various concentrations of sample in the absence of cells. This is done to observe if there are any interactions between the samples and the fluorescent indicator dyes used to determine the endpoint measurement of cell viability. The no-cell control also allows detection of some microbial contamination in whole-water samples. A positive control is also needed for each test configuration. The configuration described for 48-well plates allows eight concentrations of toxicant with four replicates including a control with four replicates. One row of the 48-well plate is dedicated to eight concentrations of a positive control toxicant including a control. The template for 96-well plates allows either twelve concentrations of toxicant including a control with six replicates each. Twelve concentrations of a positive control toxicant including a control are also included in this plate configuration. A 96-well plate can also be sub-divided into two 48-well configurations with eight concentrations of toxicant with four replicates including a control with four replicates. In this configuration two different samples can be tested on the same microwell plate. Two rows of the 96-well plate can be used for exposure to eight concentrations of a positive control toxicant.

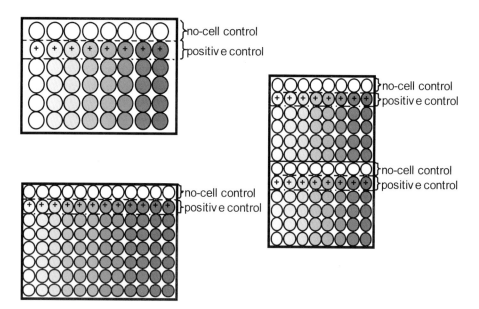

Figure 4. Suggested experimental configuration of 48-well and 96-well microplates for fish cell toxicity testing.

8.3 TEST SAMPLE CONCENTRATIONS

8.3.1 *Chemicals*

The optimal test concentration will vary depending on the chemical that is being investigated. When testing a new chemical whose toxicity is not known a good starting point is to test a broad concentration range. In order to determine the optimal concentration of the chemical to test conduct exposures on a log scale of concentrations (*i.e.,* 0.0001, 0.001, 0.01, 0.1, 1.0, 10, 100, etc.). Testing on this scale will narrow the range of exposure concentrations to be used for further experiments. Dissolving the chemicals in water or tissue culture solution is ideal and diluting in the wells is not necessary. However, when using an organic solvent, the working solutions must be at least 200 times the final concentration desired for exposure (*i.e.,* 1 µL of the 200 times concentrated test chemical into 200 µL medium in the well). A concentrated stock solution is necessary only when the chemical to be tested is dissolved in an organic solvent or carrier, which may damage the cells if the concentration of the solvent in the well is too high.

Abietic acid dissolved in L-15/ex medium can be used as a positive control for the chemical tests. This will allow the comparison of cell response between plates/runs/days. Dissolve abietic acid (85% purity, Acros Organics through Fisher Scientific) into the L-15/ex medium at a concentration of 100 µg/mL. The concentration series for the positive control should be as follows: 0, 15, 30, 45, 60, 75, 90 and 100%.

8.3.2 *Whole-water samples*

The highest concentration of a whole-water sample that can be tested is 100%. This is due to the addition of L-15/ex constituents as solids to the whole-water sample. It is recommended that a concentration series of increments of 15% be tested (*i.e.,* 0, 15, 30, 45, 60, 75, 90 and 100%) for the whole-water sample; however, other dilution series can be used such as 0, 5, 10, 30, 50, 70, 90 and 100%. These are prepared in sterile glass vials, with dilutions prepared using previously made L-15/ex. Cells that are not exposed to the whole-water sample will serve as the control cells (*i.e.,* 0%). Also a positive control should be used when exposing cells to a whole-water sample (Fig. 5; Dayeh et al., 2002). Dissolve abietic acid into the 100% whole-water sample at a concentration of 100 µg/mL. A concentration series for the positive control should follow that of the whole-water sample (*i.e.,* 0, 15, 30, 45, 60, 75, 90 and 100%).

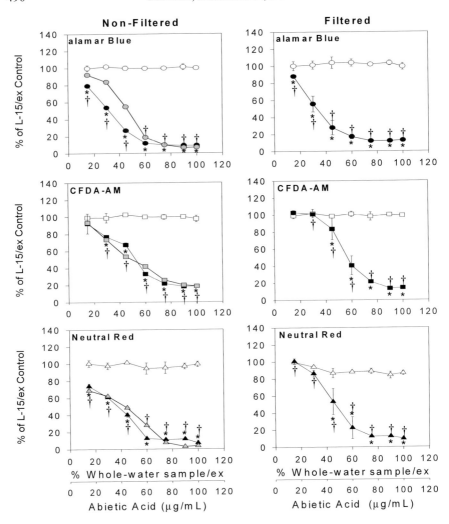

Figure 5. Viability of RTgill-W1 cultures after 24 h in abietic acid. A non-toxic whole-water sample was spiked with 100 μg/mL abietic acid in DMSO (closed symbols) or with DMSO alone (open symbols) and either filtered (right panels) or not (left panels). Abietic acid in L-15/ex was tested simultaneously (shaded symbols). Whole-water sample preparations were mixed in culture wells with various volumes of L-15/ex in order to obtain a dose-range of abietic acid-spiked whole-water sample/ex (closed symbols) or appropriate DMSO control (open symbols). Cell viability was assessed with three indicator dyes, alamar Blue (circles), CFDA-AM (squares) and neutral red (triangles). Asterisks denote the % of abietic acid-spike whole-water sample/ex that resulted in fluorescence units different than those in L-15/ex controls (one-way ANOVA followed by Dunnett's test, $\alpha = 0.05$). The † symbol indicates the % abietic acid spiked whole-water sample/ex that yielded fluorescence unit readings significantly different from DMSO-spiked control (unpaired t-test, $\alpha = 0.05$) (from Dayeh et al., 2002).

8.4 DISPENSING SAMPLE, RTgill-W1 AND EXPOSURE SOLUTIONS

8.4.1 Sample dispensing

The sample is dispensed in either a 48- or 96-microwell plates that have a confluent monolayer of RTgill-W1 cells. The cell will have been growing for approximately 3 days after plating to achieve confluency. These cultures of RTgill-W1 cells will be exposed to either dilutions of concentrated putative toxicant or the prepared whole-water sample.

8.4.2 Chemicals

(1) Turn on the vertical laminar flow hood and wipe all surfaces down with 70% ethanol solution. Place all needed equipment in the laminar flow hood and wipe each item with a 70% ethanol soaked paper towel.

(2) Remove the L-15 complete medium, which the cells have been plated in by inverting over a catch basin and blotting on a stack of paper towels.

(3) The remaining L-15 media must be removed with an L-15/ex rinse. To each well add 500 µL of L-15/ex to wells of a 48-well plate or 200 µL in a 96-well plate. Remove the L-15/ex rinse by inverting the plate over a catch basin and blotting on a stack of paper towels.

(4) The cells will be exposed to the chemicals in the L-15/ex medium. Thus, add 500 µL/well of L-15/ex to wells of a 48-well plate or 200 µL/well in a 96-well plate.

(5) Dose the cells with a 200 times concentrated working solutions of the desired test chemical when dissolved in an organic solvent. Using a positive displacement micropipette with a glass capillary tip, add 2.5 µL to wells of a 48-well plate or 1 µL to wells of a 96-well plate. When dosing, dispense the test compound above the level of the liquid in each well. The surface tension will ensure that the test chemical is dispersed evenly within the well and avoid any damage to the cells due to the organic solvent. Larger volumes for dosing can be used when the chemical is dissolved in either water, tissue culture medium or L-15/ex medium.

(6) Once all the wells have been dosed with the appropriate test compound and controls, wrap the edges of the plate with a thin strip of Parafilm M® to seal the edges and minimize evaporation.

(7) Expose the plates in the dark at 18° to 22°C for a period of 24 hours. Longer exposures up to 96 hours can also be conducted with this technique.

(8) Assess viability after the exposure using the alamar Blue/CFDA-AM and Neutral Red indicator assays as described below.

8.4.3 Whole-water samples

(1) Turn on the vertical laminar flow hood and wipe all surfaces down with 70% ethanol solution. Place all needed equipment in the laminar flow hood and wipe each item with a 70% ethanol soaked paper towel.

(2) Remove the L-15 complete medium, which the cells have been plated in by inverting over a catch basin and blotting on a stack of paper towels.

(3) The remaining L-15 media must be removed with an L-15/ex rinse. To each well add 500 µL of L-15/ex to wells of a 48-well plate or 200 µL in a 96-well plate. Remove the L-15/ex rinse by inverting the plate over a catch basin and blotting on a stack of paper towels.

(4) Using the dilution series of the prepared whole-water sample/ex, add 500 µL/well to wells of a 48-well plate or 200 µL/well in a 96-well plate. Add L-15/ex to each well that will serve as the control wells.

(5) Once all the wells have been dosed with the whole-water sample and controls, wrap the edges of the plate with a thin strip of Parafilm M® to seal the edges and minimize evaporation.

(6) Expose the plates in the dark at 18° to 22°C for a period of 24 hours. Longer exposures up to 96 hours can also be conducted with this technique.

(7) Assess viability after the exposure using the alamar Blue/CFDA-AM and Neutral Red indicator assays as described below.

9. Post-exposure observations/measurements and endpoint determinations

9.1 MICROSCOPIC OBSERVATIONS

Examine the cell cultures upon termination of the experiment. The cell cultures in the tissue culture plate can be observed using an inverted phase contrast microscope. Note the general appearance of the cultures across the various concentrations of toxicant tested taking note of any morphological changes.

9.2 MEASUREMENT ENDPOINT DETERMINATION

Viability of the cell cultures after exposure to a potential toxicant is measured using fluorescent indicator dyes. Due to the use of multiwell plates, fluorescence levels are determined using a fluorescent multiwell plate reader. There are a few manufacturers of fluorescent multiwell plate readers, which have either fixed excitation and emission filters (such as the CytoFluor, Applied Biosystems) or varying excitation and emission filters (such as the SpectraMax Gemini, Molecular Devices). These plate readers are designed to accept multiwell plates from various manufacturers. As well, the plates can be read either with or without a lid when read using multiwell plate reader configured as a bottom reader (*i.e.*, the CytoFluor 4000). However, if the multiwell plate reader is configured as a top reader (*i.e.*, the SpectraMax) the plate lid must be removed before reading the plate.

Three fluorescent indicator dyes are used to measure the viability of RTgill-W1 cultures after treatment to a toxicant. These are alamar Blue for metabolic activity, 5-carboxyfluorescein diacetate acetoxymethyl ester (CFDA-AM) for membrane integrity, and neutral red for lysosomal function. These three dyes can be used with cells in one microwell plate, this allows for three endpoint determinations on the same set of cells.

9.2.1 Alamar Blue assay

Alamar Blue (Immunocorp) is a commercial preparation of the dye resazurin (O'Brien et al., 2000) and is used to assess metabolic activity in cell cultures. Resazurin is a non-fluorescent dye that once reduced by metabolically active cells becomes the fluorescent product resorufin. It comes in pre-mixed solutions of 25 mL and 100 mL volumes ready to be prepared as a working solution to be applied to cells (alamar Blue and CFDA-AM dyes can be mixed into one working solution as these two fluorescent dyes have different excitation and emission wavelengths).

(1) Turn on the laminar flow hood and wipe all surfaces with 70% ethanol solution. Keep the light off in the flow hood.

(2) Make a 5% (v/v) working solution of alamar Blue in L-15/ex. Keep in an amber glass vessel to prevent light degradation of the dye.

(3) Remove the exposure medium from the plates. This can be done by inverting the plate over a catch basin and blotting on a stack of paper towels to drain the plates further, or careful aspiration of each well using a Pasteur pipette with a vacuum aspirator. It is recommended to invert over a catch basin if the entire plate is to be assessed at one time.

(4) Add 100 to 150 μL of the 5% alamar Blue working solution to each well of a 48-well plate, or 50 to100 μL to each well of a 96-well plate. Volumes will depend on the type of fluorescent multiwell plate reader used. In general, the bottom of the culture well must be completely covered with the working solution.

(5) Incubate the plates in the dark for 30 min at 18° to 22°C.

(6) Place plate in plate carrier of multiwell plate reader. Assess fluorescence of alamar Blue using excitation and emission filters of 530 and 590 nm respectively. Depending on the plate reader, removal of the plate lid may be necessary.

9.2.2 CFDA-AM assay

The 5'-carboxyfluorescein diacetate acetoxymethyl ester (CFDA-AM, Molecular Probes) is used to measure cell membrane integrity. CFDA-AM rapidly diffuses into cells and is converted from a non-polar, non-fluorescent dye into a polar, fluorescent dye 5'-carboxyfluoroscein (CF) by non-specific esterases present in living cells

(CFDA-AM and alamar Blue dyes can be mixed into one working solution as these two fluorescent dyes have different excitation and emission wavelengths).

(1) Turn on the laminar flow hood and wipe all surfaces with 70% ethanol solution. Keep the light off in the flow hood.

(2) Dissolve CFDA-AM in sterile DMSO to make a 4 mM stock solution. Dispense in small aliquots in sterile 0.5 mL microcentrifuge tubes to prevent degradation from thawing and refreezing. Wrap in aluminum foil to prevent light degradation. Store in a -20°C defrost cycle free freezer in a dessicator to prevent ester hydrolysis due to moisture for up to 1 year.

(3) Prepare a 4 µM working solution of CFDA-AM by diluting the 4 mM CFDA-AM stock solution 1:1000 in L-15/ex. Keep in a glass amber vessel to prevent light degradation of the dye.

(4) Remove the exposure medium from the plates. This can be done by inverting the plate over a catch basin and blotting on a paper towel to drain the plates further, or aspiration of each well using a Pasteur pipette with vacuum aspiration. It is recommended to invert over a catch basin if the entire plate is to be assessed at one time.

(5) Add 100 to 150 µL of the 4 µM working solution to each well of a 48-well plate, or 50 to 100 µL to each well of a 96-well plate. Volumes will depend on the type of fluorescent multiwell plate reader used. In general, the bottom of the culture well must be completely covered with the working solution.

(6) Incubate the plates in the dark for 30 min at 18° to 22°C.

(7) Place plate in plate carrier of multiwell plate reader. Assess fluorescence of CF using excitation and emission filters of 485 and 530 nm respectively. Depending on the plate reader, removal of the plate lid may be necessary.

9.2.3 Alamar Blue and CFDA-AM assay

As these two dyes have different excitation and emission wavelengths, they can be combined together to assess two endpoints of cell viability concurrently (Ganassin et al., 2000). To perform these two assays together, prepare a 5% (v/v) working solution of alamar Blue in L-15/ex and then dilute the CFDA-AM stock solution in DMSO (4 mM) 1:1000 in the prepared alamar Blue working solution. Add this working solution to the cells as described above.

9.2.4   Neutral red assay

Neutral red (3-amino-7-dimethylamino-2-methylphenazine hydrochloride) is a weakly basic fluorescent dye that is used to measure lysosomal function. Neutral red accumulates in acidic compartments such as lysosomes and can be applied before or after toxicant exposure (as described here) to measure neutral red release or uptake respectively. Note again that this assay can be done on a separate set of cells or on the same cells that have previously been investigated using alamar Blue and CFDA-AM. However, inasmuch as cell cultures are terminated during the NR assay, this assay always has to be carried out last.

(1) Turn on the laminar flow hood and wipe all surfaces with 70% ethanol solution. Keep the light off in the flow hood.

(2) Dissolve 3.3 mg of neutral red powder (Sigma) per mL of Dulbecco's PBS (D-PBS; Sigma or Life Technologies) in a glass amber vial. Pass dissolved neutral red through a 0.2 µm filter. Store this stock solution for up to 1 year at 4°C. Neutral red can also be purchased as a 3.3 mg/mL stock solution in D-PBS (Sigma).

(3) Prepare a 33 µg/mL working solution of neutral red by diluting the stock solution 1:100 in L-15/ex. Keep in a glass amber vessel to prevent light degradation of the dye.

(4) Remove the exposure medium from the plates. This can be done by inverting the plate over a catch basin and blotting on a stack of paper towels to drain the plates further, or aspiration of each well using a Pasteur pipette with vacuum aspiration. It is recommended to invert over a catch basin if the entire plate is to be assessed at one time.

(5) Add 100 to 150 µL of the 33 µg/mL working solution to each well of a 48-well plate, or 50 to 100 µL to each well of a 96-well plate. The bottom of the culture well must be completely covered with the working solution.

(6) Incubate the plates in the dark for 60 min at 18° to 22°C.

(7) Invert the plate over a catch basin and blot on a stack of paper towels to remove the neutral red working solution. Ensure removal of excess neutral red in each well.

(8) Rinse wells once with 100 to 150 µL to each well of a 48-well plate, or 50 to 100 µL to each well of a 96-well plate of the neutral red fixative solution: 0.5% (v/v) formaldehyde and 1% (w/v) $CaCl_2$ in deionized, distilled water; stored in the dark for up to 1 year. Remove neutral red fixative after 1 min by inverting the plate over a catch basin and blot on a stack of paper towels.

(9) Add 100 to 150 µL to each well of a 48-well plate, or 50 to 100 µL to each well of a 96-well plate of the neutral red extraction solution: 1% (v/v) acetic acid and 50% (v/v) ethanol in deionized, distilled water; stored in the dark for up to 1 year. Volumes will depend on the type of fluorescent multiwell plate reader used. In general, the bottom of the culture well must be completely covered with the working solution.

(10) Place plate on an orbital shaker and shake at ~ 40 rpm for 10 min. This will ensure solubilization of the neutral red accumulated in the lysosomes.

(11) Place plate in plate carrier of multiwell plate reader. Assess fluorescence of neutral red using excitation and emission filters of 530 and 645 nm respectively. Depending on the plate reader, removal of the plate lid may be necessary.

9.2.5 Data analysis – calculation of EC50

Upon completion of the cell viability assays the raw fluorescent units are used to evaluate the toxicity of the chemical being tested. Cell viability is expressed as a percent of non-toxicant exposed cells (% of control). For each concentration of toxicant, there is one well that has no cells in it (no-cell control) whereas all the remaining wells have cells in them. Prior to calculating % of control, subtract the fluorescent units (FU) for wells without cells from the experimental (ex.) and control (con.) values with cells. Cell viability (as % of control) can be calculated using the following formula:

$$\% \text{ of control} = (FU_{ex.cells} - FU_{ex.no\ cells}) \times 100 / (Average\ [FU_{con.} - FU_{con.no\ cells}]) \quad (1)$$

Data for each well of each concentration are expressed as a % of Control. Then, the average and standard deviation for each concentration is calculated. The data can then be plotted as % control on the y-axis versus concentration on the x-axis. These values are used to calculate the EC50 for the toxicant.

A sigmoid relationship is characteristic of dose-response data and thus can be analyzed by a nonlinear regression in most graphing software such as SigmaPlot (Jandel Scientific). The data is fitted to the four-parameter logistic function for continuous response data. The logistic function is:

$$y(d) = Y_{min} + (Y_{max} - Y_{min}) \{1 + exp[-g(ln(d) - ln\ EC50)]\}^{-1} \quad (2)$$

where y(d) is the % cell viability at the dose d, Y_{min} is the minimum % cell viability, Ymax is the maximum % cell viability, g is a slope parameter, EC50 is the dose that produces 50 % of cell viability.

Inasmuch as cell viability data are expressed on a 0 - 100 % basis, the four-parameter equation simplifies to a two-parameter equation because Y_{max} and Y_{min} are constants of 100% and 0% respectively:

$$y(d) = 0\ \% + (100\ \% - 0\ \%) \{1 + exp[-g(ln(d) - ln\ EC50)]\}^{-1}$$

10. Factors capable of influencing performance of rtgill-w1 test

10.1 EXPOSURE MEDIUM

The recommended exposure medium is L-15/ex. The RTgill-W1 cell line can survive in L-15/ex for at least 101 hours (Schirmer et al., 1997). Exposure in complete L-15 medium with or without a serum supplementation can reduce the toxicity due to the reduced bioavailability of the chemical that is being evaluated (Hestermann et al., 2000; Schirmer et al., 1997; Dayeh et al., 2003a).

10.2 DOSING METHOD FOR CHEMICALS

As numerous chemicals need to be dissolved in solvents such as DMSO or ethanol, care must be taken when dosing RTgill-W1 cultures with these solutions. Presentation of these toxicants to the cells must be conducted in such a manner as to not damage the cells due to the carrier solvent alone. This can be accomplished by using micropipettes to add small volumes (≤10 μL) of the toxicant in carrier solvent to the medium over the cells in microwells. Dispense the droplet of the carrier solution from the micropipette above the level of the medium surface and touch this droplet to the surface. This allows the surface tension to disperse the carrier solvent rapidly and evenly throughout the culture well. Failing to do this near the surface can result in a blob of DMSO falling directly onto the cell monolayer and causing the immediate death of all or part of the monolayer.

10.3 WHOLE-WATER SAMPLES

Complex samples might limit the application of whole-water samples to RTgill-W1 cultures. Complexities could include excessive microbes, precipitates, suspended particulates, and colour. Most of these problems might be overcome by adding a filtration step, which could have the detrimental effect of removing toxicants. Another problem would be hyperosmotic samples, which would necessitate diluting the sample. As mentioned in Section 11 on case studies, the full range of problems that might arise from the whole-water approach has yet to be identified.

11. Application in a case study

Testing for cytotoxicity to RTgill-W1 cells can be used to compare the toxic potency of individual chemicals and to evaluate the toxicity of whole-water samples. The methods for these two purposes are very similar. The procedure for single compounds has been presented in a previous publication (Dayeh et al., 2003b), whereas the procedure for whole-water samples is detailed in this chapter. Presented below are the chemical classes that have been examined and a discussion of a case study with paper mill effluent.

11.1 CHEMICALS

RTgill-W1 cells have been used to evaluate the cytotoxicity of PAHs, phenolics, and the surfactants, abietic acid and Triton X-100 (Schirmer et al., 1998a; Dayeh et al., 2002; 2004). An advantage of performing cytotoxicity tests in L-15/ex is being able to also test compounds for their potential to be photocytotoxic without interference from medium components. The killing of cells by concurrent exposure to a chemical and ultraviolet light (UV) is photo-cytotoxicity, and L-15/ex contains no medium components that by themselves are photo-cytotoxic. RTgill-W1 cells have been used to determine the photocytotoxicity of PAHs and creosote (Schirmer et al., 1998b; Schirmer et al., 1999). Six of sixteen PAHs were photo-cytotoxic at concentrations theoretically achievable in water (Schirmer et al., 1998b). In all these studies, toxicity has been evaluated using the alamar Blue, CFDA-AM and neutral red viability assays and calculating EC50s in order to compare the results.

11.2 WHOLE-WATER SAMPLES

The RTgill-W1 cell line bioassay has been used successfully to evaluate the toxicity of samples collected from a paper mill over a year of operation (Dayeh et al., 2002). In total, thirty-one whole-water samples were tested for their cytotoxicity to RTgill-W1 cells. Of these thirty-one samples, eleven were also tested by the conventional 96-h whole rainbow trout lethality bioassay, eighteen, by the *Daphnia* lethality bioassay. There was no correlation between the *Daphnia* and the RTgill-W1 test results. Eleven samples were toxic to *D. magna* but not to the gill cell line. Thus the Daphnia test has a greater sensitivity to something in the water samples, perhaps heavy metals, than RTgill-W1 test. Only one sample, number 28, was toxic to rainbow trout as evaluated by the 96-h lethality bioassay. This was the only sample of thirty-one that was cytotoxic to RTgill-W1 (Fig. 6). Thus the correlation between tests with rainbow trout and the rainbow trout cell line was excellent, suggesting that the fish cell line bioassay is a promising alternative to the use of whole fish in the routine toxicity testing of whole-water samples. However, this successful case study raises a number of issues and suggestions for future developments.

Firstly, the cytotoxicity of sample 28 was complex (Fig. 6, Dayeh et al., 2002). All three viability assays indicated that sample 28 was cytotoxic, but the results with neutral red had a high standard deviation and indicated more cytotoxicity than the other two assays. Surprisingly, when the sample was filtered, the neutral red assay no longer detected a decline in cell viability with an increasing % of whole-water sample/ex. On the other hand, alamar Blue detected more cytotoxicity in the filtered sample. This complexity is hard to explain, although some possible mechanisms were advanced in Dayeh et al. (2002). The results suggest two recommendations for future screens of industrial effluent. Firstly, more than one endpoint of cell viability should be tested. Secondly, both filtered and non-filtered sample should be tested. In this way, if a toxicant is removed by filtration, it should be detected with the non-filtered sample. Secondly, RTgill-W1 seems less sensitive than rainbow trout to the one toxic sample, number 28. All 10 rainbow trout died in the 96-h lethality test, whereas the reduction in RTgill-W1 cell viability was at the most only by about 55%. One possible explanation for this difference is that the toxicant(s) require more time than the 24-h of the *in vitro* tests to elicit their full toxicity. Another possibility is that the particular toxicant(s) in this sample are more potent at the organism level than the cellular level. Toxicants that target specific organ systems, such as the nervous system, might fit into this category.

Several avenues of research could be explored in the future to improve the sensitivity. One would be to expose RTgill-W1 cultures to samples for a longer period. However, as a routine practice, this is not desirable because microbial contamination is more likely to appear and overwhelm the fish cell cultures. Sensitivity might also be improved by using different or additional cellular endpoints for evaluating cell viability. A long-term solution might be to genetically engineer RTgill-W1 to be more sensitive to cytotoxicants.

Sample preparation also could be the key to sensitivity, but this will likely vary with the nature of the sample. The amount of microbes in the water sample will dictate on how essential a filtration step is. In turn, how much of a potential toxicant is adsorbed to filterable particulates will determine how filtration interferes with sensitivity. A surprising feature of the paper mill study is that the only toxic sample was from the 'clean water bypass', which is the water that is used in cooling the plant and will ultimately receive effluent, although the mechanisms behind its cytotoxicity might be complex (Dayeh et al., 2002).

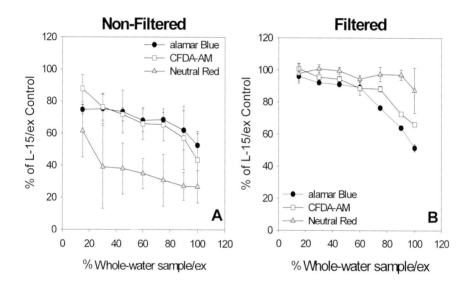

Figure 6. Viability of RTgill-W1 cultures after 24 h in whole-water sample 28 from a paper mill. The sample had been filtered through a 0.2 µm filter in Panel B but not in Panel A. Cell viability was assessed with three indicator dyes, alamar Blue, CFDA-AM and Neutral Red (from Dayeh et al., 2002).

12. Accessory/miscellaneous test information

Although a single cell viability assay might be considered as time and resources would be saved, multiple cell-viability assays are recommended because some endpoints might be less sensitive to certain toxicants than others. For example, when RTgill-W1 cells were exposed to pentachlorophenol, a dose-dependent decline in cell viability was observed with alamar Blue and neutral red, but not with CFDA-

AM (Dayeh et al., 2002). In this case pentachlorophenol seemed to impair energy metabolism more profoundly than plasma membrane integrity.

Some non-filtered samples increased alamar Blue readings as the percentage of whole-water sample increased (Dayeh et al., 2002). Although the magnitude was not large, the increase was statistically significant. These results occurred with samples that were not toxic to *Daphnia*, rainbow trout, or RTgill-W1. Why the values increased is a matter of speculation, but microbial contamination seems a likely source. Whether such increases could interfere with the detection of cytotoxicity with alamar Blue is unclear at this time.

13. Conclusions/prospects

Evaluating the toxicity of water samples by measuring their capacity to cause cytotoxicity in microwell cultures of the rainbow gill cell line, RTgill-W1, has several advantages and successes with some kinds of samples. The attractive features include cost. This method requires milliliters instead of tens of liters of effluent shipped from distant industries to central testing facilities. The time for the response of exposure to the effluent is only 24 hours as opposed to 96 hours, which in turn reflects the cost of labour. The approximate cost per sample is around $15 - 20 Canadian. Furthermore, routine cell culturing techniques done in house will keep an endless supply of cells to complete the tests instead of having to purchase rainbow trout. The use of fish cell cultures as an alternative to whole fish also satisfies the societal goal to reduce the use of animals in toxicity testing. To date, the procedure has been used successfully with paper mill samples (Dayeh et al., 2002). With these samples, the RTgill-W1 test would be a powerful tool in a program of toxicity identification evaluation (TIE).

Additional studies are needed in order to understand and validate the capability of the RTgill-W1 procedure. More samples that are toxic to rainbow trout need to be examined with the RTgill-W1 cells. In this way, enough examples could be obtained to allow a statistical test of the correlation between toxicity to rainbow trout with cytotoxicity to RTgill-W1. Different types of whole-water samples need to be examined with the RTgill-W1 procedure. Some kinds of effluents might be less successful because of the amount of particulate material or microbes or both and procedures to overcome these problems will have to be developed. As well, some effluents might need to be concentrated or extracted in order for cytotoxicity to be detected. Ultimately, with validation, the RTgill-W1 procedure could be combined with other microplate assays of this book to be part of a battery of tests to routinely appraise the quality of water samples.

Acknowledgements

The research was supported by the Natural Science and Research Council of Canada (NSERC), Centre for Research in Earth & Space Technologies (CRESTech), and the Canadian Water Network (CWN).

The authors wish to thank the copyright holder for permission to reproduce the following: Figures 3, 5 and 6, Elsevier Science, Amsterdam, Netherlands. Reproduced from: Dayeh, V.R., Schirmer, K. and Bols, N.C. (2002) Applying whole-water samples directly to fish cell cultures in order to evaluate the toxicity of industrial effluent. *Water Research* **36**, 3727-3728 (Figures 1, 4 and 6, unaltered).

References

Babich, H. and Borenfreund, E. (1991) Cytotoxicity and genotoxicity assays with cultured fish cells: a review, *Toxicology In vitro* **5**, 91-100.

Bols, N.C. and Lee, L.E.J. (1991) Technology and uses of cell cultures from the tissues and organs of bony fish, *Cytotechnology* **6**, 163-187.

Bols, N.C. and Lee, L.E.J. (1994) Cell lines: availability, propagation and isolation. in: P.W. Hochachka and T.P. Mommsen (eds.), *Biochemistry and Molecular Biology of Fishes*, Vol. 3. Elsevier Science, Amsterdam, Netherlands, pp. 145-149.

Bols, N.C., Barlian, A., Chirino-Trejo, M., Caldwell, S.J., Goegan, P. and Lee, L.E.J. (1994) Development of a cell line from primary cultures of rainbow trout, *Oncorhynchus mykiss* (Walbaum), gills, *Journal of Fish Diseases* **17**, 601-611.

Bols, N.C., Dayeh, V.R., Lee, L.E.J. and Schirmer, K. (2005) Use of fish cell lines in the toxicology and ecotoxicology of fish. Piscine cell lines in environmental toxicology, in T.W. Moon and T.P Mommsen (eds.), *Biochemistry and Molecular Biology of Fishes*, Vol. 6, Elsevier Science, Amsterdam, Netherlands (in press).

Borenfreund, E. and Puerner, J.A. (1984) A simple quantitative procedure using monolayer cultures for cytotoxicity assays, *Journal of Tissue Culture Methods* **9**, 7-12.

Castaño, A., Bols, N.C., Braunbeck, T., Dierickx, P., Halder, M., Isomaa, B., Kawahara, K., Lee, L.E.J., Mothersill, C., Part, P., Repetto, G., Sintes, J.R., Rufli, H., Smith, R., Wood, C. and Segner, H. (2003) The use of fish cells in ecotoxicology –The report and recommendations of ECVAM workshop 47, *Alternatives to Laboratory Animals* **31**, 317-351.

Dayeh, V.R., Schirmer, K. and Bols, N.C. (2002) Development and testing of methods for applying industrial effluent directly to fish cell cultures in order to evaluate effluent toxicity, *Water Research* **36**, 3727-3738.

Dayeh, V.R., Lynn, D.H. and Bols, N.C. (2003a) Assessment of metal toxicity to protozoa and rainbow trout cells in culture, in G. Spiers, P. Beckett and H. Conroy (eds.), *Sudbury 2003 Mining and the Environment Conference Programme*, Laurentian University Centre for Continuing education, Sudbury, ON, Canada, pp. 205-211.

Dayeh, V.R., Schirmer, K., Lee, L.E.J. and Bols, N.C. (2003b) The use of fish-derived cell lines for investigation of environmental contaminants, in *Current Protocols in Toxicology*, John Wiley & Sons, New York, NY, USA, pp. 1.5.1-1.5.17.

Dayeh, V.R., Chow, S.L., Schirmer, K., Lynn, D.H. and Bols, N.C. (2004) Evaluating the toxicity of Triton X-100 to Protozoan, Fish and Mammalian Cells using fluorescent dyes as indicators of cell viability, *Ecotoxicology and Environmental Safety* **57**, 375-382.

de Fries, R. and Mitsuhashi, M. (1995) Quantification of mitogen-induced human lymphocytes-proliferation - comparison of alamar Blue™ assay to 3H-thymidine incorporation assay, *Journal of Clinical Laboratory Analysis* **9**, 89-95.

Environment Canada (1989) The development document for the effluent monitoring regulation for the pulp and paper sector, Queen's Printer for Ontario, Environment Canada, pp. 1-55.

Environment Canada (1990) Biological test method: acute lethality test using rainbow trout, Environment Protection Series, EPS 1/RM/9, pp. 1-51.

Essig-Marcello, J.S. and Van Buskirk, R.G. (1990) A double-label in situ cytotoxicity assay using the fluorescent probes neutral red and BCECF-AM, *In vitro Toxicology* **3**, 219-227.

Fent, K. (2001) Fish cell lines as versatile tools in ecotoxicology: assessment of cytotoxicity, cytochrome P450, An induction potential and estrogenic activity of chemicals and environmental samples, *Toxicology In vitro* **15**, 477-488.

Freshney, R.I. (2000) Culture of Animal Cells: A Manual of Basic Technique, 4th ed. John Wiley & Sons, New York, NY, USA, pp. 1-577.

Ganassin, R.C., Schirmer, K. and Bols, N.C. (2000) Methods for the use of fish cell and tissue cultures as model systems in basic and toxicology research, in G.K. Ostrander (ed.), *The Laboratory Fish*, Academic Press, San Diego, CA, USA, Chapter 38, pp. 631-651.

Health Canada (1996) The Laboratory Biosafety Guidelines, available from http://www.hc-sc.gc.ca/pphb-dgspsp/publicat/lbg-ldmbl-96/index.html

Hestermann, E.V., Stegeman, J.J. and Hahn, M.E. (2000) Serum alters the uptake and relative potencies of halogenated aromatic hydrocarbons in cell culture bioassays, *Toxicological Sciences* **53**, 316-25.

Mochida, K. (1986) Aquatic toxicity evaluated using human and monkey cell culture assays, *Bulletin of Environmental Contamination and Toxicology* **36**, 523-526.

O'Brien, P.J., Wilson, I., Orton, T. and Pognan, F. (2000) Investigation of the alamar Blue (resazurin) fluorescent dye for the assessment of mammalian cell cytotoxicity, *European Journal of Biochemistry* **267**, 5421-5426.

Putnam, K.P., Bombick, D.W. and Doolittle, D.J. (2002) Evaluation of eight *in vitro* assays for assessing the cytotoxicity of cigarette smoke condensate, *Toxicology In vitro* **16**, 599-607.

Richardson, D., Dorris, T.C., Burks, S., Browne, R.H., Higgins, M.L. and Leach, F.R. (1977) Evaluation of a cell culture assay for determination of water quality of oil-refinery effluents, *Bulletin of Environmental Contamination and Toxicology* **18**, 683-690.

Schaeffer, W.I. (1990) Terminology associated with cell, tissue and organ culture, molecular biology and molecular genetics, *In vitro Cellular and Developmental Biology* **26**, 97-101.

Schirmer, K., Chan, A.G.J., Greenberg, B.M., Dixon, D.G. and Bols, N.C. (1997) Methodology for demonstrating and measuring the photocytotoxicity of fluoranthene to fish cells in culture, *Toxicology In vitro* **11**, 107-119.

Schirmer, K., Greenberg, B.M., Dixon, D.G. and Bols, N.C. (1998a) Ability of 16 priority PAHs to be directly cytotoxic to a cell line from the rainbow trout gill, *Toxicology* **127**, 129-141.

Schirmer, K., Chan, A.G.J., Greenberg, B.M., Dixon, D.G. and Bols, N.C. (1998b) Ability of 16 priority PAHs to be photocytotoxic to a cell line from the rainbow trout gill, *Toxicology* **127**, 143-155.

Schirmer, K., Herbrick, K., Greenberg, B.M., Dixon, D.G. and Bols, N.C. (1999) Use of fish gill cells in culture to evaluate the cytotoxicity and photocytotoxicity of intact and photomodified creosote, *Environmental Toxicology and Chemistry* **18**, 1277-1288.

Schirmer, K., Chan, A.G.J. and Bols, N.C. (2000) Transitory metabolic disruption and cytotoxicity elicited by benzo[a]pyrene in two cell lines from rainbow trout liver, *Journal of Biochemical and Molecular Toxicology* **14**, 262-276.

Schirmer, K., Tom, D.J., Bols, N.C. and Sherry, J.P. (2001) Ability of fractionated petroleum refinery effluent to elicit cyto- and photocytotoxic responses and to induce 7-ethoxyresorufin-o-deethylase activity in fish cell lines, *The Science of the Total Environment* **271**, 61-78.

Segner H. (1998) Fish cell lines as a tool in aquatic toxicology, in T. Braunbeck, D.E., Hinton and B. Streit (eds.), *Birkhäuser, Basel, Ecotoxicology*, pp. 1-38.

Wolf, K. and Rumsey, G. (1985) The representative research animal: Why rainbow tout (*Salmo gairdneri* Rich.)? *Zeitschrift fur angewandte Ichthyologie* **3**, 131-138.

Abbreviations

AB	alamar Blue
ATP	Adenosine tri-phosphate
CF	5-carboxyfluorescein
CFDA-AM	5-carboxyfluorescein diacetate acetoxymethyl ester
con.	control
D-PBS	Dulbecco's Phosphate-Buffered Saline
DMSO	Dimethyl sulfoxide
EC50	Concentration that causes an effect in 50% of the cells
ex.	experimental
FBS	Fetal Bovine Serum

FU	Fluorescent Units
IU	International Units
L-15	Leibovitz's L-15 medium
L-15/ex	L-15 exposure medium
LDH	lactate dehydrogenase
mOsmkg^{-1}	milli-osmole per kilogram
MTT	either 3-(4,5-dimethylthiazol-2-yl)-2,5 diphenyl tetrazolium bromide
NR	Neutral red (3-amino-7-dimethylamino-2-methylphenazine hydrochloride)
PAHs	Polycyclic aromatic hydrocarbons
PBS	Phosphate buffer saline
PSI	Pounds per Square Inch.

Glossary

Note to readers: Volume and chapter number(s) indicated after each **Glossary** term are those for which authors contributed a definition. They may also be found in other chapters of both volumes.

Acclimation	Adaptation to environmental conditions (usually controlled laboratory conditions). Acclimation is generally conducted over a specified period of time. Volume 1(10).
Acid volatile sulfides (AVS)	Chemical analysis that quantifies the amount of sulfides present in a sample that are assumed to be capable of forming insoluble precipitates with divalent metals. See also AVS:SEM ratio and SEM. Volume 2(10).
Acid-washing	Procedure in which laboratory articles are soaked overnight in 4% detergent (*e.g.*, Contrad 70) and rinsed five times in reverse osmosis water, soaked overnight in 10% HCl and rinsed five times in Milli-Q water and oven dried (58°C). Volume 1(6).
Activated sludge	Product that results when primary effluent is mixed with bacteria-laden sludge and then agitated and aerated to promote biological treatment, speeding the breakdown of organic matter in raw sewage undergoing secondary waste treatment (U.S. EPA, 2004). Volume 2(7).
Active biomonitoring	Use of transplanted living organisms (or part of) to assess water, air, sediment or soil quality. See also Passive biomonitoring and Biomonitoring. Volume 2(11).
Acute	Lasting a short time (test or exposure), severe enough to induce a response rapidly (stress or stimulus), having a sudden onset (effect) as opposed to chronic. Volume 1(1,2,3,5,10), Volume 2(5,8,11).
Acute effect	Overt adverse effect (lethal or sublethal) induced in test organisms within a short period of exposure to a test material. Acute effects often induce highly toxic responses (*e.g.*, mortality or assessment endpoints related to mortality). See also Acute exposure and Acute toxicity. Volume 1(1,2,3,5,10), Volume 2(5,8,11).

Acute exposure	Short period of exposure (minutes, hours, or a few days) relative to the life span of the organism (usually set at < 10% of an organism's life span). See also Acute effect and Acute toxicity. Volume 1(1,2,3,5,10), Volume 2(5,8,11). For *Selenastrum capricornutum*, whose cell numbers double every 12 h at 24°C, a contact time of 1-4 h with a test sample would correspond to an acute exposure allowing for the determination of corresponding acute toxicity effects. Measuring esterase inhibition in *S. capricornutum* after a 1-h exposure to a test chemical is another example of an acute exposure toxicity bioassay (Snell et al., 1996). Volume 1(3).
Acute toxicity	Inherent potential or capacity of a material to cause acute effects that occur rapidly as a result of a short exposure time. See also Acute effect and exposure. Volume 1(1,2,3,5,10), Volume 2(5,8,11).
Additive effect	Effect of a mixture of chemicals whereby the summation of the known effects of individual chemicals is essentially additive. For example, if individual aqueous solutions of chemical A and chemical B each yield an IC50 = 50% v/v (or 2 toxic units) for a particular biotest, their combined toxicity will correspond to an IC50 = 25% v/v (or 4 toxic units). See also Additivity. Volume 2(1,10).
Additivity	Toxicity of a contaminant mixture equal to the sum of toxic effects of the individual contaminants. See also Additive effects, Antagonism and Synergy. Volume 2(1,10).
Ad libitum	Literally means "at one's pleasure". This term is generally used with respect to feeding (see below). Volume 1(11,13).
Ad libitum feeding	Feeding with more food than the organisms are able to ingest during a period *i.e.,* until the fed organisms no longer consume food feeding or until satiation occurs. Volume 1(11,13).
Aeration of medium	Operation during which air from a compressor, which is passed through a particle and moisture filter followed by activated carbon, is directed into the aqueous solution through a Pasteur pipette to bubble the solution. The aeration period stabilizes the carbonate system (bicarbonate and CO_2) so that it is in equilibrium with air, thus preventing pH drift. Volume 1(6).
Algal fluorescence	Re-emission of light initially absorbed by chlorophyll *a* pigments in algal cells. In algal toxicity testing, it can be employed as an indirect measure of algal biomass. Volume 1(3, 6).

Algal symbiotes	Freshwater green algae which live inside the tissue of green *Hydra* in a similar symbiotic relationship to marine zooanthellae algae and corals. They are also named Zoochlorellae algae. They provide additional nutrients to *Hydra* in the form of carbohydrates via photosynthesis, while *Hydra* provides them with a protected environment. Both *Hydra* and corals can experience 'bleaching' where the symbiotic algae are expelled from the organism following some significant environmental stress, particularly increased water temperature. Volume 1(11).
Algicidal	Property of killing algae. The algicidal concentration is the lowest concentration tested which allows no net growth of the population of test organisms during either exposure to the test material or during the recovery period in the absence of test material. See also Algistatic. Volume 1(4).
Algistatic	Property of inhibiting algal growth. The algistatic concentration is the highest concentration tested which allows no net growth of the population of test organisms during exposure to the test material but permits re-growth during the recovery period in the absence of test material. See also Algicide and Algistatic effect. Volume 1(4).
Algistatic effect	Effect caused by a chemical agent which inhibits algal growth. Volume 1(3).
Algorithm	Detailed sequence of actions required for accomplishing a specific task. Volume 2(2).
Alternative assay	Biological-based assay destined to reduce the sacrifice of organisms (usually vertebrates), to reduce the cost and to replace old tests by more rapid and efficient ones. The use of fish cells is an example of an alternative for fish. Volume 1(14).
Analysis of covariance (ANCOVA)	Used to test the main and interactive effects of categorical variables on a continuous dependent variable, controlling for the effects of other selected continuous variables that covary with the dependent variable. Volume 2(4).
Antagonism	Interaction of several agents resulting in a lower effect than the one expected by addition of the individual effects. See also antagonistic effect. Volume 2(2,10).
Antagonistic effect	Toxicity of a mixture of chemicals whereby the summation of the known toxicities of individual chemicals is less than that expected from a simple summation of the toxicities of the individual chemicals comprising the mixture. Volume 2(1).

Antilogarithm	Number to which a given logarithm belongs. If $b^x = a$, then a is called the antilogarithm of x to the base b. Finding an antilogarithm is, in a sense, the inverse of finding a logarithm. Volume 2(3).
Aposymbiotic	Lacking a symbiotic organism (*e.g.*, pink *Hydra*). Volume 1(11).
Artificial sediment	Mixture of materials used to mimic the physical components of a natural sediment. See also Reference sediment. Volume 1(13).
Assessment endpoint	Effect criterion by which toxicity is estimated (*e.g.*, mortality, growth, reproduction). Volume 1(3,10).
Autolysis	Dissolution or destruction (self-digestion) of cell. Volume 1(8).
Auxinic effect	Chemical substance capable of stimulating the growth of phototrophic (micro-)organisms. Phosphorus and nitrogen are two examples of common nutrients capable of enhancing micro-algal growth. Volume 1(3).
AVS:SEM ratio	Surrogate measure of bio-availability. An AVS:SEM ratio > 1 (*i.e.*, more AVS than SEM) indicates a sample where Cd, Cu, Hg, Ni, Zn, and Pb are unlikely to be bio-available (*i.e.*, have formed an insoluble metal precipitate with sulfides). It is expressed in terms of molar differences (*e.g.*, AVS - SEM). See also AVS and SEM. Volume 2(10).
Axenic culture	A mono-specific culture of a test organism (*e.g.*, a single micro-algal species) which is devoid of other species of micro-organisms (*e.g.* other types of algae) and also free of bacterial contamination. Volume 1(3,7,8).
Bacteria	Large group of organisms that do not have organelles enclosed in cell membranes and have DNA in both a chromosome and circular plasmids. They have a protein and complex carbohydrate cell wall over a plasma membrane. Although eukaryotic and prokaryotic cells are structurally different, their basic biochemical processes are similar. Volume 1(1, 2), Volume 2(3).
Bacterial bioluminescence	Production of light by certain marine bacteria. The general consensus is that light is produced when bacterial luciferase catalyzes the bioluminescent oxidation of $FMNH_2$ and a long chain aldehyde by molecular oxygen. Volume 1(1,2).

Bacterial lyophilization	Procedure conducted under vacuum in which water is removed from bacteria (also known as freeze-drying). If the vacuum seal of the container is maintained and the bacteria are stored in the dark, they will remain viable indefinitely. Viable bacteria are activated by rehydration. Volume 1(1,2).
Bacterial reagent	In the Microtox® test, it is a standard culture of freeze-dried (lyophilized) *Vibrio fischeri*, stored in small, sealed vials which each contain about 100 million cells. Volume 1(1,2).
Basal cytotoxicity	Impairment of one or more cellular activities common to all cells. Volume 1(15).
Basal medium	In cultured fish cells assay, it is an aqueous solution of nutrients and buffering agents, such as Leibovitz's L-15, that contains a hexose, bulk ions, trace elements, amino acids, and vitamins. Volume 1(15).
Battery of (toxicity) tests	Use of several laboratory toxicity tests (at least two), usually representative of different levels of biological organization (*e.g.*, a battery composed of a bacterial, algal, micro-invertebrate and fish test) to attempt to circumscribe the full toxicity potential of a liquid or solid matrix sample. Volume 2(1,8).
Bioassay	Biological test in which the severity of the toxic effect caused by a test material is measured by the response of living organisms. Synonyms: biotest, toxicity test, toxicity assay. Volume 2(8).
Bioassay battery approach	Use of several laboratory toxicity tests (at least two), usually representative of different levels of biological organization (*e.g.*, a battery composed of a bacterial, algal, micro-invertebrate and fish test) to attempt to circumscribe the full toxicity potential of a liquid or solid matrix sample. See also battery of (toxicity) tests. Volume 2(1,8).
Biodegradability	Ability of a substance to be broken down into simpler substances by organisms such as bacteria. Volume 2(1).
Biodegradation	Process (*e.g.*, enzymatic breakdown) whereby an organic compound is transformed to a simpler carbon entity (*e.g.*, glucose to carbon dioxide). Volume 2(1,7).
Biodeterioration	Process caused by activities of living organisms whereby properties of a material are modified. Volume 2(11).

Bioindicator (biological indicator)	Measure, index of measures, or model that characterizes an ecosystem or one of its critical components. It may reflect biological, chemical or physical attributes of ecological condition. The primary uses of an indicator are to characterize current status and to track or predict significant change. With a foundation of diagnostic research, an ecological indicator can also be used to identify major ecosystem stress. Volume 2(10).
Biomagnification	Cumulative increase in contaminant body burdens up three or more trophic levels in a food chain. Biomagnification occurs when the intake of a contaminant exceeds the capacity of an organism to excrete and/or metabolize the contaminant in question. Volume 2(10).
Biomarker	Any one of a series of physiological, biochemical, behavioural or metrics measurements reflecting an interaction between a living system (tissue, organ, cell, etc.) and an environmental agent, which may be chemical, physical or biological. For example, the induction of metallothionein, a heavy metal biomarker of defense, is activated in fish hepatic tissue exposed to metals such as cadmium or mercury. Volume 1(14), Volume 2(1,10).
Biomass	Dry or wet weight of living matter. In algal tests, for example, it can be expressed in terms of mg of algae per liter. Because dry weight is difficult to measure accurately, however, surrogate measures of biomass, such as cell counts, are typically used in algal toxicity testing. Volume 1(4).
Biomonitoring	Use of resident or transplanted living organisms (or parts of) to assess water, air, sediment or soil quality. See also Passive biomonitoring and Biomonitoring. Volume 2(11).
Biosolid	Treated sewage sludge that meets US EPA regulations for land application. Volume 2(7).
Biotransformation	Ability of biological tissues to transform chemical compounds. Transformations can involve, for instance, oxidation reactions. Volume 1(14).
Bootstrap method	Re-sampling method that randomly chooses new datasets among experimental data. Volume 2(2).
Brackish	Low salinity exemplified by freshwater and seawater that are mixed near the estuary of a river flowing into the sea. Tidal flats and lagoons of low salinity are also considered as brackish areas (PIANC, 2000). Volume 2(9).

GLOSSARY

Bray-Curtis Index	Distance coefficient (*e.g.*, linked to fish and benthos field surveys) that reaches a maximum value of 1 for two sites that are entirely different and a minimum value of 0 for two sites that possess identical descriptors. It measures the amount of association between sites. Volume 2(4).
Cell line	Cells obtained from a tissue that are transferred (or passaged) to a new culture vessel and that divide readily in the culture vessel. They can be propagated *in vitro* by repeating the cycle through cell proliferation followed by transferring an aliquot of the cell population into new culture vessels, usually flasks. Volume 1(14,15).
Cell viability assay	Determined on the basis of loss of cell membrane permeability in response to a deliberate modification in culture conditions. Cell viability can be determined by either neutral red or fluorescein diacetate retention assays. Volume 1(14,15).
Chain of custody	Documented and traceable transfer of a sample from the point of collection to reception at the testing laboratory. Volume 1(10).
Chironomus	Non-biting midge with an aquatic larval stage (order Diptera). Volume 1(12,13).
Chronic	Lasting a long time (test or exposure); it can involve a stimulus or stress that is lingering or continues for a long time; it has a light onset (effect) as opposed to an acute one. Volume 1(3,5), Volume 2(2,5,11).
Chronic effects	Subtle adverse effects (lethal or sublethal) induced in the test organisms within a long period of exposure to a test material. Chronic effects often relate to growth or reproduction impairments. See also Acute exposure and Acute toxicity. Volume 1(3,5), Volume 2(2,5,11).
Chronic exposure	Long period of exposure (days, weeks or months) relative to the life span of the test organism (*i.e.*, > 10% of an organism's life span) and also relative to several life-cycle phases (*e.g.*, development, reproduction) (Férard et al., 1992). See also Chronic effects and Chronic toxicity. Volume 1(3,5), Volume 2(2,5,11).

Chronic toxicity	Inherent potential or capacity of a material to cause chronic effects that occur following long exposure times. For *S. capricornutum*, whose cell numbers double every 12 h at 24°C, a 3-d contact time with a test sample corresponds to a chronic exposure period allowing for the determination of corresponding chronic toxicity effects. See also Chronic effects and Chronic exposure. Volume 1(3,5), Volume 2(2,5,11).
CYP1A1	Gene producing cytochrome P4501A1 that biotransforms coplanar aromatic hydrocarbons (*e.g.*, benzo(a)pyrene). Volume 1 (14).
Coefficient of determination (r^2)	Measure of the closeness of fit of a scatter graph to its regression line where $r^2 = 1$ is a perfect fit. Volume 1(6).
Coefficient of variation	A statistical index of precision calculated as ([standard deviation $\times$ 100] $\div$ mean). The CV is a measure of the variability in a group of measurements. Since the CV is unitless, it can be used to compare CVs from different "experiments". It is also a quality control tool. For example, in the algal microplate toxicity test, algal cell density in control wells at the end of the test exposure period must have a CV not exceeding 20% to meet test acceptability criteria. Volume 1(1,2,3,10).
Coincident Sampling	Sampling at the same location but at different times. Volume 2(10).
Collagenase	A protease (*i.e.*, a protein enzyme that degrades other proteins) specific to collagen which is the main protein matrix that holds liver cells together. Volume 1(14).
Concordance	Total number of correct predictions (*i.e.*, presence or absence of toxic effects) between two bioassays over the total number of test samples. Volume 1(14).
Confidence interval	A range of values estimated by a sample within which the true population value is expected to fall. For example, if an LC50 and its 95% confidence intervals are estimated from a toxicity test, the true population LC50 is expected to fall within the interval 95% of the time. Volume 1(10), Volume 2(5).
Confidence limits	Upper and lower boundaries of the confidence interval. Volume 1(10), Volume 2(20).

GLOSSARY

Confined disposal	Placement of dredged material within diked nearshore or upland confined placement facilities that enclose and isolate the dredged material from adjacent waters. Confined dredged-material placement does not refer to sub-aqueous capping or contained aquatic dredged-material placement (PIANC, 2000). Volume 2(9).
Confluent monolayer	Animal cells completely covering the surface of a culture vessel. Volume 1(15).
Conspecific	Belonging to the same species. Volume 1(14).
Consumer (primary and secondary)	Heterotrophic organisms which consume other organisms and/or particulate organic matter. Primary consumers are herbivores (*e.g.*, daphnids eating micro-algae) whereas secondary consumers are carnivores (*e.g.*, hydras eating daphnids). Volume 2(1).
Contaminated dredged material	Sediments or materials having unacceptable levels of contaminant(s) that have been demonstrated to cause an unacceptable adverse effect on human health or the environment (PIANC, 2000). Volume 2(9).
Control	Treatment in an investigation or study that duplicates all the conditions and factors that might affect the results of the investigation, except the specific condition that is being studied. In an aquatic toxicity test, the control must duplicate all the conditions of the exposure treatment(s), but must contain no added test material or substance. The control is used to determine the absence of measurable toxicity due to basic test conditions (*e.g.*, temperature, health of test organisms, or effects due to their handling or manipulation). Volume 1(2), Volume 2(5).
Control Chart	Graphical plot of test results with respect to time or sequence of measurement upon which control and warning limits are set to guide in detecting whether the system is in a state of control. Volume 1(10).
Control limits	Limits or combination of limits which, when exceeded, trigger analyst intervention. These limits may be defined statistically or based on test method requirements. Control limits may be assigned to method blanks, check standards, spike recoveries, duplicates and reference samples. Most control limits for toxicity tests are based on thrice the standard deviation of the mean (*i.e.*, one in every 100 tests would be expected to exceed the control limits due to chance alone). Volume 1(10).

Corer	Hollow tubes or casings that are used to collect soil or sediment samples. Small soil corers are normally pushed into the soil or sediment by hand-held tools. See also Sediment core sample. (PIANC, 1997). Volume 2(9).
Correlation analysis	Statistical analysis that calculates the coefficient of correlation (*i.e.*, covariance divided by the product of variances) for a set of variables. Volume 2(2).
Cryovial	A two mL capacity polypropylene container with sealable screw-top lid and "V" shaped bottom. Ideal for storing dried organisms (*e.g.*, amphipods) and water samples and good for digesting small tissue samples because small acid volumes remain in contact with tissue samples. Volume 1(12).
Cryptobiotic	Relating to the dormant stage of a particular micro-organism or organism. Examples include cyst formation in micro-invertebrates such as water fleas (*e.g.*, *Daphnia magna*) or the embedding of physiologically-active algal cells (*e.g.*, *Selenastrum capricornutum*) in an alginate matrix to produce algal beads. Water fleas can later be hatched "on demand" to conduct biological testing, as can be algal cells once they are removed from their beaded matrix. Volume 1(3).
Culture	As a noun, stock of plants or animals raised under defined and controlled conditions to produce healthy test organisms. As a verb, it means to conduct the procedure of raising organisms (Environment Canada, 1999). Volume 1(7,10).
Cytogram	Bi-parametric plot of data from a flow cytometer. Each axis of the plot displays one parameter (light scatter and/or fluorescence). Data from each event (particle) analysed is represented as a dot (particle) on the cytogram. Volume 1(5).
Decomposer	Organism (*e.g.*, a bacterium) that feeds on dead or decaying plants and animals, transforming them chemically, thereby contributing to recycling (in)organic materials to the environment. Volume 2(1).
Dialysis	Removal of a small molecule from a solution with macromolecule(s) by allowing it to diffuse through a semipermeable membrane into a solvent. Volume 1(1).
Diapause	Period during which an organism does not grow, while it awaits necessary environmental conditions. Volume 1(13).

Diluent	In the Microtox® test, it is a solution of 3.5% sodium chloride in distilled or deionized water, which is prepared using reagent-grade salt. Diluent comprised of 3.5% NaCl may be used with samples of marine, estuarine, or freshwater sediment. See also "distilled water" and "deionized water". Volume 1(2).
Dilution water	Solution used to prepare the reference toxicant or effluent dilutions required for toxicity testing. Volume 1(6).
Discriminant analysis	Multivariate statistical analysis using classes of variables and calculating discriminant functions as linear combinations of the variables that maximize the inter-class variance and minimize the intra-class variance. Volume 2(2).
Dispersant	Chemical substance that reduces the surface tension between water and a hydrophobic substance (*e.g.*, oil), and thereby facilitates its dispersal via a water emulsion. Volume 1(3,7).
Dose (or concentration) response model	Function of dose (or concentration) of a chemical able to link a toxicity response to any dose (or concentration) value. Volume 2(2).
Dredged material	Material excavated from waters. The term "dredged material" refers to that which has been dredged usually from the bed of a water body, while the term "sediment" refers to material in a water body prior to the dredging process (PIANC 2000). Volume 2(9).
Dredging	Loosening and lifting earth and sand from the bottom of water bodies. Dredging is often carried out to widen the stream of a river, deepen a harbor or navigational channel, or collect earth and sand for landfill; it is also carried out to remove contaminated bottom deposit or sludge to improve water quality (PIANC, 2000). Volume 2(9).
Dredging process	A process consisting of the following three elements: 1) Excavation: this process involves the dislodgment and removal of sediments (soils) and/or rocks from the bed of the water body. A special machine – the dredger – is used to excavate the material either mechanically, hydraulically or by combined action. 2) Transport of excavated material: transporting materials from the dredging area to the site of utilization, disposal or intermediate treatment, is generally achieved by one of the following methods: in self-containing hoppers of the dredgers; in barges; pumping through pipelines; and using natural forces such as waves and currents. 3) Other, rarely used transport methods are truck and conveyer belt transport. The method of transport is generally linked to the type of dredger being used. Volume 2(9).

Duplicate	Quality control sample, often chosen randomly, from a batch of samples and undergoing separate, but identical sample preparation and analysis whose purpose is to monitor method precision and sample homogeneity. Duplicate testing also aids in the evaluation of analyst proficiency. Volume 1(10).
EC50	See ECx.
ECx	Effective concentration of a test material in the test matrix (*e.g.*, growth medium) that is calculated to exhibit a specified non-lethal or lethal effect to x% of a group of test organisms during exposure over a specified period of time. The ECx and its 95% confidence limits are usually derived by statistical analysis of responses in several test concentrations. The particular effect must be specified as well as the exposure time (*e.g.*, 48-h EC50 for immobilization). Volume 1(1,4,10).
Ecocompatibility	Situation where pollutant release from waste, when deposited in a specific physical, hydrogeological, physico-chemical and biological context, is in keeping with the acceptable pollutant level of receiving environments (Perrodin et al., 1996). Volume 2(11).
Effluent	Any liquid, gaseous or aerosolic waste discharged in the environment. Generally, it is a complex mixture. For example, wastewaters include: mine water effluent, mill process effluent, tailings impoundment area effluent, treatment pond or treatment facility effluent, seepage and surface drainage. Volume 1(9,10,14), Volume 2(2,5).
Electrophiles	Compounds representing a non reversible mode of action. Electrophilic interactions involve substitution or conjugation of electron-rich groups to nucleophilic sites in cellular macromolecules. Volume 1(8).
Elutriate	Aqueous solution obtained after adding a fixed volume of water to a solid medium (*e.g.*, waste, soil or sediment), shaking of the mixture, then centrifuging, or filtering it or decanting the supernatant. Volume 1(3,9), Volume 2(8,9).
Emulsifier	Substance that aids the fine mixing (in the form of small droplets) within water of an otherwise hydrophobic substance (Environment Canada, 1999). Volume 1(7).
Endocrine disruption	Any one of a series of effects caused by hormonally-active agents that alter the homeostatic function of hormone or physiological system under the control of hormone(s). Volume 1(14), Volume 2(1).

Endocrine disruptors	Exogenous chemicals which cause adverse health effects in organisms or their progeny as a result of changes in endocrine function. Volume 1(9, 13,14), Volume 2(1).
Endpoint	Measurement(s) or value(s) that characterize the results of a test (*e.g.*, LC50, ICp). This term might also mean the reaction of the test organisms to show the effect which is measured upon completion of the test (*e.g.*, inhibition of light production). Volume 1(2,10).
EPT Index	Total number of distinct taxa within the orders Ephemeroptera, Plecoptera and Trichoptera compared to total taxa present. Volume 2(4).
Ephippium (s.), *ephippia* (pl.)	Egg case that develops under the postero-dorsal part of the adult *Daphnia* female carapace in response to unfavorable environmental conditions. Ephippia eggs are the outcome of sexual reproduction. Volume 1(10).
Epibenthic	Characteristic of organisms that have regular contact with sediment and live just above the sediment/water interface. Volume 2(8).
Equitox parameter	Toxic unit used by the French Water Agencies. See also Toxic unit. Volume 2(2).
Esterases	Group of enzymes involved in phospholipid turnover in cell membranes. Esterase activity in algae has been shown to relate well to metabolic activity and cell viability. Volume 1(5).
Estuarine water	Coastal body of ocean water that is measurably diluted with fresh water derived from land drainage. Volume 1(2).
Eukaryotes	All organisms except viruses, bacteria and archaea. See eukaryotic cell. Volume 1(3,8).
Eukaryotic cell	Advanced cell type with a nuclear membrane surrounding genetic material and numerous membrane-bound organelles dispersed in a complex cellular structure see Eukaryotes. Volume 1(8).
Eutrophication	Excessive enrichment of waters with nutrients (essentially nitrate and phosphate), including the associated adverse biological effects (*i.e.*, aquatic plant blooms). Volume 1(3).
Exuvium (s.), *exuviae* (pl.)	Remains of the pupa, which is discarded when an insect has emerged. Volume 1(13).

Far-far field	Receiving water near an industry's effluent discharge that is more distant from the effluent outfall than the far-field and in which the effluent concentration is lower than that of the far-field. Volume 2(4).
Far-field	Receiving water near an industry's effluent discharge and located along a dilution gradient in which effluent concentration is less than or equal to 1%. Volume 2(4).
Field swipes for chemistry	Check on the quality of equipment decontamination procedures involving the "swiping" of sterile filter paper over sampling equipment after decontamination has occurred, followed by chemical analysis of the field swipe and an unused filter paper. Volume 2(10).
Fines	Sediment or soil particles which are ≤ 63 μm in size. Measurements of % fines include all particles defined as silt (*i.e.*, particles ≤ 63 μm but ≥ 4 μm) or clay (*i.e.*, particles < 4 μm). Volume 1(2).
Flow cytometer	Instrument that is capable of rapid and quantitative measurements of individual cells in a moving fluid. Thousands of cells pass through a light source (usually a laser, 488 nm) and measurements of cell density, light scatter (two parameters) and fluorescence (three or more parameters) are collected simultaneously. Volume 1(5).
Flow-through	Tests in which solutions in test vessels are renewed continuously by the constant inflow of a fresh solution, or by a frequent intermittent inflow. Synonymous term is "dynamic". Volume 1(10).
Fluorescent unit (FU)	Arbitrary unit of measurement by fluorescent plate reader. Volume 1(15).
Fluorometer	Instrument that measures the fluorescence properties of solutions. It is composed of a high-energy lamp for excitation and a phototube for emission readings. Instruments are available in either tube or microplate formats. Volume 1(14).
Foot-candle	One of several units of illumination based on units per square meter. One foot-candle = 10.76 lux. Volume 1(3).
Formulated sediment	Artificial sediment formulated from constituents such as silica sand and peat moss according to standardized recipes, intended to match the physical characteristics (*e.g.*, grain size, TOC) of the site under investigation. Volume 2(10).

Frond	Individual leaf-like structure of a duckweed plant. It is the smallest unit (*i.e.*, individual) capable of reproducing (Environment Canada, 1999). Volume 1(7).
Gamma	In the Microtox® test, it is a measure of light loss used in calculating the IC50 or ICp. It is calculated individually for each cuvette containing a filtrate of a particular test concentration. Gamma (Γ) is calculated based on the ratio between the amount of light emitted by a test filtrate and that emitted by the control solutions, as follows: $\Gamma = (I_c/I_t) - 1$, where I_c = the average light reading of filtrates of the control solutions, and I_t = the light reading of a filtrate of a particular concentration of the test material. When Gamma equals unity ($\Gamma = 1$), half of the light production has been lost. Vol. 1(2).
Genomics	Branch of genetics that studies organisms in terms of their genomes (*i.e.*, full DNA sequences). Volume 1(14).
Genotoxicity	Inherent potential or capacity of a chemical, biological or physical agent to damage the hereditary material of cells (DNA) or organ tissues (*i.e.*, causing DNA damage or alterations that can give rise to mutations, tumors and/or cancer). Volume 2(1,2).
Geometric mean	Mean of repeated measurements, calculated on a logarithmic basis. It has the advantage that extreme values do not influence the mean as is the case for an arithmetic mean. It can be calculated as the nth root of the product of the n values, and it can also be calculated as the antilogarithm of the mean of the logarithms of the n values. Volume 1(2).
Gibbosity	Fronds exhibiting a humped or swollen appearance (Environment Canada, 1999). Volume 1(7).
Groundwater	Source of water that is found below ground level. Volume 1(14).
Growth	Increase in size or weight as the result of proliferation of new tissues in a specified period of time. For example, in the duckweed test, it refers to an increase in frond number over the test period as well as the dry weight of fronds at the end of the test. Volume 1(4,7).
Growth medium	Medium promoting growth. For example, for culturing cells, basal medium plus a supplement of fetal bovine serum (FBS). Volume 1(15).
Growth rate	Rate at which the biomass increases (Environment Canada, 1999). Volume 1(7).

Hardness	Concentration of cations in water that will react with a sodium soap to precipitate an insoluble residue. Total hardness is a measure of the concentration of calcium and magnesium ions in water, usually expressed as mg/L $CaCO_3$. Volume 1(10).
Hazard	Potential for adverse effect(s) that might result from exposure to a chemical, biological or physical agent. Volume 2(8,10).
Hazard assessment	Process that evaluates the type and magnitude of adverse effect(s) caused by a stressor (such as chemical contamination). Volume 2(8,10).
Hepatocyte	Main epithelial cell in the liver. Volume 1(14).
Heterotroph	Organism that requires complex nutrient molecules as a source of carbon and energy. Volume 1(1).
Hexagenia	Burrowing mayfly (order Ephemeroptera). Volume 1(12).
Highest effect concentration (HEC)	Concentration related to the highest induced effect. In the Mutatox® test, for example, this effect refers to induced luminescence. Volume 2(11).
Histogram	Single-parameter plot of data. In flow cytometry, the horizontal axis displays the light scatter or fluorescence intensity parameter and the vertical parameter displays the number of events (*e.g.*, cell count). Volume 1(5).
Holding Time	Time elapsed between the end of sample collection or sample preparation and the initiation of analysis. Volume 1(10).
Hyalella	Amphipod crustacean (suborder Gammaridea). Volume 1(12).
Hydraulic dredgers	Dredgers using hydraulic centrifugal pumps to provide the dislodging and lifting force for sediment material removal in a liquid slurry form. Hydraulic dredging and transport methods "slurry the sediment", that is, they add large amounts of process water and thus change the original structure of sediments (PIANC, 2001). Volume 2(9).
Hydrodynamic dredging	See Hydraulic dredgers. Volume 2(9).
Hydroid	Individual *Hydra* including any attached buds. Volume 1(11).
Hydrophobic	Molecules or molecular groups that mix poorly with water (*e.g.*, hydrocarbons and fats are hydrophobic). Volume 1(3).
Hydroponic cultures	Methods of culturing plants by growing them, for example, in gravel, through which water containing dissolved inorganic nutrient salts is pumped. Volume 2(6).
IC25 or IC50	See ICp. Volume 1(4), Volume 2(4,5).

GLOSSARY

ICp (or ICx) — Inhibiting concentration for a (specified) percentage effect. It relates to a point estimate of a test sample concentration that causes a designated percent inhibition (p) compared to the control, *e.g.*, a corresponding percent reduction in a quantitative assessment endpoint such as algal growth inhibition. The ICp and its 95% confidence limits are usually derived by statistical analysis of responses in several test concentrations. Examples of frequently-reported ICps are IC50 (50% effect in relation to control organisms) or IC20 (20% effect in relation to control organisms). This term should be used for any bioassay which measures a continuously-variable effect, such as light production, reproduction, respiration, or dry weight at test end. Volume 1(2, 3, 4), Volume 2(4,5,8).

Imhoff settling cone — Cone-shaped container (1 L capacity) for measuring the volume of suspended matter in liquids. Also used as toxicity test chambers because their shape results in adequate sediment depth when using small volumes of sediment and large volumes of water. Volume 1(12).

Immobility — In the daphnid test, inability to swim during the 15 seconds following gentle agitation of the test solution, even if the daphnids can still move their antennae. Volume 1(10).

Immunotoxicity — Inherent potential or capacity of a chemical agent which specifically affects cells having immune functions (*e.g.*, heavy metals can intoxicate bivalve hemocytes and impede them from ingesting and lysing pathogenic micro-organisms which can lead to either sub-lethal or lethal infections). Volume 2(1).

Index — Single parameter summarizing several values while minimizing the loss of information and attempting to be relevant to the notion of interest (*e.g.*, toxicity). Volume 2(2).

Inhibitory concentration (IC) — See ICp. Volume 1(6).

Inter-laboratory — Among-laboratory activities. For example, inter-laboratory variability evaluates reproducibility of similar analyses by different laboratories. Estimation of inter-laboratory variability addresses a measure of quality assurance of laboratories (Environment Canada, 1999). Volume 1(10).

Interstitial water	Water occupying space between sediment particles. The amount of interstitial water in sediment is calculated and expressed as the percentage ratio of the weight of water in the sediment to the weight of the whole sediment including the pore water. It can be recovered by methods such as squeezing, centrifugation, or suction. Synonymous term is pore water. Volume 1(2,9,14), Volume 2(5,8,9).
Intra-laboratory	Within-laboratory activities. For example, intra-laboratory variability evaluates repeatability of analysis within the same laboratory system. Estimation of intra-laboratory variability of data is a principal quality control measure of a laboratory (Environment Canada, 1999). Volume 1(10).
Isogenic population	Members of a population having similar genetic make-up since they are clones of the original organisms. For example, *Hydra* use asexual budding as their prime form of reproduction, and all buds are genetic clones of the parent *Hydra*. Sexual reproduction in *Hydra* involving the production of testes and ovaries only occurs when environmental conditions become unfavorable: in this case, *Hydra* produce sperm and eggs which result in a resistant fertilized zygote being produced that can withstand dessication (drying out) and freezing. Volume 1(11).
L-15/ex	Simplified version of the basal medium L-15 that contains only galactose, pyruvate and bulk ions Volume 1(15).
Laboratory	Body or part of an organization that is involved in calibration and/or testing. Volume 1(10).
Laboratory accreditation	Formal recognition, by a registered accrediting body, of the competence of a laboratory to conduct specific functions. The process by which a laboratory quality system (*i.e.*, laboratory management system) is evaluated through regular site assessments by the accrediting body, and may include annual or twice-yearly proficiency testing rounds. Volume 1(10).
Lag phase	Stage in the growth cycle when the growth rate is changing. There may be increase or decrease in algal cell mass per unit volume of cell suspension. Volume 1(6).
Larval instar	Period of the life-cycle between molts. Volume 1(13).

GLOSSARY

LC50	Median lethal concentration of a test material in the test matrix (*e.g.*, growth medium) that is calculated to exhibit a lethal effect to 50% of a group of test organisms during exposure over a specified period of time. The LC50 and its 95% confidence limits are usually derived by statistical analysis of mortalities in several test concentrations. The duration of exposure must be specified (*e.g.*, 48-h LC50). Volume 1(1,4,10), Volume 2(5).
Leachate	Water recovered after its percolation through a solid medium (*e.g.*, soil or solid waste). Volume 1(3).
Lemna root	Part of the *Lemna* plant that assumes a root-like structure (Environment Canada, 1999). Volume 1(7).
Lentic system	Still-water aquatic system, such as a lake, a pond or a swamp. Volume 1(13), Volume 2(1).
Lethal	Causing death. Death is defined as the cessation of visible signs of all movement or other activity. For example, death of daphnids is defined as the cessation of all visible signs of movement or other activity, including second antennae, abdominal legs, and heartbeat as observed through a microscope. Volume 1(10,14), Volume 2(8).
Limnic environment	Ecological conditions (affecting the life of a plant or animal) related to lakes and other bodies of fresh standing water or (more widely) all inland water. Volume 2(9).
Linear interpolation	Statistical method used to determine a precise point estimate of the test sample (*e.g.*, toxicant solution, effluent) that produces a specific percent effect. In algal assays, for example, one would strive to determine a particular reduction (*e.g.*, 20, 25 or 50%) in algal growth by calculating ICps corresponding to IC20, IC25 or IC50. Volume 1(3).
Liquid-phase (toxicity) test	Bioassay using a biological system which measures toxic effects of the liquid/aquatic phase of a test material (*e.g.*, porewater, elutriate, leachate) and determines a response (*e.g.*, acute and/or chronic toxicity). See also Solid-phase (toxicity) test. Volume 1(2), Volume 2(9).
LOEC	Lowest observed effect concentration, that is the lowest concentration in the tested series at which a biological effect is observed (*i.e.*, where the mean value for the observed response is significantly different from the controls). It is one of the tested concentrations obtained, for example, after analysis of variance and multiple comparison statistical testing (*e.g.*, Dunnett test). Volume 1(3,4), Volume 2(8,11).

Log (logarithmic) phase	Stage in the growth cycle when the mass of microbial cells doubles over each of the successive and equal time intervals. The doubling time and, therefore, the growth rate during the entire log phase is thus constant. Volume 1(6).
Lotic system	Running-water aquatic system including rivers, brooks or streams. Volume 1(13), Volume 2(1).
LT50	Lethal time (period of exposure) estimated to cause 50% mortality in a group of organisms held in a particular test solution. The value can be estimated graphically or by regression. Volume 1(10).
Lumen	One of several units of illumination based on units per square metre. Synonymous term is lux (*i.e.*, 1 lumen = 1 lux). Volume 1(3).
Lux	One of several units of illumination based on units per square metre. One lux = 0.0929 foot-candles, and 1 foot-candle = 10.764 lux. Relationships between lux and $\mu E.m^{-2}.s^{-1}$ is variable and depends on light source, light meter used, geometrical arrangement of the exposure environment and possible light reflections, so one lux $\approx 0.015 \mu E.m^{-2}.s^{-1}$ (range of 0.012 to 0.019). Synonymous term is lumen (*i.e.*, 1 lux = 1 lumen). Volume 1(3,10).
Lyophilization	Process which extracts water from biological products or field samples, so that they remain stable over time. It is carried out using a principle called sublimation, which is the transition of a substance from the solid to the vapour state. Synonymous term is freeze-drying. Volume 1(2,3).
Lyophilized organism	Organisms which have been freeze-dried under vacuum (see above). Some bacteria, for example, can be lyophilized and stored for months at room temperature. They can then be rehydrated on demand and used to conduct bioassays. In the Microtox® test, lyophilized *Vibrio fischeri* are stored in a freezer at -20°C and will be ready for use until the expiration date, which is provided with each batch of Bacterial Reagent. Volume 1(2,3).
Macro	Computer program able to execute sequences of interactive software functions together with instructions using a programming language. Volume 2(2).
Manning sampler	Piece of equipment employed in fluid monitoring as in the collection of specific volumes of wastewater over time and commercialized by Manning Environmental Inc. Volume 2(1).

Marine water	Water coming from or within the ocean, sea, or inshore location where there is no appreciable dilution of water by natural fresh water derived from land drainage. Volume 1(2).
Matrix effect	Phenomenon occurring when toxicants interact with other effluent constituents in ways that change their toxicity. Volume 2(5).
Maximum standing crop	Algal biomass which results after cells have used up all available growth-stimulating nutrients under controlled experimental conditions. Volume 1(3).
Measurement endpoint	Numerical expression of a specific assessment endpoint or effect criterion (*e.g.*, IC50, NOEC, LOEC). Volume 1(3,10), Volume 2(8).
Mechanical dredgers	Dredgers well suited for removing hard-packed sediment material or debris and for working in confined areas (PIANC, 2001). Volume 2(9).
Mesocosm	Experimental system reflecting semi-realistic conditions. Volume 2(2).
Metallothionein	Small molecular weight protein family, rich in cysteine, that binds strongly to divalent heavy metals. The synthesis is under the control of essential metals like zinc and copper. Other metals such as cadmium, mercury and silver can induce its concentration in cells. Volume 1(14).
Milli-Q water	Reverse osmosis water which is passed through a Milli-Q Plus system (Millipore Corp.) to produce water, which meets the American Society of Testing Materials (ASTM) type 1 reagent grade water standard. Volume 1(6).
Model parameter	Constant value in a model that explains its properties. Volume 2(2).
Molting	Shedding of carapace during the growth phase. Volume 1(10).
Monitoring	Act of observing something and sometimes keeping a record of it over space and time. It can refer to the periodic (routine) checking and measurement of certain biological or water-quality variables, or the collection and testing for toxicity of samples of effluent, elutriate, leachate, or receiving water. Volume 1(7,14).
Monotonous response	Response that continuously increases (or decreases) with dose or concentration. Volume 2(2).
Mortality	Ratio of deaths in a population of cells. It is usually expressed in percentage (%). Volume 1(14).

Multiple regression method	Linear regression using several variables. Volume 2(2).
Multitrophic	Use of organisms from several different trophic levels, which can include decomposers, primary producers and (primary, secondary and tertiary) consumers. Volume 2(1).
Near-field	Receiving water adjacent to the point of industry's effluent discharge in which the water or sediment quality is potentially affected by the effluent discharge. Effluent concentration in the receiving water of the near-field will be greater than or equal to 1%. Volume 2(4)
Neat effluent sample	Undiluted or unaltered wastewater sample. Volume 2(1).
Necrosis	It indicates dead (*i.e.*, with brown or white spots) frond tissue, (Environment Canada, 1999). Volume 1(7).
Negative control sediment	Uncontaminated (clean) sediment which does not contain concentrations of one or more contaminants that could affect the performance (*e.g.*, light production) of test organisms. This sediment may be natural, field-collected sediment from an uncontaminated site, or artificial sediment formulated in the laboratory using an appropriate mixture of uncontaminated (clean) sand, silt, and/or clay. This sediment contains no added test material or substance. For example, in the solid-phase test using *V. fischeri*, it must enable an acceptable rate of light production in line with test conditions and procedures. The use of negative control sediment provides a basis for judging the toxicity of coarse-grained (< 20% fines) test sediment. See also Artificial control sediment and Reference sediment. Volume 1(2).
Neonate	Newly born organism (*e.g.*, daphnid). Volume 1(10).
NOEC	No-observed-effect-concentration, that is the highest concentration in the tested series where exposed organisms present no significant effect in relation to control organisms (*i.e.,* where the mean value for the observed response is not significantly different from the controls). It is always the next lowest concentration in the dilution series after the LOEC. Volume 1(3,4), Volume 2(3,8,11).
Non linear regression	Regression where the model is not a linear function of each parameter. Volume 2(2).
Non polar narcotics	Compounds causing baseline toxicity, *i.e.*, reversible state of arrested activity of protoplasmic structures (Bradbury et al., 1989). Volume 1(8).

GLOSSARY

Organic extract	Organic solution obtained from, for example, Soxhlet extractions, after adding an extractant (*e.g.*, dimethyl sulfoxide) to samples. Volume 2(8).
Orthogonal variables	Variables for which coefficients of correlation are inexistent. Volume 2(2).
Oxidative stress	Stress condition where oxygen (radical) reacts with internal components in cells (*e.g.*, lipids and DNA) and produces damages that eventually kill or destroy tissues. Considered as a universal mechanism of toxic damage in cells. Vol. 1(14).
Parshall flume	Specially-shaped open channel flow section device which may be installed in a canal, lateral, or ditch to measure the flow rate, such as that of an industrial effluent. Volume 2(1).
Passive biomonitoring	Use of resident living organisms (or part of) to assess water, air, sediment or soil quality. See also Active biomonitoring and Biomonitoring. Volume 2(11).
Pelagic	Aquatic organism which remains free-swimming or free-floating. Volume 2(8).
Perfusion	Pumping a liquid into an organ or tissue by way of blood vessels. Volume 1(14).
Permeability	Property of a cell or a material that can be pervaded by a liquid such as by osmosis or diffusion. Volume 1(14).
Persistence	Resistance of an organic molecule to transformation by either chemical or biological processes contributing to its longevity in the environment (*e.g.*, many organochlorine compounds are known to be persistent). Persistent organic compounds, because they are lipid-soluble, tend to accumulate in aquatic biota where they may exert adverse effects. Volume 2(1).
Petrographic analysis	Examination of a sediment sample under a high-powered microscope by trained experts in order to quantify the percentage of coal particles present. Volume 2(10).
pHi	Initial pH of an effluent sample as received by the test laboratory, before any adjustment or manipulation has been performed. Volume 2(5).
Photoperiod	Duration of light and darkness within 24 hours. Volume 1(10).
Phototrophic	Organism which must use sunlight as an energy source for nutritional purposes (*e.g.*, phytoplankton). Volume 1(3).

pH-scale	Logarithmic scale devised by Sørensen for expressing acidity or alkalinity of a solution. It is expressed numerically as the logarithm to the base 10 of the reciprocal of the hydrogen ions activity (in moles per litre). Volume 2(3).
Phytotoxicity	Potential of any agent (physical, biological, chemical) to cause adverse effects toward vegetal systems. Volume 1(3).
PLS regression	Partial least square regression: a regression method that maximizes the co-inertia of a table of independent and a table of dependent variables. Volume 2(2).
Polar narcotics	Aromatic compounds with strong electron releasing amino or hydroxy moieties, which have a narcotic mode of action (Bradbury et al., 1989). Volume 1(8).
"Polluter pays" principle	Principle stating that a polluting entity (*e.g.*, an industrial plant) should be charged the cost of restoration of the environment. Volume 2(2).
Ponar grab	Sampling device operated using a boat-mounted winch that allows collection of a relatively undisturbed surface sediment sample. Essentially, a set of "jaws" with a trigger that closes the sampling device on impact. Volume 2(10).
Pore water	See interstitial water. Volume 1(2,9), Volume 2(5,8,9).
Positive control sediment	Sediment which is known to be contaminated with one or more toxic chemicals, and which causes a predictable toxic response (for instance, inhibition of light production) with the test organisms according to the procedures and conditions of the test. This sediment might be one of the following: a standard contaminated sediment; artificial sediment or reference sediment that has been spiked experimentally with a toxic chemical; or a highly-contaminated sample of field-collected sediment, shown previously to be toxic to a (battery of) bioassay and for which its physicochemical characteristics are known. The use of positive control sediment assists in interpreting data derived from toxicity tests using test sediment. For a reference method, positive control sediment must be used as a reference toxicant when appraising the sensitivity of the test organisms and the precision and reliability of results obtained by the laboratory for that material. See also Standard contaminated sediment, Artificial sediment, Reference sediment, and Reference toxicant. Volume 1(2).
Primary consumer	Animal that eats, for example, green plants or algae in a food chain. Volume 1(8).

GLOSSARY

Primary cultures	Cells freshly extracted and isolated from an organ or tissue and plated in a defined culture medium (*e.g.*, PBS or L-15 media). During this procedure two parallel processes occur: 1) differentiated cells of the original tissue explants usually do not divide and, with time, will successively lose some of their specialized functions (dedifferentiation); and 2) decrease of number of specialized cells (*e.g.*, fibroblasts divide rapidly, and will eventually outnumber the specialized cells). Volume 1(14).
Producer (primary)	Autotrophic organisms (plants and algae) which synthesize organic matter from inorganic materials (*e.g.*, algae photosynthesize sugars from CO_2). Volume 2(1).
pT-bioassay	Bioassay belonging to a test battery for the determination of the toxicity class of a wastewater effluent. Volume 2(3).
pT-index	Numerical ecotoxicological classification of environmental samples attained with a test battery. The pT-value of the most sensitive organism within a test battery, the pT_{max}-value, determines the toxicity class of an investigated sample. Roman numerals are assigned to each toxicity class which corresponds to a pT-index. Volume 2(3).
pT-method	Procedure in accordance with a particular theory for environmental protection which includes the determination of pT-values and pT-indices. Volume 2(3).
pT-scale	A logarithmic scale for expressing aquatic toxicity with regard to a single test organism, along which distances are proportional to the pT-values. Volume 2(3).
pT-value	Numerical designation of aquatic toxicity: the highest dilution level without effect is used for the numerical designation of the toxicity with regard to a single test organism. The pT-value is the negative binary logarithm of the first non-toxic dilution factor in a dilution series in geometric sequence with a dilution factor of two. Volume 2(3).
Quantal	Toxicity test or effect endpoint for which the result can only be expressed as pass/fail or yes/no (for instance, survival/no survival). Volume 2(8).
Quantal flux	Illumination or irradiance of light in the photosynthetically effective wavelength range (400 - 700 nm), expressed in lux, foot-candles or $\mu E.m^{-2}.s^{-1}$. Volume 1(3).
Quantitative	Toxicity test or effect endpoint for which the result can be anywhere on a numerical scale (for instance, weight gained, number of young produced). Volume 2(8).

Receiving water	Surface water (*e.g*, stream, lake) receiving the effluent of a discharged waste. A representative receiving water sample should be collected upstream from the source of contamination or adjacent to the source but unaffected by it. Volume 1(6), Volume 2(5).
Reconstitution solution	Non-toxic distilled or deionized water that is used to activate a vial of Bacterial Reagent. Volume 1(2).
Reference material	Material that may consist of one or more substances whose properties are sufficiently well established to be used for the calibration of a test system. Volume 1(10).
Reference sediment	Field-collected sample of presumably clean (uncontaminated) sediment, selected for properties (*e.g.*, particle size, compactness, total organic content) representing sediment conditions that closely match those of the sample(s) of test sediment except for the degree of chemical contaminants. It is often selected from a site that is uninfluenced or minimally influenced by the source(s) of anthropogenic contamination but within the general vicinity of the site(s) where samples of test sediment are collected. A reference sediment should not produce a toxic effect (or have a minimum effect) on a test species. A sample of reference sediment should be included in each series of toxicity tests with test sediment(s). See also Artificial sediment, Positive control sediment and Test sediment. Volume 1(2,13), Volume 2(8).
Reference substance (or toxicant)	Selected chemical employed to measure the sensitivity of the test organisms in order to establish confidence in toxicity data obtained for a given test sample (or a batch of test samples). In most instances, a toxicity test with a reference toxicant is performed i) to confirm that test organisms (or cells) are in good physiological health for bioanalytical purposes at the time the test sample is evaluated, and ii) to assess the precision and reliability of results obtained by the laboratory for that reference toxicant. The toxicant selected should meet different properties as defined by Environment Canada, 1990. Volume 1(2,3,6,7,14), Volume 2(11).
Reference toxicant testing	See above. Volume 1(2,7,10).
Reference toxicity test	Test conducted using a reference toxicant in conjunction with a toxicity test to appraise the sensitivity of the organisms and the precision and reliability of results obtained by the laboratory at the time the test material is evaluated. Deviations outside an established normal range indicate that the sensitivity of the test organisms (and/or the performance and precision of the test) are suspect. Volume 1(2,7,10).

GLOSSARY

Regression	Statistical method to calculate a set of model parameters for which a model best fits the experimental data. Volume 2(2).
Regulatory authorities	Administrative or political authorities in charge of setting-up and enforcing a law or set of rules. For example, regulatory authorities implement rules to protect the aquatic environment from impairment due to the release of toxic effluents. Volume 2(2).
Residue	Difference between a modeled value and an experimental observation. Volume 2(2).
Response ratio (relative to control cells)	Amount of retained dye in cells treated to a test substance divided by the amount of retained dye in control (unexposed) cells. It indicates changes in cell viability. Volume 1(14).
Resting egg	Cyst, dormant organism or organism in a cryptobiotic stage. Volume 1(9).
Rhizosphere	Part of the ground which is located in the immediate environment of plant roots. It is very rich in micro-organisms and biological substances. Volume 2(6).
Richter scale	Logarithmic scale devised by Richter for expressing the magnitude of an earthquake from seismograph oscillations. Volume 2(3).
Risk assessment	Process of estimating the probabilities and magnitude of undesired effects resulting from the release of chemicals, other human actions or natural catastrophes. Volume 2(10).
Root exudates	Low molecular weight metabolites that enter the soil from the roots of plants. Volume 2(6).
Rotifer cyst	Encysted, diapausing embryo capable of remaining dormant for decades. Volume 1(9).
Sample	Portion of a lot or population consisting of one or more single units. Volume 1(10).
Sample preparation	All procedures applied to a sample prior to analysis; may include pre-treatment (*e.g.*, filtration, homogenization). Volume 1(10).
Sample pre-treatment	All procedures applied to a collected sample prior to sample analysis, including removal of unwanted material, removal of moisture, sub-sampling and/or homogenization. Vol. 1(10).

Sediment	Particulate material (*e.g.*, sand, silt, clay) which has been transported and deposited in the bottom of a body of water. Sediment input to a body of water comes from natural sources, such as erosion of soils and weathering of rock, or as the result of anthropogenic activities, such as forest or agricultural practices, or construction activities. The term can also describe a material that has been experimentally prepared (formulated) using selected particulate material (*e.g.*, sand of particular grain size, bentonite clay, etc.). Volume 1(2), Volume 2(5,9).
Sediment core sample	Sediment sample collected with a corer. The advantage of corers is that they preserve the vertical profile of the chemical constituents of the sediment. This allows for sediments to be sub-sampled to specific depths. Volume 2(9).
Sediment quality triad	Effects-based approach for assessing the status of contaminated sediments based on chemistry, biology and ecotoxicology. Volume 2(10).
Sediment reference area	Area with sediment that has similar physical characteristics as the site under investigation, but without elevated contaminant concentrations. Volume 2(10).
Sediment relocation	Aquatic disposal/placement of dredged material in water bodies including navigable and non-navigable waters, small lakes, lagoons and rivers (PIANC, 2000). Volume 2(9).
Sensitivity	1- Ability to detect a toxic effect at a very low concentration of test sample,
	2- In quality control, it is the slope of a concentration-response relationship,
	3- Number of toxic samples in a test system (*e.g.*, trout hepatocyte culture) divided by the number of toxic samples in another test system (*e.g.*, rainbow trout test). In the context of alternative tests, the sensitivity of fish cell methods is usually compared with the corresponding whole organism response. Volume 1(14).
Sexual dimorphism	Differences between males and females. Volume 1(13).
Sexually immature fish	Young fish that has not started its reproductive cycle with the absence of secondary sexual characteristics. Volume 1(14).
Simpson's Diversity Index	Proportion of individuals for each taxonomic group that contributes to the total individuals in a field site under study. The arithmetic mean (plus or minus the standard error, plus or minus the standard deviation), minimum and maximum for the area are also calculated. Volume 2(4).

Simpson's Evenness Index	Expressing Simpson's Diversity Index, D, as a proportion of the maximum possible value of D_s assumes individuals were completely evenly distributed among the species. Volume 2(4).
Simultaneously Extractable Metals (SEM)	Chemical analysis that quantifies the sum of Cd, Cu, Hg, Ni, Zn and Pb that can be extracted from a sediment sample. Volume 2(10).
Soil	Whole, intact material representative of the terrestrial environment that has had minimal manipulation following collection. It is formed by the physical and chemical disintegration of rocks and the deposition of leaf litter and/or decomposition and recycling of nutrients from plant and animal life. Its physicochemical characteristics are influenced by microbial and invertebrate activities therein, and by anthropogenic activities. Volume 1(2).
Solid-phase (toxicity) test	Bioassay using a biological system which measures toxic effects of solid phase of a test material (*e.g.*, bulk/whole sediment) and determines a response (*e.g.*, acute and/or chronic toxicity). It usually comprises a series of test concentrations prepared using an aliquot of the test material. See also Liquid-phase (toxicity) test. Volume 1(2), Volume 2(9).
Speciation effects	Any of a series of physical, chemical or biological factors that can cause changes in the form, bioavailability, uptake, mobility and toxicity of a chemical substance. Volume 1(3).
Stabilization pond	Relatively shallow body of wastewater contained in an earthen basin used for secondary biological treatment. Volume 2(7).
Standard deviation	Square root of the sample variance. Volume 1(10).
Standard operating procedure (SOP)	Written, authorized and controlled quality document that details instructions for the conduct of laboratory activities; SOPs are developed by laboratories when adopting a standard method or when developing laboratory-specific procedures. Volume 1(10).
Standardization	Imposition of rules permitting to check or validate the accuracy of a test using live organisms. For example, the use of a well-defined experimental procedure and the use of a reference toxicant are important rules to standardize a test. Test standardization also requires that the test be feasible by many laboratories and yield comparable results with the same test substance. Volume 1(14).

Static test	Toxicity test in which test solutions are not renewed during the test period (Environment Canada, 1999). Volume 1(3,7,10).
Static renewal test	Toxicity test in which test solutions are renewed (replaced) periodically (*e.g.*, at specific intervals) during the test period. Synonymous terms are batch replacement, renewed static, renewal, intermittent renewal, static replacement, and semi-static (Environment Canada, 1999). Volume 1(7,10).
Static replacement	See above. Volume 1(10).
Stock	Ongoing laboratory culture of a specific test organism from which individuals are selected and used to set up separate test cultures (Environment Canada, 1999). Volume 1(7).
Strain	Variant group within a species maintained in culture, with more or less distinct morphological, physiological, or cultural characteristics (Environment Canada, 1999). Volume 1(7).
Subculture	1- As a noun, laboratory culture of a specific test organism that has been prepared from a pre-existing culture, such as the stock culture. 2- As a verb, to conduct the procedure of preparing a subculture (Environment Canada, 1999). Volume 1(7).
Sublethal	Stress condition that is not immediately lethal to the organisms or below the level which directly causes death within the test period. Sublethal effects are most of the times reversible in contrast with mortality which is an irreversible condition. Volume 1(10,14).
Sum of squares	Sum of the squared residues. This sum is used as a criterion for goodness of fit in a regression procedure. Volume 2(2).
Surface water	Water column of a given water body (*e.g.*, lake, river, estuary, bay). Volume 1(14).
Surfactant	Surface tension decreasing agent that facilitates dispersion of hydrophobic materials in water. Volume 1(3).
Suspended matter	1- Fine insoluble particles originating from soil erosion, organic debris, urban wastewater or industrial effluent. Excessive levels of suspended matter lead to oxygen deficiencies in water bodies, and may have harmful effects on fauna and flora. 2- Part of the sediment load that is in suspension. Vol. 2(9).
Synergism	Interaction of several agents resulting in a greater effect than the one expected by addition of the individual effects. See also Synergistic effect. Volume 2(10).

Synergistic effect	Toxicity of a mixture of chemicals whereby the summation of the known toxicities of individual chemicals is greater than would be expected from a simple summation of the toxicities of the individual chemicals comprising the mixture. Volume 2(1).
Synoptic Sampling	Sampling at the same location at the same times; ideally, subsampling from the same original or composite sample. Volume 2(10).
Taxation principle	Guideline used to tax economic actors (*e.g.*, as a function of the load of pollutants that their activity generates). Ecological monetary taxes place pressure on polluters to limit pollution provided they are sufficiently substantial to incite clean-up action. Volume 2(2).
Test culture	Culture for providing organisms for use in a toxicity test. It can be established from organisms isolated from a stock culture. In the *Lemna* test, it refers to the 7 to 10-day old *Lemna* cultures maintained in Hoagland's medium that are then transferred to control/dilution water for an 18 to 24-h acclimation period. Volume 1(7).
Test medium	Synthetic culture medium that enables the survival or growth of test organisms during exposure to the test substance. It is prepared with deionized or glass-distilled water (*e.g.*, ASTM type-1 water) to which reagent-grade chemicals have been added. The resultant synthetic test medium is free from contaminants. The test substance will normally be mixed with, or dissolved in, the test medium. Volume 1(7).
Test sediment	Field-collected sample of whole sediment, taken from a marine, estuarine, or freshwater site thought to be contaminated (or potentially so) with one or more chemicals, and intended for use in solid-phase toxicity tests. In some instances, the term also applies to any solid-phase sample (including reference sediment, artificial sediment, negative control sediment, positive control sediment, or dredged material) used in testing. Volume 1(2).
Threshold Effect Concentration (TEC)	Value lying between the NOEC and LOEC derived by calculating the geometric mean of the NOEC and LOEC where $TEC = (NOEC \times LOEC)^{1/2}$. Volume 1(3), Volume 2(8).
Threshold Observed Effect Concentration (TOEC)	See above. Volume 1(3), Volume 2(8).

TIE Blank	During Toxicity Identification Evaluation, performance of a Phase I test on control water to determine if toxicity is added by the effluent manipulation itself. See also Control. Volume 2(5).
Toxic	Poisonous. A toxic chemical or material can cause adverse effects on living organisms, if present in sufficient amount at the right location. "Toxic" is an adjective, and should not be used as a noun, the term "toxicant" being the legitimate noun. Volume 1(2).
Toxic Threshold Effect Concentration (TTEC)	See Threshold Effect Concentration. Volume 1(3).
Toxic Unit (TU)	Inverse of the concentration of the test sample that is toxic calculated to make toxicity data directly proportional to the intensity of toxicity. For example, if a 25% dilution of a municipal wastewater has an effect on organisms, then the sample will have 100% v/v ÷ 25% v/v = 4 toxic units. Volume 1(14), Volume 2(5,8).
Toxicity	Inherent potential or capacity of a material or substance to cause adverse effect(s) on living organisms. The effect(s) can be lethal or sublethal. Volume 1(2,6,10).
Toxicity Identification Evaluation (TIE)	Iterative series of chemical manipulations (*e.g.*, pH adjustment, filtration, aeration) followed by toxicity testing designed to determine the contaminant responsible for the observed toxicity in the original sample. Volume 1(10), Volume 2(5,10).
Toxicity test	Determination of the effect of a material or substance on a group of selected organisms (*e.g.*, *Vibrio fischeri*), under defined conditions. An aquatic toxicity test usually measures either (a) the proportions of organisms affected (quantal); or (b) the degree of effect shown (quantitative or graded), after exposure to specific concentrations of test material or complex mixture (*e.g.*, chemical, effluent, elutriate, leachate, or receiving water). Volume 1(2,10).
Trickling filter	Fixed–film biological process for secondary domestic wastewater treatment. Volume 2(7).
Tubifex	Oligochaete worm, deep burrower and relatively tolerant to anoxia. Volume 1(12).
Ubiquitous	Found everywhere, present in most ecosystems around the world. Volume 1(11).

GLOSSARY

Viable cells	Cells capable of maintaining membrane permeability which is essential for the maintenance of life processes. A viable cell is able to maintain its membrane integrity to assure proper exchanges with its environment. Volume 1(14).
Vitellogenin (Vg)	Precursor of egg-yolk proteins rich in carbohydrates, lipids, phosphates and calcium. It is the principal energy reserve in oocytes. Vg expression is under the control of estradiol-17β receptors. This protein complex is produced in the liver by oviparous vertebrates and used as a biomarker to detect environmental estrogens. Volume 1(14).
Vortex mixer	Compact laboratory mixer used for stirring small sample volumes in containers (*i.e.*, test tubes, centrifuge tubes, colorimetric tubes, small flasks). Volume 1(1,3).
Warning chart	Graph used to follow changes over time, in the endpoints for a reference toxicant. Date (or number) of the test is on the horizontal axis and the effect-concentration is plotted on the vertical logarithmic scale. Volume 1(2).
Warning limits	Boundary or combination of limits which, when exceeded, may trigger analyst intervention; most toxicity laboratories use 2 X the standard deviation of the mean to create warning limits (*i.e.*, one in every 20 tests would be expected to exceed the warning limits, due to chance alone). Volume 1(2,10).
Wastewater	Water mixed with waste matter usually released by man-made activities, townships, municipal treatment plants and industries. Volume 1(14).
WaterTox Program	International network organized by the IDRC (International Development Research Centre), in collaboration with the National Water Research Institute and the Saint-Lawrence Centre of Environment Canada, to undertake bioanalytical intercalibration exercises with participating laboratories from eight different countries (Argentina, Canada, Chile, Colombia, Costa Rica, India, México and Ukraine). The battery of simple, affordable and robust tests was initially selected to detect the toxic potential of chemical contaminants in drinking water and freshwater sources. Volume 2(7).
Weak acid respiratory uncouplers	Compounds that abolish the link between substrate oxidation and adenosine triphosphate (ATP) synthesis (Cajina-Quezada and Schultz, 1990). They are generally bulky and electronegative. Volume 1(8).

Wet sediment phase	Solid phase obtained after extracting pore (or interstitial) water from whole sediment. Porewater is commonly extracted by centrifugation (*e.g.*, at 3,000 rpm, 30 min, 15°C). Volume 2(8).
Whole-water sample	Sample of water that has not been filtered or extracted. Volume 1(15).
Xenobiotics	Chemicals that have no relevant function for maintenance and reproduction of biological organisms. These compounds are usually produced by anthropogenic activity. Volume 1(14).

Glossary References

Bradbury, S.P., Carlson, R.W. and Henry, T.R. (1989) Polar narcosis in aquatic organisms, in U.M. Cowgill and L.R Williams (eds.), *Aquatic toxicology and hazard assessment: 12th Vol.*, ASTM, Philadelphia, PA, pp. 59-73.

Cajina-Quezada, M. and Schultz, T.W. (1990) Structure – toxicity relationships for selected weak acid respiratory uncouplers, *Aquatic Toxicology* **17**, 239-252.

Environment Canada (1999) Guidance documentation: Control of toxicity test precision using reference toxicants, Environmental Protection Service, Report EPS 1/RM/12, Ottawa, ON, 85 pp.

Environment Canada (1999) Biological test method: Test for measuring the inhibition of growth using the freshwater macrophyte, *Lemna minor*, Environmental Protection Service, Report EPS 1/RM/37, Ottawa, ON, 98 pp.

Férard, J.-F., Costan, G., Bermingham, N. and Blaise, C. (1992) Comparative assessment of effluents with the *Ceriodaphnia dubia* and *Selesnastrum capricornutum* chronic toxicity tests, Communication Meeting of SETAC-Europe, Postdam (Germany).

Perrodin, Y., Cavéglia, V., Barna, R., Moszkowicz, P., Gourdon, R., Férard, J.-F., Ferrari, B., Fruget, J.F., Plenet, S., Jocteur-Monrozier, L., Poly, F., Texier, C., Cluzeau, D., Lambolez, L. and Billard, H. (1996) Programme de recherche sur l'écocompatibilité des déchets: étude bibliographique, Report, ADEME, France (in French).

PIANC (1997) Handling and treatment of contaminated dredged material from ports and inland waterways, Permanent International Association of Navigation Congresses (PIANC), Report of Working Group No. 17, Supplement to Bulletin No. 89.

PIANC (2000) Glossary of selected environmental terms, Permanent International Association of Navigation Congresses (PIANC), Report of Working Group No. 3 of the Permanent Environmental Commission, Supplement to Bulletin No. 104.

PIANC (2001) Dredging: the facts, Permanent International Association of Navigation Congresses (PIANC), ISBN 90-75254-11-3.

Snell, T., Mitchell, J.L. and Burbank, S.E. (1996) Rapid toxicity assessment with microalgae using *in vivo* esterase inhibition, in G.K. Ostrander (Ed.), CRC Press/Lewis Publishers, Boca Raton, Florida, U.S.A., pp. 13-22.

US EPA, 2004, http://www.epa.gov/OCEPAterms/aterms.html.

A

Abietic acid, 302, 475, 489, 497
Acute and chronic toxicity testing with *Daphnia* sp., 337
Alamar Blue, 475, 478-495, 498, 500
Algal assay bottle test, 181
Algal microplate toxicity test, 137
    algal microplate toxicity test, 167, 171, 174
Algal microplate toxicity test suitable for heavy metals, 243
Algal preservation techniques, 152
Algal toxicity test, 181
    algal toxicity test(ing) (see also toxicity testing), 11, 17, 30, 33, 138, 140, 167-168, 171-174, 182, 184, 186, 191, 198-200, 205-206, 209, 233, 238, 274
Algicidal, 190, 194, 198-199
Algistatic, 156, 190, 194, 198-199
Alginate bead matrix, 152
Alginate-immobilized algae, 153
American Type Culture Collection (ATCC), 175, 184, 477, 482
Amphipod (*Hyalella azteca*) solid-phase toxicity test, 413
Amphipod, 112, 130-131, 413-414, 417-418, 420-421, 425-426, 428-429
Amphiporeia virginiana, 131
Anabaena cylindrica, 140
Anabaena flos-aquae, 184, 199, 244-245, 249-250
Analysis of variance, 331, 465, 490
ANOVA (see Analysis of variance)
Artemia salina, 396, 402-403
Artificial sediment, 437
    artificial and/or natural sediments, 29, 124, 129
    artificial and synthetic sediments, 28
    artificial silica sand, 441, 444
Asexually-reproducing, 395, 403
Average specific growth rate, 182-183, 195-196, 198-199, 223-224, 288
Axenic culture, 149-151, 264
Axenic strain (see also axenic culture), 143

B

Bacillus subtilis, 332
Battery of test organisms (see also test battery approaches), 25, 29, 70, 97, 101, 108, 113, 132, 239, 243-245, 263, 267-269, 271, 299, 301-302, 319, 321, 323, 325, 410, 500
Best Practicable Technology, 171
Bioaccumulation, 27, 413-418, 421, 425-429, 431-433
Bioassay battery (see also test battery approaches), 29, 302, 319, 321

Bioassay techniques, 27, 300
   development of, 1-3, 25, 27-28, 75-76, 174, 204, 239-240, 269, 342-344, 384, 410, 477-478
   initiatives promoting the use of, 31
   refinement of, 27, 29, 455
   validation of, 11, 27-29, 93, 99, 168, 192, 343, 453, 500
Bioluminescent bacteria, 69, 71-72, 74, 77, 87, 93, 97
Bioterrorism, 70
Biotransformation, 455, 457
Boot-strap methods, 263
Brachionus calyciflorus, 97, 324-328, 331-332
Brachionus plicatilis, 324, 326, 328, 332-333
Brine shrimp, 396, 402-404
Brood chamber, 348
Budding, 395-399, 402

C

Candida tropicalis, 332-333
5-carboxyfluorescein diacetate acetoxymethyl ester (CFDA-AM), 475, 478-479, 490-495, 498-499
Carbon dioxide limitation, 148, 182, 189
Carrier solvents (see solvents), 72, 80-81, 88, 93, 98, 154, 156, 172, 183, 190-191, 215, 328, 369, 401-402, 486, 489, 497
Cell density, 164-167, 204, 206, 215, 217-218, 220, 222, 224-225, 227-228, 235-236, 460, 470, 485
Cell lines, 476-478, 481, 483
   continuous cell line, 476
   piscine cell lines, 477
   mammalian cell lines, 477, 481
   RTgill-W1, 480, 496, 498
Ceriodaphnia dubia, 13, 325, 332, 338, 342, 381-382, 386
Chelating agent, 186, 249, 266
   EDTA, 147, 186-188, 210-211, 217-220, 225-227, 235, 238, 249-251, 266, 270, 276-277, 283, 287, 380, 401, 415, 425, 455, 482-483
Chemical analysis/testing, 3, 7-10, 24, 217, 219, 302, 343, 371, 380, 383, 413, 417, 424, 426
Chironomus riparius solid phase assay, 437
Chironomids, 112, 132, 417, 441, 449
Chironomus, 431-432, 438, 440, 448
Chironomus riparius, 432, 437-441, 446, 448-449
Chironomus tentans, 27, 382, 438-439
Chlamydomonas reinhardti, 140
Chlamydomonas variabilis, 140
Chlorella sp., 140, 237-238
Chlorella kessleri, 140
Chlorella pyrenoidosa, 140

Chlorella vulgaris, 140, 349, 354
Clonal cultures, 72, 344, 355
Clonal strain, 69, 76, 78
Clonal variation, 344, 377, 378
Colored samples, 30, 169, 234
Commercial chemicals, 337-339, 382, 386
Comparative studies, 28-29, 169, 244
Complex mixtures, 93, 96, 271, 454, 465, 470
Confounding factors (see also factors capable of affecting bioassay responses), 28, 88, 128, 133, 437, 439, 448
Contaminant analysis studies (see also chemical testing), 24
Contaminant bioavailability, 27, 90-91, 206, 235, 239, 266, 343, 371, 418, 429-432, 497
Control chart, 127, 167-168, 224-226, 233, 264-265, 279, 290, 377, 438, 446, 465, 467
Cost of testing, 10, 25, 69, 72, 75, 77, 97-99, 132, 173-174, 222, 239, 243, 271, 274, 303, 320-321, 323, 372, 398, 409, 428-429, 433, 449, 453-455, 473, 476-477, 500
Cost per sample (see also cost of testing), 325, 500
Critical body residue (CBR) studies, 27, 429
Cryptobiotic, 152, 326
Cyanobacteria, 19, 184, 199, 237-238, 244, 250, 257, 269, 302
Cysts (see also resting eggs), 323-328, 332, 396, 402, 404
Cytotoxicity, 454-455, 468-469, 473-478, 480, 482, 488, 497-500

D

Daphnia magna, 11, 13-14, 301, 338-339, 342, 345-350, 353, 361, 363, 365, 373, 377, 379-382, 384-385
Daphnia pulex, 339, 345-350, 353, 361, 363, 365, 380, 382
Daphnia sp., 319, 337-351, 354-356, 358-363, 372, 376-382, 384-386, 403, 498, 500
Databases, 1, 93, 344
Definitive test(ing), 81, 85, 154, 157-158, 185, 189, 190, 197, 203, 219, 226, 254, 279, 291, 300, 308-309, 312-313, 315-316, 319-320, 329, 368-369, 377
Desmodesmus subspicatus, 184
Dialysates, 71, 79, 89, 91, 98, 100
Diapause, 440
Diatoms, 184, 205, 244, 249, 349
2,6 dinitrotoluene (2,6 DNT), 85-86
Diverse types of environmental samples, 6-7
Dormancy, 441, 447
Duckweed, 5, 7, 9, 11, 14, 16, 23, 26, 273-274, 293
Dunaliella tertiolecta, 184, 199

E

Ecological relevance, 243-244, 269, 321
Edge effect, 155
EDTA (see chelating agent)
Effluents (diverse), 2, 4-5, 7-10, 33, 73, 75, 94, 98, 100, 140, 154, 166, 171, 173,181-182, 184, 189, 203, 206, 243, 278, 283-285, 299, 320, 323, 328-329, 333, 337-338, 364, 377, 382, 384, 454, 500
   industrial effluents, 2, 4-5, 7-10, 93, 141, 157-158, 169-171, 174, 217, 237, 302-303, 337, 341, 379, 386, 410, 453, 462, 465-466, 470
      composite sample of effluent, 97, 170, 216, 226-227
      pulp and paper effluents, 4, 8, 10, 171, 337, 455,
      metal plating effluent, 4, 10, 219, 337, 470
      mining effluents, 4, 7, 10, 267-269, 290-293, 337, 408-409, 455, 470
      oil refinery effluent, 4, 8, 474
      textile effluent, 4
   municipal effluents, 4, 5, 8-10, 170, 271, 453, 455, 468-470
Electronic particle counter, 138, 142, 144-145, 149, 155, 164, 185, 193, 261
Elutriate, 21-26, 154-155, 170, 243, 271, 285, 337-339, 364, 381, 383, 459
Emergence, 440, 449
Energy metabolism, 478, 480, 500
Endocrine disruption/disruptors, 34, 323, 449, 456
Endpoint(s), 16-17, 27, 29-30, 33
   Bacteria, 71-72, 76, 101, 108, 115-118, 123-124, 126, 132
   Fish cells, 454, 465, 475-478, 480, 488, 492-494, 498-499
   Microalgae, 138-140, 156-157, 163, 165-167, 169, 173-174, 183, 190, 192-193, 195-196, 198-199, 204, 207, 222, 231, 233, 235, 244, 261
   Invertebrates, 324-325, 331-332, 338, 340-341, 343-344, 362-363, 366, 375-376, 380, 383, 396-398, 410, 416, 427-430, 438-439, 445, 449,
   Plants, 272, 279, 286-289, 293
   Protozoans, 299-300, 313, 315, 321
Enrichment effect, 170
Enzyme (activity) inhibition, 97, 169, 204-207, 216, 324, 332
Ephippia, 338, 343, 344, 348, 350, 356, 358-359, 363
Erlenmeyer flasks, 141, 144, 148, 153, 182-183, 185, 208-211, 216, 276, 278, 391
Esterase(s), 140, 204, 207, 215-216, 224-225, 228-233, 235-238, 332, 463-464, 479, 493
Eutrophication studies, 143, 170, 181
Exuviae, 442

F

Factors capable of affecting bioassay responses, 28, 30
Feeding, 457

Brachionus, 323-325, 327-328, 330-332
Chironomus, 437-440, 442, 445, 447
Daphnia, 349-350, 354, 356-357, 361, 363, 373
Hyalella, 415, 420, 425
Hydra, 396, 402-403, 408-410
Fetal bovine serum (FBS), 482-485
Fish cell bioassays, 5-9, 11, 14, 19, 24, 28-29, 31, 33, 454-456, 470, 473-474, 477, 481-483, 488, 498, 500
Flask method (see algal assay bottle test)
Flow cytometry, 140, 203-208, 212-216, 218, 220, 222, 224, 227-229, 234-240
Fluorescein diacetate (FDA), 204, 207-208, 213-215, 224-225, 228-233, 235-238, 453-454, 463, 465-466
    carboxyfluorescein diacetate (cFDA), 494-495, 498-499
Fluorescent illumination, 138, 228
Fluorescence, 140, 203-207, 213-215, 221, 223-225, 228-233, 235-238, 241, 245, 259, 261-262, 325, 462-464, 466, 490, 492-494, 496
Fluorescent dyes, 205, 321, 478, 493-494
Fluorometer, 155, 247, 258, 459, 462
Fluorometry, 173
Fluorometric multiwell plate reader, 474, 478

G

Genotoxicity, 34, 101, 456
Good Laboratory Practices (GLP), 75, 145
Green algae, 140, 142, 172, 184, 203, 207, 237, 244, 268, 349-350, 354-357, 363, 399
Green *Hydra*, 395-396, 398-399, 402, 406, 409-410
Growth rate,
    Microalgae, 151, 182-183, 195-199, 203-205, 216-217, 222-224, 234-235, 238, 244-245, 261-266
    Plants, 276, 279, 281, 287-288
    Invertebrates, 338, 343, 397, 404-405, 421
Guidance documents, 32, 33, 382

H

Haemocytometer (see hemacytometer)
Health criteria, 350-351, 355-358, 378
Heavy metals (see also toxicity testing of metals), 10, 13, 17, 24, 38, 94, 130, 243-244, 266, 300, 304, 498
Hemacytometer, 138, 144-145, 152, 161, 185, 193, 247, 258, 261, 459-460, 482, 485
Hepatocyte, 453-470
Herbicides, 10, 15, 17, 95, 140, 155, 172-173, 176, 181, 198, 200, 273, 292, 341
Heterocapsa niei, 239
Hexagenia, 431-433

Hexazinone, 172, 268
Hyalella azteca, 13, 381-382, 413-433
***Hydra* population reproduction toxicity test method,** 395
Hydra sp., 14, 397-410
Hydra viridissima, 395-396, 398, 402
Hydra vulgaris, 395-396, 398, 402
Hydroids, 395-397, 399, 402-405
Hypothesis testing, 166, 196, 224, 233, 278-279, 375

I

Imhoff settling cones, 413-414, 417-418, 422, 427, 433
Immobility, 338, 340-341, 343, 362-363, 372, 374-376, 379, 386
Immunotoxicity, 34
Industrial effluents (see effluents)
Industrial wastewaters (see industrial effluents)
Ingestion test, 326-329, 332-325
Ingestion rate, 323, 325, 332
Instars, 348, 439-440
Instar larvae, 438, 440
Inter-calibration exercises, 28-29, 32, 174
Inverted phase-contrast microscope, 482, 484

L

Laminar flow hood, 247, 275, 459, 481-484, 491-495
Landfill leachates (see leachates)
Leachate(s), 8-9, 71, 93, 100, 137, 154, 203, 216, 219, 243, 271, 278, 285, 292, 299, 303, 364, 383, 453
   agricultural leachates, 8-9, 93
   industrial leachates, 4, 8-9, 93
   municipal solid waste leachates, 4, 9
   landfill leachates, 6, 7, 93-94
   solid waste leachates, 3-5, 7-9, 140, 173, 320, 453
Lemna gibba, 293
Lemna minor, 271-272, 274-276, 278-281, 284-293
***Lemma minor* growth inhibition test,** 271
Lemnaceae, 4-9, 11, 14-16, 22, 26, 28, 33, 274
Lemna spp., 273-274, 293
Life cycle reproduction test, 339, 359
Life cycle risk assessment, 449
Light scatter(ing), 128, 203-204, 206, 213, 221, 223
Linear interpolation method, 166
Line of best fit, 166
Lipophilic contaminants, 79-80
Lipophilic chemicals, 80, 89-90

INDEX 545

Liquid media toxicity assessment, 3
Logarithmic phase cells, 151, 258
Log-phase growth (see also logarithmic phase cells), 151, 256, 258
Lyophilization, 76
Lyophilized bacteria, 72, 115-116, 118
Lysosomal activity, 475, 478, 480
Lysosomal damage, 480
Luminescent bacteria, 12-13, 72, 74-78, 87, 97, 101, 108, 110-112, 123, 128, 130-133, 152
Luminous bacteria (see luminescent bacteria)
Luminometer, 69, 72-74, 77, 81-84, 97

M

Mass culture, 356-357, 359
Material Safety Data Sheets (MSDS), 79, 215, 364, 382
Matrix effect (see sediment)
Mayflies, 417
Medium, 12, 30-31
    Bacteria, 100
    Fish cells, 459, 469, 474, 477, 479, 486, 489
        L-15/ex, 462-463, 468-470, 477-479, 484, 486, 489, 491-497
        L-15 cell culture medium, 453, 460, 462
        Leibovitz's L-15 complete medium, 457, 482-486, 491
    Microalgae, 145, 153, 169, 173, 181-183, 185-186, 190-191, 194, 200, 204-205, 208, 210-211, 217-219, 225, 227, 233-235, 238, 244-245, 249, 251-252, 254-256, 258-260, 264, 266-267
        AAP algal assay procedure medium, 186-188
        ACM-1x algal culture medium, 146, 148, 153, 155
        ATEM solution algal test, 148-152, 162-163
    Invertebrates, 332, 343, 354, 371, 386, 419
        CMCP-9/9 AF medium, 386
        Hyalella medium, 420, 422
        Hydra medium, 400-401, 403
        SAM-5S medium, 420
    Plants, 272
        Hoagland's E^+ medium, 276-281
        Modified APHA medium, 278
        SIS medium, 278
    Protozoans, 300, 302, 305-307, 320
Membrane integrity, 207, 236, 475, 478-480, 493, 500
Membrane permeability, 206, 216, 236, 237, 454
Metal toxicity (see also toxicity of metals), 94, 218, 235, 243
Metallothioneins, 468, 470
Microalgal toxicity test using flow cytometry, 203
Microbiotests (see micro-scale toxicity tests)
Microcystis aeruginosa, 19, 184, 239, 244-245

Microcystis sp., 19, 256, 258, 268
Micromonas pusilla, 239
Microplates, 137-144, 148-149, 155, 167-174, 182, 206, 234, 239, 243
  24-well, 299-301, 304, 309-315, 318-320, 323, 408
  96-well, 137-139, 141, 144, 154-165, 170, 173, 244, 246-248, 258-261, 264, 269, 453-454, 459-464, 473-474, 487-488, 500
Micro-scale toxicity tests, 2
Microtox acute toxicity test, 69
Microtox solid-phase test, 127
Mining effluents (see industrial effluents)
Molting, 348
Monorhaphidium pusillum, 140
Municipal effluents (see effluents)

N

Navicula pelliculosa, 184, 199
Navicula spp., 194
N*annochloris* sp., 244-245, 256, 258, 268
Neonates, 340, 343, 348-349, 353, 355-359, 371, 373, 378
Neutral red (NR),453-454, 458, 461-463, 467, 470, 475, 478, 480, 490-496, 498-499
Nitzschia closterium, 237-238
Nitzschia sp., 244, 245, 249, 256, 258, 268
Non-monotonic/monotonous responses, 263
Number of young, 341, 348, 356-357

O

Oncorhynchus mykiss, 453-454, 456-457, 480
Organic chelator, 249
Osmolality, 474, 478, 486
Overview of contemporary toxicity testing, 1

P

Paramecium aurelia, 332
Parthenogenesis, 348
Persistent (lipophilic) compounds (see also lipophilic contaminants/chemicals), 26
Pesticides, 16-18, 34, 93, 95-96, 142, 181-182, 184, 285, 301, 319, 323, 353, 382, 383, 397, 448, 456
  biopesticides, 18-19
Petroleum products, 95, 96
Phaeodactylum tricornutum, 184, 237, 239
Phytotoxicity, 13-14, 137, 139-143, 148, 151, 154-157, 163, 165, 168-172, 174, 176, 195, 203, 205, 243, 246, 267, 273

Photobacterium phosphoreum, 77, 78, 113
Photocytotoxicity, 477, 480, 498
Photometer, 76, 109-110, 114, 116-118, 120, 122-123, 155
Pink *Hydra*, 395-396, 398-399, 402, 404, 408, 410
Polychlorinated biphenyls (PCBs), 16, 95-96, 130-131, 397
Population growth, 151, 181-182, 189, 194, 205, 395, 397-398, 404-405
Porewater (see water)
Primary culture, 399, 453-455, 459, 465, 467-470, 476-477, 481
Prokaryotic cells, 78
Protective wear (see safety)
Protozoa (see also toxicity testing with protozoans), 5, 7-11, 14, 18-19, 22-23, 26, 28-29, 31, 175, 299-321, 364
Pseudokirchneriella subcapitata, 143, 184, 207, 245
Pulp and paper effluents (see industrial effluents)

Q

Quantitative structure-activity relationships (QSAR), 17-18, 301
QA/QC program, 79, 345
Quality assurance, 32, 78, 220, 224, 233, 342, 345
Quality control, 32, 72, 78, 113, 127, 132, 167, 172, 224-226, 233, 246, 265, 279, 289, 302, 316, 342, 345, 465-467

R

Rainbow trout hepatocyte toxicity test, 453
Rainbow trout gill cell line microplate cytotoxicity test, 473
Range-finding test, 154, 157-158, 189, 191, 219, 309-311, 329, 366, 368-369, 445
Reference sediment, 88, 108, 123-124, 126-130, 426, 430, 437
Reference toxicants, 33, 79-80, 97, 108, 123, 127, 132, 138, 157, 167-168, 220, 224-227, 229-231, 233, 244-245, 252-254, 259-260, 264-265, 267, 272, 279, 289, 291, 300, 316-319, 324, 340, 343, 360, 376-378, 381, 385, 396, 408, 427, 438, 446, 454, 458, 465-467, 470, 475
Resting eggs, 323, 326
Resazurin, 478-479, 493
Raphidocelis subcapitata, 143, 184
Rotifers, 97, 323-333
Rotifer Ingestion Test, 323
Round-robin testing (see inter-calibration exercise)

S

Safety, 75, 79, 132, 154, 172, 173, 215, 308, 319-320, 363, 364, 382, 481
   protective equipment, 75, 79, 132, 173, 363

Screening test, 101, 173, 300, 308-310, 315, 319
Scenedesmus quadricauda, 140
Scenedesmus subspicatus, 140, 184
Sediment, 3, 7, 18, 20-28, 30-33, 70-71, 79, 88-89, 91, 94, 97, 100-101, 108-110, 112-113, 115-117, 122-133, 137, 154-155, 170, 173, 203, 205-206, 237-240, 321, 326, 342, 381-382, 413-433, 437-449, 453-459
Sediment porewater (see water)
Sediment toxicity, 20, 27-29, 31, 33, 110, 112, 126, 129-131, 133, 239, 416-417, 427-433, 437, 439, 441, 447
Sediment toxicity assessment, 20, 25, 26, 32, 33, 94
   of areas of concern, 20-24
   of oil spills and flooding events, 20
Selenastrum capricornutum, 97, 137-175, 184, 203-204, 207-208, 210, 212, 214-225, 232, 234-240, 244-245, 249, 256, 258, 265, 267-268, 349, 354, 357, 373
Semipermeable membrane device (SPMD), 89-92, 94, 98
Sexual dimorphism, 447
Silicate, 249-251, 257, 444
Single species tests, 6, 10-11, 25-26, 206, 239
Skeletonema costatum, 183-184, 189, 191-192, 194, 197
Small-scale tests (see micro-scale toxicity tests)
Solid-phase test for sediment toxicity using the luminescent bacterium, *Vibrio fischeri*, 107
Solid waste leachates (see leachates)
   municipal solid waste leachates (see leachates)
Solubilizing agent, 154, 156
Solvents, 16, 31, 80-81, 88, 93, 154, 190-191, 215, 308-309, 369, 477, 497
   acetone, 72, 80-81, 154, 186, 191, 208, 215, 228-229, 308-309, 328, 369
   ethanol, 72, 80-81, 215, 309, 386, 482-484, 486, 491-495, 497
   DMSO, 72, 80-81, 88-89, 91, 96, 98, 308, 453-454, 458-460, 462, 464, 466, 486, 490, 494, 497
   methanol, 81, 154, 308-309, 369, 444, 458, 462
SOS Chromotest, 173
Spectrophotometer, 459, 462-463
Spectrophotometry, 425
Spiked sediments, 417, 422, 426-427, 430, 444
***Spirostomum ambiguum* acute toxicity test, 299**
Spirostomum ambiguum, 299, 300, 302-304, 307, 314
Spirotox, 299-303, 308-310, 315-316, 318-321
Standardization, 11, 28-29, 32, 69, 99, 112-113, 118, 168, 174, 200, 221, 273, 279, 338, 343, 354, 376, 382, 384, 470
Standardized test methods (TMS), 32-33, 341, 384
Surface water (see water)
Surfactants, 15-17, 154, 301, 475, 497
Synchronization (life cycle), 440

T

Teratogenicity, 34
Test battery approaches (TBA), 3-11, 13-16, 19, 21-26, 137, 174, 203, 453
    algal test battery, 268
Test method development (see bioassay techniques)
Test validity (conditions for) (see also validity criteria), 167, 170, 261, 279, 284, 316, 331, 465
Thalassiosira pseudonana, 184, 188
Tissue culture facility, 481
Toxic effects, 2, 87, 132, 139, 148, 155, 163, 170-171, 174, 215-216, 269, 271, 278, 283-284, 301, 309, 316, 431, 447, 453, 468
    additive, 477, 154
    antagonistic, 17, 477
    synergistic, 17, 154, 477
Toxicant potentiality, 88
Toxicity
    identification evaluations (TIEs), 70, 133, 169, 500
    reduction evaluations (TREs) (1), 70
Toxicity testing (TT), 4-9, 12-16, 19, 21-24, 26
    of biological contaminants, 18-19
    of organic substances/compounds, 15-16, 269, 448
        dimethylphenol, 332
        methanol, 81, 154, 309, 369
        naphthol, 332
        PAHs, 27, 96, 130-131, 239, 447, 474, 477, 497-498
        pentachlorophenol, 16, 95, 324, 332, 340, 377, 499, 500
        phenol, 10, 16, 73, 79, 85-86, 88, 91, 94-95, 156, 169, 302, 332, 497
    of various classes of (in)organic chemicals, 12-16, 19, 173
    of metals, 12-14, 16-17, 27, 94, 209, 243-269, 278, 291-292, 300, 304, 318, 323, 332, 344, 377, 378-379, 382, 397-398, 401, 408, 423, 431, 447-448, 498
        copper, 13-14, 16-17, 165-166, 226, 235-239, 267, 292, 300, 304, 331-332, 353, 378, 380, 397, 409, 421, 446
        cadmium, 27, 91, 267-268, 292, 318, 332, 353, 378, 397-398, 421, 431
        mercury, 13, 301, 332, 353, 429
    of metals, ions and oxidizing agents, 12-14
    of pesticides, 16-19, 95-96, 181, 301, 323, 353, 382, 397
        diazinon, 16, 332, 421
        chlorpyrifos, 17, 332
    with algae, 4-9, 11-12, 15-16, 19, 21-24, 26, 28, 137, 181, 203, 243
    with bacteria, 4-9, 11-12, 15-16, 19, 21-24, 26, 28, 69, 107
    with fish, 4-9, 11, 15-16, 19, 21-24, 26, 28, 453, 473
    with invertebrates, 4-9, 11-16, 19, 21-24, 26, 28, 323, 337, 395, 413, 437
    with plants, 4-9, 11, 15-16, 22-23, 26, 28, 271
    with protozoans, 4-9, 11, 14, 19, 22-23, 26, 28, 299
    with seeds, 4-9, 11, 16, 19, 22-23, 26

Toxicity tests,
  developed and applied at different levels of biological organisation, 2
Trophic levels, 10, 25, 33, 97, 174, 374
Trophic-level specificity, 10
Trypsin, 277, 482-484
Tubifex, 431-433
Tubificid worms, 417

U

Ultrasonic dispersion, 154

V

Validity criteria, 170, 244-245, 263, 279, 289, 291, 340, 362-363
Versene, 482-484
Vibrio fischeri, 11, 69, 77-78,100, 108, 112-113, 118, 130, 408
Vitellogenin, 470
Volatile substances, 169, 246, 365

W

Wastewater (see water)
Wastewater treatment plants (WWTP), 70
Water,
  brackish, 326
  drinking water, 32, 70, 98, 299, 308, 323
  groundwater, 6, 9, 71, 93, 137, 140, 173, 203, 319-320, 350, 354, 361-362, 453
  lake/limnic, 6, 18, 89, 91, 274, 326, 345, 381, 410, 416, 418, 431-432, 457
  pore water, 21-26, 79, 100, 110, 115, 127-128, 130, 173, 203, 205, 299, 321, 323, 325, 328-329, 332-333, 364
  receiving waters, 17, 34, 70, 154, 170, 174, 182, 216, 269, 271, 278, 282, 337-338
  river/stream, 6, 9-10, 20, 24, 89, 91, 93-94, 143, 237-238, 300, 326, 345, 380, 410, 447-48, 457, 468
  surface water, 7, 32, 70-71, 170, 173, 181, 203, 205, 238, 320, 323-324, 327-329, 333, 337, 350, 361-362, 381, 386, 427, 453, 468-469
  urban creeks, 332
  wetland, 6, 9
  wastewater (also see industrial effluent), 3-10, 18, 21-23, 32, 70-71, 73, 93-94, 100, 137, 169, 171, 174, 203, 205, 227, 235, 267, 274, 278-279, 282, 284, 286, 290, 293, 337-339, 364-367, 375, 381
Water fleas, 345
Water renewal, 413, 416-417, 431, 433, 441-442

Whole water samples, 474-475, 477-478, 486-492, 497-500

Y

Young per brood, 350, 356-357
Young production, 338, 340, 343, 357-359, 363, 372-375